KEY SCIENCE
THIRD EDITION

Biology

DAVID APPLIN MSc DIC PhD FRES

Tutor, St Andrew's Cambridge
Formerly Head of Science, Chigwell School

First published in 1994 by:
Stanley Thornes (Publishers) Ltd
Second edition 1997

This edition published in 2002 by:
Nelson Thornes Ltd
Delta Place
27 Bath Road
CHELTENHAM
GL53 7TH
United Kingdom

02 03 04 05 / 10 9 8 7 6 5 4 3 2 1

A catalogue record for this book is available from the British Library

ISBN 0 7487 6241 8

Illustrations by Barking Dog Art, Peters and Zebransky, Harry Venning, Cauldron Design Studio, Tech Set Ltd
Page make-up by Tech Set Ltd, Gateshead, Tyne and Wear

Printed and bound in Spain by Graficas Estella

Related titles
Key Science: Physics (0 7487 6243 4)
Key Science: Chemistry (0 7487 6242 6)
Key Science: Biology Extension File (0 7487 6256 6)

Acknowledgements

Andy Trott: 8.1A; Action Plus: 14.1D; Ardea: p.31 (bottom – John Clegg), 3.6I (R M and Valerie Taylor), 3.7B (Pascal Goetghleluck), 11.3C, 15.2A (P. Morris), 26.1F (b – Jack Swedburg), 26.2A; Associated Press: 8.1Q (Ricardo Mazalan); Australian Information: 3.1C; Allsport: p.202 (Clive Brunskill); AES Education: 24.4A (Marlow Foods), 24.4B (Marlow Foods); Axon Images: 25.2D (right – John Bailey); Biophoto Associates: 2.3D (a), 4.1A (top), 15 (bottom), 16, 17 (middle), 18, 24, 25 (bottom), 26 (top), 28, 29 (top), 3.7A (e) (f), 9.2A (a) (b), 9.4D (bottom), 10.1G, 10.2B, p. 180 (top), 12.3A, 15.2F (a) (b), 15.2H (a), 15.2L (a) (b), 16.18, 19.3C, 23.1E (a) (b), 23.2A (b); Blackpool Tourist Office: 8.10; Bristol City Council: 8.3A (c); British Diabetic Association: 24.2B; Bruce Coleman: 1.3B (b – Joe Van Warmer), p.20 (Jane Burton), 1.3A, p.29 (bottom – Pacific Stock), 8.2D (a – Steven Kaufman); Cheltenham General Hospital Medical Photography: 22.5A; Corbis: p.50; Corel (NT): 8.2B, 8.3D, 10.1H, p.168, 18.2A, 340, 21.2C, 22.5B; David Applin: 2.1B (c), p.86, 5.1A, 8.3A (a), 8.3A (d), 23.4F, 26.1J; Dr. Eileen Ramsden: 10.1A (a), 10.2C (a) (b), 10.3A; Digital Vision (NT): 5.1C, 6.3A, 6.6B, 14.4G; Digital Stock (NT): 22.5C; Environmental Picture Library: 8.2A; Ecoscene: 7.3A (a – Frank Blackburn) (b – Sally Morgan), 20.1A (d – Diana Aldam), 24.bB (bottom: Ian Harwood); Forensic Science Service: 24.2A; Frank Lane Picture Agency: 5.1B (J C Allen), 8.1J (a – W Broadhurst), 24.3F (R Anstring), 26.5D (David Hosking), 23.1A (left – Peter Dean), 26.5L (Walter Rondich), 26.5M (Foto Natura Stock), 26.5P (centre left – Minden Pictures), 26.1F (a – Martin Withers); Gene Cox: 3.7A (c), 10.1F, 9.4A; Geoscience Features Picture Library: 2.1B (a); Getty Images: p.138 (Spike Walker), 16.1C (Andrew Syred), 285 (top – Will and Demi McIntyre); Grant Heilman Photography: 13.2A (top left – Runk/Schoenburger); Greenpeace: 8.1G (Hodson); Glasshouse Crops Research Centre: 11.1D; Gonville and Cains College, Cambridge: p.383; Holt Studios International:

4.1A (bottom), p.88 (bottom – Julia Chalmers), 5.aA, 5.5A, 13.1A (a), 13.2B, 21.1B (b), 23.1A right, 24.3D – Nigel Cattlin), 5.6B, 11.2A, 11.3D (Dick Roberts), 13.1A (b), 13.2C (Rosemary Mayer), 21.2E, 21.3F (Inga Spence), 4.2B (adult), 14.3C (a), 24.5E; Hulton Getty: 6.1A (Topical Press), 26.1H (Hulton Deutsch); ICCE: 11.1A (Chris Rose), 3.7A (b – Cath Wadforth), 3.7A (a – Manfred Kage), 3.7A (d – E M Unit/CVL Webridge), 5.2C; Impact Photos: 14.2B (Jeremy Nicholl), 23.2D (Julian Calder); Janine Weidel Photolibrary: 17.1K; John Innes Institute: 13.2A (top right), 23.2A, 24.3C (a) (b); John Reader: p.448 (left); Leslie Garland Picture Library: p.34 (Andrew Lambert; Mark Boulton: 22.6B; Martyn Chillmaid: 14.2S (left) (right), 14.3D, p 251 (top), 22.2D, 22.7A, 24.1B, 24.2C, 25.1B; Mary Evans Picture Library: p.287, p.405, 26.1D, 26.1E, 26.1I, 26.2C, 26.3A; Mike Peters: p.418; Natural History Museum: 26.5F (bottom – Orbis), 426, 26.5F (a); Natural History Photographic Agency: 26.1G (b – Phillipa Scott), 26.5N (b – ANT); News International: 14.2R (left and right – The Times); Natural Visions: 22 (bottom), 29 (middle), 2.1A (bottom right), 2.3D (b), 2.4A, 3.7D, 10.2A, 14.4E, 20.1A (c), 21.1B (a), 21.2B (top) (bottom), 26.5N (a); Oxford Scientific Films: 15 (middle – G W Willis), 17 (top), 26 (middle), 27 (top) (bottom – Mike Birkhead), 19 (bottom), 2.3B (b – JAL Cooke), 2.3C (S R Morris), 2.6A, 3.6G (ab – Kim Westerskov), 4.2B (top centre and bottom left – Tim Shepherd), 4.4A (Alastair Shay), 5.5B (left – Stan Osokriski) (right – Deni Brown), 8.2D (b – Mike Hill), 8.3C (M Leach), 9.4D (top – Carolina Biological Supply Co. Ltd), 13.1A (c – G I Bernard) 17.4F (left and right – David Thompson/London Scientific Films), 17.4G (Rudie H Kuiter amd Scott Camazine), 17.4H (Breck P Kent/Animals Animals/London Scientific Films), 20.1A (e-Mantis Wildlife Films), 20.2E (London Scientific Films), 20.2D (Irving Cushing), 21.1C (Richard Packwood), 21.3I (Harold Taylor), 26.1G (a – Peter Ryley), 17.1I, 20.2F, 20.2G, 20.2H, 26.4C (Peter Parks); Panos Pictures: 8.1R (P Wolmuth); Papilio: 4.2B (bottom right – Robert Picket); Pictor International: 26.5E; Popperfoto: 17.1L (b), 10.4I (Erik Hill); PA News: 23.4B; PA Photos: 23.4E (Joyce Naltchayan); Rex Features: 16.1F (Sipa Press), 17.5D (Nils Jorgensen), 25.2B (Andrew Dunsmore); Robert Harding Picture Library: 7.1A, 14.3A (right – N A Callow); Roslin Institute, Edinburgh: 24.3E (Dr John Clark); Royal Botanic Gardens: 8.3E; Royal Army Medical College: 4.3A (Dept. of Military Entomology; Russel Kighyley Media: 16.1H; Science Museum: 9.1A (top and bottom – reproduced by permission of the Trustees of the Science Museum, London); Science Photo Library: p.13 (Dr Steve Patterson), p.14, 8.3I (top) (middle-London School of Hygiene and Tropical Medicine), p.17 (left – Dr D Patterson) (right – Claude Nuridsany and Marie Perennou), 26B (Eye of Science), 27 (left – J. Berger, Max-Planck Institute), 2.1B (b), 9.3A (Dr Gopal Murti), 46 (top left – Jan Hinsch), 3.7A (g – Dr P Marazzi), 4.2B (pupa – John Burbage), 8.1M (NASA), 6.6B, 10.4E (Oxford Molecular Biophysics Laboratory), 10.4H (A Barrington Brown), 10.1E (Andrew Syred), 14.1H (Biophoto Associates), 14.1C, 12.2A, 12.2E (a) (b), 14.3A (left – Eric Grave), 14.3C (b – A B Dowse) p. 302 (John Paul Key/ Peter Arnold Inc.), 18.3E (Dept of Clinical Radiology, Salisbury District Hospital), 14.5H, 15.2H (b – Jonathon Ashton), 15.3A (J C Revy), 15.1C (K R Porter), 16.1G, 16.3C (professor P M Motta/A Gaggiati/GMachiarelli), 16.4A (b – BSIP/VEM), 17.1D, 17.3A, 17.4B (Mark Clark), 19.2A (Alfred Pasieka), 20.1A (Eric Grave), 21.1I (Juergen Berger, Max-Planck Institute), 21.3G, 22.2F (Prof. P Motta, Univ. 'La Sapienza', Rome), 22.3D (Dr Vorgos Nikas), 23.2F (a – Jomes Holmes, Cellmark Diagnosis) (b – Saturn Skills), 23.2A (a – Richard Hutchings), 23.4C (Peter Menzel), 23.4D (Simon Fraser), 24.5G (Dept of Medical Photograph, St Stephen's Hospital, London), p.407 (top – C C Studio), 24.6B (top – Andrew Syred), 24.5I (Jackie Lewin, Royal Free Hospital), 26.1A (Space Telescope Science Institute, NASA), 26.2E (top – Tom Myers/Agstock), 26.5A (William Ervin), 26.5C (Sally Benusen), 26.5G (Phillipe Plailly, Eurelios); Stockmarket: 17.1L (a); Still Pictures: 8.1I (a – David Drain) (b – Nigel Dickenson); Stone: 8.2C (Sylvain Grandadam); St. Mary's Hospital Medical School, London: 24.5D; Telegraph Colour Library: 16.1A (David Grossman); Topham Picturepoint: 8.1H, 24.2D; TRIP: 11.1E; University of Bristol: 18.2K (Dr Jeremy Rayner); Wellcome Medical Picture Library: 4.2B (person). Photo research by johnbailey@axonimages.com

The third edition of *Key Science Biology* has been written with the aim of bringing the latest advances in the Biological Sciences to students. The revision of the text and illustrations of the previous edition is extensive and reflects the rapid pace of change which is trickling down to the specifications for GCSE (and equivalent examinations) Science and Biology. It is a pleasure to acknowledge the colleagues who have contributed to the work. In particular, Eileen Ramsden and Jim Breithaupt, authors respectively of *Key Science Chemistry* and *Key Science Physics*, who have given wholehearted support and encouragement, Simon Read and Michael Cotter who have provided the means and context for the work to progress, Sarah Ryan whose patience and editorial skills have resulted in the final product, John Bailey who in a real sense visualised the illustrations which are a vital, integral part of the book, Debra Goring who sympathetically dealt with the new text and ensured its seamless insertion into the main body of existing material, and Diane Freeborn whose meticulous indexing allows ease of access to all of the exciting information which is the Biology for GCSE. Christopher Hilton's invaluable research of the internet-based sources of data and information helped to ensure that *Key Science Biology* is up to date, thereby providing students with the resources that include the advances of the "new biology". Finally, the unstinting support of my family helped to ensure a successful conclusion to the project.

David Applin

Contents

Preface   1

Theme A ▶ Living things in the environment

Topic 1 Variety and classification 3
1.1 ▶ Earth as a place to live   3
1.2 ▶ Variety of life   4
1.3 ▶ What's in a name?   7
1.4 ▶ Classification   9
1.5 ▶ Five kingdoms   11

Topic 2 Ecology 35
2.1 ▶ What is the environment?   35
2.2 ▶ Ecosystems   40
2.3 ▶ Habitats, niches and the community   42
2.4 ▶ Community structure   45
2.5 ▶ Ecological pyramids   48
2.6 ▶ Decomposition and cycles   53
2.7 ▶ The nitrogen cycle   55
2.8 ▶ The carbon cycle   56

Topic 3 Populations 58
3.1 ▶ Population growth   58
3.2 ▶ Checks on population increase   60
3.3 ▶ Population size   61
3.4 ▶ Human populations   63
3.5 ▶ Competition   65
3.6 ▶ Predators and prey   68
3.7 ▶ Parasitism and other associations   72

Topic 4 Insects and humans 76
4.1 ▶ Beneficial and harmful insects   76
4.2 ▶ Malaria and mosquitoes   78
4.3 ▶ Houseflies spread disease   80
4.4 ▶ Honeybees   82

Theme questions 85

Theme B ▶ Human impact on the environment

Topic 5 Food production 87
5.1 ▶ Intensive farming   87
5.2 ▶ Soil: the lifeblood of farming   89
5.3 ▶ Natural fertilisers: manure, slurry and compost   93
5.4 ▶ Synthetic fertilisers   94
5.5 ▶ Herbicides   94
5.6 ▶ Insecticides   96
5.7 ▶ Energy   99

Topic 6 Air pollution 102
6.1 ▶ Smog   102
6.2 ▶ The problem   103
6.3 ▶ Sulphur dioxide   103
6.4 ▶ Acid rain   105
6.5 ▶ Carbon monoxide   108
6.6 ▶ Chlorofluorohydrocarbons   109
6.7 ▶ The greenhouse effect   110

Topic 7 Water pollution 113
7.1 ▶ Pollution by industry   113
7.2 ▶ Thermal pollution   115
7.3 ▶ Pollution by agriculture   115

Topic 8 Conservation 119
8.1 ▶ The world problem   119
8.2 ▶ What has it got to do with us?   132
8.3 ▶ Conservation in action   134

Theme questions 142

Theme C ▶ Cell biology

Topic 9 The cell 145
9.1 ▶ Seeing cells   145
9.2 ▶ Movement into and out of cells   148
9.3 ▶ Making proteins   152
9.4 ▶ Cell division   154
9.5 ▶ Cells, tissues and organs   157

Topic 10 The chemicals of life 162
10.1 ▶ Carbohydrates   162
10.2 ▶ Lipids   164
10.3 ▶ Proteins   167
10.4 ▶ Nucleic acids   170

Theme questions 174

Theme D ▶ Green plants as organisms

Topic 11 Photosynthesis and mineral nutrition 177
11.1 ▶ Leaves are adapted for photosynthesis   177
11.2 ▶ Mineral nutrition   183
11.3 ▶ Growth and support   185

Topic 12 Transport and water relations 188
12.1 ▶ The transport system of plants   188
12.2 ▶ Xylem transports water and mineral ions   189
12.3 ▶ Phloem transports food   192

Contents

Topic 13 Hormonal control of plant growth 195

13.1 ▶ Growth movements 195
13.2 ▶ Using growth substances 198

Theme questions 201

Theme E ▶ Humans (and other animals) as organisms

Topic 14 Nutrition 203

14.1 ▶ Food – why do we need it? 203
14.2 ▶ Diet and food 211
14.3 ▶ Keeping food fresh 225
14.4 ▶ Teeth 229
14.5 ▶ Digestion and absorption 235

Topic 15 Respiration 243

15.1 ▶ What is respiration? 243
15.2 ▶ Exchanging gases: lungs 247
15.3 ▶ Exchanging gases: gills and other surfaces 256

Topic 16 Circulation and transport 259

16.1 ▶ Blood and its functions 259
16.2 ▶ Understanding immunology 267
16.3 ▶ The blood system 270
16.4 ▶ Understanding heart disease 278

Topic 17 Responses and co-ordination 283

17.1 ▶ Senses and the nervous system 283
17.2 ▶ The ear 292
17.3 ▶ The eye 295
17.4 ▶ Hormones 300
17.5 ▶ Maintaining a constant internal environment 306

Topic 18 Support and movement 314

18.1 ▶ Skeletons and muscles 314
18.2 ▶ Locomotion: how different animals move 317
18.3 ▶ The skeleton 325

Topic 19 Disease 331

19.1 ▶ What is disease? 331
19.2 ▶ Influenza: a case study 334
19.3 ▶ Fighting disease 336

Theme questions 339

Theme F ▶ Reproduction

Topic 20 Outlines of reproduction 341

20.1 ▶ Asexual reproduction 341
20.2 ▶ Sexual reproduction 342

Topic 21 Plant reproduction 345

21.1 ▶ Flowers 345
21.2 ▶ Seeds and fruits 351
21.3 ▶ Asexual reproduction in plants 355

Topic 22 Human reproduction 360

22.1 ▶ Getting to know you 360
22.2 ▶ The human reproductive system 361
22.3 ▶ How do sperms meet eggs? 364
22.4 ▶ Fertilisation and development of the embryo 366
22.5 ▶ Birth 368
22.6 ▶ Looking after baby 370
22.7 ▶ Controlling fertility 371
22.8 ▶ Sexually transmitted diseases 374

Theme questions 376

Theme G ▶ Genetics and biotechnology

Topic 23 Inheritance and genetics 379

23.1 ▶ Gregor Mendel 379
23.2 ▶ Human genetic disease 385
23.3 ▶ The growth of modern genetics 390
23.4 ▶ Discovering the human genome 392

Topic 24 Biotechnology 397

24.1 ▶ Introducing biotechnology 397
24.2 ▶ Using biotechnology 399
24.3 ▶ Biotechnology on the farm 403
24.4 ▶ Single-cell protein 408
24.5 ▶ Biotechnology and medicine 410
24.6 ▶ Treating sewage 417

Theme questions 423

Theme H ▶ Variation and evolution

Topic 25 Genes and the environment 427

25.1 ▶ Continuous and discontinuous variation 427
25.2 ▶ Sources of variation 428

Topic 26 Evolution 432

26.1 ▶ Evolution: the growth of an idea 432
26.2 ▶ Artificial selection 437
26.3 ▶ *The Origin of Species* 440
26.4 ▶ Evolution in action 441
26.5 ▶ The vertebrate story 445

Theme questions 454
Index 456

Preface

Key Science: Biology is a comprehensive and up-to-date textbook designed to meet the requirements of the different examining groups for all of their science specifications for GCSE and overseas examinations. The textbook can be used for the biology component of the Single or Double Award and for all GCSE Science: Biology specifications.

Topics are differentiated into core material for Single or Double Science and extension material for Science: Biology (blue/grey margin). The examining groups have chosen different extension topics in addition to the Programme of Study for Key Stage 4; therefore teachers and students need to be familiar with the specific requirements of the specification. *Key Science: Biology* contains the extension topics for all boards.

Key Science: Biology is organised in 8 Themes covering 26 major topics. Each topic is sub-divided into numbered sections covering all aspects of the topic. Each section includes special features, usually located in the margin:

- **First Thoughts**, setting the scene and reminding students of the background knowledge to the section,
- **It's a Fact**, sharpening interest and enhancing text,
- **Key Scientist**, providing a historical perspective on how scientific ideas have developed,
- **A key** 🔑 , in the margin indicates an opportunity to exercise one or more of the key skills: application of number (AoN), communication (Comm) and information technology (IT). These opportunities are listed in the Teachers' Guide,
- **A star**, ☆ in the margin suggests the topic can be used for the development of scientific ideas and the weighing of evidence. These suggestions are listed in the Teachers' Guide,
- **A computer mouse**, 🖱 in the margin indicates an appropriate opportunity to use computer software or hardware. These opportunities are listed at www.keyscience.co.uk.
- **Commentary and Summaries**, within each section of a topic and at the end of each section, helping students review their work and highlighting the key points to learn and understand,
- **Checkpoints**, testing knowledge and understanding through sets of short questions at regular intervals at the end of each section,
- **Theme Questions**, based on examination questions.

Key Science: Biology is extended and supported by an Extension File which contains Teachers' Guide, and a bank of photocopiable material. The extra resources include over 90 photocopiable Activities for Sc1 and Assignments for Sc2 and Science: Biology topics. There are 50 Exam File questions for homework and revision. Mark Schemes are provided for all questions. Answers to the Checkpoints in *Key Science: Biology* are also included.

Student's Note

I hope that Key Science: Biology provides all the information and practice needed to enable you to do well in your science examinations at GCSE and similar courses. However, although examinations are an important goal they are not the only reason for studying science. I hope also that you will develop a real and lasting interest in this fascinating area of human endeavour. Finally, I hope that you will enjoy using the book. If you do, all the effort will have been worth while.   *David Applin*

If I have seen further ... it is by standing upon the shoulders of giants.

This quote is attributed to Isaac Newton, but the English clergyman Robert Burton who died before Isaac Newton was born wrote:

A dwarf standing on the shoulders of a giant may see further than the giant himself.

Living things in the environment

Throughout the universe we know of only one planet where there is life – planet Earth. Scientists have described about 5 million types of living thing. Millions more await discovery and millions once lived on Earth but are now extinct. The phrase 'The balance of nature' refers to the delicate relationships that exist between living things and the environments in which they live. People upset this balance with their need for food, homes, hospitals, schools and their demands for a wide range of manufactured goods. Now people are becoming aware that the human race is responsible for preserving the environment and that each one of us can play a part in protecting planet Earth.

Topic 1

Variety and classification

1.1 ▶ Earth as a place to live

Earth is home to millions of different kinds of living things. It seems likely that none of the other planets in the solar system support life.

Figure 1.1A 🔺 The Earth from space. Water covers 75% of the planet's surface. It is home for the majority of living things, a solvent (see *Key Science Chemistry* p. 151) for chemical reactions n cells and a coolant for warm bodies. Without water in liquid form, life on Earth would not exist.

It's a fact!

The first living things on Earth were probably simple bacteria-like organisms. They originated about 4000 million years ago. The human race appeared about 2.5 million years ago.

Make sure you can answer an exam question which asks you to distinguish between respiration and gaseous exchange.

rritability is another word used to describe the sensitivity of living things to changes in their surroundings.

What is it about Earth that enables it to support life? One factor is the distance between Earth and the Sun. Earth is **close enough** to the Sun for its surface temperature to be in the range in which life can exist. The energy of sunlight is converted by living things into the energy which they need for their life processes.

The size of Earth is a factor. It is **massive enough** to have a sufficiently large gravity to hold down an atmosphere. The gases oxygen, nitrogen, carbon dioxide and water vapour are essential for living organisms and are present in the air on Earth's surface.

Several factors combine to keep the **temperature at the surface** of Earth at an average of 30 °C.

- The distance of Earth from the Sun.
- The layer of ozone which surrounds Earth prevents too much ultraviolet light from the Sun reaching Earth. (Too much ultraviolet light destroys living organisms.)
- The layer of carbon dioxide and water vapour which blankets Earth prevents too much heat radiating from Earth into space. Without it, the temperature at the surface of Earth would be −30 °C.

There are at least 35 million different kinds of living organism on Earth. Their variety of appearance is enormous, as you will see in later sections of this topic. All living things, however, share certain characteristics, which non-living things do not. These are the **characteristics of life**.

- **Movement** – animals are able to move from place to place because of the action of **muscles** (see Topic 18). Plants do not move from place to place; they move by growing (see Topic 13).
- **Respiration** – occurs in cells and releases energy from food for life's activities. **Aerobic** respiration uses oxygen to release energy from food; **anaerobic respiration** does not use oxygen to release energy from food (see Topic 15). The oxygen needed for aerobic respiration comes from the environment. We breathe in air and the oxygen it contains. Aerobic respiration in cells produces carbon dioxide which is breathed out. Breathing in and breathing out are the visible signs that we exchange oxygen and carbon dioxide between ourselves and the environment. **Gaseous exchange** describes the process (see Topic 15).
- **Sensitivity** – allows living things to detect changes in their surroundings and to respond to them (see Topic 13 and Topic 17).
- **Growth** - leads to an increase in size. **Development** occurs as young change and become adult in appearance (see Topic 11 and Topic 17).
- **Reproduction** – produces new individuals (see Topic 21 and Topic 22).
- **Excretion** – removes the waste substances produced by the chemical reactions (called **metabolism**) taking place in cells (see Topic 17).
- **Nutrition** – makes food (called **photosynthesis** – see Topic 11) or takes in food for use in the body (see Topic 14).

Notice the first letter of each of the characteristics of life. In order, they spell the mnemonic **MRS GREN**, helping you remember the characteristics which distinguish living things from non-living things.

CHECKPOINT

▶ **1** **Mrs Gren** will help you remember the characteristics of living organisms. Complete the names of the processes which occur in living things.
M _____   R _____   S _____
G _____   R _____   E _____   N _____

▶ **2** What would happen to Earth's water if Earth were (a) nearer to and (b) further from the Sun?

▶ **3** The Moon is about the same distance from the Sun as Earth is. The Moon is not massive enough to hold down an atmosphere. How does this affect the possibility of finding life on the Moon?

▶ **4** An aeroplane moves, it has to be 'fed' with fuel, and it gives out waste products. Do these characteristics prove that an aeroplane is alive? Explain your answer.

1.2 ▶ # Variety of life

FIRST THOUGHTS

Have you ever thought about how many different living things there are in a garden? As you read about the variety of life think about:
● the large number of species of living things that have been named,
● the even greater estimate of the number of species awaiting discovery,
● using biological keys to identify living things.

There are probably millions of species of living things awaiting discovery.

It's a fact!

Over 600 new species of beetle have been discovered living in just one type of tree that grows in the rainforests of Costa Rica.

Will we ever know all the species (a species is a particular type of living thing) of life on Earth? So far we know of about 5 million species but this figure is only a fraction of the millions of species that biologists estimate are awaiting discovery.

For example, techniques of collecting insects from the tops of trees in tropical rain forests lead us to believe that over 30 million new insect species are living in the leaves, branches, flowers and fruit (Figure 1.2A).

Figure 1.2A ⬆ The lightweight matress dropped by helicopter is strong enough to support people using it as a sampling platform. Notice the scientists taking samples of wildlife living in the rain forest canopy through perforations in the mattress. Other scientists are measuring environmental conditions (humidity/light intensity) and taking photographs

Figure 1.2B 🔺 A deciduous wood. 'Deciduous' means that the leaves fall from the trees each year.

 See www.keyscience.co.uk
for more about keys.

Using keys

How many kinds of living thing can you find in Figure 1.2B? You can see
from the picture that leaves from trees litter the ground with a thick
carpet of dead organic material. Leaf litter is a rich source of food and
home for many small animals. Specimens collected from leaf litter can be
grouped according to their appearance and structure. Each description is
a clue that helps to identify a particular group of animals found in leaf
litter (Figure 1.2C). We call a set of clues a **key**.

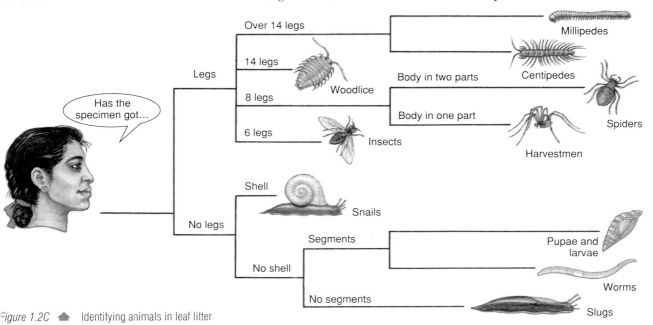

Figure 1.2C 🔺 Identifying animals in leaf litter

5

It's a fact!

False scorpions are common in leaf litter but they are well hidden and very small (about 1–2.5 mm long) and little is known about them. False scorpions are relatives of true scorpions (together with spiders, harvestmen and mites) but they do not have a sting in the tail.

The easiest type of key to use is called a **dichotomous key**. 'Dichotomous' means branching into two. Each time the key branches you have to choose between the two statements. Eventually you will track down the identity of your specimen.

A key in biology, therefore, is a guide to a name. Different keys are used to name different living things.

Try this

Naming names

Amphibians are **vertebrates** (animals with backbones) that spend part of their lives on land and the rest of the time in water. Identify each of the illustrated specimens using the key on the next page, then draw a table by putting the correct letter against each name for all of the specimens.

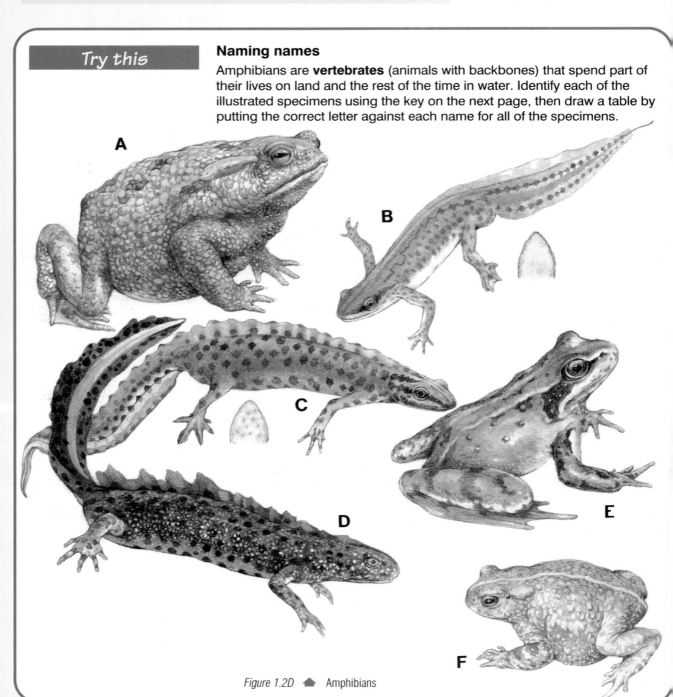

Figure 1.2D ⬆ Amphibians

■ Key to Amphibians

1	The animal has a tail	**Newts**	go to 2
	The animal has no tail	**Frogs and toads**	go to 3
2	The newt has a rough warty skin, is dark brown with dark spots on top and has a yellow or orange belly blotched with black markings	**Great crested (warty) newt**	
	The newt has a smooth skin, is green or brownish in colour with or without dark spots; belly is yellow or orange and may be spotted; 10 cm or less in length	**Smooth or palmate newt**	go to 4
3	The animal has a smooth, moist skin and a dark flash behind the eye	**Common frog**	
	The animal has a warty skin and no dark flash behind the eye	**Toads**	go to 5
4	The throat is whitish and spotted	**Smooth newt**	
	The throat is pinkish and unspotted	**Palmate newt**	
5	The toad has a yellow stripe running down the middle of the back	**Natterjack toad**	
	There is no yellow stripe running down the middle of the back	**Common toad**	

SUMMARY

Although about 5 million species of living thing have been described, millions more are estimated to await discovery, especially in tropical rain forests. Making a collection of living things in the environment will give an idea of life's variety. Biological keys help us to identify them.

CHECKPOINT

▶ **1** Look at Figure 1.2A. Briefly describe the method being used to investigate the wildlife living in tropical treetops.

▶ **2** What is leaf litter?

▶ **3** (a) What is a biological key used for?
   (b) Suggest why certain characteristics like size, exact colour and mass are not suitable for keys.

1.3 ▶ What's in a name?

FIRST THOUGHTS

What's in a name? This section emphasises the importance of biological names. As you read it think about:

● the confusion caused by having the same name for different living things and different names for the same living thing,

● Carolus Linnaeus who cleared up the confusion by defining each type of living thing with a species name,

● the meaning of 'species'.

Sometimes a living thing has several different names which describe it. For example, the plant shown in Figure 1.3A is called 'cuckoo pint', 'lords and ladies', 'parson-in-the-pulpit' and 'wake-robin' in different parts of the country.

On the other hand different living things are sometimes given the same name. For example, the robin in the USA is different from the robin in the UK. They are shown in Figure 1.3B. *List some of the differences between the two birds.*

If living things were known only by their everyday names, you can imagine the confusion there would be when British and American ornithologists (an ornithologist is someone who studies birds) spoke to each other about 'robins'.

Figure 1.3A 🔺 Cuckoo pint

Figure 1.3B 🔺
(a) The North American robin

(b) The British robin

The individual is identified by its species name.

SUMMARY

The Swedish naturalist Carolus Linnaeus introduced the idea of the species. Each type of living thing is given a species name which distinguishes it from all other types of organism.

How can people describe living things to each other and know that they are talking about the same organism?

The Swedish scientist Linnaeus tackled this question in 1735. Linnaeus' system was to give each living organism a name consisting of two parts. The first part of the name is the name of the **genus** to which the organism belongs. The second part of the name is the name of the **species** to which the organism belongs. Since each name has two parts, the Linnaeus method of naming is called the **binominal system** (from the Latin: *bis* – twice, *nomen* – a name). The genus name begins with a capital letter; the species name begins with a small letter.

To illustrate how the binominal system works look at the biological name for human beings – *Homo sapiens*. Our species name is *sapiens* (Latin for 'wise'). Our genus name is *Homo*. The other species of our genus are now extinct.

So the binominal system clears up confusion over names. For example, ornithologists can easily distinguish between the robins mentioned earlier when the birds are called by their binominal names which are *Turdus migratorius* (common name North American robin) and *Erithacus rubecula* (common name British robin).

What, then, is a species? Most biologists agree that if individuals can sexually reproduce offspring which are themselves able to reproduce, then they belong to the same species. Sometimes members of closely related species mate and produce offspring called **hybrids**. However, animal hybrids are usually sterile, that is they cannot reproduce.

Different species of plant also form hybrids between themselves. Unlike most animal hybrids, hybrid plants are often fertile and able to produce offspring. This throws open the question 'What is a species?', to which there is no entirely satisfactory answer.

CHECKPOINT

▶ 1 Who was Linnaeus? Explain how his binominal system of biological names works.

▶ 2 What is a hybrid? Use a named example in your explanation.

1.4 ▶ Classification

Classification organises living things into groups. Each group consists of individuals which have features in common.

 See www.keyscience.co.uk for more about classification.

Linnaeus' system is more than just a list of names cataloguing the variety of life on Earth. Linnaeus believed that as many characteristics as possible should be used to describe species, rather like the way we use characteristics to describe a type of car. Some characteristics are unique to the species but others are shared with other species. This means that species which have characteristics in common can be put together into groups.

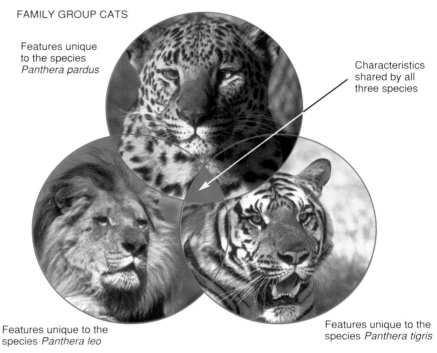

FAMILY GROUP CATS

Features unique to the species *Panthera pardus*

Characteristics shared by all three species

Features unique to the species *Panthera leo*

Features unique to the species *Panthera tigris*

Figure 1.4A ⬆ Unique features describe each species: shared features unite cats into a larger family group

Organising living things into groups means that we don't need to know everything about every species to be able to identify them. For example, we do not have to know about every insect species to know what an insect looks like!

Groups within groups

Groups can be combined to form larger groups. The cats (Figure 1.4A) are a **family** which belong to the **order** called carnivores. Also included in the order carnivores are dogs, bears, seals and sealions. The carnivores and other orders are members of the **subclass** placentals, which is one of the **class** mammals.

Organising things into groups is called **classification**. Figure 1.4B shows the classification of cats. It also shows that cats are part of a still larger group called a **phylum**. In this phylum mammals are grouped with fish, amphibians, reptiles and birds. All these animals have a feature in common – a backbone. Animals with backbones are called **vertebrates**.

Phyla (plural of phylum) come together in the largest group of all used in classification – the **kingdom**. For example, phylum 'vertebrates' along with 32 other phyla are grouped in the animal kingdom.

Phylum	Class	Subclass	Order	Family	Genus	Species

Vertebrates
Backboned animals with a skeleton made of bone and/or cartilage.

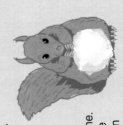

Skeleton of the cat *Felis*. Notice the rib cage and the arching backbone.

Mammals
The female feeds her young with milk produced from milk glands called mammary glands which give the class its name. Many mammals have fur that helps to retain body heat.

Placentals
Mammals whose young complete their development inside the mother attached to her by a placenta. Food and oxygen pass from mother to young and wastes pass from young to mother across the placenta.

Carnivores

	Cats	*Felis*	*Felis silvestris* Wildcat
			Felis catus Domestic cat
			Felis lynx Lynx
		Panthera	*Panthera pardus* Leopard
			Panthera tigris Tiger
			Panthera leo Lion

Dogs

Bears

Seals and sealions

Plus six more families

Primates*
Monkeys and apes

Cetaceans
Whales and dolphins

Lagomorphs
Rabbits and hares

Plus twenty-two more orders

Marsupials
Mammals that give birth to young which complete their development attached to a milk nipple inside the mother's pouch (marsupium). Most species, like the kangaroo, are found in Australia though some come from South America.

Monotremes
Mammals that lay eggs. Includes the duck-billed platypus from Australia and three species of spiny anteater; one species from Australia and the other two from New Guinea.

Birds
Animals with feathers that help to retain body heat. Feathers also make it possible for birds to fly.

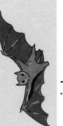

Reptiles
Animals with dry horny scales that waterproof their body.

Amphibians
Animals with a soft skin that quickly loses water in a dry atmosphere. They can live in water and on land.

Fish
Animals with flat, wet scales which make the body smooth (bony fish) or with tiny, backward sloping spines (cartilagenous fish, e.g. sharks).

Figure 1.4B ◀ Classification of cats. Groups within groups: notice monotremes and placentals belong to the larger and inclusive group, mammals, etc. Group names increase in size indicating that more and more species are included in groups to the left of the figure. In-between groups are named by adding the prefixes sub or super to any of the main groups for a more detailed classification.

*Our order – humans are primates and close relatives of apes.

CHECKPOINT

▶ 1 The diagram shows a plan for organising living things into groups. Copy the diagram and use the words provided to fill in the blank spaces. Kingdom and Family have been filled in to show the idea.

species class order genus phylum

▶ 2 With the help of Figure 1.4A explain why organising living things into groups helps us to understand them.

▶ 3 What does the word 'arthropods' mean literally?

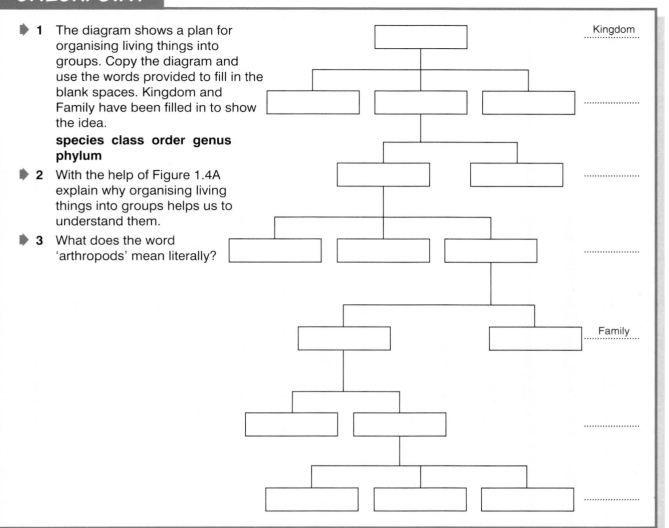

1.5 ▶ Five kingdoms

FIRST THOUGHTS

The largest grouping of animals is the kingdom. Here we review the five kingdoms. As you read, think about:
- each kingdom, representing a way of life shared by all its members,
- the size of living things, ranging from the very small to the very large,
- how living things are adapted for their way of life.

The living world is often organised by biologists into five kingdoms. Individuals grouped within a kingdom share characteristics in common. For example, plants make food by photosynthesis (see p. 177). Each kingdom, therefore, represents a way of life which all of its members share. This helps us to learn about the wide variety of living things. (See Figure 1.5A.)

The organisms pictured in Figure 1.5B are not drawn to scale. They vary in size from single-celled bacteria only just visible under a light microscope to the giant redwoods (*Sequoiadendron giganteum*) which tower into the sky. The wide range of size of living things, from the very small to the very large is shown.

The Five Kingdoms

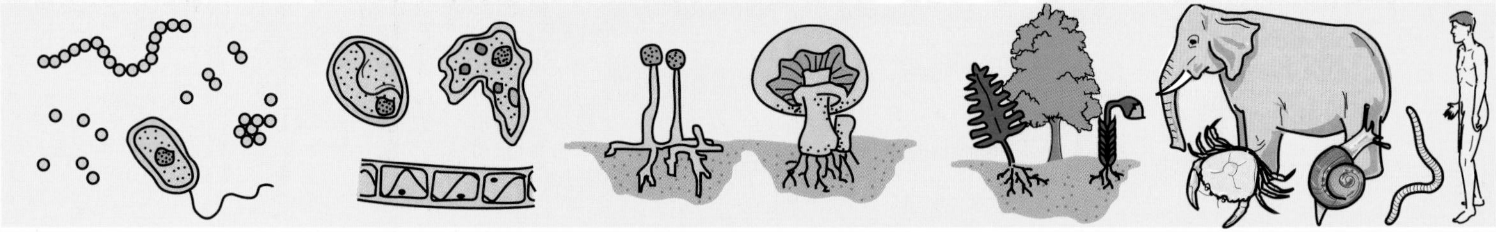

		Bacteria	Protists	Fungi	Plants	Animals
Cell Structures	Nucleus	Absent.	Present.	Present.	Present.	Present.
	Cell wall	Made of different polysaccharides (not cellulose).	Of different types in different species.	Made of polysaccharide (mainly chitin).	Made of polysaccharide (mainly cellulose).	Absent.
	Chloroplasts	Absent (although chlorophyll is found in some species).	Present in some species.	Absent.	Present.	Absent.
Bodies		Single-celled (unicellular) although clusters or chains of identical cells may form.	Most species are single-celled though some are made of many cells of different types (multicellular).	Nearly all species are multicellular although a few are unicellular.	All species are multicellular.	All species are multicellular.
Food		Some species make their own food by photosynthesis. Some species are parasites; they live off other living things. Some species are saprobionts. They secrete enzymes which digest plant and animal material and then absorb the digested food. Saprobionts cause decay and decomposition.	Some species make food by photosynthesis. Some species take in food from their environment in different ways and digest it in the body. A few species make food by photosynthesis and take it in from their surroundings.	Most species are saprobionts. Hyphae secrete enzymes on to organic material which once digested is absorbed. This causes decay and decomposition. Some species are parasites.	All species make their own food by photosynthesis.	All species take in food from the environment and digest it inside the body. Exceptions are parasites like the tapeworm which absorbs the semi-digested food of its host.

Figure 1.5A ◀ Characteristics of the five kingdoms

(kilometre, km)	10^3 m		
	10^2 m	Giant redwood	
	10^1 m	Indian python	
(metre, m)	10^0 m	Dog	
	10^{-1} m	Locust	
	10^{-2} m	Woodlouse	
(millimetre, mm)	10^{-3} m	Mite	
	10^{-4} m	Human egg cell	
	10^{-5} m	Red blood cell	
(micrometre, µm)	10^{-6} m	Bacterium	

Figure 1.5B ▲ The scale and size of living things

Careful study of pp. 13–32 will help you to understand the way of life of different organisms from each kingdom. Some of the organisms will be familiar to you; others will not. Many have been selected because they are easy to obtain and study. Each of the selections on particular kingdoms and phyla contains a fact file, body file and food file on representative organisms of that group. The files list important characteristics of the organisms shown, give details of their appearance and describe how they feed.

Are viruses alive? It is difficult to say.

Many biologists do not accept the five kingdoms scheme of classification. They think that more than five kingdoms should be recognised.

To reproduce (**replicate**), viruses must enter a living cell. Usually the virus attaches itself to the plasma membrane (see p. 147). Its nucleic acid (DNA or RNA – see p. 170) is inserted into the cell. Once inside the cell, the nucleic acid becomes part of the cell's genetic material. It makes the cell form new virus particles (**replication**) which escape when the cell bursts. Depending on the type of virus, the escape of particles may give rise to the symptoms of disease.

Viruses are not cells. They do not seem to need food as a source of energy and they cannot reproduce independently. Crystallised and stored like minerals, some viruses can remain unchanged for years until moistened and provided with a favourable environment for renewed activity. *Are viruses alive? Check p. 3 for the characteristics of living organisms and try to decide for yourself.*

Consisting of a strand of nucleic acid (DNA or RNA – see p. 170) enclosed in a coat of protein, a virus can only be seen using an electron microscope (see p. 146). Viruses infect and take over the biological machinery of living cells to reproduce new viruses. Some types of virus cause diseases in plants and animals.

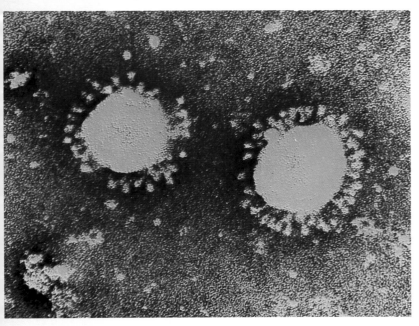

Bacterium kingdom

■ Food facts

Most bacteria are **saprobionts** —so called because they are responsible for decay and decomposition (see p.18). Bacterial cells secrete enzymes which convert dead organic matter and dead organisms into food which the cells absorb. Some bacteria are parasites, that is they live by feeding on the tissues of other organisms. Bacteria which cause disease belong to this group. Other bacteria can obtain their food through photosynthesis or by making use of chemicals like sulphur and hydrogen sulphide which they extract from the environment.

Cocci bacteria are spherical in shape and may stick together in pairs, chains or clusters as illustrated below. The photograph illustrates *Staphylococcus aureus* which causes boils and food poisoning.

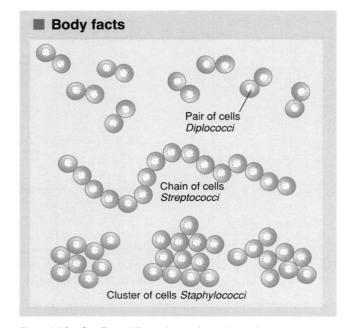

■ Body facts

Pair of cells
Diplococci

Chain of cells
Streptococci

Cluster of cells *Staphylococci*

Figure 1.5C ◆ Three different forms of cocci bacteria

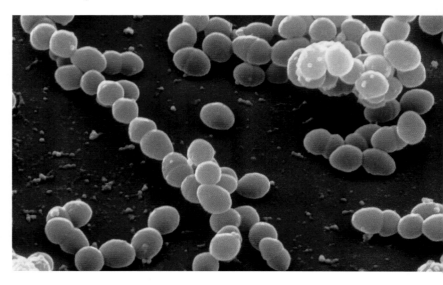

Bacilli bacteria are rod-shaped and can form into filaments of cells that look like the hyphae of fungi. The photograph shows *Escherichia coli* which is found in the human intestine. More is known about *E.coli* than any other living thing because it is used by scientists to investigate a wide range of biological problems.

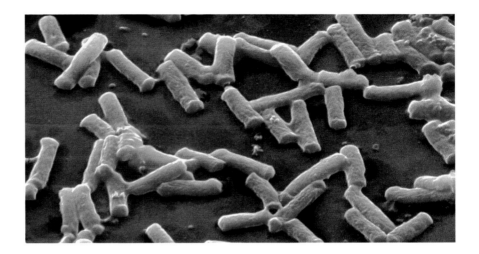

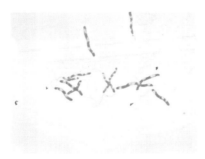

Spirilla bacteria look like tightly coiled springs. The photograph shows *Spirillum volutans* which is found in stagnant water.

Vibrios are comma-shaped bacteria. The photograph shows *Desulfovibrio desulfuricans* which releases sulphur into the environment. If the bacteria are living in water rich in iron compounds, the sulphur they release reacts to form iron sulphide which glitters like gold (iron sulphide is called 'fool's gold').

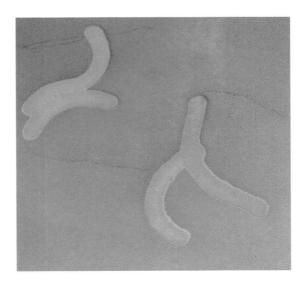

Protist kingdom

Amoeba lives in ponds and ditches. Flowing extensions (called **pseudopodia**: literally meaning 'false feet') of its single-celled body change its shape. Movement occurs when they flow in a particular direction. The name *Amoeba* describes all single-celled organisms that have pseudopodia. The bodies of some are surrounded by delicate shells of **calcium carbonate**. Some types of amoeba are **parasites** (see Topic 3.7) and can cause **dysentery** in their animal hosts.

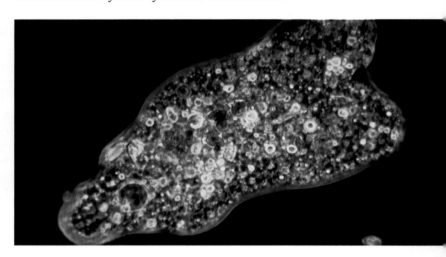

■ Body facts

Contractile vacuole removes water from the cell

Plasma membrane forms body surface

Nucleus

Food vacuole – space where food is digested

Pseudopodium – changes the shape of the cell

Figure 1.5D ⬆ *Amoeba*

■ Food facts

The diagram shows how *Amoeba* captures its food. Pseudopodia flow around the food particle which *Amoeba* detects by sensing the chemicals released into the water by the food. The pseudopodia join around the food particle and a food vacuole forms in the cell body. Digestion takes place in the food vacuole. The method of feeding by capturing food in this way is called **phagocytosis**.

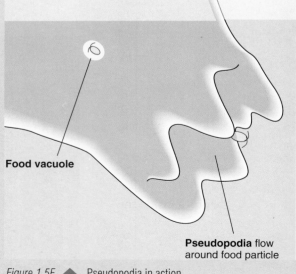

Food vacuole

Pseudopodia flow around food particle

Figure 1.5E ⬆ Pseudopodia in action

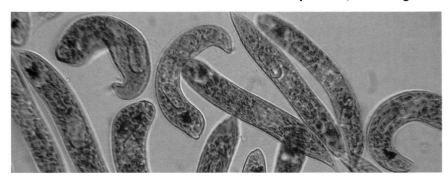

Euglena is a single-celled organism that lives in ponds and puddles, especially where there is a high level of nitrogen compounds. Notice the whip-like flagellum which lashes to and fro, driving the cell through the water. *Euglena* has chloroplasts and therefore can photosynthesise its food. However, like *Amoeba*, it can gather food from the environment. Single cells closely related to *Euglena* do not have chloroplasts and obtain food only by feeding. Other close relatives develop chloroplasts in bright light but lose them when they are kept in the dark. *Euglena* and closely related organisms display both plant and animal characteristics. This makes it difficult to put them into either the animal or plant kingdom and they are therefore included within the protists.

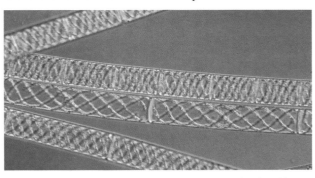

Spirogyra forms filaments of cells joined end to end. The filaments make a green tangled mat on the surface of ponds. Each cell is surrounded by a wall made of cellulose which in turn is surrounded by a layer of slimy substance called mucilage. A chloroplast winds in a spiral against the inner wall of each cell, giving *Spirogyra* its name.

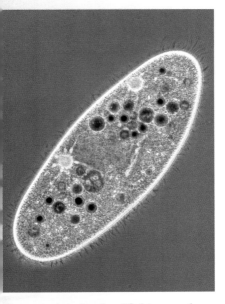

Paramecium is plentiful in ponds. The surface of the cell is covered with fine hair-like extensions called **cilia** which beat against the water, propelling *Paramecium* along. *Paramecium* slowly rotates as it swims. It takes in food particles through the gullet and encloses them in **food vacuoles** inside the cell. The vacuoles slowly move around the cell (**cyclosis**), and the particles inside are digested by enzymes. The digested food is absorbed into the cell.

Bladder wrack is a brown seaweed, common along the sea shore. Its body is made of branched, flat fronds. The **holdfast** at the end of the short stalk holds the seaweed in place on exposed rocks. Pairs of air-filled sacs (called bladders, hence the common name) help the organism to float when it is covered in water. The whole organism is limp and slimy to touch. These features protect it against the action of waves. The limp seaweed can bend and sway in the moving water so that water flows freely over the slimy surface of the fronds.

- Fungi are made up of slender tubes called **hyphae** (singular, hypha).
- The mass of hyphae which form an individual is called the **mycelium** (plural, mycelia).
- **Fruiting bodies** produce **spores**. Each spore can grow and develop into a new mycelium.

Fungus kingdom

Field mushroom: the fruiting body is all that we see above ground. The word 'mushroom' refers to the field mushroom which can be eaten. However, many types of so-called 'toadstool' make a good meal while others are very poisonous. So, 'mushroom' and 'toadstool' have little meaning. Mushrooms and some toadstools are edible; other toadstools are not. A few species are fatal if eaten! NEVER collect fungi for the table unless you are certain of their identification – BE SAFE NOT SORRY.

■ Body facts

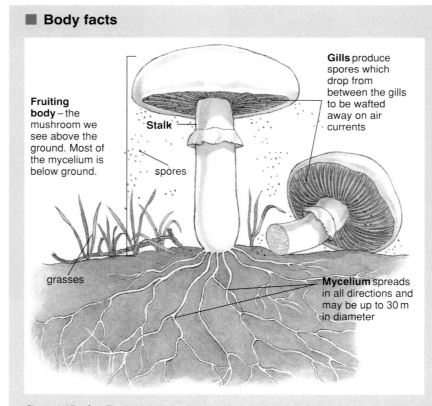

Figure 1.5F The fruiting body and mycelium of a field mushroom

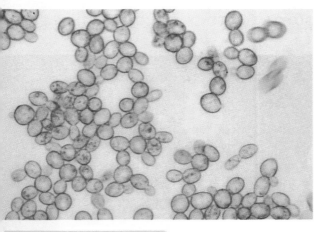

Yeasts do not grow hyphae but live as single cells. Their name *Saccharomyces* means 'sugar fungi' and they are commonly found wherever sugar occurs. For thousands of years yeast has been used to ferment the sugar in rice and barley to produce beer, and the sugar in grapes to make wine. Yeast added to dough ferments the sugar in the dough to produce carbon dioxide. This makes the dough rise.

EXTENSION FILE
ACTIVITY

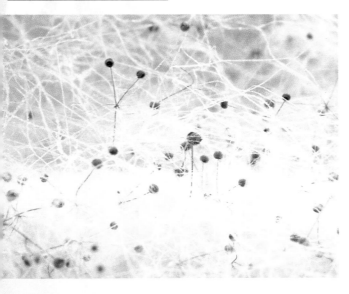

Fungi (and bacteria) are responsible for the decay and decomposition of dead organic material.

Bread mould hyphae spread over and into uncovered food, spoiling it. They form a mat-like mycelium. Hyphae (called aerial hyphae) poke up into the air, each carrying a capsule-shaped fruiting body. In the picture you can see that some are black, a sign that they contain ripe spores. Soon they will burst open, releasing a cloud of spores which are carried away on the air currents. If the spores land in a suitable place they will grow into new mycelia.

Other types of mould also spoil food but some add flavour. For example, blue cheeses such as Gorgonzola, Stilton and Roquefort owe their taste to the moulds which create the blue veins in them.

The mould *Penicillium notatum* produces a substance which can destroy bacteria. The substance, called **penicillin**, is an antibiotic used to fight diseases caused by bacteria (see p. 413).

It's a fact!

A lichen is a close partnership between an alga and a fungus. The algal cells grow in the fungal mycelium. They make food by photosynthesis and some of it is passed to the fungus. The fungus cannot survive without the alga. However, the alga does quite well on its own and grows even faster when free of the fungus.

■ Food facts

Hyphae secrete enzymes which convert dead organic material in the soil into food which the hyphae absorb. Obtaining food in this way is called saprobiontism (see p. 14). Most fungi are saprobionts but some are parasites, feeding on the living tissue of plants and animals (see p. 73).

FACT FILE

- The plants reproduce by means of spores.
- Water is needed for the sperm to swim to the eggs.
- Spore-producing plants live on land in damp habitats.

Mosses live in damp places because:
- the plant body quickly loses water in dry air,
- water is needed for sperm to swim to eggs.

Plant kingdom 1: spore-producing plants

Mosses quickly lose water in a dry atmosphere, so the plants live in damp places such as the banks of streams and the woodland floor where the air is moist. They do not have tissues to draw water from the ground as ferns and seed-producing plants do. They rely on capillary action to soak up water like a sponge. Closely fitting leaves help the capillary movement of water. This mechanism for water collection limits the plant's size. The tallest species of moss is only about 40 cm high and most are much shorter.

In summer, stalks grow from the moss plants, each carrying a capsule which contains spores. When ripe the capsule opens and the spores are shaken out to be carried away on air currents. If spores land on damp soil, they develop into new moss plants.

■ **Body facts**

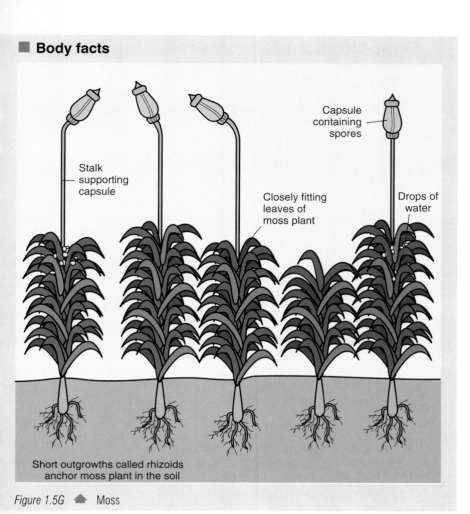

Capsule containing spores

Stalk supporting capsule

Closely fitting leaves of moss plant

Drops of water

Short outgrowths called rhizoids anchor moss plant in the soil

Figure 1.5G ◆ Moss

■ **Food facts**

Moss leaves are thin and delicate, each made up of only a few layers of cells. Chloroplasts in the cells of the leaves make food by photosynthesis.

Ferns have clumps of leaves called **fronds** which grow near the top of a thick stem. Roots sprout from the base of each clump, anchoring the plant firmly into the soil. Each leaf is built up of leaflets called **pinnae** which stand out on either side of a sturdy rib that runs through the centre of the leaf. Each leaflet is further subdivided into lobe-shaped **pinnules**. Young leaves are coiled in a bud.

A double row of greenish-white patches develops on the underside of the pinnae. Each patch consists of a group of spore-producing capsules which darken as the spores mature. When ripe, the capsules break open and the spores shoot out to be carried away on air currents.

A waxy layer waterproofs the plant's surfaces, reducing water loss. Ferns can live in a drier environment than moss plants, although they grow best in damp, shady places. Tissues in the roots, stem and leaves draw water from the soil and carry it to all parts of the plant. They also strengthen and support the plant which is much bigger than the moss plant, growing to a height of a metre or more.

- Plants reproduce by means of seed which protects the embryo plant from drying out on land.
- The sperm of seed-producing plants is transported to the eggs by pollen grains.
- Very successful land plants are able to live in dry, hot places where there is little water.

It's a fact!

Some trees live for a very long time. For example the Cedar of Lebanon lives for up to 1000 years. However, Bristlecone pines may be much older. They grow high in the mountains of eastern California and some are believed to be over 4000 years old.

Plant kingdom 2: seed-producing plants

Beech trees are widespread on gentle slopes and low-lying land. They are **woody** plants with massive trunks covered with smooth bark and bearing a crown of thick branches which carry leaves. New wood grows every year, increasing the girth of the tree. A cross-section of a tree trunk shows rings for each year's growth. Counting the **annual rings** will tell you the age of the tree. The beech tree's growth is greatest in the spring. Every autumn its broad leaves are shed. Plants that lose their leaves in this way are described as **deciduous**. Beech trees are **perennial**; that is they keep on growing and producing seeds for many years. A beech tree can live for 200 years or more.

Scots pine trees grow naturally in cool climates. They cover hill slopes and valleys with thick forests. The Scots pine is a large tree with a crown of branches covered with rough scaly bark. Short shoots growing from the branches carry clusters of needle-like leaves which are lost and replaced throughout the year. This is why the Scots pine and trees like it are called **evergreens**: they are covered with leaves all year round. **Cones** containing seeds form on young branches that grow in spring. If the tree is damaged a sticky resin oozes out from the wound and plugs it, preventing infection by disease-carrying organisms.

Buttercup grows in meadows and unimproved grassland. It is a non-woody **annual**; that is it grows from seed to maturity and produces new seeds all within one growing season. It then dies with the onset of the first frosts, leaving the seed to lie dormant through the winter ready to grow when conditions improve the following spring. Non-woody plants are called **herbaceous** plants. **Biennials** are plants that grow during one season and produce flowers and seeds in the next before dying. They often have tough woody stems to survive harsh weather between growing seasons.

A waxy layer waterproofing the surfaces means that seed-producing plants (and ferns) can live in a drier environment than mosses. Fern plants still need water for sperm to swim to eggs; seed-producing plants do not.

■ Body facts

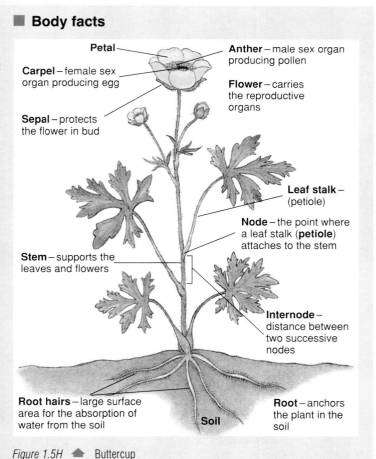

Petal

Anther – male sex organ producing pollen

Carpel – female sex organ producing egg

Flower – carries the reproductive organs

Sepal – protects the flower in bud

Leaf stalk – (petiole)

Node – the point where a leaf stalk (**petiole**) attaches to the stem

Stem – supports the leaves and flowers

Internode – distance between two successive nodes

Root hairs – large surface area for the absorption of water from the soil

Soil

Root – anchors the plant in the soil

Figure 1.5H ▲ Buttercup

■ Food facts

Buttercup leaves are complex structures. Each leaf is made up of different types of cell. The cells nearest the upper surface receive most light and have the most chloroplasts for photosynthesis.

Animal kingdom 1: *Hydra* and its relatives

- The body shape shows **radial symmetry:** that is it has no front or rear and its parts are arranged evenly 'in the round'.
- Tentacles surround an opening which is both mouth and anus.
- Stinging cells are used to capture prey.

Hydra lives in ponds and water-filled ditches. It hangs from water plants and other firm surfaces and trails its tentacles in the water. The tentacles carry sting cells which can paralyse small organisms which touch them. *Hydra* can then feed on its defenceless prey.

The body of *Hydra* is very simple in structure.

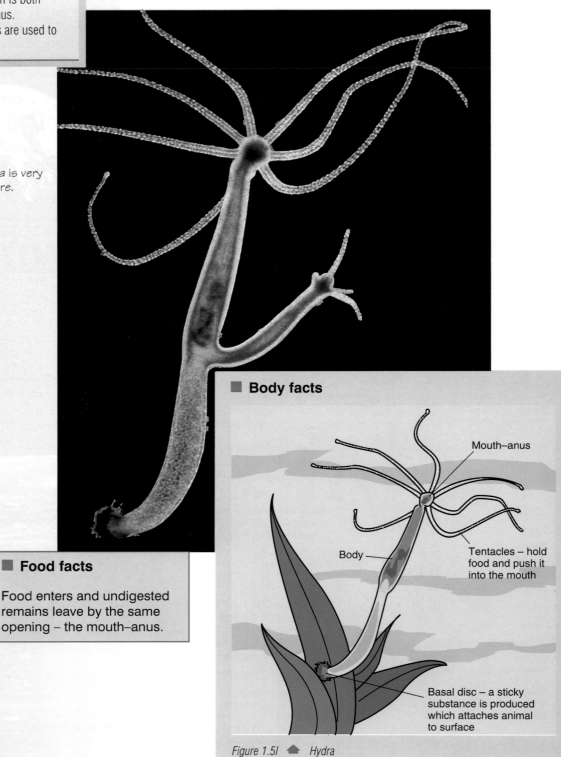

■ Body facts

Mouth–anus

Body

Tentacles – hold food and push it into the mouth

Basal disc – a sticky substance is produced which attaches animal to surface

Figure 1.5I ⬆ *Hydra*

■ Food facts

Food enters and undigested remains leave by the same opening – the mouth–anus.

Sea anemone. The specimen in the picture is called the beadlet anemone. It lives attached to rocks in pools that are left as the tide moves down the sea-shore.

Coral is made from limestone which hydra-like animals secrete around themselves. It builds up into reefs in warm sunlit seas. The Great Barrier Reef off the northeast coast of Australia is 2000 kilometres long and the world's largest reef.

Jellyfish usually hang, bell-like, in water with the tentacles floating around the mouth. When the bell pulses in and out the water jet which is produced propels the animal through the water. Most types of jellyfish live in the sea, but a few live in ponds and streams.

It's a fact!

The sea blubber *Cyanea* is the largest jellyfish in the world. Its tentacles are more than 30 metres long and the bell measures over 3 metres in diameter.

It's a fact!

Hydra takes its name from a mythical water-snake with nine heads. An ancient Greek legend tells that the hero Hercules was sent to kill the monster. Hercules cut off each of Hydra's heads but the ninth was immortal so he rolled a huge rock over it, trapping it for ever.

FACT FILE

- Flatworms have flattened bodies with upper (**dorsal**) and lower (**ventral**) surfaces.
- Their body shape shows **bilateral symmetry**.
- Flatworm bodies have a front end (**anterior**) and a rear end (**posterior**).
- Many of the different types of flatworm are parasites.

Animal kingdom 2: flatworms

Planaria is a flatworm which is common in ponds and streams where it lives under stones and plants. Notice the eyespots at the front end. They do not form images like our eyes but are sensitive to changes in light and shade. Other sense organs help the animal to test new environments as it moves forward.

■ Food facts

In *Planaria* the digestive cavity branches to all parts of the animal and its outline is just visible through the body wall. A part called the pharynx pokes out from the mouth–anus and captures small animals or breaks off fragments of dead larger ones. Most non-parasitic flatworms obtain their food in this way. Most parasitic flatworms feed either like the tapeworm or the fluke shown below.

■ Body facts

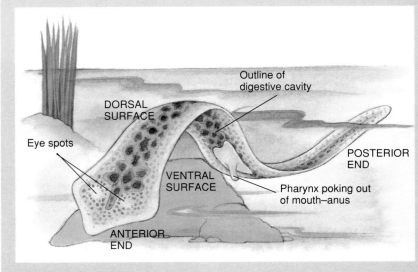

Eye spots

DORSAL SURFACE

Outline of digestive cavity

VENTRAL SURFACE

POSTERIOR END

Pharynx poking out of mouth–anus

ANTERIOR END

Figure 1.5J ⬅ *Planaria*

Flukes are common flatworm parasites. The specimen illustrated lives in the livers of sheep and cattle. It feeds on the tissue and blood of the liver causing great damage and even death. Notice the branching digestive cavity, and the suckers at either end which anchor the worm inside the **host** sheep or cow. Part of the life cycle of the fluke occurs in snails which are called the **secondary host** (or **vector** – see Topic 3.7). Flukes are transferred to sheep and cows feeding on grass to which infected snails are attached.

Tapeworm is a common flatworm parasite that lives in the intestines of vertebrates surrounded by the host's semi-digested food. The body is long and flat and, unlike *Planaria*, made up of sections called **proglottids**. The front end is called the **scolex** and has hooks and suckers which fasten the tapeworm to the inner lining of the host's intestine. The tapeworm does not have a digestive cavity. It absorbs some of the host's semi-digested food through its body wall. Tapeworms are particular about where they live. For example, a dog tapeworm is not usually found in humans or other animals; we say that tapeworms are 'host specific'.

Animal kingdom 3: larger worms

■ Food facts

The worm has a mouth at the head end and an anus at the rear end. The digestive cavity is an open tube that runs the length of the body. The means of taking in food and the means of passing out the undigested remains of a meal are separated. Food can therefore be digested and processed more efficiently than in flatworms which have only one opening to the digestive cavity.

It's a fact!

Nematodes are called roundworms because they are circular in cross-section. Found in vast numbers in water and soil, some species are parasites of vertebrates including humans (see p. 72). Other species parasitise plants.

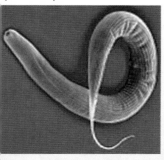

FACT FILE

- The body of the worm is long and thin with a distinct head end.
- The body is made up of many rings, called **segments.**
- The body is bilaterally symmetrical.

Earthworms are so called because they live in soil. One hectare of fertile grassland may support more than seven million earthworms! A worm makes its burrow by taking in soil through its mouth. The soil is finely ground up in the intestine where any organic material is extracted for food. The remains are passed out through the anus to form a **cast** above ground. Notice the **saddle**.

■ Body facts

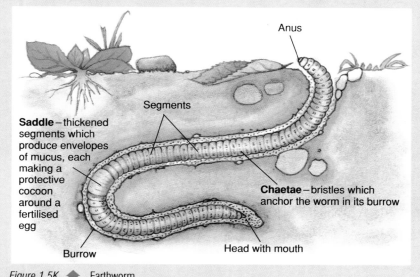

Anus

Segments

Saddle – thickened segments which produce envelopes of mucus, each making a protective cocoon around a fertilised egg

Burrow

Chaetae – bristles which anchor the worm in its burrow

Head with mouth

Figure 1.5K ◆ Earthworm

Medical leeches live in ponds and streams and feed on the blood of different vertebrate hosts. The leech has suckers which fix it to the host's body. It makes a small cut in the host's skin. Blood flows freely with the help of an anticoagulant which the leech produces. The anticoagulant stops the host's blood from clotting. A leech may take in up to five times its own weight of blood in a single meal and may not need another one for several months. When full, it drops from the host and digests the blood. Even until the nineteenth century doctors used to think that taking blood from a person who was ill would help them to recover. They used leeches to 'let blood' from the patient. This is why doctors were given the nickname 'leeches'.

It's a fact!

Although surrounded by a shell, snails are food for a variety of animals. For example, the picture shows a thrush hitting a snail against a stone to smash open the shell to get at the juicy flesh inside. The stones they use are called 'thrushes' anvils'.

EXTENSION FILE
ACTIVITY

FACT FILE

- The body of a mollusc is made up of a head, a foot, a mantle and a visceral mass, which contains the digestive cavity and other organs.
- The mantle of a mollusc produces the shell. This is outside the body of some molluscs but inside the body of others.
- The body shape shows bilateral symmetry.
- There is no segmentation in the body of a mollusc.

Animal kingdom 4: molluscs

Garden snails live under stones, rotting plants and in other damp, cool places. The **foot** of the snail moves the animal along on a trail of slimy mucus. If disturbed, it withdraws into its protective **shell**. This also helps to reduce the loss of water from the body. In dry weather the snail may retreat into its shell and secrete a film of mucus (called the **epiphragm**) across the opening. By sealing the animal inside further water loss is avoided. When it rains the epiphragm is cast off and the snail becomes active again.

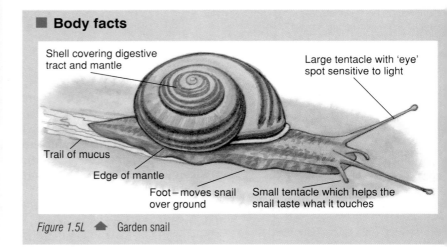

■ Body facts

Shell covering digestive tract and mantle

Large tentacle with 'eye' spot sensitive to light

Trail of mucus

Edge of mantle

Foot – moves snail over ground

Small tentacle which helps the snail taste what it touches

Figure 1.5L Garden snail

■ Food facts

Figure 1.5M is a cross-section through the head of a garden snail. It shows the **radula** which is a file-like tongue that the snail uses to rasp fragments of food into its mouth. All molluscs have a radula except for the molluscs that have two shells, such as mussels.

Figure 1.5M Cross-section through the head of a garden snail

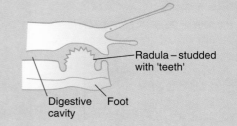

Radula – studded with 'teeth'

Digestive cavity

Foot

Chiton lives on the sea-shore attached to rocks and stones. The shell is made of eight overlapping plates and fits over the animal. When large waves threaten a chiton's grip on its rocky platform, the animal draws the shell down around itself, sealing itself away from danger.

Common mussels live in shallow water near the sea-shore. The mussel attaches itself by the foot to rocks and stones. Two shells close round the animal. Water is drawn through the crack where the shells meet and **filtered** for food. Mussels and their close relatives, oysters, are fished to provide delicacies for the dinner table.

It's a fact!

The diagram shows the two shells of a mussel prised open and the arrangement of body parts within. The gills are covered with beating cilia which draw a current of water containing microscopic organisms between the shells. The mussel filters the microscopic organisms from the water by trapping them in mucus that covers the mantle and gills. The mucus and entrapped food are then drawn into the mouth.

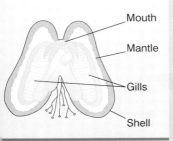

Mouth

Mantle

Gills

Shell

The octopus is a close relative of the squid and the cuttlefish. Its shell is inside the body. Some species live in shallow water near the sea-shore, half buried in sand during the day. At night they come out to feed on crabs, fish, snails and prawns. Other species inhabit deep water where they catch particles of food drifting down from above. The **siphon**, or funnel, is able to produce a water jet that can propel the animal through the water in any direction at high speed. When threatened, an octopus can squirt an inky liquid into the water. Under the black cloud it creates, the octopus can escape unseen by predators.

- The body of an insect is covered by a hard outer layer called the **exoskeleton**.
- Jointed limbs are adapted variously as legs for walking, paddles for swimming and for many other uses.
- The head and other regions of the body are formed from segments of the exoskeleton fused together.
- The insect head is well developed and has **compound** eyes.

Insects and their relatives are classified as arthropods because they have jointed legs and a body enclosed by an exoskeleton.

Animal kingdom 5: insects and their relatives

The locust shows the body features typical of most insects. It has two pairs of **wings** and is a strong flier. Its exoskeleton has a **waxy**, waterproof layer which helps to reduce the loss of water from the locust's body. This feature is so successful that locusts live in hot dry places where few other types of animal survive. All insects are waterproofed in this way, which is why there are so many of them living on land.

■ Food facts

Figure 1.50 shows the specialised mouthparts of a locust. The **mandibles** have saw-like edges which work like scissors. They cut up and chew plant food which the other mouthparts guide into the mouth. The **palps** help the locust to taste what it touches. In the butterfly the mouthparts form a tube-like **proboscis**. When feeding from flowers, nectar is sucked up the proboscis like water up a drinking straw. When not in use the proboscis is coiled up under the head like a clock spring. The way aphids (green fly and black fly), mosquitoes and houseflies feed is described in Topic 4.

■ Body facts

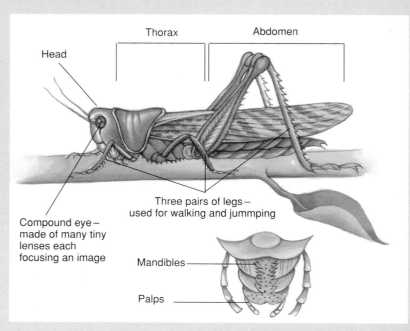

Figure 1.5N ◆ Locust

Figure 1.5O ◆ The specialised mouthparts of a locust

The garden spider spins sticky **webs** in which it catches its prey. Not all spiders spin webs; some chase their prey, others lie hidden in wait to ambush their victims. Like that of insects, the exoskeleton of the spider is waterproofed. They can live, therefore, in hot dry places. There are two regions to the body of a spider, a fused head and thorax with four pairs of legs and an abdomen. Spiders do not have wings.

The centipede feeds on insects and other small animals and lives on land. Like the woodlouse, it does not have a waterproof exoskeleton and is restricted to living in damp places. The millipede, a close relative of the centipede, feeds on dead leaves and other plant remains. The body consists of a head and trunk which is divided into segments. Each segment carries a pair of legs.

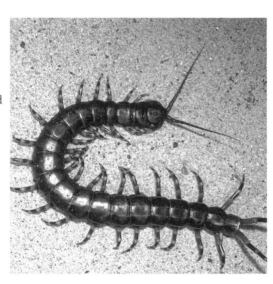

EXTENSION FILE
ACTIVITY

Crayfish live in clean, clear streams. Their body is divided into two regions, an abdomen and a fused head and thorax, with pairs of limbs adapted for swimming, catching food and other uses. Two pairs of antennae are attached to the head. Crabs, shrimps and prawns are close relatives of crayfish that live in the sea. The woodlouse is another close relative that lives on land. The woodlouse does not have the waterproof layer present in the exoskeleton of insects and spiders and is therefore restricted to living under stones, rotting wood and in other damp places.

- All vertebrates have a **vertebral column** (backbone) which runs along the dorsal (top) surface of the body.
- The backbone extends to form a tail.
- Vertebrates have internal skeletons, called **endoskeletons** which are most often made of bone.
- Limbs are variously adapted for swimming, walking and flying.

Animal kingdom 6: vertebrates

The carp is a species of bony fish. Its skeleton is made of **bone**. It lives on the bottom of muddy rivers and lakes. Different species of sea-living bony fish, like cod, plaice and haddock, are important sources of food. Bony fish have a **swim-bladder**, a gas-filled sac that helps the fish to control its depth in the water.

Fish without a swim bladder, for example sharks and skates, use their fins to control their depth. In such fish the skeleton is made of **cartilage**: they are cartilaginous fish. Stroking the body of a cartilaginous fish from tail to head feels like stroking sandpaper, but **scales** covering the body of a bony fish feel smooth. The **gill flap** is another feature which distinguishes the two types of fish; bony fish have them but cartilaginous fish do not.

The flap covering the gill slits of bony fish is called the operculum.

The frog is an amphibian. The adult frog lives on land but breeds in water-filled ditches and ponds. Fertilised eggs hatch into swimming tadpoles which grow and develop into adults. The change from tadpole to adult is called **metamorphosis**. Toads and newts are also amphibians. Toads, like frogs, divide their time between water and land, but newts rarely leave the water even as adults. The frog's skin is **soft** and wet and in dry air the body quickly loses water. So frogs, toads and most other amphibians live in wet environments.

■ Body facts

Dorsal surface

Eye – contains a single lens
which focuses a sharp image

Legs

Tail

Teeth

Lower jaw

Figure 1.5P ⬆ Lizard

Figure 1.5Q ⬆ The sharp, cone-shaped
teeth of a lizard

■ Food facts

The cone-shaped teeth of a lizard grip
and chew food. In different mammals
teeth are different shapes adapted for
cutting, piercing or grinding food.

Lizards are reptiles and are fully adapted for life on
land. Their skin is dry and covered with **horny scales**
which restrict water loss from the body. All reptiles
are waterproofed like this and so can live in hot, dry
places. They lay eggs in which the young develop,
protected by a **hard shell**. Water, therefore, is not
necessary for breeding.

The pigeon is a domestic bird which is descended
from the rock dove which makes its home on cliffs
and craggy hills. In towns and cities, tall buildings
substitute for cliffs (as far as pigeons are concerned).
Birds are covered with **feathers** which make flying
possible and keep in body heat. Feathers are covered
in oil which keeps out water and prevents them
from getting waterlogged in the rain. The oil is
produced from a special gland commonly called the
'parson's nose'. Instead of teeth, birds have a **beak**
which is adapted differently in different species to
deal with various types of food. Birds lay eggs
protected by a **hard shell**, usually in a nest. Parent
birds **care** for their young.

Homo sapiens is a species of mammal. Us! Humans have mammalian features and characteristics. We **care** for our young, which the female feeds in early life with **milk** from her **mammary glands** (breasts). We have **hair**, which helps to conserve body heat, though we do not have as much as other mammals. We have a small tail bone called the **coccyx** at the base of our backbone. The ear tube (called the **pinna**) funnels sound waves down the ear tube. Most mammals move by using all four limbs, but we stand upright and our forelimbs are adapted as hands, able to carry out different tasks. Our large brain and dexterous hands mean that we have more control over our environment than the members of any other species.

SUMMARY

Living things are organised into five kingdoms which represent different ways of life. Some living things are very small and only visible under the microscope, others are very large. All are adapted for their way of life.

CHECKPOINT

▶ 1 Look at Figure 1.5A. With the help of named examples, summarise the different ways of life represented by the five kingdoms.

▶ 2 Compare the appearance of yeast, bread mould and the field mushroom.

▶ 3 Why can fern plants live in a drier environment than mosses?

▶ 4 Using the buttercup and beech tree as examples, explain the meaning of the words annual and perennial. State an important function of wood.

▶ 5 (a) What kind of symmetry is shown by the body of (i) a mussel and (ii) a jellyfish?
  (b) Explain how the type of symmetry suits the way each animal spends its life.

▶ 6 Snails and frogs live in water and on land. What prevents the animals losing too much water when on land?

▶ 7 List the features that make humans mammals. Which ones help us to have more control over our lives and the environment than any other kind of living thing?

Topic 2   Ecology

2.1 ▶ What is the environment?

FIRST THOUGHTS

Ecology is the study of how living things interact with each other and with their environment. The scientists involved in the work are called ecologists.

Planet Earth (see Figure 1.1A) is our home and the home of all life as we know it. Elsewhere in the Universe there may be other planets with life forms on them, but we have not yet discovered them. Figure 2.1A shows that high mountains, slopes and flat ground, coastal waters and ocean depths are all homes for living things. These and all other places on Earth where there is life form the **biosphere**.

Living things are not found just anywhere in the biosphere. For example, fish swim in the sea, trees are rooted in soil, and so on. Each organism is suited (**adapted**) to the place where it lives. This place is its **environment**. It consists of a non-living (**abiotic**) part of air, soil or water and a living (**biotic**) part of plants, animals and microorganisms. *How many environments does Figure 2.1A show?*

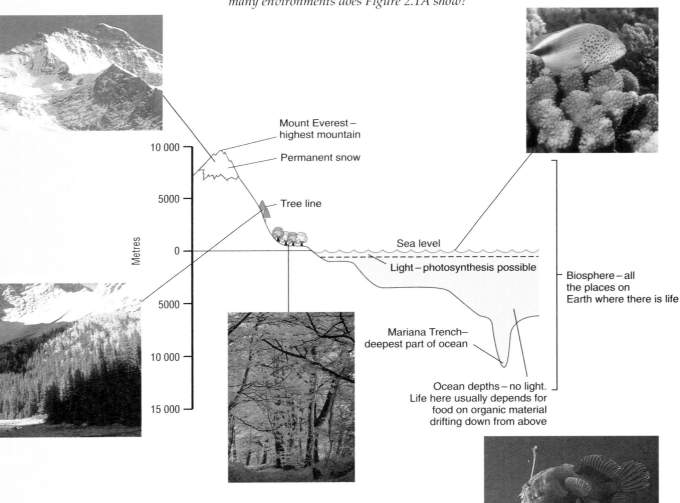

Figure 2.1A ⬆ Different environments in the biosphere. Little can live in permanent snow, apart from a few simple organisms. Light fades gradually as it passes through water, giving way to darkness below 200 m

The physical environment: soil

Soil, air and water are the physical (**abiotic**) environments in which organisms live (see Figure 2.1B) Figure 2.1C shows how soil is formed. The weather breaks down rocks into small particles. The roots of plants push rocks apart, cracking and splitting them into fragments.

Figure 2.1B ⬆ (a) Soil forms a thin layer covering most of the Earth's land surface. Notice the layered arrangement of the soil. Each layer is called a horizon

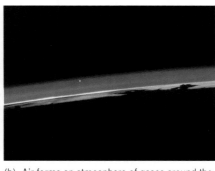

(b) Air forms an atmosphere of gases around the Earth

(c) Water (fresh and salt) covers 75% of the Earth's surface

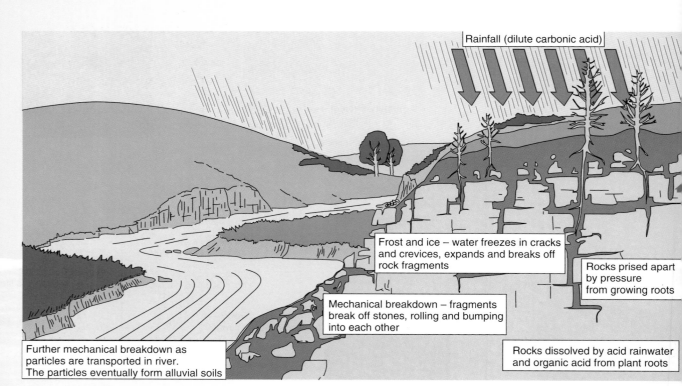

Rainfall (dilute carbonic acid)

Frost and ice – water freezes in cracks and crevices, expands and breaks off rock fragments

Rocks prised apart by pressure from growing roots

Mechanical breakdown – fragments break off stones, rolling and bumping into each other

Further mechanical breakdown as particles are transported in river. The particles eventually form alluvial soils

Rocks dissolved by acid rainwater and organic acid from plant roots

Figure 2.1C ⬆ Soil formation

◼ Humus

Rock particles form the mineral part of soils. The breakdown of wastes and remains of dead organisms forms **humus**. Humus formation depends on the activities of a variety of organisms. Figure 2.1D shows how some of them work in the dead wood of a fallen tree.

Humus is a dark, fibrous material (Figure 2.1E). It contains the decomposed remains of plants and animals and a mixture of complex substances produced by the **decomposers** (see p. 46) themselves as they break down dead organic matter.

Saprobiontic fungi and bacteria feed on the dead wood causing decay and decomposition

Woodlice, wood-boring beetles and other wood-eating animals break up the tree's remains into crumbly pieces called frass. This increases the surface area of the wood exposed to attack by fungi and bacteria

When dead material has decomposed to the point where the original organism is unrecognisable, it is called **humus**

Earthworms pull dead leaves and other organic fragments into their burrows for food

Decomposition releases gases, minerals and water from the dead wood into the soil

Nutrients for the growth of new plants

Figure 2.1D ▲ Humus is formed through the feeding activities of decomposers, so called because they break down dead organic matter. Soil fertility depends on their activities

Humus improves soils by:

- absorbing large amounts of water. This increases the soil's water content
- reducing evaporation from the soil in dry weather
- absorbing heat and warming the soil
- providing food for organisms such as earthworms, which help keep the soil in good condition
- improving soil texture: in wet, heavy clay humus helps soil particles stick together into larger crumbs; in dry, light, sandy soils humus helps improve water and mineral content by absorbing water and reducing drainage (see p. 89).

Texture is an important property of soil. It affects:

- the distance between soil crumbs
- the amount of air in the soil
- drainage (see also p. 89).

All this in turn affects the plants growing in the soil and the animals and microorganisms underground which help keep the soil in good condition.

Figure 2.1E ▲ Beech leaves take a long time to decompose. They contain high levels of tannin. This stops or slows down the action of fungi or bacteria which break down dead organic matter.
- In the top layer, leaves from the previous autumn are easily recognisable; there is little decomposition. This is the **litter** layer.
- In the deeper layers, leaves begin to lose their shape as decomposers break them down. This is the partly decomposed **fermented** layer.
- The bottom layer consists of leaves which fell four to five autumns ago. They are unrecognisable as leaf material. This is the completely decomposed **humus** layer.

SUMMARY

Soil is formed from small particles of broken-down rock. Humus contains the decomposed remains of plants and animals; it improves soil.

CHECKPOINT

▶ **1** List the physical (abiotic) environments in which organisms live.

▶ **2** How is soil formed?

▶ **3** What is humus. How does it improve soil?

▶ **4** Explain the meaning of 'soil texture'.

The physical environment: air and water

■ Air

The thin blanket of gases surrounding the Earth is called the **atmosphere**. Without it, life could not exist. The atmosphere can be divided into layers according to temperature.

The bottom layer of the atmosphere, called the **troposphere**, extends for about 16 km above sea level. The air temperature drops as it gets further away from the Earth's surface. That is why there is always snow on top of high mountains like Mount Everest (see Figure 2.1A). The troposphere is important for living things. It contains most of the atmosphere's gases, water vapour and dust particles (Table 2.1). Oxygen is essential for the **respiration** of most organisms (see Topic 15.1), carbon dioxide is needed for **photosynthesis** (see Topic 11.1).

Above the troposphere is the **stratosphere**. Here oxygen molecules (O_2) break down and recombine to form a layer of **ozone** (O_3). This screens out harmful ultraviolet rays which would kill most forms of life if they reached the Earth.

Table 2.1 ▼ Gases of the troposphere make up the air we breathe

Gas	%
Nitrogen	78
Oxygen	21
Carbon dioxide	0.03
Other gases, e.g. water vapour, argon and xenon	<1

■ Water

98% of the Earth's water is in the seas, oceans, rivers and lakes. The remaining 2% is locked up as ice, in the soil, in the bodies of living organisms or is vapour in the atmosphere.

The sun's heat evaporates water from seas, oceans, rivers and lakes and turns it into vapour. In the atmosphere the vapour cools, condenses and falls to earth as snow or rain. Some of the water that falls on soil passes down (**percolates**) to the bedrock beneath to become part of lakes and rivers which eventually flow into the sea. Living organisms are part of the **water cycle**. Plants absorb water and this passes to animals when they feed and drink. They also give out water in **transpiration** (see Topic 12.2).

Life processes such as **respiration** (see Topic 15.1). return water to the environment. Water, therefore, is used again and again and the amount in circulation stays more or less constant (see Figure 2.1F).

Figure 2.1G (a) and (b) represent sections through a pond. They show some of the physical and chemical conditions of water which affect where pond organisms live.

The physical and chemical conditions at the surface and bottom of a pond are different. Over 24 hours, conditions change at the surface but remain more or less the same (except

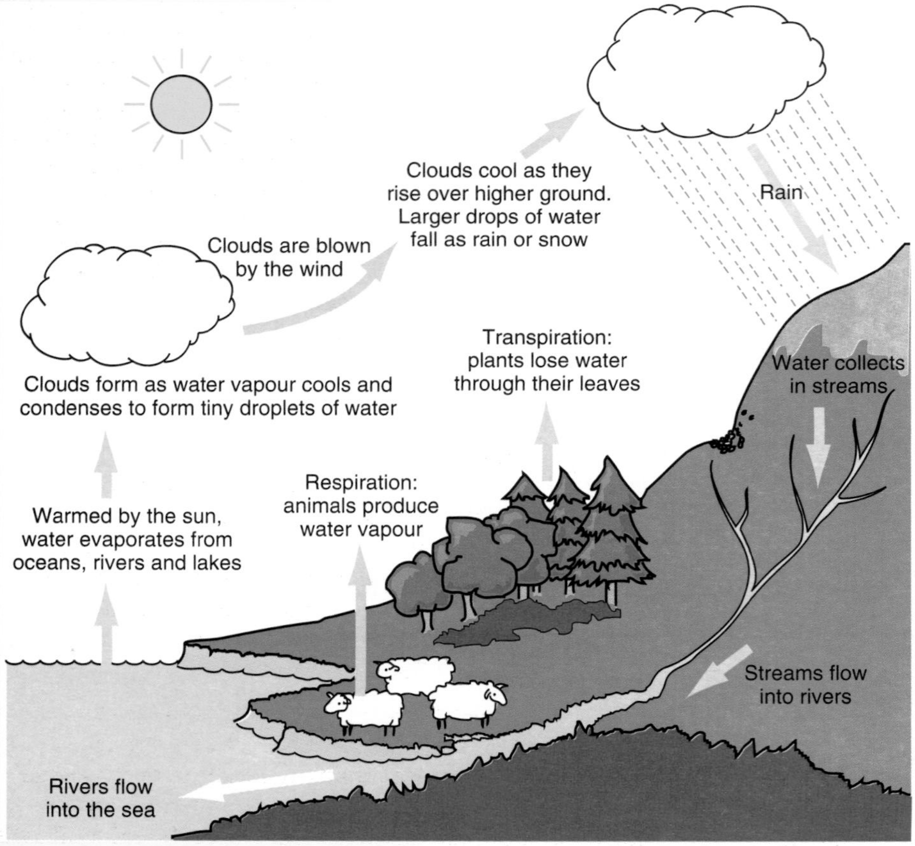

Clouds cool as they rise over higher ground. Larger drops of water fall as rain or snow

Clouds are blown by the wind

Rain

Clouds form as water vapour cools and condenses to form tiny droplets of water

Transpiration: plants lose water through their leaves

Water collects in streams

Warmed by the sun, water evaporates from oceans, rivers and lakes

Respiration: animals produce water vapour

Streams flow into rivers

Rivers flow into the sea

Figure 2.1F ◆ The water cycle

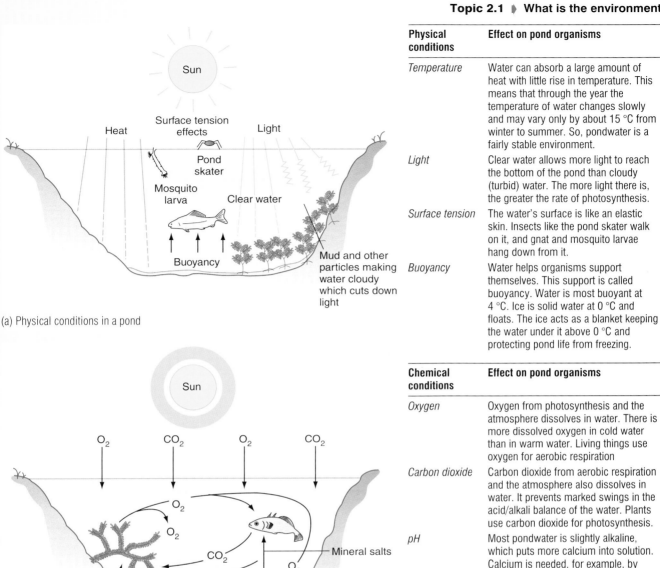

(a) Physical conditions in a pond

Physical conditions	Effect on pond organisms
Temperature	Water can absorb a large amount of heat with little rise in temperature. This means that through the year the temperature of water changes slowly and may vary only by about 15 °C from winter to summer. So, pondwater is a fairly stable environment.
Light	Clear water allows more light to reach the bottom of the pond than cloudy (turbid) water. The more light there is, the greater the rate of photosynthesis.
Surface tension	The water's surface is like an elastic skin. Insects like the pond skater walk on it, and gnat and mosquito larvae hang down from it.
Buoyancy	Water helps organisms support themselves. This support is called buoyancy. Water is most buoyant at 4 °C. Ice is solid water at 0 °C and floats. The ice acts as a blanket keeping the water under it above 0 °C and protecting pond life from freezing.

Chemical conditions	Effect on pond organisms
Oxygen	Oxygen from photosynthesis and the atmosphere dissolves in water. There is more dissolved oxygen in cold water than in warm water. Living things use oxygen for aerobic respiration
Carbon dioxide	Carbon dioxide from aerobic respiration and the atmosphere also dissolves in water. It prevents marked swings in the acid/alkali balance of the water. Plants use carbon dioxide for photosynthesis.
pH	Most pondwater is slightly alkaline, which puts more calcium into solution. Calcium is needed, for example, by snails – for building their shells.
Mineral salts	Mineral salts (especially phosphates, nitrates and sulphates) dissolve in water. They are nutrients for plant growth and healthy life.

O_2 = oxygen
CO_2 = carbon dioxide

Figure 2.1G ⬆ (b) Chemical conditions in a pond

SUMMARY

The atmosphere creates an environment suitable for life. Water is recycled in the environment. The physical and chemical conditions of water affect where pond organisms live.

for light) at the bottom. At night-time, for example, the surface temperature drops but the temperature at the bottom of the pond remains fairly steady. So at night, water at the bottom of the pond is warmer than water at the top. This switch in temperature is called a **temperature inversion**. At night-time, too, there is less oxygen in the water, particularly at the surface, because plants no longer release oxygen from photosynthesis but aerobic respiration, which uses oxygen, continues. Figure 2.1H shows these changes.

Figure 2.1I shows annual changes in the amount of mineral salts in solution. In autumn and winter decomposers break down the dead remains of plants and animals (**detritus**) lining the bottom of the pond. This releases minerals and gases. In spring and early summer, longer daylight hours bring a burst of plant growth which uses the mineral salts built up over the winter months.

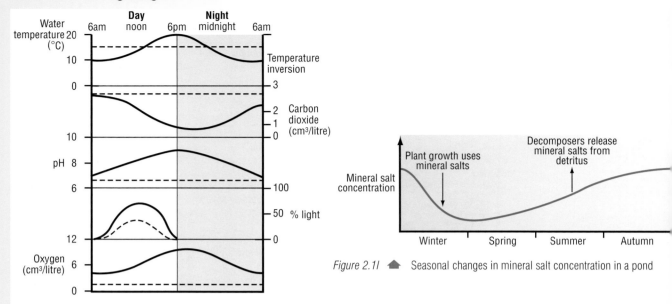

Key
—————— Surface conditions
------------ Conditions at bottom

Figure 2.1H Changes in some physical and chemical conditions at the surface and bottom of a pond

Figure 2.1I Seasonal changes in mineral salt concentration in a pond

CHECKPOINT

▶ **1** How does the water cycle work?

▶ **2** Why can pond skaters walk on water?

▶ **3** Describe the seasonal changes in mineral salt concentration in a pond. Why do these occur?

▶ **4** What is meant by 'the biosphere'?

2.2 ▶ Ecosystems

FIRST THOUGHTS

Ecosystems are the basic units of study in ecology. Each one is made up of a community of living things in non-living surroundings.

See www.keyscience.co.uk for more about ecology.

An ecosystem consists of a community of organisms living in a particular environment.

Ecology involves studying the relationships between organisms and between organisms and their environment. The **ecosystem** is an important idea in ecology. It describes a more or less self-contained part of the biosphere like a pond or an oakwood. 'Self-contained' means that each ecosystem has its own characteristic organisms not usually found in other ecosystems. They are the living (**biotic**) part of the ecosystem. The physical environment is the non-living (**abiotic**) part where they live. Ecosystems exist in great variety as Figure 2.2A shows. Each one is unique.

There is always overlap where the boundaries of ecosystems meet. Such boundaries are really only imaginary lines for our convenience, but they are based on natural limits. For example, the boundary of an oakwood is where the trees thin into grassland; the boundary of a pond is at the water's edge.

As Figure 2.2B shows, a frog may leap from its pond (one ecosystem) on to land (a different ecosystem) and back again. However, two-way traffic between ecosystems is limited because organisms are usually adapted to the particular conditions of only one ecosystem. For example, although the frog can survive on land it lives in damp places, because its body rapidly loses water in dry air. This would soon kill it. The frog must also return to water to breed.

Figure 2.2A 🔺 Ecosystems

SUMMARY

The subject matter of ecology is the relationships between organisms and between organisms and their environment. An ecosystem consists of organisms (the biotic part) which live in a physical environment (the abiotic part). Organisms are adapted to the particular conditions of an ecosystem.

Figure 2.2B 🔺 The boundary between neighbouring ecosystems is not always clear-cut

41

2.3 ▶ Habitats, niches and the community

FIRST THOUGHTS

A community consists of groups of living things. Each group lives in the community in its habitat. Its role within the habitat is called its niche.

Organisms are the living (biotic) part of an ecosystem. They form the **community**. The place where a group of organisms live in the community is called its **habitat**. Figure 2.3A shows a pond community and its main habitats.

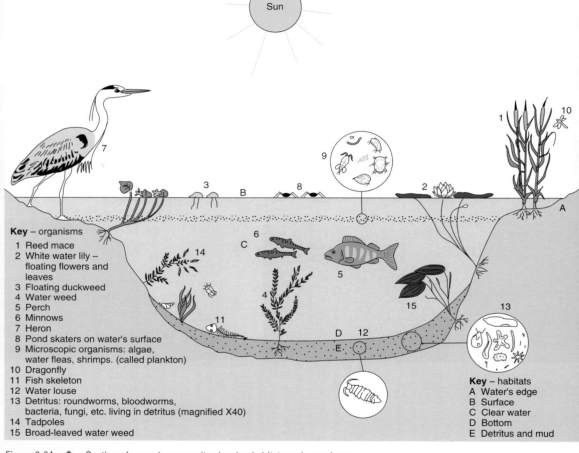

Key – organisms

1 Reed mace
2 White water lily – floating flowers and leaves
3 Floating duckweed
4 Water weed
5 Perch
6 Minnows
7 Heron
8 Pond skaters on water's surface
9 Microscopic organisms: algae, water fleas, shrimps. (called plankton)
10 Dragonfly
11 Fish skeleton
12 Water louse
13 Detritus: roundworms, bloodworms, bacteria, fungi, etc. living in detritus (magnified X40)
14 Tadpoles
15 Broad-leaved water weed

Key – habitats
A Water's edge
B Surface
C Clear water
D Bottom
E Detritus and mud

Figure 2.3A 🔺 Section of a pond community showing habitats and organisms

EXTENSION FILE
ACTIVITY

The habitat of water-lily flowers and duckweed is the water's surface (Figure 2.3B). The habitat of water weeds is the bottom of the pond, where it is rooted in the mud (Figure 2.3C). In turn, these plants are habitats for groups of animals. Water spiders and shrimps (Figure 2.3D(a)) hide under the floating duckweed and water lily leaves. Water lice (Figure 2.3D(b)) walk on the bottom and among the water weeds.

When pond organisms die they sink to the bottom. They form a layer of **detritus** along with leaves, twigs and other bits and pieces that fall into the pond. This layer is the habitat of many kinds of organisms (Figure 2.3E). The best way of understanding the pond community is to make a practical study of it.

Figure 2.3F shows the oakwood community and its main habitats. The plants are arranged in layers. Different birds live in different layers and each layer is a habitat for a variety of living things. An oak tree itself is a habitat for a range of species as Figure 2.3G shows.

Figure 2.3B 🔺 (a) Water lily flowers and leaves float on the water's surface. Insects are attracted to the flower's yellow pollen. The plant is rooted in the mud at the bottom of the pond. It closes at night and sinks below the water's surface

(b) Duckweed covers a pond with a bright green carpet of thousands of tiny plants. Each one is a single leaf-like disk with a root hanging down just under the water's surface

(a) Shrimps like clean, well-oxygenated water. The shrimp swims on its side and eats mainly decaying plants. It is eaten by fish

Figure 2.3C 🔺 Water weed rooted in gravel and mud at the bottom of a pond

Figure 2.3D 🔺 (b) The water louse is large and easily seen, climbing about among water weeds. Unlike the shrimp, the water louse tolerates dirty water with a low oxygen content

Figure 2.3E 🔺 Detritus at the bottom of a pond

Key – habitats
A Canopy layer
B Shrub layer
C Field layer
D Ground layer
E Detritus

Key – organisms
1 Oak tree
2 Hazel
3 Holly
4 Bluebell
5 Wood anemone
6 Primrose
7 Moss
8 Pigeons, rooks living in canopy
9 Blue tits, woodpeckers living further down tree
10 Great tits, warblers living in shrubs
11 Wrens, blackbirds living on ground
12 Toadstools on rotting log
13 Woodlice in detritus
14 Earthworm pulling leaf into burrow
↓ Falling leaves

Look back to p. 42 and remind yourself of the meaning of the word 'habitat'. Now look at Figure 2.3G again. *Do you think an oak tree is a habitat in itself or many habitats?* Give reasons for your answer, briefly explaining your understanding of a habitat.

Figure 2.3F ◆ Section of an oakwood community, showing the layered arrangement of plants. Each layer is a habitat

It's a fact!

The **species diversity index** is a measure of the number of different types (species – see p. 8) of organism in an ecosystem (see p. 40). Most biologists believe that the more species there are, the more stable is the ecosystem. **Biodiversity** is the term used to refer to the number of different species in an ecosystem.

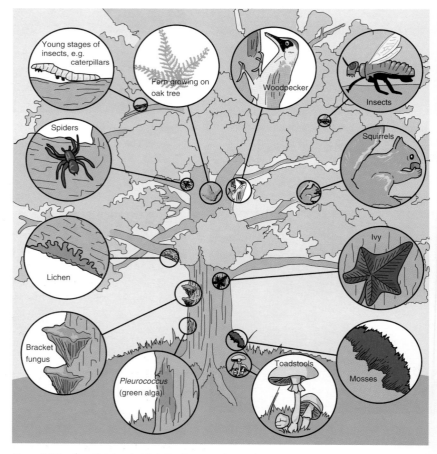

Figure 2.3G ◆ Oak tree habitat. The fern growing on the tree trunk is an example of an **epiphyte**. Epiphytes are plants that use other plants for physical support

■ Niches

■ Niches

Within the habitat each species has its own role to play. This role is called its **niche**. Information about the niche tells us what a species feeds on, what feeds on it, where it rests and how it reproduces.

Clearly, food is an important part of the niche idea. For example, the niche of a water spider is feeding on water fleas and other small animals living in the pond; the niche of a woodpecker is feeding on insects with its long, pointed beak.

Detritus is a rich source of food. In a wood, for example, earthworms, woodlice and different insects break up detritus into small pieces which are then food for fungi, microscopic animals and bacteria (Figure 2.3F). Fungi and bacteria are called decomposers, because their feeding activities cause decay and decomposition (see Topic 2.6). Their niche is that of recycling nitrogen and other elements locked up in the detritus to be used as nutrients for the growth of new plants.

SUMMARY

Organisms form the community. The place where a group of organisms lives in the community is its habitat. The niche is the role of a species in its habitat.

Make sure that you can distinguish between the community, its habitats and the niches of living things within habitats.

CHECKPOINT

▶ 1 Explain, in your own words, what you understand by 'ecology'.
▶ 2 Distinguish as clearly as possible between the following:
   (a) Biosphere and ecosystem
   (b) Community and environment
   (c) Habitat and niche
▶ 3 (a) What is detritus?
   (b) Compare the detritus in a pond with the detritus of a wood.
▶ 4 Using examples briefly explain the niche of decomposers.

2.4 ▶ Community structure

Producers and consumers

FIRST THOUGHTS

Finding out the feeding relationships between producers and consumers is an important way of describing a community. Feeding relationships can be shown as food chains or food webs. Most communities ultimately depend on photosynthesis (see p. 245).

In any community, the organisms can be classified by their method of feeding. Basically, either they can make their own food or they can't.

Autotrophs are able to make their own food from simple substances such as carbon dioxide and water by photosynthesis (see p. 177). Green plants are autotrophs. They use light as a source of energy to make sugars. Algae and many types of bacteria are also autotrophs. Some obtain their energy from light and others obtain it from simple chemical reactions. Since autotrophs ultimately provide food for all the other members of the community, we call them **producers** (see Figure 2.4A).

Heterotrophs cannot make their own food so they have to eat it. For this reason they are called **consumers**:

● **Primary consumers** are herbivores. They eat producers.
● **Secondary consumers** are carnivores. They eat herbivores.
● **Tertiary consumers** are carnivores that eat secondary consumers.

Each category of feeding is known as a **trophic level** (from the Greek *trophos* which means 'a feeder'). *But how do we know what an animal feeds upon?* We can get a good idea by looking at **teeth** (see p. 229) and other feeding structures. We can also investigate the contents of the animal's gut or study its feeding behaviour.

Figure 2.4A ⬆ Green plants trap the radiant energy of sunlight and use it to make sugars by photosynthesis

Make sure you can answer an exam question which asks you to distinguish between autotrophs and heterotrophs.

Decomposers

Decomposers are saprobionts (see pp. 14 and 19). They are consumers that obtain energy from dead material which is broken down in the process. Bacteria and fungi (see Topic 1.5) are decomposers. As we see on p. 54 they perform an important role in the cycling of nutrients. By breaking down dead material, they make it available as nutrients for the growth of new plants. (See Figure 2.4B.)

Figure 2.4B Decomposition in progress

Food chains and food webs

Feeding methods are a useful way of showing the relationships that exist in a community. We can highlight these by means of **food chains** which show the passage of food (and therefore energy) from one organism to another. Food chains always start with a producer. The food then passes to a primary consumer, next to a secondary consumer and so on. The number of links (producer and consumers) in a food chain may vary, but is seldom more than five. (See Figure 2.4C.)

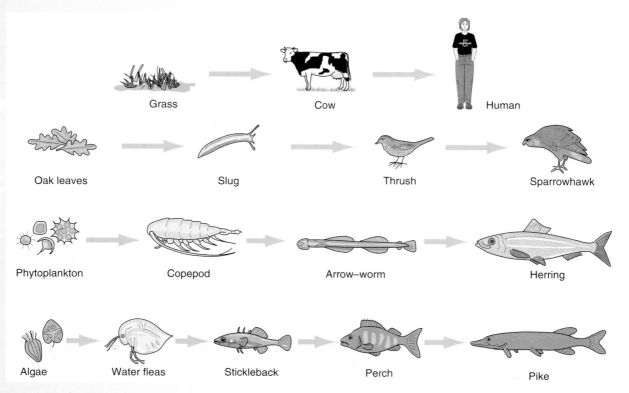

Grass → Cow → Human

Oak leaves → Slug → Thrush → Sparrowhawk

Phytoplankton → Copepod → Arrow–worm → Herring

Algae → Water fleas → Stickleback → Perch → Pike

Figure 2.4C Food chains

SUMMARY

Producers are able to make their own food. They include green plants and some bacteria. Producers provide food for primary consumers (herbivores). These in turn provide food for secondary consumers (carnivores). Decomposers feed upon dead and decaying material. These feeding relationships can be shown by the use of food chains and food webs.

Different food chains exist in a community and link up to form a **food web** (see Figure 2.4D) because the diet of an animal consists of a number of different species. A food web represents a more complete picture of the feeding relationships in a community.

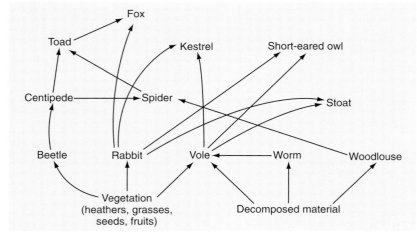

Figure 2.4D ⬆ Food web for a moorland community

Remember why a food web is usually a more accurate description of feeding relationships in a community than a food chain.

CHECKPOINT

▶ **1** Study the food web and answer the questions which follow.

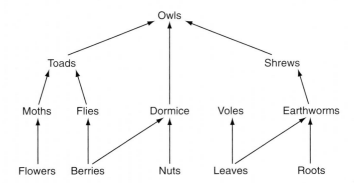

(a) From the diagram name (i) two primary consumers, (ii) one secondary consumer.

(b) Construct, using the food web above, two different food chains, each with four links. The arrows should point from eaten to eater.

(c) Explain why every food chain must begin with a green plant.

(d) If all the owls were killed, what would happen to the size of the populations of (i) dormice, (ii) earthworms?

(e) In the food web shown, the moths depend upon the flowers and the dormice depend upon the hazel tree for food. Give one example to show how (i) the flowers may depend on the moths, (ii) the hazel tree may depend upon the dormice.

▶ **2** The diagram below shows a food web.

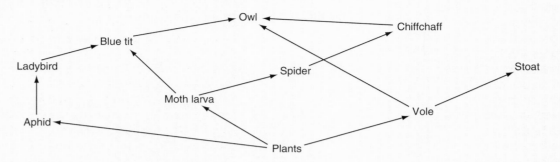

(a) Name the ultimate source of energy for all the organisms in this food web.

(b) From this food web name one example of a secondary consumer.

(c) Construct a food chain which includes examples of animals from four different feeding levels in this food web.

(d) If all the spiders were killed by disease, suggest the effects this could have on the food web shown.

2.5 ▶   # Ecological pyramids

FIRST THOUGHTS

Ecological pyramids are another way of describing feeding relationships in a community. There are three types: numbers, biomass and energy.

Food chains and food webs describe the feeding relationships that occur in a community. However they do not tell us about the numbers of individuals involved. It takes many plants to support a few herbivores. Similarly there must be far more prey than there are predators. (See Figure 2.5A.)

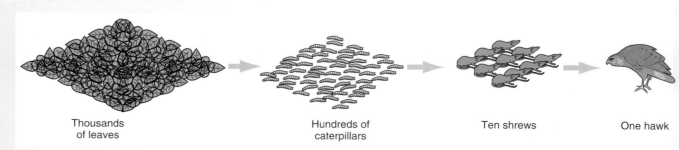

Thousands of leaves   Hundreds of caterpillars   Ten shrews   One hawk

Figure 2.5A ◆ More plants than herbivores – more prey than predators. Caterpillars and shrews are prey. The hawk is a predator.

Pyramids of number

Pyramids of number give information about the numbers of individuals in each trophic level. They are drawn up by counting the number of individuals in a certain area, say one square metre. These numbers are then plotted like bar charts but horizontally (see Figure 2.5B). The producers are at the base of the pyramid. Above them are the primary consumers. Next are the secondary consumers and so on. The numbers of individuals decrease as we go up the pyramid.

Figure 2.5C shows a pyramid of numbers drawn from data collected for a grassland community.

Pyramids of number do not take into account the size of the organisms at each trophic level. For instance one oak tree will support many more herbivores than one grass plant. Figure 2.5D shows an example. The pyramid appears top heavy if parasites are included. Many parasites feed on the secondary consumers (ladybirds, for example). The pyramid is said to be **inverted**. Figure 2.5E illustrates the point.

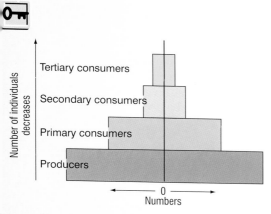

Figure 2.5B ⬆ How to plot a pyramid of numbers. Half of the number of organisms in each trophic level is plotted on one side of the vertical line, the other half is plotted on the other

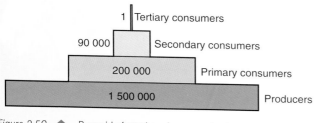

Figure 2.5C ⬆ Pyramid of numbers for a grassland community in 0.1 hectare

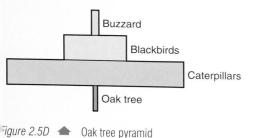

Figure 2.5D ⬆ Oak tree pyramid

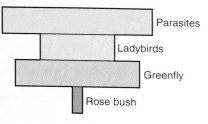

Figure 2.5E ⬆ Inverted pyramid

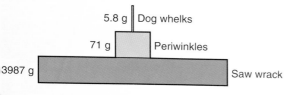

Figure 2.5F ⬆ Biomass pyramid for a rocky shore (m⁻²)

Pyramids of biomass

One way of overcoming the problem of the size of organisms, is to chart the amount of organic material (**biomass**) at each trophic level. A representative sample of the organisms at each trophic level is weighed. The mass is then multiplied by the estimated number in the community. In practice the **dry mass** is used since fresh mass varies so much with water content. The sample is heated in an oven at $110\,°C$ until there is no further change in mass. Figure 2.5F shows a biomass pyramid for a rocky shore community.

Biomass pyramids also have their drawbacks:

- Some organisms grow at a much faster rate than others, for example grass does not have a large biomass, but it grows at a fast rate.

- The biomass of an individual can vary during the year – a beech tree will have a much greater biomass in June than it has in November. *Why do you think this is so?*

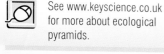

See www.keyscience.co.uk for more about ecological pyramids.

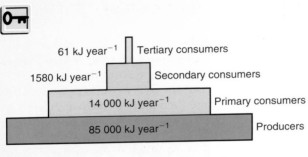

61 kJ year⁻¹ — Tertiary consumers

1580 kJ year⁻¹ — Secondary consumers

14 000 kJ year⁻¹ — Primary consumers

85 000 kJ year⁻¹ — Producers

Figure 2.5G ← Pyramid of energy for a river (m^{-2})

Check the advantages and drawbacks of each type of ecological pyramid as a way of representing feeding relationships in a community.

It's a fact!

In the inky blackness of the ocean depths, communities of organisms flourish around openings in the sea floor called **vents**. No light means no photosynthesis. *How do the communities survive?* Larva and gases from the liquid rock of the Earth's mantle escape through the vents. Figure 26.1B on p. 433 shows you the idea. The extreme heat (volcanic activity) and pressure (depth of water) provide the conditions for chemical reactions in the sea water, producing hydrogen sulphide. Different types of bacteria oxidise the sulphide, releasing the energy needed for them to make sugars from carbon dioxide (in solution) and water. These bacteria are the producers upon which the food webs of the communities in the vent depend.

Energy is lost in the transfer of food from one trophic level to the next.

Pyramids of energy

Pyramids of energy provide the most accurate representation of the feeding relationships in a community. They give information about the amount of new tissue at each trophic level over a certain period of time. They make it possible to compare different communities accurately. They can also be used to compare the production of food by different methods of farming. Figure 2.5G shows an energy pyramid for a river.

Energy flow through producers

The energy in most ecosystems has its origin in sunlight. Green plants (and algae and some bacteria) absorb sunlight and use its energy to convert carbon dioxide and water, through the chemical reactions of photosynthesis, into sugars (see Topic 11.1). This is why food chains, food webs and the trophic levels of ecological pyramids begin with plants (or algae/bacteria).

Plants absorb very little of the sunlight that falls on them. Figure 2.5H shows you why. *How many kilojoules (kJ) of light energy falling onto 1 m^2 of plants each year are used for photosynthesis?*

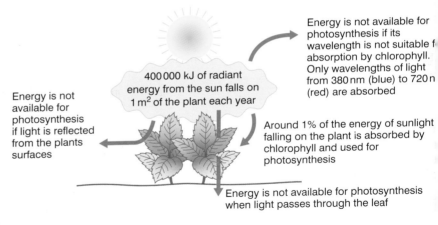

Energy is not available for photosynthesis if its wavelength is not suitable for absorption by chlorophyll. Only wavelengths of light from 380 nm (blue) to 720 nm (red) are absorbed

400 000 kJ of radiant energy from the sun falls on 1 m^2 of the plant each year

Energy is not available for photosynthesis if light is reflected from the plants surfaces

Around 1% of the energy of sunlight falling on the plant is absorbed by chlorophyll and used for photosynthesis

Energy is not available for photosynthesis when light passes through the leaf

Figure 2.5H ← Energy flow through a green plant

The plant needs energy to stay alive. The energy comes from the sugars it makes through photosynthesis. Some is used to drive the chemical reactions of photosynthesis. Some is also released through aerobic respiration (see Topic 15.1) to fuel metabolism (see p. 3). The rest is used in growth and repair which increases the biomass (see p. 49) of the plant. The increase in biomass represents the food energy available to primary consumers, and indirectly to secondary consumers when they eat primary consumers. Energy is transferred from the plant when fruits and seeds are dispersed. Transfer also occurs to decomposers (see p. 46) when the plant loses its leaves and, eventually, when it dies.

Energy flow through consumers

Scientists estimate that for every 100 g of plant material eaten only 10 g ends up as herbivore biomass. In other words, approximately 90% of the energy transferring between trophic levels is lost. The transfer of energy between trophic levels is inefficient because:

- some of the plant material is not digested and passes out of the herbivore's body as faeces
- the herbivore uses energy to stay alive
- when the herbivore dies its body represents 'locked up' energy, some of which transfers to decomposers.

Similar transfers of energy occur between higher trophic levels:

primary consumers ———→ secondary consumers ———→ tertiary consumers

Figure 2.5I shows the energy intake and output of a cow. Notice that more than 50% of the energy content of the grass that the cow eats is transferred via its urine and faeces. Energy is also transferred from the body as heat produced in respiration (see Topic 15.1).

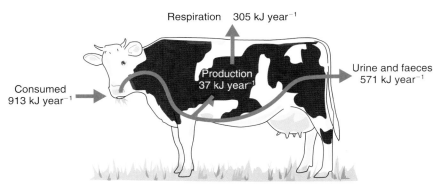

Respiration 305 kJ year^{-1}

Production 37 kJ year^{-1}

Consumed 913 kJ year^{-1}

Urine and faeces 571 kJ year^{-1}

Figure 2.5I ▲ The energy intake and output of a cow

The energy flow through a cow can be summarised in the following equation:

$$\begin{matrix} \text{Energy} \\ \text{intake} \end{matrix} = \begin{matrix} \text{Energy transfer} \\ \text{in respiration} \end{matrix} + \begin{matrix} \text{Energy transfer} \\ \text{in production} \\ \text{of cow's biomass} \end{matrix} + \begin{matrix} \text{Energy} \\ \text{in urine} \end{matrix} + \begin{matrix} \text{Energy} \\ \text{in faeces} \end{matrix}$$

Shortening the food chain

Feeding transfers energy from one trophic level to the next (between the links in a food chain). However, the transfer is never 100% efficient as we have seen. The energy lost from each trophic level through life's activities means that the amount of food energy in a trophic level is less than the one below it. As a result the amount of living material (biomass) in a trophic level is less than the one below it. *What happens when food energy dwindles to nothing?* The trophic levels of pyramids and links in food chains can no longer exist. Now you know why an ecological pyramid tapers off and why its trophic levels (and links in a food chain) are limited in number.

What is the message for feeding people?
The fewer the trophic levels (or links in a food chain) in the farm ecosystem, the *less* energy is transferred from the system and so the *more* food is available for the consumer. In other words, eating a vegetarian diet is less wasteful of the food energy available to people than eating meat.

By eliminating links in the food chain, more individuals at the end of the food chain can be fed. *Why do you think that the diets of people in underdeveloped countries tend to be made up mostly of plants?* In the Western developed countries people have a varied diet including large amounts of poultry, fish, lamb, beef and pork. *What does this tell you about the economies of these countries?*

With the huge increase in the world's population, what is likely to happen to the price of meat and what may be the inevitable change in human diets in the future?

Feeding transfers available food energy from one trophic level to the next.

It's a fact!

Meat is *easier* to digest and has a *higher* energy value than plant material. As a result the transfer of energy from primary consumer (herbivore) ———→ secondary consumer (carnivore) is *more efficient* (20%) than producer ———→ primary consumer (5–10%).

The loss of energy in the transfer of food between trophic levels limits the number of links in a food chain.

SUMMARY

Pyramids of number tell us about the numbers of individuals at each trophic level. Biomass pyramids give information about the mass of material present. More accurate are energy pyramids that show the transfer of energy through a community. A short food chain can support far more people than one with many links in the chain.

A vegetarian diet reduces the inefficiency of energy transfer between trophic levels because the food chain is short.

CHECKPOINT

▶ **1** Look at Figure 2.5l.

    (a) Would you say that the cow is efficient at converting grass into biomass (production)? Explain your answer.

    (b) What percentage of the energy intake is present in faeces and urine? Faeces provide a rich food source for a number of organisms. Name some of them.

    (c) What percentage of the energy intake is used up in respiration?

    (d) Cows spend a great deal of their time grazing. In view of your other answers, why do you think this is?

    (e) How do you think the energy budget of a secondary consumer would compare with the example here?

▶ **2** Look at the figure below and answer the following questions (the size of the arrows represents the relative amounts of energy passed on at each stage).

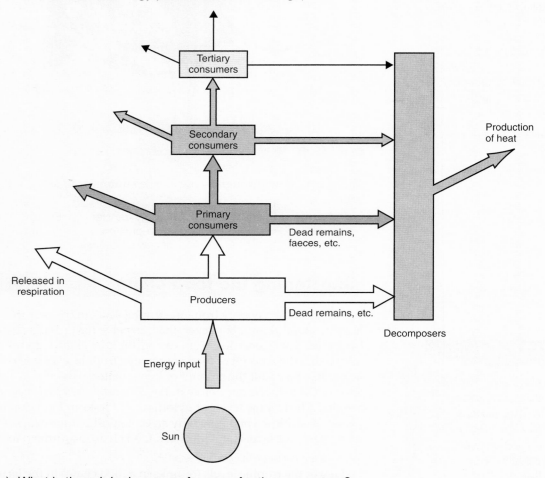

    (a) What is the original source of energy for the ecosystem?

    (b) How is energy eventually lost from the ecosystem?

    (c) Would you say that the transfer of energy through the ecosystem was a cyclical process where the energy is reused, or a flow in one direction? Give reasons for your choice.

    (d) In what ways is energy lost from (i) the producers and (ii) the consumers?

    (e) Predict what might happen to the biomass of the primary consumers when the rate of photosynthesis increases.

    (f) How might an increase in the number of primary consumers affect
        (i) the number of producers,
        (ii) the number of secondary consumers?

▶ **3** From every one square metre of grass it eats, a cow obtains 3000 kJ of energy. It uses 100 kJ for growth, 1000 kJ are lost as heat and 1900 kJ are lost in faeces.
   (a) What percentage of the energy in one square metre of grass
      (i) is used in growth,   (ii) passes through the gut as food which is not absorbed?
   (b) If beef has an energy value of 12 kJ per gram, how many square metres of grass are needed to produce 100 g of beef?

2.6 ▶ Decomposition and cycles

FIRST THOUGHTS

Biological cycles circulate materials between organisms and the environment. Bacteria and fungi circulate nutrients through the processes of decomposition.

Decomposition

Eventually all plants and animals die. You may have seen the remains of a blackbird caught by the family cat, a hedgehog run over and left at the side of the road or a dead gull washed up on the shore (see Figure 2.6A). In autumn dead leaves form a carpet on the ground (see Figure 2.6B). All this dead material eventually disappears. *Where does it go?*

Figure 2.6A ▶ Animal decomposition

Figure 2.6B ▲ Plant decomposition

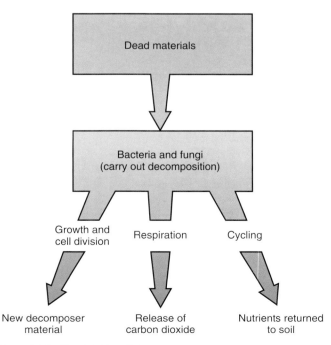

Figure 2.6C ▲ Activities of decomposers

It's a fact!

In the UK decomposition is quicker in the summer than in the winter. Maggots scavenge up to 80% of small mammal and bird carcasses, speeding up decomposition. In the winter, there are few maggots around.

There are many animals that feed upon dead remains, we call them **scavengers**. Crows, maggots, woodlice and vultures all make a living by feeding upon dead material. However scavengers cannot account for the removal of all dead material and even without them it would slowly disappear anyway owing to the action of decomposers. These are the fungi and bacteria that break down dead material and also make food go rotten (see p. 225). They release enzymes on to the dead material, digesting it (saprobiontism – see pp. 14 and 19). The products of digestion are absorbed by the decomposers and used for their own growth (see Figure 2.6C).

Decomposition releases nutrients essential for the growth of new plants. Animals obtain nutrients by eating plants and/or other animals.

Cycles

Not all of the products of decomposition are used to make new fungi and bacteria: most are returned to the soil as nutrients. These nutrients are then made available to plants and in turn to animals for growth (see p. 183). This cycling of nutrients is vital if life is to continue. Most living matter (95%) is made up of just six elements: carbon, hydrogen, nitrogen, oxygen, phosphorus and sulphur.

A constant supply of nutrients containing these elements is essential if living organisms are to continue to make important compounds like proteins, fats and carbohydrates. Figure 2.6D shows the pattern of cycling nutrients which takes place in all ecosystems: on land, in freshwater and in the sea.

Remember the mnemonic 'CHNOPS'. Each letter is the chemical symbol for the most common elements which make up living matter.

SUMMARY

Decomposers break down dead remains and release nutrients into the ecosystem. These nutrients are absorbed by plants and passed along food chains. Nutrients such as carbon, nitrogen and sulphur are continuously being cycled through ecosystems.

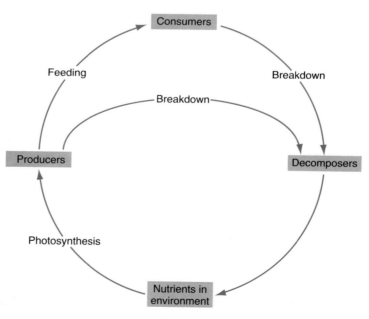

Figure 2.6D Cycling of nutrients in an ecosystem

2.7 ▶ The nitrogen cycle

FIRST THOUGHTS

The atmosphere is a vast reservoir of nitrogen (78% by volume). Nitrogen is an essential part of protein. Animals obtain compounds containing nitrogen from the food they eat. Plants absorb nitrogen in the form of dissolved nitrates from the soil or water in which they live.

Nitrogen is an essential element in **proteins**. Some plants have nodules on their roots which contain **nitrogen-fixing** bacteria. These bacteria **fix** gaseous nitrogen, that is, convert it into nitrogen compounds. From these nitrogen compounds, the plants can synthesise proteins. Members of the **legume** family, such as peas, beans and clover, have nitrogen-fixing bacteria (see Figure 2.7A).

Plants other than legumes synthesise proteins from nitrates. *How do nitrates get into the soil?* Nitrogen and oxygen combine in the atmosphere during lightning storms and in the engines of motor vehicles during combustion. They form nitrogen oxides (compounds of nitrogen and oxygen). These gases react with water to form nitric acid. Rain showers bring nitric acid out of the atmosphere and wash it into the ground, where it reacts with minerals to form nitrates. Plants take in these nitrates through their roots. They use them to synthesise proteins. Animals obtain the proteins they need by eating plants or by eating the flesh of other animals.

Ammonium salts are present in the excreta of animals and the substances released during the decay of dead organisms by **putrefying** bacteria. **Nitrifying** bacteria in the soil convert ammonium salts into nitrates. Both nitrates and ammonium salts can be removed from the soil by **denitrifying bacteria**, which convert the compounds into nitrogen. To make the soil more fertile, farmers add both nitrates and ammonium salts as fertilisers. The balance of processes which put nitrogen into the air and processes which remove nitrogen from the air is called the **nitrogen cycle** (see Figure 2.7B).

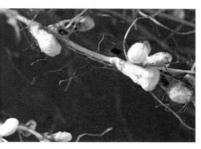

Figure 2.7A 🔺 Nodules containing nitrogen-fixing bacteria on the roots of a legume

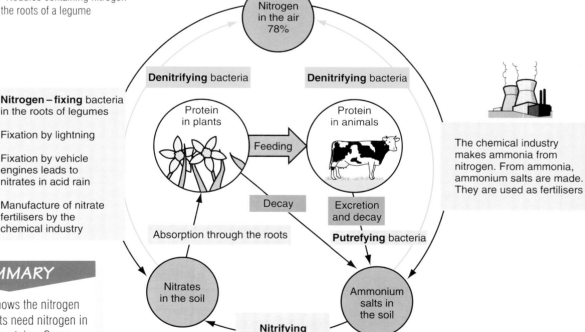

Nitrogen–fixing bacteria in the roots of legumes

Fixation by lightning

Fixation by vehicle engines leads to nitrates in acid rain

Manufacture of nitrate fertilisers by the chemical industry

The chemical industry makes ammonia from nitrogen. From ammonia, ammonium salts are made. They are used as fertilisers

Figure 2.7B 🔺 The nitrogen cycle

SUMMARY

Figure 2.7B shows the nitrogen cycle. All plants need nitrogen in order to build proteins. Some plants have bacteria which can fix gaseous nitrogen. Other plants use nitrates. Nitrates and ammonium salts are added to soil to make it more fertile.

2.8 ▶ The carbon cycle

Plants need carbon dioxide, and animals, including ourselves, need plants. Plants take in carbon dioxide through their leaves and water through their roots. They use these compounds to **synthesise** (build) sugars. The reaction is called **photosynthesis** (photo means light). It takes place in green leaves in the presence of sunlight. Oxygen is formed in photosynthesis (see Topic 11.1).

Photosynthesis in plants

$$\text{Sunlight + Carbon dioxide + Water} \xrightarrow{\text{catalysed by chlorophyll in green leaves}} \text{Glucose + Oxygen}$$
(a sugar)

The energy of sunlight is converted into the energy of the chemical bonds in glucose.

Animals eat foods which contain starches and sugars. They inhale (breath in) air. Inhaled air dissolves in the blood supply to the lungs. In the cells, some of the oxygen in the dissolved air oxidises sugars to carbon dioxide and water. These reactions release energy. This process is called **respiration**. Plants and other living things also respire to obtain energy (see Topic 15.1).

Respiration

$$\text{Glucose + Oxygen} \rightarrow \text{Carbon dioxide + Water + Energy}$$

The processes which take carbon dioxide from the air and those which put carbon dioxide into the air are balanced so that the percentage of carbon dioxide in the air stays at around 0.03%. The balance is achieved through the processes of the **carbon cycle** (see Figure 2.8A).

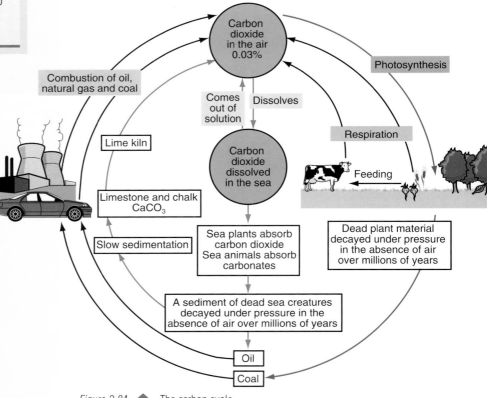

Figure 2.8A ◀ The carbon cycle

CHECKPOINT

▶ **1** Name two processes which add carbon dioxide to the atmosphere.

▶ **2** *Process A* Nitrogen-fixing bacteria convert nitrogen gas into nitrates.

Process B Nitrifying bacteria use oxygen to convert ammonium compounds (from decaying plant and animal matter) into nitrates.

Process C Denitrifying bacteria turn nitrates into nitrogen.

(a) Say where Process A takes place.

(b) Say what effect the presence of air in the soil will have on Process B.

(c) Say what effect waterlogged soil which lacks air will have on Process C.

(d) Explain why plants grow well in well-drained, aerated soil.

(e) A farmer wants to grow a good crop of wheat without using a fertiliser. What could he plant in the field the previous year to ensure a good crop?

(f) Explain why garden manure and compost fertilise the soil.

▶ **3** Suggest a place where you would expect the percentage of carbon dioxide in the atmosphere to be lower than average.

▶ **4** Suggest two places where you would expect the percentage of carbon dioxide in the atmosphere to be higher than average.

▶ **5** Explain why nitrogen is used in (a) food packaging (b) oil tankers (c) hospitals and (d) food storage.

Topic 3

Populations

3.1 ▶ Population growth

Figure 3.1A ◀ Ants swarming on a branch

A herd of bison

A colony of penguins

A **population** is a group of individuals of the same species living in a particular habitat at the same time. Examples of populations are shown in Figure 3.1A.

One of the reasons ecologists study populations is to gain information about the rate at which their numbers change. Such studies are of practical use if the species is a pest or if it causes disease. For instance, population studies of the locust are important since a swarm of locusts can devour a harvest. Studies of mosquito populations can help scientists to control the spread of malaria (see p. 80). Investigations into the human population explosion are very important for the future of mankind.

Why do individuals live in populations? There may be a number of reasons:

- The habitat provides food, shelter and other vital factors.
- Individuals come together to breed.
- Individuals may gain more protection as a population; for example a shoal of fish or a colony of gulls.

However, overcrowding may result in **competition** (see Topic 3.5) within the population. Individuals may compete for food, space, light and other resources. Inevitably some will not survive this competition and numbers may decline.

Think about a plant or animal species colonising a new area. A few individuals enter the new habitat. Provided there are no predators and that there is no shortage of food, the individuals will thrive and reproduce. At first their numbers increase slowly. Later, as the population grows, so does the rate of increase (the number of individuals added to the population in a given time). The sequence of numbers gives you the idea: 2 4 8 16 32 64 128 256 512 1024 … and so on. Notice that the next number in the sequence is double the previous number.

If a population increases in a similar fashion, then each generation would be double the size of the previous generation. The increase is said to be **exponential**. The exponential increase in numbers of a population only takes place when environmental conditions are ideal (for example, predators and disease do not kill significant numbers of individuals).

FIRST THOUGHTS

Individuals of the same species living in a particular place at the same time form a population. The balance between births, deaths, immigration and emigration affect the size of a population. Growth of a population does not go on for ever. Limiting factors stop it. The world's human population is growing rapidly.

Check the meaning of 'species' on p. 8 and 'niche' on p. 45. Can you work out the relationship between species, niche and population ?

See www.keyscience.co.uk for more about population growth.

EXTENSION FILE ACTIVITY

Limiting factors control the rate of growth of a population.

It cannot go on for long. Eventually **limiting** factors such as lack of food and overcrowding cause the rate of increase to slow and eventually to level off. Numbers become stable and then vary in a fairly regular way around an average (see Figure 3.6B). The description of population growth holds for many different species and is illustrated in Figure 3.1B.

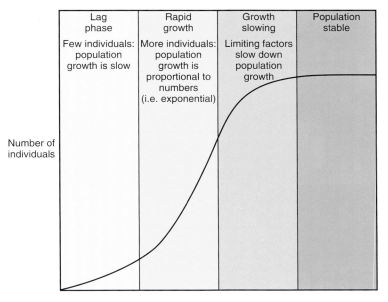

Lag phase	Rapid growth	Growth slowing	Population stable
Few individuals: population growth is slow	More individuals: population growth is proportional to numbers (i.e. exponential)	Limiting factors slow down population growth	

Number of individuals

Figure 3.1B ▲ Population growth curve

Rabbits and rhododendrons

The rabbit was introduced to Britain from the Mediterranean following the Norman conquest of 1066. It soon established itself adapting to landscape, climate and predators (see Topic 3.6) such as the fox, weasel and stoat. Indeed the rabbit was so successful that in the 1950s its numbers were estimated at around 100 million animals.

Rabbits were also very successful following their introduction into Australia in 1859. They had no natural enemies and conditions were ideal for their increase. Within a few years rabbits were a national nuisance, eating grass intended for sheep and stripping vegetation so that soil erosion often followed. Before the introduction of the myxomatosis virus in 1950 (see p. 332) the wild rabbit population of Australia fluctuated at around 500 million animals (see Figure 3.1C).

Figure 3.1C ▲ The few rabbits imported into Australia were the start of a population explosion in the absence of limiting factors. Notice that the vegetation has been eaten by the hungry rabbits

The rhododendron is another example of a species introduced to Britain – this time from south-east Europe. Imported originally as an ornamental plant in large gardens, it became widespread and abundant in the nineteenth century. The fleshy leaves and woody stems of the rhododendron help the plant to survive cold winters and account for its success. The rhododendron is a pest when growing wild. It shades out other plants and may dominate the landscape unless cut back and rooted out (see Figure 3.1D)

Rabbits and rhododendrons are examples of introduced species where the natural checks and balances that control numbers in their native environment are absent. Their new surroundings provide an ideal environment for the rapid growth of populations with the result that the new arrivals may soon become pests. After a time a new balance between the introduced species and its new environment is established as it adapts to its surroundings. Numbers stabilise as limiting factors (see Figure 3.1B) establish themselves and cause the rate of increase to slow and eventually level off.

Figure 3.1D ⬆ Where rhododendron flourishes other plants are shaded out

---

3.2 ▶ Checks on population increase

Figure 3.2A ⬆ Plants of the tropical rainforest compete for light

What prevents populations increasing? Different environmental factors limit the growth of populations, preventing them from becoming too large. They are called **limiting factors** and include

- **food, water and oxygen**. Individuals in the population compete for these resources. If the resources are in short supply, some individuals die and the numbers in the population level off or decline

- **light** which is necessary for plants to carry out photosynthesis. Plants compete for light in areas of dense vegetation (see Figure 3.2A)

- **lack of shelter.** Animals compete for shelter in order to avoid predators or harsh weather.

- **overcrowding** which often leads to unhygienic conditions, favouring the spread of disease. Overcrowding can also stress the members of a population.

- **predators** which are a natural check on the population of their prey. The size of a predator population depends on the numbers of prey. An increase in the number of predators means that more prey will be caught and so their numbers will fall. Hence the predators' food supply is reduced, leading to a drop in the number of predators (see Topic 3.6)

- **accumulation of toxic wastes** including carbon dioxide and nitrogenous waste (such as ammonia and urea). The waste is toxic. As a population increases in size, more waste is produced and its toxicity may restrict further growth.
- **diseases** can spread very quickly through large populations. Monocultures of arable crops (see Topic 5.1) are vulnerable to the rapid spread of disease.
- **climate.** Harsh weather can reduce the numbers in populations. Animals and plants may die if the temperature is very high or very low. Drought, flood, storms and gales all take their toll on populations.

Sometimes populations decrease dramatically in a short space of time but then **crash** instead of levelling off. The crash may occur because the population runs out of food or is overcome by disease. Population crashes are often temporary or seasonal affairs. For example, the large numbers (**blooms**) of algae (see p. 115) on the surfaces of lakes and ponds may crash as the nutrients in solution needed for reproduction and growth are used up (see p. 115).

It's a fact!

Lemmings are mouse-like animals that feed on grasses and roots in the mountains of Norway. Every few years the population grows rapidly owing to increased breeding.

Overcrowding occurs and large numbers emigrate to the lowlands. Many die or are killed. Those that reach the sea attempt to swim out and drown in the process.

3.3 ▶ Population size

The birth rate and death rate help to determine the size of a population. Some species are capable of phenomenal increases in numbers. For example, different species of insect lay hundreds of eggs, frogs lay thousands and fish may produce millions.

Few animals stay in one place, they usually move into and out of the habitat in which the population lives. We call the movement of individuals into a population **immigration** and the movement of individuals out of a population **emigration** (see Figure 3.3A). In a stable population

When the factors that add individuals to a population (births and immigration) are balanced by the factors that remove individuals from a population (death and emigration), then the population is stable.

$$\text{Birth rate} + \text{Immigration rate} = \text{Death rate} + \text{Emigration rate}$$
$$(B) \qquad\qquad (I) \qquad\qquad\qquad (D) \qquad\qquad (E)$$

If the population is increasing then

$$B + I \text{ will be more than } D + E$$

If a population is declining then

$$D + E \text{ will be more than } B + I$$

Estimating the size of a population

Ecologists often need to estimate the size of a particular population when they are carrying out a scientific study. In the case of plants there is usually little difficulty. A **quadrat** is laid down on the vegetation (a quadrat is a sampling device which consists of a metal or wooden rectangle, usually one metre square). The number of individuals of a particular plant occurring within the quadrat is then counted. The number of quadrats needed to cover the habitat is estimated and multiplied by the number of plants per quadrat to give the total population.

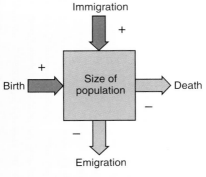

Figure 3.3A ▲ Factors affecting the size of a population

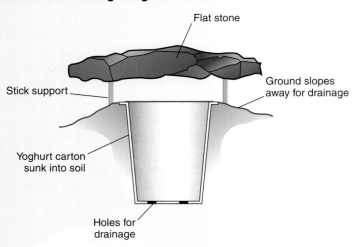

Flat stone

Stick support

Ground slopes
away for drainage

Yoghurt carton
sunk into soil

Holes for
drainage

Figure 3.3B  A pitfall trap is one way of
trapping small animals prior to counting

The capture/recapture method
of estimating population size is
also called the Lincoln Index,
after the American who worked
it out.

EXTENSION FILE
ACTIVITY

Sampling animals is more difficult. Animals
tend to move about – most do not stay in a
quadrat. Also many animals only come out at
night so they have to be trapped before they can
be counted (see Figure 3.3B). Ecologists use a
method called mark, release, recapture to
estimate many animal populations. The method
for estimating the numbers of a particular
species works as follows:

- **collect** a number of individuals from the
  habitat
- **mark** the captured individuals. Ecologists try
  to make the mark in some way that is
  harmless and does not make the individuals
  conspicuous to predators; for example, a
  small dot in a dull colour on a beetle's body
- **release** the marked individuals into the habitat they came from and
  leave them for a day or two to mix freely with the rest of the
  population
- **make a second collection** of a similar number of individuals from the
  same habitat. Note the number of marked individuals in this second
  sample
- **the total population** of the species can be estimated by using the
  equation

$$\text{Total population} = \frac{\text{Number in first catch} \times \text{Number in second catch}}{\text{Number of marked individuals in second catch}}$$

Worked example: 20 ground beetles were caught in a pitfall. They
were each marked with a small dot of paint and when it was dry, they
were released. After two days a second sample of 18 beetles was caught.
Of these, six were marked individuals from the first occasion. Estimate
the total population of beetles.

Solution
Number in first catch = 20
Number in second catch = 18
Number of marked individuals in second catch = 6
$$\text{Total population} = \frac{20 \times 18}{6} = 60$$

CHECKPOINT

▶ 1 By using a net to sweep long grass, 90 froghoppers (small insects) were collected. Each was
marked and released back into the grass. The next day a second sweep produced 80
froghoppers and of these six had a mark on them. Estimate the total population of froghoppers in
the field.

▶ 2 It is important that the handling and marking of animals causes them no harm. Explain why it is
also important that:
(a) the mark does not rub off too soon,
(b) the mark does not make the animal conspicuous.

▶ 3 On the seashore, periwinkles are excellent animals for a mark, release, recapture exercise. They
are easy to find under rocks and seaweed and are most active when the tide is in. A sample of 18
periwinkles was marked and released and after a period of four days, 26 were collected. Of these,
12 were marked individuals. Estimate the periwinkle population for the stretch of shoreline.

3.4 ▶ Human populations

See www.keyscience.co.uk for more about human populations.

A word of warning: the chart shows the numbers of people predicted for 2025 and 2050. However, the numbers are just that – predictions. What will happen if trends change between now and then? Will the rate of human population growth increase or slow down? No-one can say for sure.

For thousands of years the rate of growth of the world's human population has been relatively slow. During the last 300 years, however, it has increased dramatically. In the nineteenth century and the first part of the twentieth century, the human populations of Europe and North America (the **developed nations**) grew exponentially. Now populations in these countries are levelling off. Population growth in Latin America, Asia and Africa (the **developing nations**) is still exponential and the worldwide trend is upwards as Figure 3.4A shows.

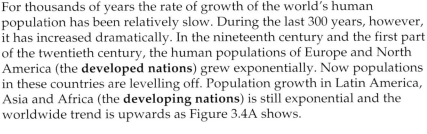
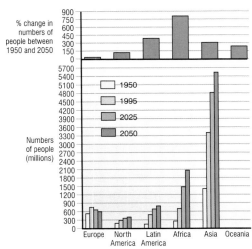

Figure 3.4A ◀ Where in the world human populations are growing
Source: UN Population Division

There are many factors which contribute to population increase. Improvements in

- agriculture has meant greater food production
- public health. For instance clean drinking water and better sanitation are available to some people
- medical provision. Vaccines and new drugs have been developed to combat disease (see Topic 24.5)
- the control of disease-causing organisms.

All of these factors contribute to lower infant mortality and increased life expectancy. The average life expectancy has increased, in Europe and North America, to 75 for men and 79 for women. In India, the figures have for a long time been less than half these ages because of the high death rate among infants. However, with improved health and living conditions, the average life expectancy in India is increasing.

The rate of population growth is affected by the number of young people in the population, particularly women of child-bearing age. For example, Figure 3.4B shows the **population pyramids** of India, China and Britain. The base of the pyramids represents the percentage of children in each country's population; the top represents the percentage of each country's oldest inhabitants. The percentage of all other age groups are represented in five-year steps in between. Each pyramid therefore shows the percentage of different age groups that make up each country's total population. If we compare countries (although the size of each country's population is different, the comparison is possible because the pyramids show age groups as a percentage of each population), the bulge at the bottom of the graph for India and China means that these countries have a greater percentage of young people than Britain. Table 3.1 shows why this is a problem.

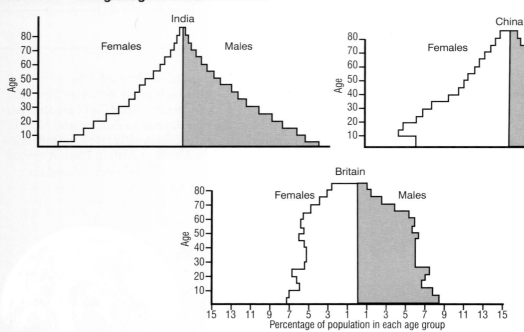

Figure 3.4B 🔺 Pyramids showing the age structure of the populations of India, China and Britain

SUMMARY

Populations grow and change in particular ways. Different factors limit growth and control eventual size. The effects of limiting factors such as disease and food availability on the human population have been reduced by improvements in medicine and agriculture

Table 3.1 🔻 Problems of a young population

Problem	Result
Present birth rate	High, adding to the rate of population increase
Future birth rate	High, as children in the population 'bulge' grow older and have their own children, adding further to the rate of population increase
Financial	Young children do not contribute to the economy
Social services	Huge numbers of children put a great strain on the educational system, medical services and housing

Like India and China, a large percentage of the present world population is young, so the problems are worldwide, even though there are differences between the developed and developing countries, as Figure 3.4B shows. Britain's population, for example, has an increasing proportion of old people. What problems do you think this might bring?

CHECKPOINT

Refer to Figure 3.4B and answer the following questions.

▶ 1 (a) (i) Which country has the higher number of children below the age of 5 years?
(ii) Which country has the higher number of old people over 80 years?
(b) Which country has (i) the higher birth rate (ii) the lower death rate?
(c) Explain your answers to (a) and (b) above.

▶ 2 (a) What forecast can you make about the numbers of child-bearing women in each country in 15 years time?
(b) What will happen to the birth rate in each country as a result?

▶ 3 (a) What is likely to happen to the number of adult workers in relation to elderly dependants as the average age of the population increases?
(b) What implications does this have for (i) health and social services and (ii) employment in this country?

3.5 ▶ Competition

FIRST THOUGHTS

Organisms compete for resources in the community. Competition ensures that resources are not over-exploited and may result in natural selection.

What is the outcome of competition?

Competition between individuals of the same species is a form of natural selection. It is part of the mechanism of evolution (see p. 437) and it also helps to control population size.

Competition between individuals of different species may result in the competitive exclusion of the unsuccessful species and its eventual extinction.

We all have an idea of what is meant by competition. In a race, all the competitors strive as hard as they can to be ahead, but at the end there is only one winner. In nature, competition means much the same thing. Individuals compete for resources that are in short supply. The resources may include food, space, light and other things that are vital for life. Clearly only those individuals able to compete successfully will survive, so competition restricts the size to which a population will grow. Competition between individuals is of two types:

- Competition between individuals of the same species.
- Competition between individuals of different species.

■ Competition between individuals of the same species

Living things tend to produce far more offspring than the habitat can support. Overproduction of offspring ensures the survival of the species and enables the population to colonise new areas. As a result, there is competition between individuals. Only those that are the best adapted survive to breed. They are the ones who pass on their genes to the next generation. The competition results in the 'survival of the fittest'. The 'struggle for existence' ensures that only the best adapted individuals survive to breed and pass on their characteristics to the next generation. The competition which weeds out less fit individuals is a form of **natural selection** (see Figure 3.5A and Topic 26).

Figure 3.5A ▲ Competition between individuals of the same species

■ Competition between individuals of different species

There will be competition for scarce resources between the different species in a community. Weeds compete successfully with crops. They germinate rapidly and grow quickly to establish themselves before the crop matures. They occupy space and take light, water and minerals from the soil, which could have been used by the crop. By the time the crop plants have grown to their full height, the weeds will already have flowered and set seed, so ensuring their survival.

Animals of different species quite often compete for food. For instance predators like owls and weasels compete for the same prey, for example shrews. Different species of fly larvae will compete for the limited food available within a cow pat. Animals also compete for territory, for shelter and for nesting sites. Many different species are able to co-exist in a particular habitat because they occupy different **niches** (see Topic 2.3).

Competition is most intense when different species attempt to fill the same niche. For example, in Scotland two species of barnacle are commonly found on rocky shores (see Figure 3.5B). The species belonging to the genus (see p. 8) *Chthalamus* is found higher up the shore than the other species belonging to the genus *Balanus*. The two species form zones which stand out clearly, as Figure 3.5B shows. Normally *Balanus* grows faster and crowds out *Chthalamus* from the lower zone. In an experiment, when *Balanus* was cleared from an area of rock, *Chthalamus* settled and flourished. However, when an area of *Chthalamus* was cleared in the upper zone, *Balanus* could not live there and did not move up shore to take its place. So *Chthalamus* is driven from the lower zone through competition with *Balanus*, but lives higher up the shore where *Balanus* cannot survive.

The removal of one species from an area through competition with another is called **competitive exclusion**. It means that only one species occupies a given niche (see p. 45) at any particular time. With the barnacles, *Chthalamus* was forced into another niche higher up the shore.

Figure 3.5B Chthalamus and Balanus occupy different zones on the seashore

Diagram labels: Splash zone; High tide; Chthalamus can withstand desiccation; zone of overlap; Average sea level; Balanus (out-competes Chthalamus); Low tide

Because niches are separate from each other, the available food, water, etc. can be shared between different species. This means that limited resources will not be overexploited.

Grey squirrels and red squirrels

Competition between grey squirrels and red squirrels is probably an example of competitive exclusion.

Soon after the introduction to Britain of grey squirrels from North America in the 1870s, the number of red squirrels declined (see Figure 3.5C). Until the end of the nineteenth century red squirrels lived in broad-leaved (oak, beech) and coniferous (fir tree) woods throughout Britain. As grey squirrels extended their range red squirrels have retreated to coniferous woods and to isolated areas which grey squirrels have not colonised. *Why?*

Apparently reds and greys compete for food. The grey squirrel wins because it is better adapted to British trees (its food source) than the red. For example, grey squirrels exploit hazel nuts (fruit containing the seeds of hazel trees) so efficiently that hazel is one of the most threatened of British trees. It seems that competitive exclusion is at work. The niches (see p. 45) of the two squirrels are so similar that only one can survive in any particular habitat.

Figure 3.5C (a) Red squirrels are a species native to (originated in) Britain

(b) Grey squirrels are a species introduced to Britain. Larger than red squirrels, they damage the bark of trees and take eggs and young birds

Succession

What was it like when plants first colonised a habitat? We can get a good idea if we look at an area of ground that has been disturbed, perhaps by developers, by farming or by fire. Within a few weeks some weed and grass species will have grown. These species are known as **pioneers**, since they are the first to colonise. With the passage of time the pioneer plants may be replaced by a wider range of taller herb species. These may in turn be replaced by other more competitive scrub vegetation. This gradual change in the mix of species is called **succession**.

We can see succession taking place if arable land is neglected or on a smaller scale, if we do not mow the lawn. Other plant species soon appear: in the lawn, daisies and dandelions, on arable land, poppies and thistles. In most lowland areas of Britain the process of succession would eventually end as deciduous woodland (see p. 22) if we were to let it, but it would take over 150 years. By then, the numbers of different species would have increased and the community as a whole would have become more complex and more stable. This sort of community changes very little and is known as a **climax community** (see Figure 3.5D). Other examples of climax communities include tropical rain forests and coral reefs. They contain a huge variety of different species so it is vital that we conserve them by protecting them from destruction (see Topic 8).

SUMMARY

Competition exists between members of the same species and between members of different species. Only those individuals that are the best adapted will survive the competition to fill a niche, others will be excluded. The sequence of changes in species, ending in a climax community is called succession.

Figure 3.5D ⬤ A climax community – deciduous woodland

CHECKPOINT

▶ 1 What is competition?

▶ 2 Briefly explain why a given niche is occupied by only one species.

▶ 3 Why does mowing the lawn hold back succession?

▶ 4 What is a climax community?

3.6 ▶ Predators and prey

Predators are animals that catch and kill other animals called **prey**. Thus predation has a major effect upon the number of the prey population. It is also true to say that the prey will have an effect upon the predator population. If the prey is scarce then some predators will starve. Figure 3.6A shows the relationship between the numbers of a predator and its prey. Following the numbers will help you to understand the events taking place.

1 Prey breed and increase in numbers if conditions are favourable (e.g. food is abundant).

2 Predators breed and increase in numbers in response to the abundance of prey.

3 The rate of predation increases and the number of prey declines.

4 Predator numbers decline in response to the shortage of food.

5 With fewer predators, prey numbers increase ... and so on.

The relationship is self-regulating (see p. 306). Notice that since there are fewer predators than prey and that predators tend to reproduce more slowly than their prey:

● fluctuations in predator numbers are less than fluctuations in prey numbers

● fluctuations in predator numbers lag behind fluctuations in prey numbers.

Other factors affect the size of both populations. The incidence of disease, harsh climate and lack of shelter all play their part. The most widely documented example of a predator–prey cycle is that of the snowshoe hare and its predator, the lynx. Records of their numbers were kept by the Hudson Bay Fur Company in Canada between 1845 and 1935. The fluctuations in the numbers of the predator and its prey are shown in Figure 3.6B. Counts each year of hare and lynx skins brought to the Hudson Bay Fur Company give the figures for the graph.

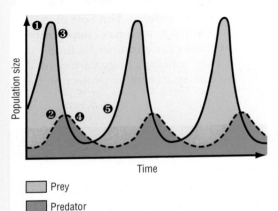

Prey
Predator

Figure 3.6A ⬆ Predator–prey relationship

Predation affects the numbers of the prey population which affects the numbers of the predator.

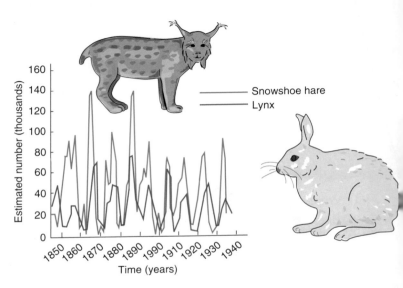

Figure 3.6B ⬆ The lynx and the snowshoe hare predator–prey cycle

What makes a successful predator?

A predator must be able to hunt and kill efficiently. *But how can a predator make life easier for itself?* By:

The strategies of predator and prey are different:
- Predators are adapted to catch prey.
- Prey is adapted to avoid capture.

- **eating a variety** of prey species, reducing the risk of starvation should one prey species decline in numbers
- **catching young**, old and sick prey. The predator uses less energy if its prey can be caught easily. A result is the weeding out of weaker individuals in the prey population. Those that remain are the best adapted individuals, who pass on their genes to the next generation
- **catching large prey** which provides more food for the predator per kill
- **migrating** to areas where prey is more plentiful.

However well adapted the predator becomes, it should not be too successful or else it risks wiping out its food source.

The prey's guide to survival

Prey species are adapted to avoid capture although they are never completely successful.

- Many try to **out-run**, **out-swim** or **out-fly** the predator.
- **Large groups**, for example herds of antelope and shoals of fish, distract the predator from concentrating on one particular individual (see Figure 3.6C).
- Some potential prey, such as bees and wasps, **sting** to avoid capture. Others just **taste** horrible, for instance ladybirds, so the predator thinks twice before eating one again.
- Some prey species possess **warning colours** that tell the predator to 'keep clear'. The hoverfly has striking yellow and black stripes that remind the predator of a wasp (see Figure 3.6D). A hungry frog that has been stung previously by a wasp will think again before it attempts to eat a hoverfly. In fact, the hoverfly has no sting, so it is quite harmless.
- **Camouflage** is often used by prey to escape capture. However predators may also be camouflaged to avoid detection when stalking prey. Many animals blend in with their surroundings. Others look like something else, twigs or leaves for instance (see Figure 3.6E).
- Some prey species use **shock tactics**. They startle their predators into abandoning their attack. The eyed hawkmoth can quickly open its wings to reveal a pair of huge, coloured eye-spots. This often has the effect of scaring off a hungry bird (see Figure 3.6F).

Figure 3.6C ⬆ Hiding in the crowd

Figure 3.6D ⬆ A hoverfly displaying the 'danger' colours of a wasp

Figure 3.6E ⬆ Two camouflaged moths

Figure 3.6F ⬆ The eyed hawkmoth

Humans as predators

Mesh? – the size of gaps in fishing nets.

Can you think of a modern day example of how humans hunt and kill for food? Most of our food is produced by farming (see Topic 5), but fishing is an activity where humans are still hunters – albeit on a large scale (see Figure 3.6G).

Figure 3.6G
a) A deep sea stern trawler in action

b) The skipper watching a large school of fish (the red band on screen) on the trawler's echo sounder

Overfishing shows that humans can be so efficient as predators that stocks of fish are threatened.

In the last 50 years the rise in fish landings around the world has been remarkable. Large catches of fish are possible through improved technology.

- Fishing boats use **big nets** made of strong, light, transparent, plastic materials. Transparency means that fish do not see the nets to avoid them.
- **Sonar** makes it possible to find shoals of fish so that nets can be positioned to make a maximum catch.
- **Factory ships** receive the catch from the fishing boats and refrigerate it. In this way fishing fleets can stay at sea for longer periods, extending the time available for fishing.

The increased efficiency of fishing methods means that for some species, more fish are caught than are replaced by reproduction (**overfishing**). As a result, populations dwindle and in some cases the size of a particular fish stock is so reduced that it is no longer possible to fish. Figure 3.6H shows the annual catches of different fish stocks from the North Sea. Notice the sharp decrease in the annual catch of herring between the mid-1970s and the 1980s. Financial hardship for the people working the herring boats followed in the wake of overfishing. It was no longer profitable or allowed to fish for herring for a number of years, so stocks have been able to recover to a point where it is financially worthwhile to catch herring once more. The challenge is to balance fishing effort against the natural ability of herring (and other fish stocks) to reproduce themselves. Striking the right balance means that fishing can continue in a **sustainable** way.

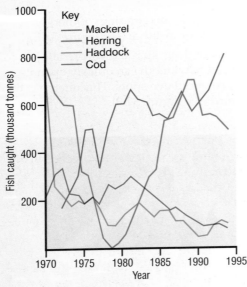

Figure 3.6H ⬆ Total catches of fish in the North Sea
Source: DEFRA

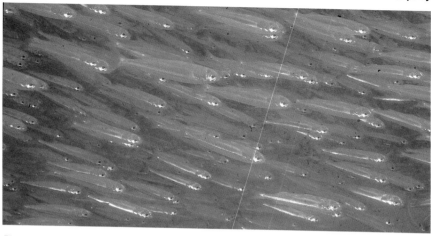

Figure 3.6I ◀ A shoal of herring

SUMMARY

The relationship between predator and prey regulates their respective numbers. Different strategies ensure predator success and prey survival. Human predation endangers the survival of different food resources (especially fish).

The management of fish stocks highlights the conflict between short-term human needs (food, economical use of expensive equipment) and the ecological health of a particular environment. Using living things for food (and as a source of different products) only at a rate that allows them to be naturally replaced (sustainable use – see p. 137) helps the environment to provide resources in the long-term. In the case of fishing stocks from the North Sea:

● reducing the intensity of fishing helps balance the numbers of fish spawned with the numbers of fish caught

● increasing the mesh size of nets allows small fish and young fish to escape capture. As a result the fish that escape have more time to grow, providing more food for the same effort of catching them when they were small

● encouraging the catching of species little-used for food means that the intensity of fishing is spread over more fish stocks, reducing the intensity on individual species.

The different countries of the EU operate a *Common Fisheries Policy* in an attempt to achieve the goal of fishing in a sustainable way. A system of *Total Allowable Catches* sets a maximum catch (measured in tonnes of fish landed) for each species of commercially important fish. The intention is to preserve fish stocks in the long-term to benefit the environment and the thousands of people who depend on fishing for a living.

Whaling is another example of the effects of humans as predators on the numbers of different species, in this case whales, which are prized for their meat and oil. Figure 3.6J shows that following a peak of catches in 1960, blue whales and fin whales had been hunted almost to the point of extinction and attention turned to hunting sperm whales and pilot whales instead. Overhunting has led to a decline in their numbers … and so on to the present. A few nations (especially Norway and Japan) still insist that whaling remains a significant source of jobs and income. *Do you think the argument put forward by Norway and Japan is reasonable?* Explain your response to the question.

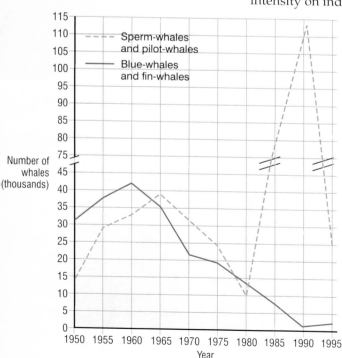

Figure 3.6J ◀ The decline in numbers of whale species due to overhunting
Source: FAO

3.7 ▶ Parasitism and other associations

FIRST THOUGHTS

Associations between living things:

Parasitism – an association between a species that benefits (parasite) and another which is harmed (host)

Mutualism – two or more species benefit from their close relationship

Commensalism – an association between a species that benefits (commensal) and another which is unaffected either way.

Parasites often disgust us because of the harm that they can cause. In fact, probably every living animal has some of these uninvited guests. Some parasites cause disease and death and therefore reduce the size of a population (see Figure 3.7A).

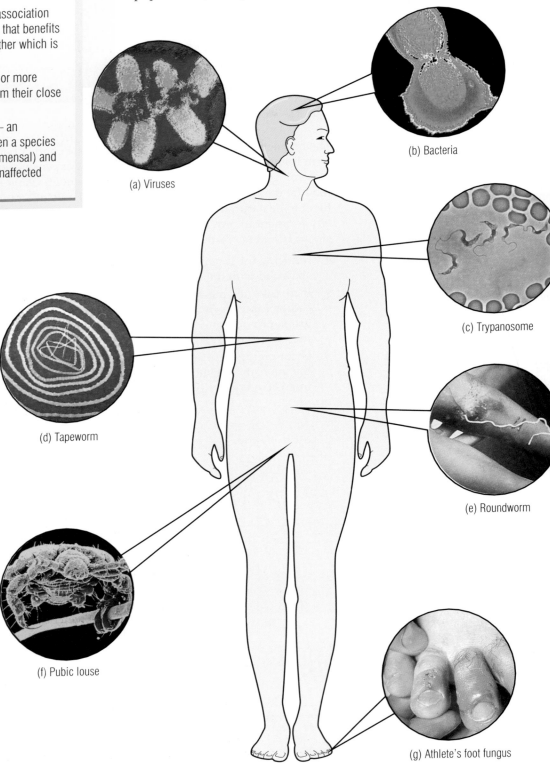

(a) Viruses

(b) Bacteria

(c) Trypanosome

(d) Tapeworm

(e) Roundworm

(f) Pubic louse

(g) Athlete's foot fungus

Figure 3.7A ⬆ Some of the parasites of humans

Parasitism is one of a number of different associations that can exist between two organisms. **Symbiosis** is a general term for close associations in which either one of the partners benefits to the detriment of the other or both benefit or neither appears to benefit. Benefits can be food, shelter or protection (see Table 3.1).

Table 3.1 ▼ Some associations

Association	How each participant is affected
Competition	Neither population benefits, they inhibit each other
Predation	The predator benefits at the expense of the prey
Parasitism	The parasite benefits to the detriment of the host
Commensalism	One participant, the commensal, benefits whilst the other, the host, is unaffected
Mutualism	The association benefits both partners

Parasitism

Parasitism is a one-sided relationship between two organisms. The parasite obtains food at the **host's** expense, either by consuming the host itself or by consuming the host's food supply. Parasites of humans include fungi, bacteria, viruses, protists, worms and arthropods. Microorganisms that cause disease are termed **pathogens**. When an animal is healthy, it has little difficulty supporting all of its parasites with food. However, if the animal is weakened, by starvation or drought for instance, then parasites can have a fatal effect upon it.

There is a great variety of parasites. The degree to which they are tied to the host varies. **Ectoparasites** attach themselves to the outside of the host and many only form temporary associations. For example, a tick attaches itself to a cow long enough to take a blood meal and then drops off. An **endoparasite** lives inside the host's body which becomes the parasite's habitat, giving it food, shelter and protection. The malarial parasite only leaves the human body when it is carried to a new host by a mosquito (see Topic 4.2).

■ Parasitic adaptations

Parasites have very specialised life-styles, particularly endoparasites which live inside a host that may well react against their presence. Often endoparasites have poorly developed organs of:

- locomotion. They move around very little inside the host
- digestion. For example tapeworms absorb the digested food of the host through their body wall
- sense. Receptors for sight, hearing and smell are often absent.

However endoparasites often have:

- well-developed organs for attachment to the host; for example, the hooks and suckers of the tapeworm (see Figure 3.7A)
- an ability to resist the host's **immune reactions** (see Topic 16). For example, parasitic roundworms have a thick resistant cuticle
- an ability to survive where there is little oxygen. The human intestines where tapeworms may live is an example of an oxygen-poor environment.

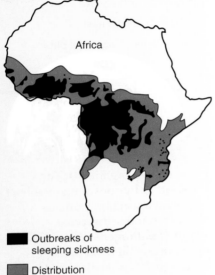

Africa

■ Outbreaks of
sleeping sickness

■ Distribution
of tsetse flies

Figure 3.7B 🔺 Tsetse fly and sleeping
sickness. Notice that the fly's abdomen is swollen
with blood.

■ How do parasites find new hosts?

The offspring of parasites must infect new hosts, otherwise the species will die out. Different strategies ensure infection of new hosts. Endoparasites in particular:

- may produce hundreds of thousands of eggs every day so that some find their way to a new host
- each may have male and female reproductive organs (**hermaphrodite**). Consequently when an individual mates with a partner, both donate sperm and twice as many fertilised eggs are produced
- may become very attached – literally. The male and female of a particular species link up. For example the male of the species of fluke that causes the disease **bilharzia** carries the female in a groove in his body so that she is easily available when ready for mating
- may use another organism (called a **secondary host**) to transmit their offspring to the main (**primary**) host. The secondary host is called a **vector**. Different insect species act as vectors in this way. For example, tsetse fly transmits the protist (see p. 16) that causes sleeping sickness from one person to another. The flies feed on blood. Each has piercing mouthparts that penetrate skin like a hypodermic needle and then inject the protist parasite into a new host (see Figure 3.7B). When feeding on a human being, the fly injects the protist parasite into the victim's blood stream. The person is the parasites new primary host. In the case of the pork tapeworm a human is the primary host and a pig the secondary host. The tapeworm lives in the human intestine. Its millions of eggs are passed out in the person's faeces and some may be eaten by a pig. The eggs hatch and the larvae find their way into the muscle of the pig. If the muscle is eaten as undercooked pork, then larvae develop inside the person who becomes the tapeworms' new primary host (see Figure 3.7C)

Figure 3.7C 🔻 Life cycle of the pork tapeworm

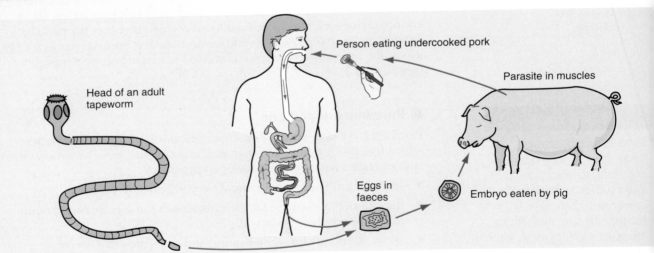

Head of an adult
tapeworm

Person eating undercooked pork

Parasite in muscles

Eggs in
faeces

Embryo eaten by pig

Mutualism

Mutualism is an association between organisms of different species in which each partner benefits. Sometimes the partners become so dependent upon each other that they are unable to exist separately.

The roots of legumes such as peas, beans and clover, have lumps on them called **root nodules**. These contain **nitrogen-fixing bacteria**. The bacteria

SUMMARY

Parasites live in or on their hosts to which they cause harm. In order to be successful, parasites have evolved many adaptations. Some parasites use a secondary host as a vector to ensure the spread of their offspring.

can convert nitrogen from the air into compounds which the plant can use to make protein (see Topic 11.3). In return, the bacteria obtain carbohydrates from the plant's roots. Many bacteria and protists live in the large intestines of mammals. They break down the cellulose in plant material to sugars. They are especially important to herbivores in helping them to digest the cellulose which forms a large part of their diet. In return, the microorganisms get some of the food and a suitable environment in which to live (see p. 241).

The mutualism between oxpecker birds and buffaloes is less tight-knit. The oxpecker removes ticks and other ectoparasites from the skin of the buffalo and often alerts it if predators approach. Cleaner-fish do a similar job for larger fish, removing lice and other ectoparasites.

Commensalism

An association between two organisms of different species, in which one benefits without harming the other is called **commensalism**. The sea anemones shown in Figure 3.7D are hitching a ride on the crab. The crab is a messy feeder and scraps of food float towards the sea anemones which pick them up in their tentacles. The sea anemones are the commensals; their presence is of no advantage or disadvantage to the crab.

Figure 3.7D ▲ Sea anemones attached to a whelk shell taken over by a hermit crab

CHECKPOINT

▶ 1   Why are undernourished people more likely to be affected by parasites than people who are well fed?

▶ 2   Look at the figure below. It shows the life cycle of the malarial mosquito. The female mosquito is the vector of the protozoon, *Plasmodium*, that causes malaria.

    (a) Suggest ways in which the eggs, larvae and pupae of the mosquito are well adapted for living in water.

    (b) Attempts to control the parasite aim to break the life cycle of the vector. Suggest ways in which each of the stages in the life cycle could be eradicated.

    (c) How can the adult mosquito be prevented from biting people and so spreading the disease?

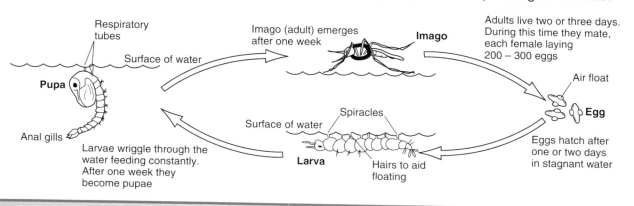

Topic 4

Insects and humans

4.1 ▶ Beneficial and harmful insects

FIRST THOUGHTS

There are more insect species and more insects living than all other animal species put together. Many insects are beneficial; some are harmful. This section introduces you to them.

See www.keyscience.co.uk for more about biological pest control.

Their method of feeding makes aphids ideal vectors of viruses which cause diseases in plants.

It's a fact!

Houseflies, tsetse flies, screw-worm flies and mosquitoes are examples of true flies. They belong to the group called Diptera which literally means 'two-winged'. Most insects have four wings.

EXTENSION FILE
ASSIGNMENT

See www.keyscience.co.uk for more about insects.

Blackfly and greenfly are everyday names for **aphids** although they are not flies at all. They feed on the sugar-rich **sap** flowing through the **sieve tubes** of plants (see Topic 12). The hollow needle-like mouthparts (called stylets – see p. 194) are tapped into a stem. Sap, under pressure in the sieve tubes, is forced through the mouthparts and into the aphid gut.

Aphids spend a lot of time feeding. Large colonies can quickly build up and cover a plant, which wilts and may die through loss of food and water.

Aphids also transmit **virus diseases**. While aphids are feeding, their mouthparts inject viruses into the sieve tubes. The sap stream transports the viruses around the plant, which sickens and dies. Virus-free seed potatoes are grown in Scotland where aphids are generally scarce. The risk of aphid-borne viruses is therefore less than elsewhere in the United Kingdom. Potato growers can reduce the risk of devastation by raising their crops from Scottish seed potatoes.

Controlling aphids

Crowded colonies of aphids are a juicy target for insect predators (see Figure 4.1A). Despite the attentions of predators, however, aphids are major pests of agricultural crops. As well as potatoes, broad beans, barley, wheat and sugar beet are just some of the crops which suffer serious damage.

Varying the sowing date, the number of plants sown (crowded plants are less vulnerable to aphid attack than plants spread apart) and the use of varieties of plant which are resistant to attack help to control the impact of aphids. However, insecticides may have to be used if damage is to be kept to acceptable levels.

Systemics – so called because the chemicals pass into the soil in solution to be absorbed by the roots of the plant – are the type of insecticide most often used to control aphids. While feeding, the aphid takes in a large dose of insecticide and is killed. *Why do you think systemics are an efficient way of controlling pests that feed on plant sap? What advantage does their use have for wildlife?*

Controlling whitefly

Whitefly is a common pest of greenhouse crops. The moth-like insect covers the undersides of leaves, feeding on the sugary liquid transported in the phloem tissue of the plants (see Topic 12.3). The sticky juice which the whiteflies excrete slows crop growth and makes fruit dirty and unattractive.

Encarsia formosa is a small parasitic wasp (see Topic 3.7). It lays its eggs in whitefly. The eggs hatch and the wasp larvae feed on the tissues of the whitefly and kill them. Adult wasps emerge from the dead whitefly. The wasp gives reliable **biological control** (see Table 4.1) of whitefly, providing the control of temperature in the greenhouse is accurate. If the temperature is too low, the parasite is sluggish and the whitefly gains the upper hand. If the temperature is too high, the parasite overwhelms the whitefly leaving itself without hosts for the next generation of wasps. *What would then happen to the wasp population?*

Table 4.1 ▼ Advantages and disadvantages of biological control

Advantages	**Disadvantages**
Harmless – programmes of biological control do not harm people, wildlife or the environment	**Slow** – biological control acts slowly
Low cost – biological control is relatively cheap	**Unpredictable** – changes in the environment may cause sudden outbreaks of pest problems
Specific – control agents (predators and parasites) usually attack only the pest species and do not harm other wildlife	**Pests survive** – a pest must persist even if at a low level for the control agent to survive long-term
No resistance – pests are unlikely to develop resistance to controlling agents	**No insecticides** – if a crop has several pests then biological control against one means insecticides cannot be used against the others. *Why? Briefly summarise your opinions*

Figure 4.1A ▲ Aphid predators

Make sure that you can distinguish between the biological control and chemical control of pests (see Topic 5).

SUMMARY

Different predators reduce the numbers of aphids. However, insecticides are needed to control aphids and minimise damage to crops. Biological control and sterile male techniques successfully reduce the damage to crops and livestock caused by other types of insect pest.

Controlling screw-worm fly

Screw-worm flies are a serious pest of cattle. An insect bite, even a small scratch, attracts the female fly to lay her eggs around the edge of the wound (called **strike**). After a few days the larvae (**maggots**) hatch and eat into the tissue increasing the size of the wound. Victims of attacks quickly lose condition, and unless they are treated with insecticides to kill the maggots they eventually die.

Treatment of livestock for strike is difficult where animals are scattered over large areas. In the 1950s scientists in North America wondered if releasing sterile males into the environment would control screw-worm fly. They knew that:

- female screw-worm flies only mate once
- exposing male flies to X-rays causes lethal mutations in their sperm, sterilising them
- sterile males are as effective as wild fertile males in finding mates
- eggs fertilised by sperm from sterile males fail to hatch.

Success with trial experiments on the island of Curaçao off the coast of Venezuela encouraged scientists to repeat their efforts in North America. Beginning in Florida in 1957, the campaign gathered pace, and in 1958 50 million flies were sterilised and released every week. Invasion by wild flies across the Mexican border was countered by moving the campaign westward. A factory was set up producing millions of sterilised pupae which were boxed for airlifting and dropping by parachute into the countryside.

By the mid-1960s, screw-worm fly ceased to be a problem. Later outbreaks, particularly in Texas, were soon brought under control, but gave a warning that in the long term, screw-worm fly is still a threat.

4.2 ▶ Malaria and mosquitoes

The parasite and the disease

There are four species of the protist *Plasmodium* responsible for human malaria. The parasite infects liver cells and red blood cells (see Topic 16.1).

Once the parasite is in the blood stream, flu-like symptoms are followed up to 30 days later by high fevers, teeth-chattering chills and heavy sweating. The timing of the attacks depends on the species of *Plasmodium*. *Plasmodium vivax* is the most common species. It causes a mild form of the disease. *Plasmodium falciparum* is the most dangerous. It may cause red blood cells to become sticky, blocking blood vessels to the kidneys, brain, intestines and lungs. Unless treated the victim dies within a few days.

The geography of malaria

Plasmodium needs warmth to reproduce inside the mosquito. Malaria therefore flourishes in the warmer tropical and equatorial regions of the world. Poorer countries are particularly vulnerable. They are short of expert help and also find it difficult to finance measures for treatment and prevention.

Mosquitoes

Mosquitoes suck sugary liquids. Nectar and honey are favourite foods. Females need a blood meal as well before their eggs can develop. Most vertebrates are hosts for female mosquitoes (see Topic 1.5).

Only female *Anopheles* mosquitoes transmit the malaria parasite to humans. About 30 different species world-wide are responsible. As in the tsetse fly (see p. 74), the female's pointed mouthparts pierce capillary blood vessels beneath the surface of the skin and take up blood like a hypodermic needle (Figure 4.2A). If the person is infected with *Plasmodium*, then the parasite is sucked up as well and then passed on when she feeds on another person.

FIRST THOUGHTS

Female mosquitoes (genus: *Anopheles*) transmit the protist parasites (genus: *Plasmodium*) that cause the disease malaria in humans. Think about these facts:

- 2400 million people are at risk from malaria
- 280 million people are infected with malaria parasites
- 110 million people develop the disease each year
- 2 million people die from malaria each year.

How can malaria be treated or prevented? What is the role of mosquitoes in its transmission? Read this section and find out.

Their method of feeding makes *Anopheles* mosquitoes ideal vectors of the malaria parasite *Plasmodium*. **Remember** that vectors (often insects) are organisms which transmit parasites from host to host. The vector is usually not affected by the parasite.

It's a fact!

The mosquitoes *Culex*, *Aedes* and *Mansonia* are vectors for different diseases but not malaria. They transmit the tiny worms that cause **filariasis**. The worms block lymph vessels and lymph glands (see p. 273). As a result limbs may swell to alarming proportions.

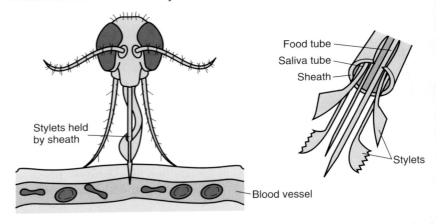

Figure 4.2A 🔺 Female mosquitoes feed on blood. Their mouthparts are long and sharply pointed forming stylets. They are surrounded by a sheath. Inside is a tube through which saliva passes and a food tube. When feeding, the sheath bends back and holds the stylets in place as they pierce the victim's skin until a blood vessel is reached. The food tube is then pushed in and saliva pumped down through the salivary tube into the wound

The malaria life cycle

Since *Plasmodium* infects blood and female mosquitoes feed on blood, female mosquitoes are efficient vectors (see Topic 3.7) of the parasite. Female mosquitoes are unaffected by the parasite, but people are. Figure 4.2B shows the role of mosquitoes and people in the malaria life cycle.

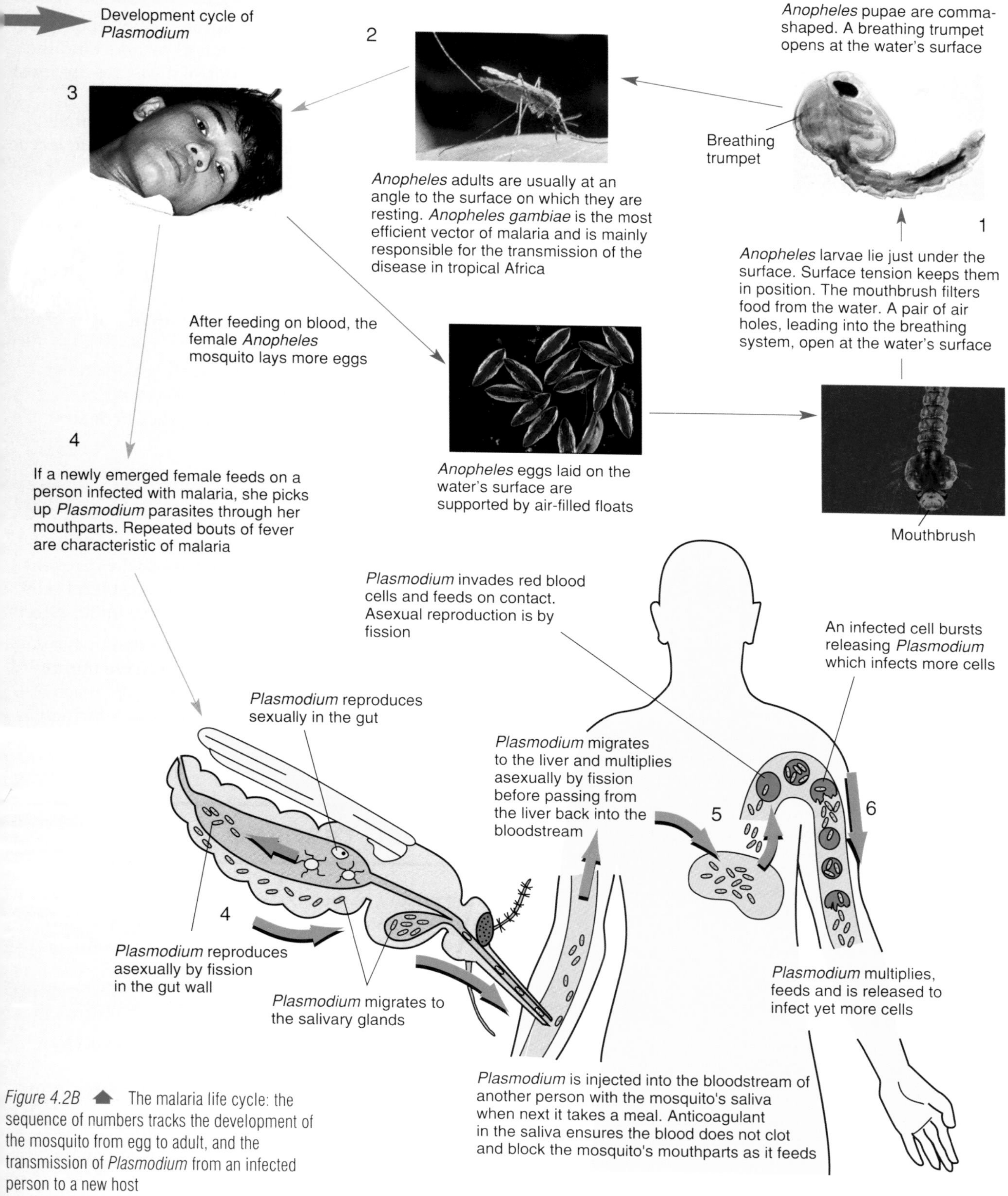

Life cycle of *Anopheles* mosquito

Person to person contact by mosquito showing route for transmission of *Plasmodium*

Development cycle of *Plasmodium*

2

3

Anopheles pupae are comma-shaped. A breathing trumpet opens at the water's surface

Breathing trumpet

1

Anopheles adults are usually at an angle to the surface on which they are resting. *Anopheles gambiae* is the most efficient vector of malaria and is mainly responsible for the transmission of the disease in tropical Africa

Anopheles larvae lie just under the surface. Surface tension keeps them in position. The mouthbrush filters food from the water. A pair of air holes, leading into the breathing system, open at the water's surface

After feeding on blood, the female *Anopheles* mosquito lays more eggs

4

If a newly emerged female feeds on a person infected with malaria, she picks up *Plasmodium* parasites through her mouthparts. Repeated bouts of fever are characteristic of malaria

Anopheles eggs laid on the water's surface are supported by air-filled floats

Mouthbrush

Plasmodium invades red blood cells and feeds on contact. Asexual reproduction is by fission

An infected cell bursts releasing *Plasmodium* which infects more cells

Plasmodium reproduces sexually in the gut

Plasmodium migrates to the liver and multiplies asexually by fission before passing from the liver back into the bloodstream

5

6

4

Plasmodium reproduces asexually by fission in the gut wall

Plasmodium migrates to the salivary glands

Plasmodium multiplies, feeds and is released to infect yet more cells

Plasmodium is injected into the bloodstream of another person with the mosquito's saliva when next it takes a meal. Anticoagulant in the saliva ensures the blood does not clot and block the mosquito's mouthparts as it feeds

Figure 4.2B ▲ The malaria life cycle: the sequence of numbers tracks the development of the mosquito from egg to adult, and the transmission of *Plasmodium* from an infected person to a new host

Repeated invasions of and release from red blood cells of *Plasmodium* are tracked by a cycle of chills and fevers. The victim's body temperature soars when the red cells burst releasing parasites into the blood stream. Eventually enormous numbers build up, causing serious and sometimes fatal illness.

Treatment and prevention of malaria

Fighting malaria depends on destroying the mosquito vector, the parasite or preventing contact between the vector and people. Find the numbers on Figure 4.2B. They indicate where control measures are most effective. Control measures include:

- **draining** marshes, ponds and ditches, thus preventing the female mosquito from laying eggs, and the eggs from developing into larvae
- **biological control** in the form of fish which eat mosquito larvae (see Topic 4.1)
- **insecticides** sprayed on the water's surface killing mosquito larvae and pupae
- **drugs** like quinine, mefloquine and primaquine, which prevent fission in the liver cells or red blood cells of the host
- **vaccines,** which are being developed to inhibit different stages in the parasite's life cycle (see Topic 16.1)
- **bed nets** soaked with insecticide to protect the resting person not only from mosquitoes but from other biting insects as well
- **chemical repellents** sprayed on to the skin and clothes to deter mosquitoes from landing on the body.

Natural immunity

In many parts of Africa people are partly immune to malaria. The body produces antibodies in response to *Plasmodium*. The antibodies prevent infection in the liver cells and red blood cells, or destroy red blood cells which are already infected. Other immune responses have similar effects.

The children of an infected mother are given some protection by her antibodies, which cross the placenta before birth. This **passive** immunity disappears after 4–6 months. However if the child survives to reach 5–6 years, then repeated infections result in **acquired** immunity which gives a high level of protection (see Topic 16.1).

4.3 ▶

Houseflies spread disease

Houseflies feed on most organic matter. Look at the fly's trail shown in Figure 4.3A. How many sources of food for flies can you find in Figure 4.3B? Using both figures as evidence, explain why flies are a health hazard.

The problem

Figure 4.3C shows the mouthparts of the housefly. Digestive juices are released through the fine tubes on to the food on which the fly has landed. The digested food is sucked up. In the process, microorganisms on the mouthparts and in the digestive juice are transferred from the fly to the food. If the fly's food is also human food, then people are in danger from the disease-causing microorganisms the fly leaves behind (see Figure 4.3D).

Fly-borne diseases

The bacteria (*Shigella spp*) or protists (*Entamoeba histolytica*) that cause **dysentery** are transferred to human food by houseflies. Dysentery results in diarrhoea. Contractions of the gut (peristalsis – see p. 240) are more frequent than normal, and food and water are not absorbed through the gut wall. The faeces are liquid. The patient loses water and feels very weak. Children are particularly vulnerable and may die through dehydration (loss of water).

Dysentery appears wherever people are crowded together in unhygienic conditions. During the Second World War (1939–1945), the British Army in North Africa kept outbreaks of dysentery to a minimum by strict attention to hygiene and the control of flies. The picture was different in the Crimean War of 1854 when filth provided ideal breeding conditions for flies, and 8000 British soldiers suffered from dysentery. More than 2000 died of the disease.

re 4.3A ⬆ The petri dish is covered with in layer of blood agar (a jelly-like substance which bacteria grow). A housefly has walked r the agar surface. The petri dish was sealed then incubated at 37°C for 4 days. *Why was petri dish sealed and then incubated?*

gure 4.3B ⬆ Faeces and rotting organic matter are homes for microorganisms which cause disease. Houseflies pick up and transfer microorganisms as they move from one food source to another

It's a fact!

Up to the 1950s, sticky fly-papers hung from the ceiling were used to catch flies in the home. In summer a character nicknamed 'Flypaper Joe' called each week selling fly-papers from a band worn around his top hat.

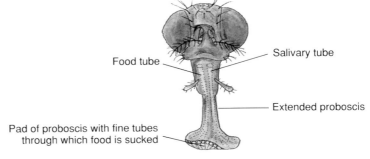

Food tube — Salivary tube

Extended proboscis

Pad of proboscis with fine tubes through which food is sucked

Figure 4.3C ⬆ The proboscis of the housefly has a two-lobed pad at its tip. In the pad are many thin tubes. During feeding, digestive juices are pumped down these tubes on to the food. The juices make the food liquid. The liquid food is then sucked up the tubes and into the mouth

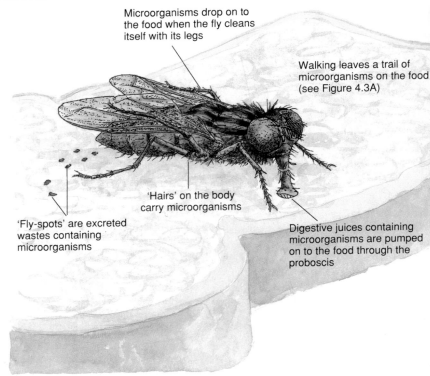

Microorganisms drop on to the food when the fly cleans itself with its legs

Walking leaves a trail of microorganisms on the food (see Figure 4.3A)

'Hairs' on the body carry microorganisms

'Fly-spots' are excreted wastes containing microorganisms

Digestive juices containing microorganisms are pumped on to the food through the proboscis

Flies spread disease-causing organisms through their feeding habits.

Figure 4.3D ⬆ The housefly is a vector for microoganisms that cause disease.

SUMMARY

Houseflies transfer disease-causing microorganisms from organic matter to human food as they move from one food source to another. Strict hygiene controls flies and drugs help to treat the diseases they spread.

Typhoid fever is caused by the bacterium *Salmonella typhosum* carried in contaminated water. Houseflies also transfer the bacterium to human food. After about fourteen days, the infected person develops a fever which grows steadily worse. Delirium and uncontrollable shakes appear about two weeks later, and the victim may die unless treated with antibiotic drugs.

Cholera bacteria (*Vibrio cholerae*) are also carried in contaminated water and transferred to food by houseflies. Symptoms of cholera include vomiting and passing large amounts of liquid faeces. Loss of salts from the body causes severe cramps in the limbs and, unless treated with antibiotics, the patient may die from the effects of dehydration.

Improvements in treatment, health, personal hygiene and sanitation have helped to reduce the extent and effects of fly-borne diseases. However, remember that flies carry microorganisms. Infections are reduced by covering food to prevent flies from landing and feeding.

4.4 ▶ Honeybees

FIRST THOUGHTS

'Busy as a bee' is an old saying, but how busy are bees? What are they busy doing and why? Reading this section will help you to answer these questions.

Figure 4.4A shows honeybees collecting **nectar** from a flower. The bees pass from flower to flower and then fly away. If you were in the field following them, the bees would probably lead you to their hive. Landing near the entrance, 'your' bees would join hundreds of others coming and going in seemingly ceaseless activity.

Figure 4.4A 🔺 Honeybee collecting nectar. Notice the pollen sac on the back leg

EXTENSION FILE
ASSIGNMENT

Adaptations for collecting nectar and pollen

Bees visit flowers to collect nectar from which they make honey. Nectar is a dilute solution of various natural sugars. Figure 4.4B shows that the nectar is sucked through the mouthparts into the **honey sac**. Here chemical reactions which change the nectar into honey begin. The reactions are catalysed by the enzyme **invertase** (see p. 168).

The bee uses some of the mixture as food for herself. The rest she brings up and transfers to a 'house' bee on her return to the hive. Further processing occurs as the 'house' bee passes the liquid on to other bees. Water evaporates from the mixture and the end result is honey which is stored in the honey pot cells of the comb. Further evaporation reduces the water content of the honey to less than 20%. The bees then seal the cells with cappings of wax.

Honeybees collect **pollen** as well as nectar. Pollen is a source of protein. Grains of pollen stick to the 'hairs' of the bees' bodies when they visit flowers. The pollen grains are combed from the body with a row of bristles (**pollen comb**) on each of the front legs. The pollen is mixed with a little nectar to make two balls of pollen paste which are then passed to the **pollen sac** on each back leg (see Figure 4.4A). When bees return to the nest they push the pollen paste into food-storage cells in the comb. 'House' bees pack down the pollen and when the cells are full seal them with caps of wax. Pollen and honey are eaten in the winter when no other food is available.

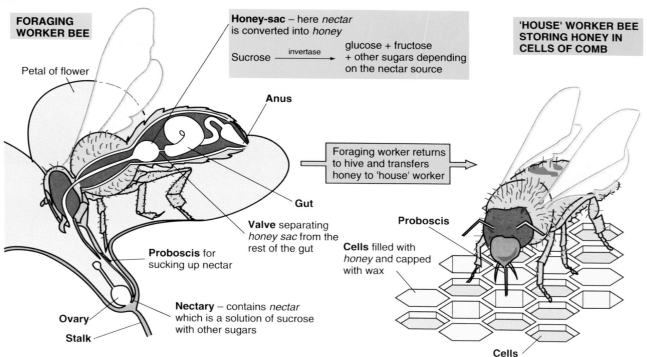

Figure 4.4B 🔺 Worker bees making and storing honey

EXTENSION FILE
ASSIGNMENT

Honeybees make honey and other useful products. They are also useful pollinators.

The value of honeybees

Honey was used as a sweetener long before sugar became popular. Its flavour depends on the source of nectar, which varies from month to month. Of all the sources, rape is the most important. Much farmland in the UK is covered with its bright yellow flowers in late spring and early summer. Bees love rape flowers and eagerly visit them to forage on the nectar and pollen. The honey produced is distinctive and strong in flavour.

Propolis is a dark brown substance which bees obtain from plants like the horse chestnut tree. Its sticky resin is used to seal and repair cracks in the honey combs and to block up holes. Bees also smear the frames and wax surfaces of the cells of the honeycombs with propolis. This prevents the growth of bacteria and moulds which would otherwise flourish in the warm humid environment of the hive. The value of honey, propolis and other honeybee products is illustrated in Figure 4.4C.

Although we usually think of honeybees as producers of honey, their value as pollinators of food crops, wild plants and garden flowers outstrips their value as a source of food. Farmers pay bee-keepers to take their hives into fields and orchards so that the bees can pollinate the crop in flower.

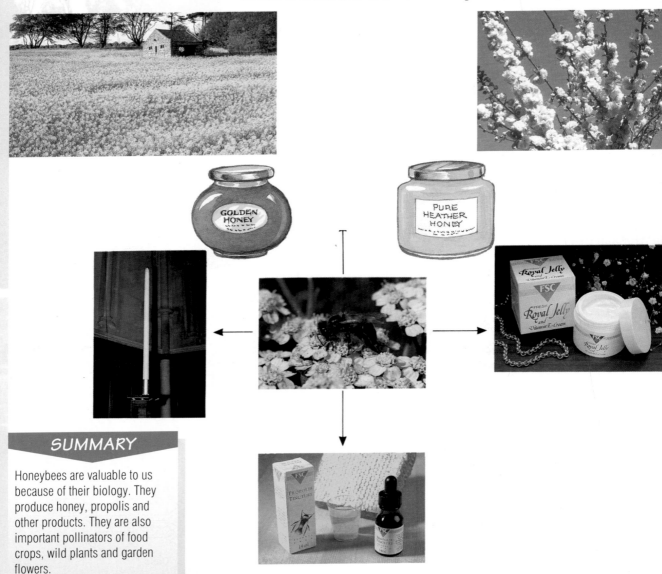

SUMMARY

Honeybees are valuable to us because of their biology. They produce honey, propolis and other products. They are also important pollinators of food crops, wild plants and garden flowers.

Figure 4.4C ⬆ Honeybee products

Theme Questions

Dogs may have several kinds of parasitic worms in their bodies. The life cycle of one kind of tapeworm is shown in the diagram. Some stages of the life cycle of this worm are spent inside a flea, which is itself a parasite of the dog.

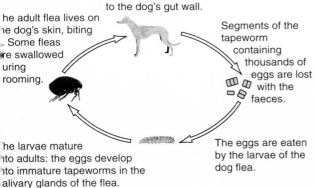

The tapeworm undergoes final development once attached to the dog's gut wall.

The adult flea lives on the dog's skin, biting. Some fleas are swallowed during grooming.

Segments of the tapeworm containing thousands of eggs are lost with the faeces.

The larvae mature into adults: the eggs develop into immature tapeworms in the salivary glands of the flea.

The eggs are eaten by the larvae of the dog flea.

(a) State **two** features of parasites shown by **both** the flea and the tapeworm.

(b) State **one** difference between the ways in which the two animals (flea and tapeworm) parasite the dog.

(c) Suggest how regular use of flea powder may reduce the incidence of tapeworms in dogs.

(AQA)

2. The diagram shows the passage of energy along a food chain.

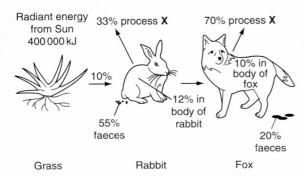

Radiant energy from Sun 400 000 kJ

33% process **X**

70% process **X**

10%

12% in body of rabbit

55% faeces

10% in body of fox

20% faeces

Grass   Rabbit   Fox

(a) Only about 1% of the energy which falls on grass is used to produce sugar in photosynthesis. State **two** possible ways in which the plant uses the sugar. What happens to the rest of the energy falling on the leaf?

(b) When rabbits feed on grass only about 10% of the energy stored in the grass is passed on. What happens to the rest of the energy trapped by the grass leaves?

(c) 400 000 kJ of energy fall on 1 m^2 of grass over 1 year. How much of this energy will be passed on to rabbits in a year? Show your working.

(d) 55% of the energy taken into the rabbit is lost in faeces. The rest of the energy is lost by process **X**. What is process **X**?

(e) Suggest why the fox loses far less of its energy intake as faeces compared with the rabbit.

(f) Explain in terms of energy efficiency, why food chains are rarely of more than 4 or 5 links.

(OCR)

3. Many insects feed using a tube called a proboscis.
(a) The mosquito feeds by sucking human blood. How is the proboscis of the mosquito adapted to its function?

(b) The diagram shows a section through the proboscis of a housefly.

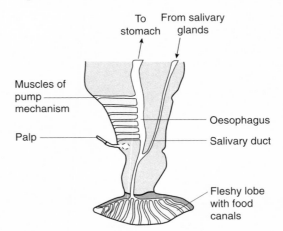

To stomach   From salivary glands

Muscles of pump mechanism

Palp

Oesophagus

Salivary duct

Fleshy lobe with food canals

(i) Describe how the proboscis of the housefly is adapted to the way in which it feeds.

(ii) Describe how houseflies can transmit diseases to humans through their feeding habits.

(SEG)

Human impact on the environment

L and is the environment of millions of kinds of plants and animals including humans. We use part of the land for agriculture, to grow crops and to raise livestock. Modern methods of agriculture are intensive: they are mechanised and require extensive use of fertilisers and pesticides. Conflicts arise between utilising the land for the maximum benefit of humans and allowing plants and animals to flourish in their natural habitats. In this theme you will be able to weigh up some of the pros and cons of different farming methods.

A modern industrial society makes plenty of goods to keep us free from squalor and starvation. Plenty has led to pollution. The air is polluted by objectionable gases, dust and smoke from our cars and factories. Rivers and lakes are polluted by waste from factories, fields and our own bodies. The land is polluted by rubbish tips. Urgent measures are needed to combat pollution. If we take the necessary steps, we can avoid disaster. We can reduce the pollutants in our atmosphere; we can make our rivers and lakes clean again; by making use of much of our rubbish, we can conserve the Earth's resources.

Topic 5   Food production

5.1 ▶ Intensive farming

FIRST THOUGHTS

There are more than twice as many people in the world as there were in 1950. Farmers have to supply more food to meet the needs of the growing world population (see Topic 3.4).

See www.keyscience.co.uk for more about intensive farming.

The population of the world is increasing (see Topic 3.4).

When the human population began to rise sharply in the nineteenth century, the need for food increased. Farming methods had to become more **intensive**, that is, to produce as much food as possible from the land available for raising crops (arable farming) and animals (livestock farming).

Figure 5.1A ⬆ Intensive cereal production

The field of wheat in Figure 5.1A has been grown by intensive methods. It is called a **monoculture** because one type of plant is being grown over a large area. Chemical **fertilisers** have been applied to produce stronger, larger plants in a shorter time. Chemical pesticides have been used: **insecticides** have been sprayed to kill insect pests and **herbicides** have been applied to kill weeds. The use of all these chemicals costs the farmer money. The farmer has to balance the expense against the increased yield of wheat grain. Most of our bread is made from cereals which have been grown by intensive methods.

Most of the pork, chicken and beef we eat is produced by intensive methods. For example, pigs are kept indoors in pens and provided with specially prepared food which is not produced on the farm but bought in (Figure 5.1B). As a result of the lack of exercise and the specially prepared food, the pigs gain weight more quickly than pigs allowed to roam outdoors (**free range**) and so are ready for slaughter earlier.

In the following sections we shall look at the various factors that contribute to intensive farming. Sometimes the new methods have an unwanted effect on the environment. We have to balance the advantages of the new methods against the impact which they have on the environment.

Figure 5.1B ⬆ Intensive pork production – pigs feeding indoors. Food is supplied to the circular trough through the chute. Although kept 'indoors' intensively reared animals indirectly require 'land' for production of the food on which they are fed

Figure 5.1C shows intensive egg production. Hens are kept in **battery** cages, so called because the cages are stacked in tiers. Control of the environment and their food intake ensures that the hens each lay on average 300 eggs in a year.

Fairness for animals

The intensive rearing of animals indoors is often called **factory farming**. *Is it fair or even humane to keep animals in such conditions?* For example, the battery hens shown in Figure 5.1C cannot stretch their wings and may peck their feathers. The vice is checked by removing the tips of the birds' beaks (**de-beaking**). Many people think that the system is cruel but others point out that intensively reared animals are kept safe from predators, free of parasites, warm and well fed. Table 5.1 summarises the arguments for and against factory farming.

Figure 5.1C ⬆ Battery hens are kept in artificial light. Heat, light, humidity, ventilation and availability of water are controlled. Up to a million hens may be kept in long poultry sheds. In such crowded conditions birds are regularly given antibiotics in their food to prevent (or control) the outbreak of disease

Intensive farming maximises the amount of food produced from the land available.

It's a fact!

Around 15% of the eggs sold in the UK are free range. To be called free range, hens must be able to move around in runs during the day which are mainly covered with plants. Also, the number of hens should not exceed 1000 per hectare.

Table 5.1 ⬇ Arguments for and against factory farming

For	Against
Food – people's demand for high quality meat can only be met by raising animals intensively.	**Food** – people need less meat than was previously believed. The health benefits of eating less animal fat have also been established.
Efficiency – animals confined indoors do not waste energy in exercise and keeping warm. Therefore, they convert their food energy into growth (meat) or milk or egg production more efficiently.	**Efficiency** – although outdoor (free-range) methods of rearing animals is very inefficient, it is possible to devise systems that allow animals to live 'natural' lives and yet be productive.
Breeding – animals may be bred for their productivity without worrying about the characteristics that enable them to look after themselves or their offspring outdoors.	**Quality** – the meat, milk and eggs of animals raised on a monotonous diet is less tasty than the products from free-range animals.
Labour – one person can look after large numbers of animals indoors.	
Cost – increased efficiency, better breeding and reduced labour costs mean producing intensively reared milk, meat and eggs is much cheaper than the equivalent free-range products.	**Cost** – although non-intensively reared animals are more expensive to keep than those reared intensively we do not need to eat as much meat as once was thought (see above). The proportionate cost of meat, therefore, to our food bill should not increase.
Animal welfare – intensive methods protect animals from parasites, harsh weather and shortages of food.	**Animal welfare** – cruelty to animals is not just a matter of pain or physical discomfort. Boredom and frustration cause great suffering which often shows in bizarre behaviour (see above for the example of feather pecking).

The fact that intensively reared livestock must be content because they eat well and are extremely productive is an argument often heard in support of factory farming. However, we know that overeating, leading to obesity (see p. 221) is a common sign of depression in people. Evidence suggests that intensively reared animals suffer similar problems.

SUMMARY

Intensive farming produces food (crops and livestock) in great quantities. Arguments for and against the intensive farming of livestock identify cost advantages and high productivity set against possible cruelty to the animals.

New laws that allow animals to follow more 'natural' lives while remaining productive are being introduced. The changes are to be phased in gradually. For example, under current proposals keeping hens in the conditions shown in Figure 5.1C will not be allowed. Ideas for a more humane system of keeping hens intensively include proposals for an 'enriched cage' which will be double the size of the battery cage shown above and provide each bird with a nest, perch and scratching strip. For hens an enriched cage should provide a more interesting place to live.

Apart from new laws, pressure for change comes from people's increasing awareness of the suffering that can be caused by factory farming. Products that come from free-range animals are gaining in popularity as a result.

5.2 ▶ Soil: the lifeblood of farming

Soil is formed of small paticles produced by the **weathering** of rocks (see p. 36).

Differences in particle size means that different soils have different properties (see Figure 5.2A and Table 5.2).

Table 5.2 ▼ The properties of sandy soils and clay soils

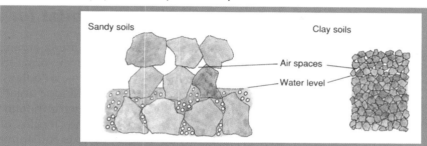

	Sandy soils	Clay soils
Particle size	Large, > 0.2 mm	Small < 0.002 mm
Air spaces	Large	Small
Drainage	Rapid, leaving a dry soil	Slow, leaving a wet soil
Temperature	Fluctuates, tending to be higher	More consistent, tending to be lower
Cultivation	Easy to dig and plough because they are dry and loose	Difficult to dig and plough because they are wet and sticky
For plant growth …	Plants may suffer from lack of water. Minerals may be leached (washed) from the soil by rain	Plant roots may lack oxygen if the soil becomes waterlogged. The mineral content is high since mineras tend to stick to clay particles.

Soil crumbs form when sand and clay particles stick together
A film of water binds to the soil crumbs
Excess water drains out of the air spaces so that the soil does not become waterlogged
Air spaces provide oxygen for the growth of roots and for useful microbes
Air spaces make the loam less dense and easier to cultivate

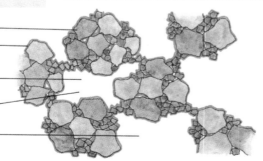

Figure 5.2A ▶ The crumb structure of loam

Most soils are a mixture of particles of different sizes. They therefore combine the properties listed in Table 5.2, and the advantages and disadvantages are balanced. Such a mixture is called **loam** and is ideal for growing plants.

The 'feel' of a soil depends on the proportion of large and small particles in it. This 'feel' is the texture of the soil. Its texture decides the amount of air in the soil and the speed at which water drains from the soil. These factors affect the plants which grow in the soil and also the animals and microorganisms which keep the soil in good condition.

Part of the soil is called **humus**. It is formed from the decayed remains of dead organisms. A variety of organisms take part in the formation of humus (see pp. 36–37).

Water table

The roots of a plant anchor it in the ground and extract nutrients from the soil. Many plants have enough room for their roots in a layer of soil 15 cm deep: some plants need a layer of soil as deep as 4 m.

The natural level of water is called the **water table** (see Figure 5.2B). Plants need to have their roots above the water table, so that they can draw water from it without becoming waterlogged. Roots should not be immersed in water the whole time. By means of a good drainage system, it is possible to lower the water table. Farmers are careful to ensure good drainage in their fields. Sometimes, they install underground pipes to help drainage. Good drainage helps air to enter the soil; it helps seeds to start growing, and it helps to control plant diseases.

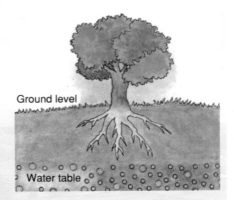

Ground level

Water table

Figure 5.2B ⬆ The water table

Soil as a habitat for animals

Soil is teeming with living organisms. One hectare ($10\,000\,m^2$) of soil contains a million earthworms and millions of mites, springtails, millipedes, centipedes, woodlice, fly larvae, beetles, ants and microorganisms. Some soil organisms are pests, for example, wireworms burrow into crops such as potatoes and ruin them. Others play a vital part in maintaining the fertility of the soil. Some bacteria decompose the dead remains of plants and animals, helping to make humus. Other bacteria fix nitrogen; that is, convert nitrogen in the air into nitrogen compounds which plants need to make proteins (see p. 185). Earthworms improve the texture of the soil. Moving through it, they open channels for roots to spread through. As well as mixing the soil, they help air to enter and help water to drain from the soil (see p. 27).

Water, air and nutrients

Plants obtain water from the soil. Plants lose water constantly as water vapour passes out through openings (**stomata**) in the leaves into the atmosphere. The process is called **transpiration** (see p. 189). Plants take in water through their roots to replace that lost by transpiration. Dissolved in the water are many nutrients the plants need. During a dry spell, plants need to be watered (see Figure 5.2C).

Soil must be aerated (contain air) so that plant roots can obtain air. Many of the microorganisms in the soil also need air. If soil is packed down by tramping feet as hikers walk across the land, or by heavy vehicles, such as tractors, it will contain little air.

Plants need certain chemical elements (nutrients). In photosynthesis (see Topic 11.1), plants obtain carbon and oxygen from air, and hydrogen and oxygen from water. Land plants obtain the rest of the elements which they need from the soil. Aquatic plants obtain their nutrients from water (see Topic 11.2).

The natural cycle is for plants to die and decay, returning chemicals to the soil (see Topic 2.6). When crops are harvested, the chemicals they

Figure 5.2C ◀ Many plants need to be watered during dry spells

 See www.keyscience.co.uk for more about Soil Analysis (program).

contain are not returned to the soil. **Fertilisers** are needed to replenish the soil with the nutrients which the crops have taken out of it (see Topics 5.3 and 5.4).

pH

A soil may be acidic, alkaline or neutral (measured as pH values – see *Key Science Chemistry* Topic **9.5**). For each type of crop there is a range of soil pH over which the plants can be grown successfully (Table 5.3).

Table 5.3 ▼ Crop growth and pH

Crop	Potatoes	Swedes	Oats	Wheat	Sugar beet	Kale	Barley
pH	2	2.7–5.6	4.8–6.3	6.0–7.5	7.0–7.5	9–11	12

Farmers have a choice. They can measure the pH of their soil and then choose a crop which will grow well on it. Alternatively, they can alter the pH of the soil to suit the crop which they want to grow. Often soils are too acidic. Rainwater is weakly acidic and therefore neutralises bases which are present in topsoil. When soils are too acidic, farmers add lime (calcium oxide) which neutralises the excess acid. Occasionally, soils are too alkaline. Then iron(II) sulphate is added. The salt is weakly acidic in solution and neutralises the excess alkali in the soil (see *Key Science Chemistry* Topic **9.4**).

Erosion

The top few centimetres of soil (**topsoil**) contains the most nutrients. It can be washed away by high rainfall or blown away by strong winds (**eroded** – see Figure 5.2D). When farmers convert grassland into arable fields, erosion may follow. A field of winter wheat can lose 2.4 kg m^{-2} of soil or more to erosion by wind and rain each year. Loss of soil leads to loss of growth of crops and therefore less food.

Good farming methods and conservation help prevent soil erosion.

- Grass prevents erosion: its roots keep the soil in place.
- Hedges reduce erosion by reducing the speed of the wind.
- Forests reduce erosion as the roots of the trees hold soil in place.
- Farmers plough land so that furrows lie across the natural slope of the land; rainwater will then not run down the furrows and cause erosion.
- A terraced hillside is less likely to be eroded than a natural hillside. By slowing down the flow of rain down a hillside, terraces give the water time to soak in and nourish the crop.

EXTENSION FILE
ASSIGNMENT

Figure 5.2D Soil erosion

Cultivation

Cultivation breaks up soil into small pieces. A gardener digging the garden and a farmer ploughing the land are both culivating the soil. The benefits of cultivation are:

- uprooting weeds
- mixing fertilisers and decaying plants with the soil
- introducing air into the soil
- giving plants enough depth for their roots to grow.

SUMMARY

Soil contains small particles of weather rock and humus – decayed organic matter. Plants need enough soil for their roots to grow in. They also need water, air and certain chemical elements as nutrients. The soil must have a suitable pH. Topsoil can be eroded by rain and by wind. Soil needs to be cultivated if it is to grow crops.

CHECKPOINT

▶ 1 (a) Why are cereals so important for the diet of the human population? (See Topic 14.)

(b) Use the bar chart opposite to work out the ratio: Cereal production in 1999/Cereal production in 1970.

(c) What changes in farming methods have made the increase possible?

(d) Which block shows the greatest percentage increase over the preceding figure?

▶ 2 Give two reasons in each case why the following are important features of soil: good drainage, cultivation, microorganisms, the level of the water table.

Chart: Cereal production (thousand million tonnes) vs Year, with y-axis from 1.0 to 2.2 and x-axis labelled 1965, 1975, 1985, 1995.

5.3 ▶ Natural fertilisers: manure, slurry and compost

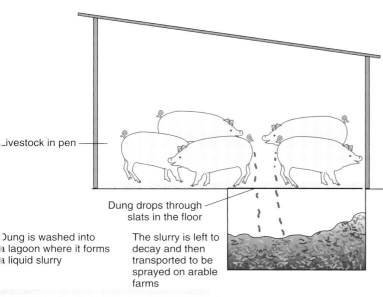

Livestock in pen ——

Dung drops through slats in the floor

Dung is washed into a lagoon where it forms a liquid slurry

The slurry is left to decay and then transported to be sprayed on arable farms

Figure 5.3A ▲ Collecting dung on an intensive farm

A substance which adds nutrients to the soil is called a **fertiliser**. In a mixed arable and livestock farm, the dung of farm animals is collected, left for a few months to decay and then used as **manure**, a natural fertiliser. When a herd of animals is reared indoors on an intensive farm the dung has to be disposed of. Most intensive farms specialise, and a livestock farm may have no arable land to spread manure on. Figure 5.3A shows how the dung is collected.

A serious hazard to the environment arises if slurry leaks from the lagoon. Then slurry may seep into rivers and lakes, where it will be broken down by bacteria. As the bacteria multiply, they use up dissolved oxygen in the water (see Topic 7.3). The result is that fish and other forms of aquatic life die. The slurry may even find its way into ground water and contaminate drinking water.

Compost

Many gardeners build a **compost heap** to recycle vegetable scraps and garden waste. Any rottable material can be used. Worms break it up and bacteria and fungi decompose it, releasing nutrients which can be used by plants. Figure 5.3B shows you how to make a compost heap.

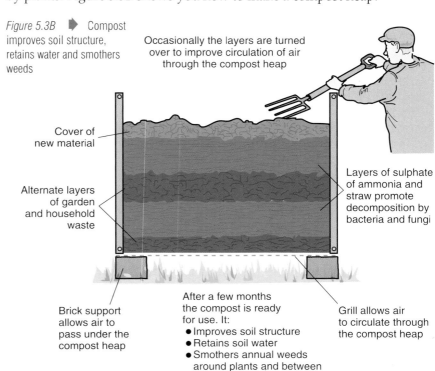

Figure 5.3B ▶ Compost improves soil structure, retains water and smothers weeds

Occasionally the layers are turned over to improve circulation of air through the compost heap

Cover of new material

Alternate layers of garden and household waste

Layers of sulphate of ammonia and straw promote decomposition by bacteria and fungi

Brick support allows air to pass under the compost heap

After a few months the compost is ready for use. It:
● Improves soil structure
● Retains soil water
● Smothers annual weeds around plants and between rows of vegetables

Grill allows air to circulate through the compost heap

5.4 ▶ **Synthetic fertilisers**

FIRST THOUGHTS

Most modern farms do not produce enough manure for their needs and use synthetic (man-made) fertilisers.

In the present century, we have seen the dramatic effects of synthetic fertilisers in the **green revolution**. This is the way fertilisers have turned barren areas in many parts of the world into green, fertile agricultural land. In India, the rice yield has increased by 50%; in Indonesia, the maize crop has doubled (see Figure 5.4A).

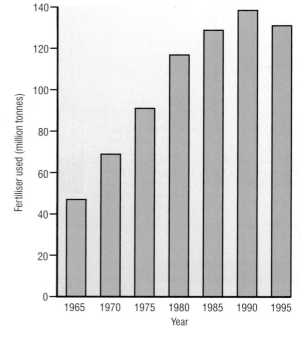

Figure 5.4A ▲ With and without nitrogen fertilisers; one plot of wheat has received fertiliser, the other has not

Figure 5.4B ▲ The worldwide use of NPK fertilisers
Source: FAO

It's a fact!

There are islands off the coast of South America which are inhabited by large flocks of sea birds. Mounds of their droppings, called guano, accumulate. Since sea birds eat a fish diet, which is high in protein, guano is rich in nitrogen compounds. For a century, European farmers imported guano from South America to use as a fertiliser.

How much fertiliser?

Most fertilisers supply the necessary nitrogen (N), phosphorus (P) and potassium (K). **NPK fertilisers** are consumed in huge quantities: in 1995 world consumption was over 130 million tonnes (see Figure 5.4B). The fertilisers cost a lot of money. Farmers and market gardeners need to invest in fertilisers. They want good crops, but they do not want to spend more than they need. They can obtain expert advice from the Department of the Environment, Food and Rural Affairs. Agricultural chemists at the Department will advise them on the type and quantity of fertiliser to apply and the best season of the year for applying it. Every farmer and grower has a different problem. The agricultural chemists must weigh up the type of crop and the type of soil before they can recommend the most suitable treatment.

5.5 ▶ **Herbicides**

Fertile soil is a good growing ground for weeds as well as for the crops that farmers sow. Weeds are plants that compete with crops for space, light and nutrients. At one time, farmers employed many workers, and one of the jobs they did was pulling out weeds by hand. Now, farmers employ fewer people and save labour by using chemical weedkillers –

Farmers and gardeners have to cope with pests. These include weeds which compete with crops for nutrients, animals which eat crops, and disease-causing organisms like harmful fungi which attack plants and livestock. Substances which are used to kill pests are **pesticides**. There are three main types: **herbicides**, which kill plants, **insecticides**, which kill insects (Topic 5.5) and **fungicides**, which kill fungi.

EXTENSION FILE
ASSIGNMENT

Weeds are plants growing where they are not wanted

herbicides. By using a **selective herbicide**, they can kill weeds and leave the crop untouched (see Figure 5.5A). The most common selective herbicides are 2,4-D and 2,4,5-T. They kill broad-leaved plants (dandelions, thistles and nettles).

Figure 5.5A ⬆ Sugar beet: part of the crop has been treated with herbicide to kill weeds; the rest of the crop is untreated

They do not harm plants with narrow leaves (grasses and cereals – see Figure 5.5B). These selective herbicides are organic compounds which contain chlorine. Other weedkillers, such as paraquat and sodium chlorate(V), are **non-selective:** they kill all plants.

Figure 5.5B ⬆ Broad-leaved and narrow-leaved plants

Chemical weedkillers are of great benefit to us. It is difficult to see how scientific farming could continue without them. There have also been some tragic results of using herbicides.

Agent Orange

From 1961–71, there was a war in Vietnam between the USA and the communist Vietcong forces. The US planes found it difficult to carry out bomb attacks on the Vietcong because the Vietcong positions were hidden by dense jungle vegetation. US planes dropped a herbicide called Agent Orange, which killed all the trees and other vegetation where it landed. It destroyed the forests which sheltered Vietcong

It's a fact!

Vietnam received 72 million litres of herbicides during the war. About half of this was Agent Orange.

SUMMARY

Selective weedkillers are of enormous benefit to agriculture. Nevertheless, one of them, 2,4,5-T, is now banned because it contains an impurity called dioxin, which has teratogenic effects and genetic effects.

fighters and their bases and killed their crops. Agent Orange contained 2,4,5-T and an impurity, **dioxin**. At the time, no-one knew that dioxin has **teratogenic** effects (effects on unborn babies). Terribly deformed babies were born to Vietnamese mothers. Dioxin also has genetic effects. When US servicemen returned home and started families, some of them found to their dismay that they had become the fathers of deformed babies. Agent Orange had affected the fathers' genes. As a result of these tragedies, 2,4,5-T has been banned in many countries. The method of manufacture always produces some dioxin as an impurity.

CHECKPOINT

▶ **1** Explain what is meant by (a) herbicide and (b) selective herbicide.

▶ **2** What are the advantages of selective herbicides over non-selective herbicides?

5.6 ▶ Insecticides

FIRST THOUGHTS

The previous section dealt with weeds – plant pests. Insects are the major animal pests, although many insects are beneficial. The pesticides which are used to kill insects are called insecticides.

It's a fact!

A family in the USA became very ill after eating a pig which they had raised. One of the children suffered permanent brain damage. The cause was the grain which they had fed to the pig. It was labelled 'Seed grain. Not for consumption. Contains mercury salts as fungicide.' But the family could not read. Mercury salts had been used to treat the seed grain so that it would not be attacked by fungi before the time came to plant it the following spring.

Insects, weeds and plant diseases destroy about one-third of the world's crops. Aphids attack the flowers of many fruits and vegetables (see Topic 4.1). Cutworms chew off plants at soil level. Wireworms bore into potatoes and other root vegetables. Grain weevils eat stored grain. Unchecked, insects would consume most of the food we grow, and humans would starve (see Figure 5.6A). On the other hand, many insects are helpful. Some pollinate flowers, and some eat harmful insects.

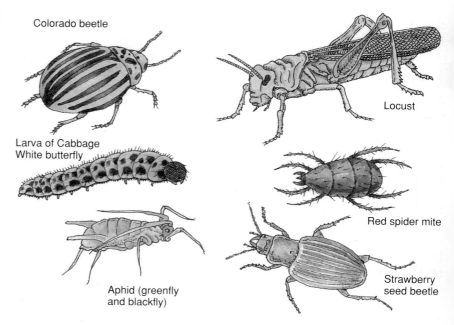

Colorado beetle

Larva of Cabbage White butterfly

Aphid (greenfly and blackfly)

Locust

Red spider mite

Strawberry seed beetle

Figure 5.6A ◆ Pests – which one is not an insect?

SUMMARY

Insecticides are used to protect crops against being eaten by insects. Insecticides have played a big part in increasing food supplies worldwide.

Pesticides help increase food production:

Herbicides kill weeds (unwanted plants) which compete with crops for space, light, water and nutrients

Insecticides kill insects which eat crops and damage farm animals.

Science at work

DDT came into production in time to be used in the Second World War (1939–1945). In wartime, the breakdown of normal hygienic living conditions means that both troops and civilians may suffer from lice, typhus and other afflictions. In the Italian city of Naples in 1943, the spread of lice caused a massive epidemic of typhus. The whole population of Naples was dusted with DDT, which killed the lice, and the epidemic was halted.

It's a fact!

According to a recent report as many as 2 million people worldwide are poisoned – 40 000 of them fatally – by pesticides each year. Pesticide controls are lax in some countries. Manufacturers recommend that agricultural workers should wear protective boots, gloves, face masks and hoods while spraying pesticides. But workers in developing countries often cannot afford this protective clothing.

The first chemicals which were used as insecticides were general poisons, such as compounds of arsenic, lead and mercury. These compounds are poisonous to all animals which swallow them. Children and family pets were sometimes accidentally poisoned. These poisons are stable: they remain in the soil for many years.

The DDT story

In 1939, a scientist called Paul Mueller discovered that a substance called DDT was very poisonous to houseflies. His experiments soon showed that DDT was a very useful substance indeed, killing many insects such as lice and mosquitoes and agricultural insect pests. The exciting aspect of DDT was that it left species other than insects unharmed. It is a **broad spectrum insecticide**. That is, it kills many types of insect. DDT is a **petrochemical** (see *Key Science Chemistry* Topic 23); making it costs relatively little. Together with similar compounds, it is often referred to as an 'organochlorine'.

DDT has saved more people than any other single substance. It has saved million of lives and prevented billions of illnesses by killing insects which carry disease. It has saved people from hunger and starvation by killing insects which eat crops. Thanks to the use of DDT, millions of people in the developing countries of the tropics now no longer live in constant fear of malaria. DDT has saved many lives. It is also used to kill the insect vectors (see p. 74) of other diseases such as cholera (see p. 82).

DDT has also contributed to agriculture. For example in the semi-desert regions of North Africa, crops are often consumed by swarms of locusts. The result is famine, and people face starvation. DDT has reduced the harm done by locusts.

It was predicted that DDT would exterminate all insect pests. In 1948, Paul Mueller was awarded the Nobel Prize for his discovery. No sooner had he received the prize than people began to discover some worrying effects of DDT. In 1962 Rachel Carson wrote a book called *Silent Spring*. She suggested that killing insects would deprive birds of their natural food, and birds would die out, giving a spring without birdsong. Many of her fears have been borne out. Since DDT does not break down easily, an application of DDT will remain in soil and water for many years (see Figure 5.6B). It is an example of a **persistent** insecticide.

Figure 5.6B ⬆ Crop spraying today. In countries where DDT is banned, safer, often biodegradable chemicals are used

p.p.m. stands for parts per million and is a measure of how much of a particular substance is dissolved in a known volume of water. For example 250 p.p.m. of DDT means that there are 250 parts of DDT per million of all parts (DDT plus water).

It's a fact!

What can be done about malaria? Every 20 seconds a young child somewhere in the world contracts malaria. The people of developing countries cannot afford modern drugs and so depend on cheap insecticides like DDT to kill the mosquitoes which transmit the malaria parasite to humans (see Topic 4.2). Some people want the ban on the use of DDT in the USA and Europe to be extended worldwide. However, for many countries a ban on DDT to help the environment could condemn many thousands of people to death. Their risk of contracting malaria would increase in the absence of effective control of mosquito numbers.

It's a fact!

Central Florida is the fruit basket of the USA. To combat fruit fly, boxes of oranges and lemons are sprayed with an insecticide before they are sold. Alternatives to spraying are refrigeration and irradiation (see p. 228). Both of these cost much more than spraying.

In Topic 7, Figure 7.3B shows what happened when Clear Lake in California was sprayed with DDT to combat mosquitoes. People were surprised when aquatic birds, such as grebes, died. The concentration of DDT in the lake water was only 0.02 p.p.m. However, DDT is soluble in fat and therefore difficult to excrete. Along the food chain, the level of DDT built up: microorganisms 5 p.p.m., fish 250 p.p.m. and at the top of the food chain, grebes 1600 p.p.m., a lethal dose.

Because it is persistent, DDT has been carried through food chains to the remotest parts of Earth, far from the places where it was used. Figure 5.6C shows how DDT can build up along a food chain on land. A problem is that many insects are now resistant to DDT (and other insecticides – see p. 442). As a result, higher concentrations must be used in spraying.

1 An elm tree is treated with DDT to kill beetles which carry the fungus causing Dutch elm disease
2 A worm eats dead leaves which contain the pesticide
3 A robin eats the worm
4 The robin is eaten by a bird of prey

Figure 5.6C ⬆ The robin's place in a food chain shows how DDT is carried from plants to predators

Persistence means that the amount of DDT in the world is steadily increasing. We know that it kills insects and that, in sufficiently large amounts, it kills fish and birds. A human being may eat fish and birds contaminated with DDT over a long period of time. We are not sure what the results of the build-up of DDT in the body will be. Cancer victims in the USA, for example, have twice as much DDT in their fatty tissues as the rest of the population. Worries over the effects of DDT on health and the environment have led to a ban on the use of DDT in the USA and Europe. Safer insecticides are used. One choice is organic compounds of phosphorus, which in time are broken down to safe compounds.

Other methods of controlling insects

Scientists are looking for other ways of controlling insects. Some methods are described as **biological control** (see Topic 4.1). One such method is to breed **predator insects**. These are insects which kill the insects that are eating the crops. Of course, the predator insects must not eat the crops! Another idea is to use radioactivity to sterilise male insects so that they cannot breed. Insects give out aromatic chemicals called **pheromones** when they want to attract a mate. Research workers have

SUMMARY

Insecticides kill insects which cause disease and damage crops. However, it would benefit the environment if we could find ways of cutting down on the use of insecticides. Research is being carried out on the use of predator insects, sterilisation and trapping.

worked out a method of extracting pheromones from insects and using the extract to bait a trap. The insects which are lured into the trap are then killed.

Great Spruce Bark beetles lay their eggs under the bark of spruce trees. When the larvae hatch, they feed on the tree's nutrients, thus killing the tree. Belgian beetles are predators of Great Spruce Bark beetles. Belgian beetles also lay their eggs under the bark of spruce trees. When the larvae of the Belgian beetles hatch, they eat the larvae of Great Spruce Bark beetles, but they do not harm the tree.

CHECKPOINT

▶ **1** In the spring of 1988, North Africa was threatened by the worst plague of locusts for 30 years. The locust control programme financed by the United Nations uses short-lived insecticides such as fenitrothion. These do not last long enough to lie in wait for advancing swarms of locusts. An area sprayed with fenitrothion is safe for locusts after only three days. With long-lasting insecticides, such as dieldrin, strips of desert can be sprayed to prevent the advance of locusts. Experts say that dieldrin would do no harm to a desert environment. However, dieldrin has been shown to cause genetic damage. Which would you use – fenitrothion or dieldrin – if you were in charge of the locust control programme? Explain your views.

▶ **2** The Mediterranean fruit fly is an insect pest. Large numbers of male flies were irradiated by a cobalt-60 source (see Topic 4.1) and then released. They mated but they produced no offspring.
   (a) Why were fruit farmers pleased with the use of radioactivity?
   (b) What is the advantage of this technique over the use of chemical pesticides?

5.7 ▶ Energy

FIRST THOUGHTS

An ecosystem consists of a community of organisms living in a particular part of the physical (non-living) environment. Ecosystems are more or less self-supporting. The animals, plants, soil, water, etc. which make up a farm are an ecosystem.

Farms are ecosystems (see Topic 2.2) which are managed so as to produce as much food as possible. The amount of food produced can be assessed in terms of the energy content of the food produced. For example, the energy content of beef is 1050 kJ per 100 g; the energy content of oatmeal is 170 kJ per 100 g. The amount of food produced depends on the amount of energy which enters the farm ecosystem and also on the efficiency with which the energy is converted into the energy content of plant and animal tissue. The energy which enters the system is the **input**, and the energy content of the food produced is the **output**.

Figure 5.7A shows the input and output of a typical intensive farm of 460 hectares in southern England. The farm is chiefly arable but raises some livestock. Figure 5.7B shows the input and output on a farm of the same size and type in the 1820s. The figures have been worked out from historical records. The unit is the gigajoule, GJ (1 GJ = 10^9 J).

The values in Figures 5.7A and B do not include the energy of sunlight as input. It is assumed to be the same today as it was in the 1820s. The rounded values in the two figures are summarised in Table 5.3).

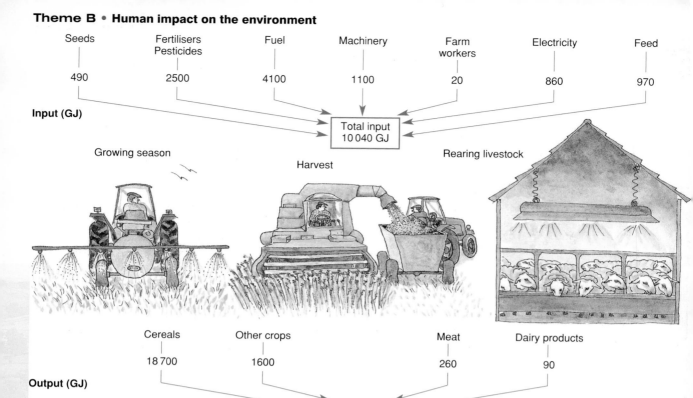

Input (GJ)

Seeds	Fertilisers Pesticides	Fuel	Machinery	Farm workers	Electricity	Feed
490	2500	4100	1100	20	860	970

Total input
10 040 GJ

Growing season Harvest Rearing livestock

Output (GJ)

Cereals	Other crops	Meat	Dairy products
18 700	1600	260	90

Total output
20 650 GJ

Figure 5.7A ⬆ Energy input and output on a 460 hectare farm in the 1990s. Note that fuel is used to power machinery, in the generation of electricity and in the manufacture of fertilisers, pesticides, etc. In total, fuel accounts for nearly 99% of energy input. Manual work accounts for less than 0.2% of energy input

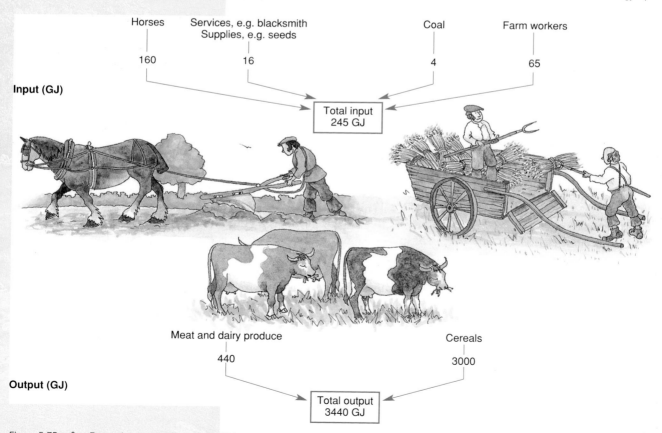

Input (GJ)

Horses	Services, e.g. blacksmith Supplies, e.g. seeds	Coal	Farm workers
160	16	4	65

Total input
245 GJ

Output (GJ)

Meat and dairy produce	Cereals
440	3000

Total output
3440 GJ

Figure 5.7B ⬆ Energy input and output on a 460 hectare farm in the 1820s. Note that a horse uses 8 MJ hour^{-1} at work, while a farm worker uses 0.8 MJ hour^{-1} (MJ = megajoule, 1 MJ = 10^6 J). More than 98% of the work is done by horses and people

Table 5.3 ▼ A comparison of intensive and traditional farms. (Energy values are in GJ.)

Energy	Intensive farm	Traditional farm	Ratio: Intensive farm / Traditional farm
Input	10 000	245	41
Output	20 700	3440	6
Ratio: Output/Input	2	14	

The increase in **productivity** (output per hectare) (see Topic 5.1) from the 1820s to the 1990s was achieved with a decreasing number of farm workers per hectare. Today's farm worker produces 60 times more food than a farm worker in the 1820s. As you see from Table 5.3, the sixfold increase in output per hectare is achieved by means of a 40-fold increase in the input of energy. This energy comes almost entirely from oil. Each year a farm worker uses the energy from over 11 tonnes of oil to enable him or her to produce so much food per worker. This brackets the energy demands of agriculture with those of the steel industry and the ship building industry. When the world's supplies of oil dwindle, the **energy crisis** will affect intensive farming.

CHECKPOINT

▶ **1** (a) How much greater is the output of the intensive farm compared with that of the traditional farm? Look at Figure 5.6.

(b) How is the increase in productivity achieved? Name three factors.

(c) How many times more energy has to be supplied to the intensive farm to achieve the increased productivity?

(d) To achieve the same level of productivity (the same output of energy), how many times more energy would have to be supplied to the intensive farm than to the traditional farm?

Topic 6

Air pollution

6.1 ▶ Smog

See www.keyscience.co.uk for more about air pollution.

Smog forms when smoke combines with fog.

Smog is a combination of smoke and fog. Fog consists of small water droplets. It forms when warm air containing water vapour is suddenly cooled. The cool air cannot hold as much water vapour as it held when it was warm, and water condenses. When smoke combines with fog, smog prevents smoke escaping into the upper atmosphere. Smoke stays around, and we inhale it. Smoke contains particles which irritate our lungs and make us cough. Smoke also contains the gas sulphur dioxide. This gas reacts with water and oxygen to form sulphuric acid, H_2SO_4. This strong acid irritates our lungs which produce a lot of mucus that we cough up.

During December 1952 in London, the weather was calm, moist and cold for a few days. The cold weather meant that plenty of coal fires were lit, producing clouds of polluting sooty smoke. The smoke and the water droplets mixed, so smog was formed. The calm weather meant that the smog did not blow away. Traffic came to a halt because of poor visibility adding exhaust fumes to the thickening smog (see Figure 6.1A). Thousands of people developed breathing problems. Figure 6.1B shows that many people died.

The Government did very little about the cause of smog until 1956. Then there was another killer smog. A private bill brought by a Member of Parliament (the late Mr Robert Maxwell, the newspaper owner) gained such widespread support that the Government was forced to act. The Government introduced its own bill, which became the Clean Air Act of 1956. The Act allowed local authorities to declare smokeless zones. In these zones, only low-smoke and low-sulphur fuels can be burned. The Act banned dark smoke from domestic chimneys and industrial chimneys. People started using gas and electricity.

Figure 6.1A ⬆ London's smoggy streets

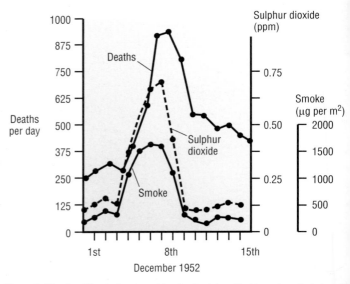

Figure 6.1B ⬆ The death rate and levels of sulphur dioxide and smoke in London smog of 1952. About 4000 more people than expected died over two weeks

6.2 ▶ The problem

All the dust and pollutants in the air pass over the sensitive tissues of our lungs. Any substance which is bad for health is called a **pollutant**. The lung diseases of **cancer**, **bronchitis** and **emphysema** (see Topic 15.2) are common illnesses in regions where air is highly polluted. From our lungs, pollutants enter our bloodstream to reach every part of our bodies. The main air pollutants are shown in Table 6.1.

Table 6.1 ▼ The main pollutants in air. (Emissions are given in millions of tonnes per year in the UK.)

Pollutant	Emission	Source
Carbon monoxide, CO	100	Vehicle engines and industrial processes
Sulphur dioxide, SO_2	33	Combustion of fuels in power stations and factories
Hydrocarbons	32	Combustion of fuels in factories and vehicles
Dust	28	Combustion of fuels; mining; factories
Oxides of nitrogen, NO and NO_2	21	Vehicle engines and fuel combustion
Lead compounds	0.5	Vehicle engines

In this topic, we shall look at where these pollutants come from, what harm they do and what can be done about them.

6.3 ▶ Sulphur dioxide

EXTENSION FILE
ACTIVITY

■ Where does sulphur dioxide come from?

Worldwide, emissions of sulphur dioxide from human sources are estimated at around 80 million tonnes per year. Almost all the sulphur dioxide in the air comes from industrial sources. The emission is growing as countries become more industrialised. Some of the output of sulphur dioxide comes from the burning of coal. Most of the coal is burned in power stations. All coal contains between 0.5 and 5 per cent sulphur.

$$\text{Sulphur (coal)} + \text{Oxygen (air)} \rightarrow \text{Sulphur dioxide}$$
$$S(s) + O_2(g) \rightarrow SO_2(g)$$

Industrial smelters, which obtain metals from sulphide ores, also produce tonnes of sulphur dioxide daily.

■ What harm does sulphur dioxide do?

Sulphur dioxide is a colourless gas with a very irritating smell. Inhaling sulphur dioxide causes coughing, chest pains and shortness of breath. It is poisonous; at a level of 0.5%, it will kill. Sulphur dioxide is thought to be one of the causes of bronchitis and lung diseases.

■ What can be done about it?

After the Clean Air Acts of 1956 and 1968, the emission of sulphur dioxide and smoke from the chimneys of houses decreased. At the same time, the emission of sulphur dioxide and smoke from tall chimneys increased. Tall chimneys carry sulphur dioxide away from the power stations and factories which produce it (see Figure 6.3A). Unfortunately it comes down to earth again as acid rain (see Figure 6.4B on p. 105).

Figure 6.3A ▲ Smoking chimneys

SUMMARY

Sulphur dioxide causes bronchitis and lung disease. The Clean Air Acts have reduced the emission of sulphur dioxide from low chimneys. Factories, power stations and metal smelters send sulphur dioxide into the air. In the upper atmosphere, sulphur dioxide reacts with water to form acid rain.

■ Detecting sulphur dioxide in the atmosphere

It is possible to monitor the emission of sulphur dioxide from the chimneys of factories and power stations from a distance. A van can carry equipment using infra-red radiation to detect sulphur dioxide. The readings are automatically recorded by a computer.

Another method of monitoring makes use of the sensitivity of some organisms to the level of sulphur dioxide in the air. Lichens fit into this category. They are a combination of a fungus and a green alga and grow on trees, stones and roof tops. Figure 6.3B shows three types of lichen: shrubby, leafy and slightly leafy. The tolerance of each type to sulphur dioxide pollution is stated. *Which type of lichen would you not expect to see in the middle of a large city?*

Different surveys show that the number of different species of lichen varies according to the amount of sulphur dioxide in the atmosphere. The further the distance from city centres where the level of sulphur dioxide pollution is high, the greater the variety of lichen species. Figure 6.3C shows you the idea.

Figure 6.3B 🔺 The tolerance of different lichens to sulphur dioxide pollution (A) Shrubby: highly intolerant (B) Leafy: medium tolerance (C) Slightly leafy: highly tolerant

EXTENSION FILE
ASSIGNMENT

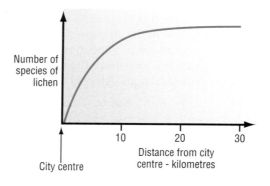

Figure 6.3C ◀ The variety of lichen species increases when the level of sulphur dioxide pollution decreases

CHECKPOINT

▶ **1** Since the 1960s, emissions of sulphur dioxide have decreased from low chimneys and increased from high chimneys.

    (a) Explain why there has been a decrease in sulphur dioxide emission from low chimneys.

    (b) Who benefited from the decrease in emission from low chimneys?

    (c) Explain why tall chimneys are not a complete answer to the problem of sulphur dioxide emission.

▶ **2** Since 1970, emissions of sulphur dioxide in the UK have decreased from 6.4 million tonnes a year to 3.1 million. What has been the percentage decrease in the emissions of sulphur dioxide since 1970? Briefly explain the reasons for the decrease.

6.4 ▶ Acid rain

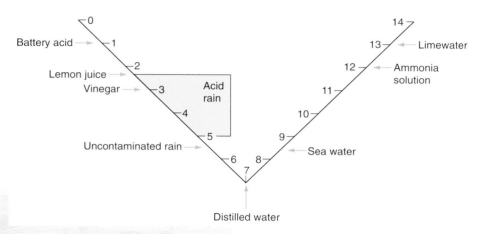

Figure 6.4A ▲ The pH values of some solutions

It's a fact!

Emissions of sulphur dioxide and oxides of nitrogen from tall chimneys in the UK and Germany fall as acid rain over Sweden and Norway.

Rain water is naturally weakly acidic. It has a pH of 5.4. Carbon dioxide from the air dissolves in it to form the weak acid, carbonic acid, H_2CO_3. What we mean by acid rain is rain which contains the strong acids, **sulphuric acid** and **nitric acid**. Acid rain has a pH between 2.4 and 5.0 (see Figure 6.4A).

How do sulphuric acid and nitric acid get into rain water? Tall chimneys emit sulphur dioxide and other pollutant gases, such as oxides of nitrogen. Air currents carry the gases away. Before long, the gases react with water vapour and oxygen in the air. Sulphuric acid, H_2SO_4, and nitric acid, HNO_3, are formed. The water vapour with its acid content becomes part of a cloud. Eventually it falls to earth as acid rain or acid snow which may turn up hundreds of miles away from the source of pollution (see Figure 6.4B).

Acid rain which falls on land is absorbed by the soil (see Figure 6.4C). At first, the nitrates in the acid rain fertilise the soil and encourage the growth of plants. But acid rain reacts with minerals, converting the metals in them into soluble salts. The rain water containing these soluble salts of calcium, potassium, aluminium and other metals trickles down through the soil into the subsoil where plant roots cannot reach them. In this way, salts are **leached** out of the topsoil, and crops are robbed of nutrients. One of the salts formed by acid rain is aluminium sulphate. The salt damages the roots of trees. The damaged roots are easily attacked by viruses and bacteria, and the trees die of a

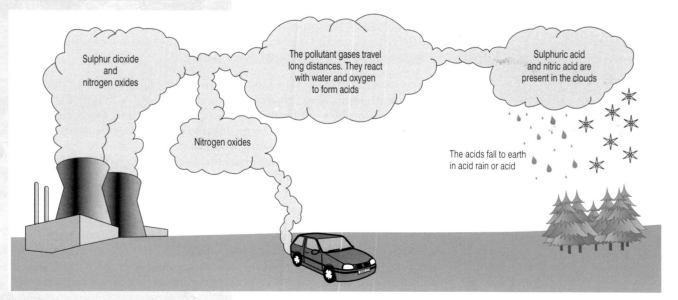

Figure 6.4B ▲ Where acid rain comes from

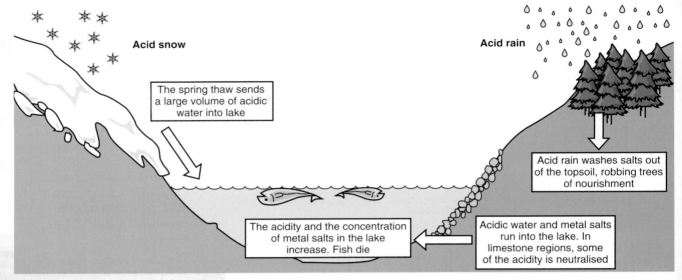

Acid snow

The spring thaw sends a large volume of acidic water into lake

Acid rain

Acid rain washes salts out of the topsoil, robbing trees of nourishment

The acidity and the concentration of metal salts in the lake increase. Fish die

Acidic water and metal salts run into the lake. In limestone regions, some of the acidity is neutralised

Figure 6.4C 🔺 Where acid rain goes

Figure 6.4D 🔺 The effects of acid rain

combination of malnutrition and disease. The Black Forest is a famous beauty spot in Germany which makes money from tourism. About half the trees there are now damaged or dead (see Figure 6.4D). Pollution is an important issue in Germany. The Green Party is a major political party. It is campaigning for the reduction of pollution. There are dead forests also in the Czech Republic and Poland. The Forestry Commission has reported that damage to spruce and pine trees in the UK fluctuates, and that identifying long-term trends is difficult. In fact, the condition of oak and beech trees has improved due to reduced insect damage compared with recent years.

The acidic rainwater trickles through the soil until it meets rock. Then it travels along the layer of rock to emerge in lakes and rivers. Lakes are more affected by acid rain than rivers are. They become more and more acidic, and the concentrations of metal salts increase. Fish cannot live in acidic water. Aluminium compounds, e.g. aluminium hydroxide, come out of solution and are deposited on the gills. The fish secrete mucus to try to get rid of the deposit. The gills become clogged with mucus, and the fish die. An acid lake is transparent because plants, plankton, insects and other living things have perished.

Thousands of lakes in Norway, Sweden and Canada are now 'dead' lakes. One reason why these countries suffer badly is that acidic snow piles up during the winter months. In the spring thaw, the accumulated snow melts suddenly, and a large volume of acidic water flows into the lakes. Acid rain is partially neutralised as it trickles slowly through soil and over rock. Limestone, in particular, keeps damage to a minimum by neutralising some of the acidity. There is not time for this partial neutralisation to occur when acid snow melts and tonnes of water flow rapidly down the hills and into the lakes.

The effects of acid rain on fresh water were first noticed in the early 1980s. In West Wales the River Twyi was particularly badly affected. The Twyi feeds Llyn Brianne reservoir which supplies most of West Wales with drinking water. The river is also very important for its salmon and trout fisheries. However, in the 1980s young fish could not survive its acidified waters.

Acid rain contains sulphuric acid and nitric acid. It has a pH between 2.4 and 5.0.

The cost of dealing with the problem of acid rain should be balanced against the cost of the damage caused by acid rain.

In 1991 treatment of the reservoir started: 500 tonnes of calcium hydroxide (powdered limestone) was distributed from boats over the water's surface every 6 months. More recently limestone has been added, from dosing silos to the Twyi and other streams feeding the reservoir. The method of treatment is more efficient and economic. The project has proved its worth, as large numbers of young salmon and trout have returned to the Twyi, re-establishing the fisheries to the benefit of the local economy.

■ What can be done about acid rain?

There are three main methods of attacking the problem of acid rain. They all cost money, but then the damage done by acid rain costs money too.

1 **Low-sulphur fuels** can be used. Crushing coal and washing it with a suitable solvent reduces the sulphur content by 10 to 40 per cent. The dirty solvent must be disposed of without creating pollution on land or in rivers. Oil refineries could refine the oil which they sell to power stations. The cost of the purified oil would be higher, and the price of electricity would increase.

2 **Flue gas desulphurisation**, FGD, is the removal of sulphur from power station chimneys after the coal has been burnt and before the waste gases leave the chimneys. As the combustion products pass up the chimney, they are bombarded by jets of wet powdered limestone. The acid gases are neutralised to form a sludge. The method removes 95 per cent of the acid combustion products. FGD is fitted to existing power stations. One of the products is calcium sulphate, which is sold to the plaster board industry and to cement manufacturers.

3 **Pulverised fluidised bed combustion**, PFBC, uses a new type of furnace. The furnace burns pulverised coal (small particles) in a bed of powdered limestone. An upward flow of air keeps the whole bed in motion. The sulphur is removed during burning. The PFBC uses much more limestone than the FGD method: one power station needs 1 million tonnes of limestone a year (4 times as much as the FGD method). The PFBC method also produces a lot more waste material, which has to be dumped.

CHECKPOINT

▶ 1 What is the advantage of building a power station close to a densely populated area? What is the disadvantage?

▶ 2 Why do power stations and factories have tall chimneys? Are tall chimneys a solution to the problem of pollution? Explain your answer.

▶ 3 Why does acid rain attack (a) iron railings (b) marble statues and (c) stone buildings?

▶ 4 (a) Why does Sweden suffer badly from acid rain?
  (b) Why do lakes suffer more than rivers from the effects of acid rain?

▶ 5 A country decides to increase the price of electricity so that the power stations can afford to use refined low-sulphur fuel oil. In what ways will the country actually save money by reducing the emission of sulphur dioxide?

6.5 ▶ Carbon monoxide

■ Where does it come from?

Worldwide, the emission of carbon monoxide is 350 million tonnes a year. Most of it comes from the exhaust gases of motor vehicles. Vehicle engines are designed to give maximum power. This is achieved by arranging for the mixture in the cylinders to have a high fuel-to-air ratio. This design leads to incomplete combustion. The result is the discharge of carbon monoxide, carbon and unburnt **hydrocarbons** (see *Key Science: Chemistry*, Topic 23).

■ What harm does carbon monoxide do?

Oxygen combines with **haemoglobin**, a substance in red blood cells (see Topic 16.1). Carbon monoxide is 200 times better at combining with haemoglobin than oxygen is. Carbon monoxide is therefore able to tie up haemoglobin and prevent it combining with oxygen. A shortage of oxygen causes headache and dizziness, and makes a person feel sluggish. If the level of carbon monoxide reaches 0.1% of the air, it will kill. Carbon monoxide is especially dangerous in that, being colourless and odourless, it gives no warning of its presence. Since carbon monoxide is produced by motor vehicles, it is likely to affect people when they are driving in heavy traffic. This is when people need to feel alert and to have quick reflexes.

Carbon monoxide is dangerous because it combines much more readily with haemoglobin than does oxygen. The body is therefore starved of oxygen.

■ What can be done?

Soil contains organisms which can convert carbon monoxide into carbon dioxide or methane. This natural mechanism for dealing with carbon monoxide cannot cope in cities, where the concentration of carbon monoxide is high and there is little soil to remove it. People are trying out a number of solutions to the problem.

- Vehicle engines can be tuned to take in more air and produce only carbon dioxide and water. Unfortunately, this increases the formation of oxides of nitrogen (see Table 6.1 on p. 103).
- Catalytic converters are fitted to the exhausts of many cars. The catalyst helps to oxidise carbon monoxide in the exhaust gases to carbon dioxide.
- New fuels may be used in the future. Some fuels, e.g. alcohol, burn more cleanly than hydrocarbons.

SUMMARY

Carbon monoxide is emitted by vehicle engines. It is poisonous. Catalytic converters fitted in the exhaust pipes of cars reduce the emission of carbon monoxide.

CHECKPOINT

▶ 1 Which types of people are likely to breathe in carbon monoxide? Is there anything they can do to avoid it?

▶ 2 How does carbon monoxide act on the body?

▶ 3 (a) What are the products of complete combustion of petrol?
(b) What harm do these products do?
(c) What conditions lead to the formation of carbon monoxide?
(d) What harm does it do?

▶ 4 A family was spending the weekend in their caravan. At night, the weather turned cold, so they shut the windows and turned up the paraffin heater. In the morning, they were all dead. What had gone wrong? Why did they have no warning that something was wrong?

6.6 ▶ Chlorofluorohydrocarbons

When the pressure is released, the propellant liquid vapourises and forces the polish out of the can

Mixture of propellant and useful liquid, e.g. polish or insecticide, under pressure

ALL BRITE

Figure 6.6A ▲ An aerosol can

The ozone layer

There is a layer of ozone, O_3, surrounding the Earth. It is 5 km thick at a distance of 25–30 km from the Earth's surface (see Topic 2.1) The ozone layer cuts out some of the ultraviolet light coming from the Sun. Ultraviolet light is bad for us and for crops. Long exposure to ultraviolet light can cause skin cancer. This complaint is common in Australia among people who spend a lot of time out of doors. If anything happens to decrease the ozone layer, the incidence of skin cancer from exposure to ultraviolet light will increase. An excess of ultraviolet light kills **phytoplankton**, the minute plant-like life of the oceans which are the primary food on which the life of an ocean depends (see Topic 2.4).

■ The problem

Ozone is a very reactive element. If the upper atmosphere becomes polluted, ozone will oxidise the pollutants. Two pollutants are accumulating in the upper atmosphere. One is the **propellant** from aerosol cans (see Figure 6.6A).

Many of the propellants are **chlorofluorohydrocarbons (CFCs)**. They are very unreactive compounds. They spread through the atmosphere without reacting with other substances and drift into the upper atmosphere. There they meet ozone, which oxidises CFCs and in doing so is converted into oxygen.

Ozone + CFC → Oxygen + Oxidation products

Another pollutant found in the upper atmosphere is **nitrogen monoxide**, NO. It comes from the exhausts of high-altitude aircraft. Ozone oxidises nitrogen monoxide to nitrogen dioxide:

Ozone + Nitrogen monoxide → Oxygen + Nitrogen dioxide
$$O_3(g) + NO(g) → O_2(g) + NO_2(g)$$

■ What should be done?

Is it happening? Is the ozone layer becoming thinner? In June 1980 the British Antarctic Expedition discovered that there was a 'hole' in the ozone layer over Antarctica during spring. Recent findings show that the 'hole' is still deepening and that springtime values of ozone levels have fallen to less than 40% of the values recorded in the 1960s. Also, the decline in ozone levels has extended into summertime, increasing the amount of harmful ultraviolet light reaching the surface of Antarctica. Other parts of the world are affected as well (see Figure 6.6B). Scientists have discovered that the ozone layer of northern Europe is thinner than it has been in the past.

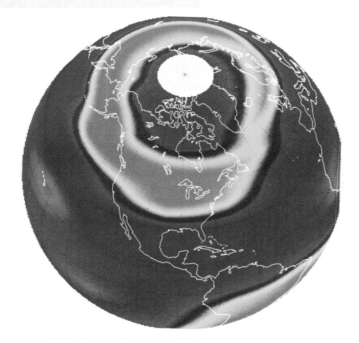

Figure 6.6B ▲ The ozone hole over the Arctic

SUMMARY

The ozone layer protects animals and plants from ultraviolet radiation. As it reacts with pollutants in the upper atmosphere, the ozone layer is becoming thinner. CFCs and nitrogen monoxide from high altitude planes are the culprits. The use of CFCs is being reduced.

Knowing that the danger had appeared over more populated regions of the globe spurred many countries to take action. At a meeting in Montreal in 1987 many countries agreed to phase out CFCs and replace them with less harmful substances. Since 1987, many countries have agreed to speed up their programme of phasing out CFCs. Aerosols containing CFCs have been banned in the USA since 1988. In 1988 makers of toiletries in the UK agreed to stop using CFCs by the end of 1989. They are now using spray cans with different propellants, which they label 'ozone-friendly', or pump-action cans. These changes have resulted in a slight decrease in some of the more harmful CFCs. However, levels of other ozone-depleting chemicals continue to increase overall despite international regulation. New agreements aim to tighten further the controls on their use and their release into the atmosphere. Only then will the ozone layer have a chance to recover, reducing the risk to the environment and human health of too much ultraviolet light reaching ground/sea-level.

CHECKPOINT

▶ 1 Look round your kitchen, bathroom and garage. How many products in aerosol cans do you buy? How convenient is it to have each of these products in an aerosol can? What inconvenience would you suffer if aerosol cans were banned? How many of the aerosol cans are labelled 'ozone-friendly'? What does this mean?

▶ 2 The Prince of Wales has banned aerosols from his household. He said that some members of his household had difficulty in finding a suitable alternative hairspray.
  (a) What concern led the Prince of Wales to take this step?
  (b) What properties must the propellant in the hairspray possess to work effectively and to be safe in use?
  (c) What substitute can you suggest for an aerosol hairspray?

▶ 3 How does their lack of chemical reactivity make CFCs (a) useful and (b) dangerous?

6.7 ▶ The greenhouse effect

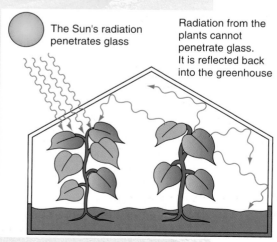

The Sun's radiation penetrates glass

Radiation from the plants cannot penetrate glass. It is reflected back into the greenhouse

Figure 6.7A 🔺 A greenhouse

The Sun is so hot that it emits high-energy radiation. The Sun's rays can pass easily through the glass of a greenhouse. The plants in the greenhouse are at a much lower temperature. They send out infra-red radiation which cannot pass through the glass. The greenhouse therefore warms up (see Figure 6.7A).

Radiant energy from the Sun falls on the Earth and warms it. The Earth radiates heat energy back into space as infra-red radiation. Unlike sunlight, infra-red radiation cannot travel freely through the air surrounding the Earth. Both water vapour and carbon dioxide absorb some of the infra-red radiation. Since carbon dioxide and water vapour act like the glass in a greenhouse, their warming effect is called the **greenhouse effect**. Without carbon dioxide and water vapour, the surface of the Earth would be at $-40°C$. Most of

EXTENSION FILE
ASSIGNMENT

Remember that there is a natural greenhouse effect. If there were not the surface of the Earth would be at −40 °C. Human activities release different gases into the air, increasing the greenhouse effect, especially that due to carbon dioxide. The increase in the greenhouse effect is the cause of global warming.

the greenhouse effect is due to water vapour. The Earth does, however, radiate some wavelengths which water vapour cannot absorb. Carbon dioxide is able to absorb some of the radiation which water vapour lets through (see Figure 6.7B).

The surface of the Earth has warmed up by 0.75 °C during the last century. The rate of warming up is increasing (**global warming**). Unless something is done to stop the temperature rising, there is a danger that the temperature of the Arctic and Antarctic regions might rise above 0 °C. Then, over the course of a century or two, polar ice would melt and flow into the oceans. If the level of the sea rose, low-lying areas of land would disappear under the sea.

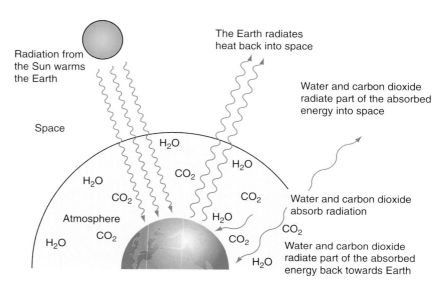

Figure 6.7B ▲ The greenhouse effect

■ Why so much gas?

One reason for global warming is that we are putting too much carbon dioxide into the air. The combustion of coal and oil in our power stations and factories sends carbon dioxide into the air. The second reason is that we are felling too many trees. In South America, huge areas of tropical forest have been cut down to make timber and to provide land for farming (see Topic 8.1). In many Asian countries, forests have been cut down for firewood. The result is that worldwide there are fewer trees to take carbon dioxide from the air by photosynthesis. The percentage of carbon dioxide is increasing; Figure 6.7C shows the trend.

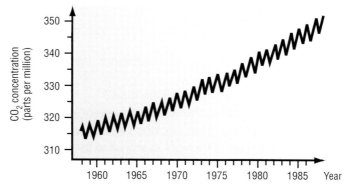

Figure 6.7C ▲ Atmosphere concentration of carbon dioxide. (These measurements were made at the Mauna Loa Observatory in Hawaii)

Methane, nitrogen oxides and CFCs (see Topic 6.6) are gases which also contribute to the greenhouse effect. Nitrogen oxides are formed from burning petrol in engines (see p. 105). Methane is given off by cattle as a waste product and by rice growing in paddy fields. As more cattle are raised for meat and milk production (see Topic 5) and more rice is grown, so more methane is produced. Nitrogen oxides are on the increase as more cars are produced and used. Also, levels of CFCs continue to increase overall, although the Montreal agreements (see p. 110) should eventually bring about improvements.

The weather could change

Global warming could upset normal weather patterns. The amount of cloud cover, the force and direction of winds, and rainfall patterns could all change. A recent study suggests that areas of the world with high rainfall may experience even more rain, while dry areas may become even drier. This means that dry areas which cannot afford to lose rain, lose what water they have and turn to desert.

■ Making changes

In 1997, at a meeting of scientists and politicians in Kyoto, Japan, targets were set for countries to reduce their carbon dioxide emissions; 2010 was the year agreed for targets to be met. The developed countries produce and use the most vehicles and manufacture the most goods. They burn the most fossil fuel to do so. The burden of cuts in carbon dioxide emissions therefore required their economies to adjust to the lower targets. Have the countries responded? In 2000, at a follow-up meeting in the Hague, Holland, it was announced that some countries were on course to meet their targets. For example, in the UK emissions had reduced by 9% since the Kyoto meeting against a target of 12.5% reduction by 2010. However, in the USA emissions had *increased* by 21.8% against a target of 7% reduction by 2010, and in 2001 President George W. Bush decided to pull the USA out of the Kyoto agreement. The disagreement between nations and between scientists and politicians shows how difficult it is to decide on how best to protect the environment without some sacrifice of national self-interest.

SUMMARY

Carbon dioxide and water vapour reduce the amount of heat radiated from the Earth's surface into space and keep the Earth warm. Their action is called the greenhouse effect. The percentage of carbon dioxide in the atmosphere is increasing, and the temperature of the Earth is rising. If it continues to rise, the polar ice caps could melt.

CHECKPOINT

▶ 1   The amount of carbon dioxide in the atmosphere is slowly increasing.
   (a) Suggest two reasons why this is happening.
   (b) Explain why people call the effect which carbon dioxide has on the atmosphere the 'greenhouse effect'.
   (c) Why are some people worried about the greenhouse effect?
   (d) Suggest two things which could be done to stop the increase in the percentage of carbon dioxide in the atmosphere.

▶ 2   *Selima*   Did you hear what Miss Sande said about the greenhouse effect making the temperature of the Earth go up?
   Joshe   I don't know what she's worried about. We wouldn't be here at all if it weren't for the greenhouse effect.
   (a) What does Joshe mean by what he says?
   (b) Is Joshe right in thinking there is no cause for worry?

▶ 3   The burning of fossil fuels produces 6000 million tonnes of carbon in the form of carbon dioxide per year. Carbon dioxide is thought to increase the average temperature of the air. It is predicted that the effect of this increase in temperature will be to melt some of the ice at the North and South Poles. What effects could polar melting have on life for people in other parts of the world?

Topic 7 Water pollution

7.1 ▶ Pollution by industry

FIRST THOUGHTS

What's the problem? We need clean drinking water – nothing could be more important. We need industry, and industry needs to dispose of waste products; in the process our water is polluted.

See www.keyscience.co.uk for more about pollution.

The National Rivers Authority controls pollution of inland rivers. It does not regulate the discharge of pollutants into tidal rivers, estuaries and the sea. The estuaries in the UK are heavily polluted by industry and by sewage.

Controls

Why do you think many industrial firms are located on the banks of large rivers? The firms can get rid of waste products by discharging them into the water flowing by (see Figure 7.1A). Until 1989 the quantities of waste which industries were allowed to discharge were controlled by the water authority of each region. Each water authority followed rules laid down in the 1974 *Control of Pollution Act*, but the rules were often broken. Wastes poured into long stretches of Britain's largest rivers, making them dirty and lacking in sufficient oxygen to keep fish alive. Recent changes in legislation bring together our different approaches to the pollution of water (including the discharge of sewage – see p. 420), air and land into an integrated package of prevention and response. The *Integrated Pollution Prevention and Control Directive* came into force in 2000. The law sets the conditions required of industry to protect the environment from the many harmful industrial activities. The law also requires sites damaged or destroyed by industrial activities to be put right when the activity stops (see p. 122)

Figure 7.1A ⬆ Industrial works on the banks of a polluted river

Why are river estuaries favourite places for the location of industry?
Estuaries are where rivers flow out to sea. In theory, the large volumes of water should carry wastes away from the land and dilute noxious substances to harmless levels. Unfortunately, experience does not always match theory, as the examples described on pages 121 and 128 show.

Recent surveys of major estuaries in England suggest that there has been some progress. Tests were used to measure how poisonous water samples are to particular species of aquatic wildlife. The results show improvements in the quality of water taken from the rivers Tyne, Tees and Mersey. Their estuaries have been the focus for the location of different industries for many years: wastes have poured out from oil refineries, chemical works, steel plants and paper mills. Today, tighter

SUMMARY

Industrial firms find it convenient to discharge wastes into rivers and estuaries. Laws set limits to the quantities discharged, but rules are often broken. However, new laws set stricter standards for the discharge of wastes. The improving water quality suggests that firms are responding to the new regulations.

Mercury is a nerve poison.

Remember that DDT is a substance that persists in the environment and is concentrated in food chains (see pp. 97, 98 and 117).

controls on the discharge of wastes means that overall water quality has improved, with the exception of a few pollution hotspots.

Mercury and its compounds

A well-known case of industrial pollution is the tragedy of Minamata, a fishing village on the shore of Minamata Bay in Japan. A plastics factory started discharging waste into the bay in 1951. By 1953, a thousand people in Minamata were seriously ill. Some were crippled, some were paralysed, some went blind, some became mentally deranged, and some died. The cause of the disease was found to be the mercury compounds which the plastics factory discharged into the Bay. Although the level of mercury compounds in the Bay water was low, mercury is a persistent substance (see p. 90). It is concentrated through **food chains** (see Figure 7.1B). The level of mercury in the fish in the Bay was high, and fishermen and their families became ill through eating the fish.

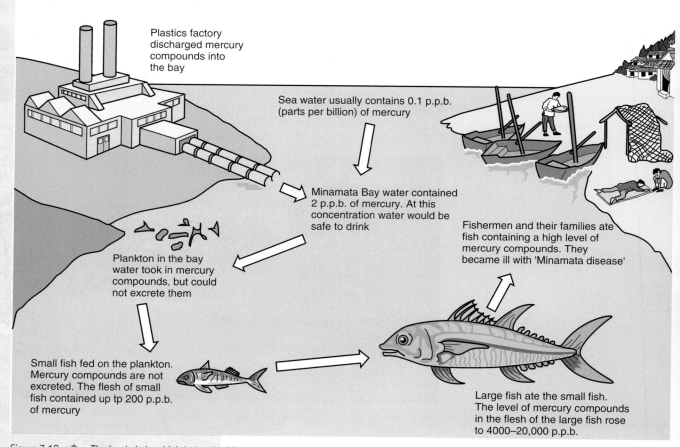

Plastics factory discharged mercury compounds into the bay

Sea water usually contains 0.1 p.p.b. (parts per billion) of mercury

Minamata Bay water contained 2 p.p.b. of mercury. At this concentration water would be safe to drink

Plankton in the bay water took in mercury compounds, but could not excrete them

Small fish fed on the plankton. Mercury compounds are not excreted. The flesh of small fish contained up tp 200 p.p.b. of mercury

Fishermen and their families ate fish containing a high level of mercury compounds. They became ill with 'Minamata disease'

Large fish ate the small fish. The level of mercury compounds in the flesh of the large fish rose to 4000–20,000 p.p.b.

Figure 7.1B The food chain which led to the Minamata disease

SUMMARY

Mercury and its compounds are poisonous. If mercury gets into a lake or river, it is converted slowly into soluble compounds. These are likely to accumulate in fish and may be eaten by humans.

Other countries have experienced the results of mercury pollution. In the 1960s, many lakes and rivers in Sweden were found to be so contaminated by mercury that fishing had to stop. In the 1970s, high mercury levels were found in hundreds of lakes in Canada and the USA. As late as 1988, the ICI plant on Merseyside discharged more mercury than the permitted level. Now that the danger is known, the polluting plants have taken care to reduce spillage of mercury. The danger is still there, however. Mercury from years of pollution lies in the sediment at the bottom of lakes. Slowly it is converted by bacteria into soluble mercury which may enter food chains.

CHECKPOINT

▶ **1** When a car engine has an oil change, the waste oil is sometimes poured down the drain. What is wrong with doing this?

▶ **2** Does it matter whether rivers are clean and stocked with fish or foul and devoid of life? Explain your answer.

▶ **3** The Minamata tragedy happened when Japan was building up its industry after the Second World War which ended in 1945. In spite of Japan's experience, Sweden, Canada and the USA found an excess of mercury in their lakes many years later. Why had they not learned from Japan's mistake?

7.2 ▶ Thermal pollution

FIRST THOUGHTS

What's wrong with warming up the water?

SUMMARY

Thermal pollution means warming rivers and lakes. It reduces the concentration of oxygen dissolved in the water.

Industries use water as a coolant. A large nuclear power station uses 4000 tonnes of water a minute for cooling. River water is circulated round the power station, where its temperature increases by 10 °C, and is returned to the river. If the temperature of the river rises by many degrees, the river is **thermally polluted**. As the temperature rises, the solubility of oxygen decreases. At the same time, the **biological oxygen demand** increases (see Topic 24.6). Fish become more active at the higher temperature, and need more oxygen. The bacteria which feed on decaying organic matter become more active and use more oxygen.

7.3 ▶ Pollution by agriculture

FIRST THOUGHTS

Farmers need to use fertilisers. What happens when a crop does not use all the fertiliser applied to it? There can be pollution, as this section explains. Persistent insecticides are also a problem. They accumulate in food chains.

Fertilisers

A lake has a natural cycle. In summer, algae grow on the surface, fed by nutrients which are washed into the lake. In autumn the algae die and sink to the bottom. Bacteria break down the algae into nutrients containing carbon, hydrogen, oxygen, nitrogen and phosphorus. Plants need nutrients (see Topic 11). Water always provides enough carbon, hydrogen and oxygen; plant growth is limited by the supply of nitrogen and phosphorus. Sometimes farm land surrounding a lake receives more fertiliser than the crops can absorb. Then the unabsorbed nitrates and phosphates in the fertiliser wash out of the soil into the lake water. When fertilisers wash into a lake, they upset the natural cycle. The algae multiply rapidly to produce an **algal bloom**. The lake water comes to resemble a cloudy greenish soup. When the algae die, bacteria feed on the dead material and multiply. The increased bacterial activity consumes much of the dissolved oxygen (increasing **biological oxygen demand** – see Topic 24.6). There is little oxygen left in the water, and fish die from lack of oxygen. The lake becomes difficult for boating because masses of algae snag the propellers. The name given to the consequences of this accidental fertilisation of lakes and rivers is **eutrophication** (see Topic 5.2).

Until recently many parts of the Norfolk Broads have suffered from eutrophication. The economic prosperity of the Broads depends on

Figure 7.3A ▲ (a) Part of the Norfolk Broads covered with a bloom of algae (b) The result of steps taken to reduce eutrophication

tourism. Action has been taken to restore water quality and the variety of wildlife to its eutrophic lakes. Figure 7.3A shows the result of progress so far.

■ Case study: Lough Neagh

Lough Neagh in Northern Ireland is the UK's biggest inland lake. It is a major source of drinking water. It is also one of the most eutrophic lakes in the world. A major culprit is phosphates. Before 1980 large amounts entered the lake in the waters discharged from sewage works. Treatment to reduce phosphate levels from sewage discharge began in 1981. It was the first time that phosphate reduction treatment had been used in the UK and phosphate levels dropped, improving the water quality of the lake. Today the process removes 100 tonnes of phosphorus as phosphates each year. A switch from the use of phosphate-based detergents and other cleaning agents has also helped. Clothes may be a little less sparkling white but the phosphate load entering the lake from these sources is reduced.

However, run-off from the soil into the lake of phosphorus in the form of the phosphate component of agricultural fertilisers (see p. 94) has steadily increased since 1974. The increase has been more than enough to cancel out the benefits of reducing phosphates from other sources (sewage, detergents and other cleaners). Today, fertilisers and animal feeds are the largest single source of the phosphates entering Lough Neagh. The phosphates accumulate in the soil, even though the amount of fertiliser spread over fields has not increased since 1974. There seems to be little hope of improving the water quality of Lough Neagh until the run-off of phosphates from the soil into the lake is substantially reduced.

The problem is not restricted to Lough Neagh. A recent survey has shown that the waters of hundreds of small lakes in Northern Ireland suffer overloading from agricultural phosphates. The reasons are not clear and research continues to try to solve the problem.

Fertiliser which is not absorbed by crops can be carried into the ground water (the water in porous underground rock). Groundwater provides one-third of Britain's drinking water. The EU has set a maximum level of nitrates in drinking water at 50 mg dm^{-3}. Some of the water companies in England and Wales have drinking water which exceeds the permitted nitrate level.

There is a health worry over nitrates. Nitrates are converted into nitrites (salts containing the NO_2^- ion). Some chemists think that nitrites are converted in the body into **nitrosoamines**. These compounds cause cancer. The level of nitrites in drinking water permitted by the EU is 0.1 mg dm^{-3}. Nitrate-stripping equipment helps reduce the level of nitrates (and therefore nitrites) in drinking water. However, the level of nitrates in ground water, from which we obtain much of our drinking water, is rising. Pollution can be reduced by cutting down the application of fertilisers and by omitting phosphates from detergents.

Insecticides

Other pollutants which cause worry are the insecticides dieldrin, endrin and aldrin (sometimes called the 'drins'). Chemically similar to DDT, their use is banned in North America and Europe. However, they persist in the environment for a long time (see p. 97). They cause liver cancer and affect the central nervous system. The European Union (the UK is a member) sets a maximum level of 5×10^{-9} g dm^{-3} for 'drins'.

Half the water in the UK exceeds this level. The danger with the 'drins' is that fish take them in and do not excrete them. The level of 'drins' in fish may build up to 6000 times the level in water.

Figure 7.3B shows what happened when DDT was used to spray Clear Lake in California to get rid of mosquitoes. It is another example of the concentration of insecticides through a food chain (see p. 98).

SUMMARY

When a crop receives more fertiliser than it can use, the excess washes into lakes and rivers stimulating the growth of plants and algae. When the plants and algae die, bacterial decay of the dead material uses oxygen (increasing biological oxygen demand). The resulting shortage of dissolved oxygen kills fish.

Insecticides are serious pollutants, especially when concentrated through a food chain.

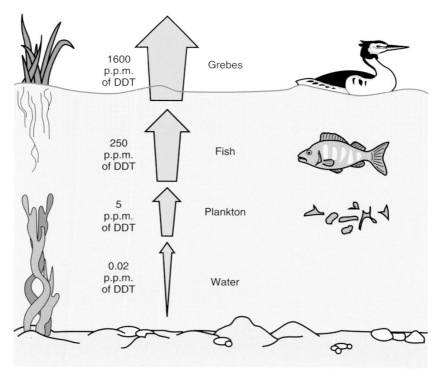

Figure 7.3B A food chain in Clear Lake, California

117

CHECKPOINT

▶ 1 (a) Why do some lakes develop a thick layer of algal bloom?

(b) Why is algal bloom less likely to occur in a river?

(c) What harm does algal bloom do to a lake that is used as (i) a reservoir (ii) a fishing lake (iii) a boating lake?

▶ 2 The concentration of nitrates in groundwater is rising. Explain:

(a) why this is happening,

(b) why some people are worried about the increase.

▶ 3 Water companies can tackle the problem of high nitrate levels by:

- blending water from high-nitrate sources with water from low-nitrate sources
- closing some sources of water
- treating the water with chemicals
- ion exchange
- microbiological methods

(a) Say what you think are the advantages and disadvantages of each of these methods.

(b) Which do you think would be the most expensive treatments? How will water companies be able to pay for the treatment?

Topic 8   Conservation

8.1 ▶   The world problem

 See www.keyscience.co.uk for more about science and the environment.

 **EXTENSION FILE**
ACTIVITY

EXTENSION FILE
ACTIVITY

 **EXTENSION FILE**
ASSIGNMENT

Our effect upon ecosystems is highlighted daily in the newspapers and on television (see Figure 8.1A). Not a day goes by without our hearing of a new environmental crisis: the pollution of our air and water, soil erosion and the creation of deserts, the population explosion and the shortages of food and shelter, destruction of natural habitats such as tropical rain forests and vanishing wildlife.

Figure 8.1A ⬆ Environmental disaster – An explosion of an oil refinery causes widespread damage and sends a thick cloud of oily smoke billowing into the air

The depletion of the ozone layer, acid rain, the increase in the greenhouse effect have all brought home to us the fact that our planet is fragile. Disasters such as oil spills and chemical leaks bring environmental problems to our attention, but less serious incidents happen every day. *What is our response?* It is easy to think that these disasters are far too vast and are impossible to cope with. *But why?* Over the past fifty years we have made great strides in reducing poverty. Improvements have come in health: infant mortality has been reduced and many diseases can now be controlled. Advances have also been made in combatting food shortages, improving welfare and education. Some nations have begun to reduce their stocks of arms. If all nations did so there would be more money to spend on undoing the damage which we have done to the Earth. One thing is sure, we cannot continue to use the Earth as a rubbish tip. This would only be storing up unsolvable problems for future generations. We have only one Earth.

Pressure on the environment

The first human beings appeared on Earth around 2.5 million years ago (see Topic 26.1). It was not until about 10 000 years ago that human activities began to have a significant impact on the environment. Think about the following timetable:

Figure 8.1B ⬆ Reconstruction of a campsite showing a family group of *Homo erectus* (an early type of human)

1.5 million years ago early humans probably moved from place to place in search of food. They hunted animals and gathered plants. Their impact on the environment was no more than that of other medium-sized animals (see Figure 8.1B).

Figure 8.1C ⬆ Archaeologists investigating an old village in modern Syria have discovered stone blades and sickles, and stones for grinding corn into flour

10 000 years ago about 12 million people lived in the world. In the Middle East they harvested wild wheat and other grains. When the grain was ripe a family could probably gather over a year's supply in just a few weeks. People had little impact on the environment beyond their village (see Figure 8.1C).

Figure 8.1D ⬆ Perhaps farming began when people watered wild crops, then sowed seeds and kept animals. Increased food production meant that some people were free to develop manufacturing skills and to barter goods, perhaps for surplus food

2000 years ago people had started to farm. Not everybody was needed to produce food. Skills in crafts and tool-making developed so that goods were made that the community needed, possibly in exchange for food. Villages became larger and some grew into towns. People had a much greater impact on the environment: farming the land, using raw materials (see Figure 8.1D).

Figure 8.1E ⬆ City life – the high standard of living uses raw materials often obtained from developing countries where environments may be stripped of resources

Today around 6000 million people live in the world. In developed countries most people live in big cities. Food is produced by relatively few people. Industry and technology offer improved living standards and leisure time. Environments in developed countries are destroyed through pressure for living space, roads, railways, airports and other services (see Figure 8.1E).

Topics 5, 6 and 7 highlight a range of human activities which put pressure on the environment. Here, we shall form an overview of the world problem.

Resources

Raw materials are needed to satisfy human demands for food, homes, hospitals, schools, cars and other manufactured goods. They fall into two main categories: **renewable resources** and **non-renewable resources**.

■ Renewable resources

These are replaced as fast as plants and animals can reproduce and grow. If a resource is over-used it will decline. Damage to the environment also limits the reproduction of renewable resources.

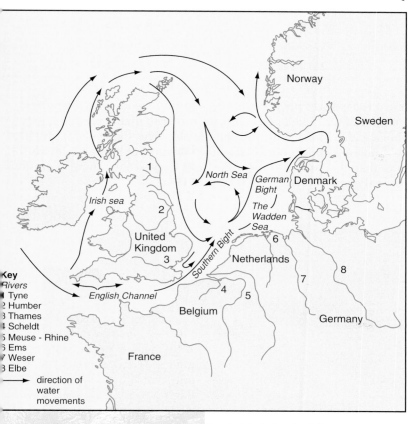

Key
Rivers
1 Tyne
2 Humber
3 Thames
4 Scheldt
5 Meuse - Rhine
6 Ems
7 Weser
8 Elbe

→ direction of water movements

Figure 8.1F ▲ Water movements in the North Sea. Anticlockwise circulation pushes water east towards the European mainland and Scandinavia

Case study: fishing in the North Sea

Catches of fish are reduced because of overfishing and pollution. The greater efficiency of modern fishing methods means that more fish are caught than are replaced by reproduction. The problem is discussed in Topic 3.6.

Causes of pollution are:

- Nutrients (e.g. nitrogen and phosphorus) from sewage works and surplus synthetic fertilisers enter rivers which discharge into the North Sea.

- Pesticides used to protect crops enter rivers which discharge into the North Sea.

- Metals (e.g. mercury, cadmium, copper) from different industrial processes are dumped or enter the North Sea through outfall pipes from the shore.

You might think that the enormous volume of water in the North Sea would dilute pollutants to harmless levels but Figure 8.1F shows the problem. Water movements carry wastes from Britain's shores across to the German Bight and Wadden Sea. The same water movements trap the outflow of wastes from the European Continent in the same area. The result is that polluting chemicals concentrate around the eastern coast of the southern North Sea.

Diseased fish are more common in the German Bight and Wadden Sea than elsewhere in the North Sea (Figure 8.1G). The area is one of the most important breeding grounds for fish in the North Sea. For example, over 80% of the plaice caught in the North Sea for food begin life in the Wadden Sea. If we pollute it to the point where wildlife can no longer survive there, we may seriously damage the ecology of the North Sea as a whole.

Figure 8.1G ◀ Diseased fish are becoming more common in the North Sea. Problems include skin cancer, ulcers, damage to the gills and abnormalities of the skeleton

■ Non-renewable resources

These cannot be replaced when used up. For example, there are only limited amounts of fossil fuels (coal, oil, natural gas) and metals.

Case study: world reserves of metals

Table 8.1 shows the estimated world reserves of some metals vital for the manufacture of goods and technological development. *How long will the reserves of each metal last at present rates of use?*

Table 8.1 ▼ Estimated world resources of some metals (billion = thousand million)

Metal	Annual use (tonnes)	Reserves (tonnes)
Iron ore	1 billion	800 billion
Copper	10 million	1.6 billion
Tin	226 thousand	12 million

Some of the world's poorest countries supply the metals which the economies of the developed countries turn into manufactured goods. As this means that many countries sell the resources they need for their own development to other countries whose standard of living is already greater than their own. At the same time, large-scale mining operations damage the environment (see Figure 8.1H).

Make sure that you can distinguish between renewable and non-renewable resources.

It's a fact!

Reclaiming land following mining operations takes time, is expensive and is often hindered by environmental problems. Since the 1980s, and despite difficulties, tens of thousands of hectares of derelict land in the UK have been reclaimed. More than 50% of the reclaimed land is used for agriculture.

Figure 8.1H ▲ Copper mining in Zaire

The picture repeats itself if we look at the energy used to turn resources into goods: 70 per cent of the world's population lives in developing countries, but these people use only 15 per cent of the world's energy output. The remaining 85 per cent is used by the 30 per cent minority of people who live in developed countries.

Land use

Habitats are destroyed, driving thousands of species of plants and animals to the verge of extinction because of our demand for land. Pressures are:

● economic development

● growing human population

● the increasing need for food to feed people

All these increase the use of land.

Case study: land use in the United Kingdom

There are 24 million hectares of land in the UK. Around 19 million hectares are used for agriculture. Building, quarrying and waste disposal are some of the other human activities that account for the rest (Figure 8.1I).

Housing covers 1.7 million hectares of land in the UK

Nearly 95 per cent of household waste in the UK is dumped into large holes or pits in the ground. The operation is called landfill

Each year quarrying in the UK produces around 300 million tonnes of gravel, limestone, sand and sandstone for concrete and other building materials

Figure 8.1I ◆ Impact on the environment of building, quarrying and waste disposal

Dumping waste from long-term mining activities has an obvious visible impact on the environment. Figure 8.1J shows that pit heaps remodel the landscape and that power stations produce vast quantities of pulverised ash.

Estimates put the amount of waste lying in pit heaps in UK coalfields at around 200 million tonnes

Coke/coal-fired power stations in the UK each year produce 10 million tonnes of pulverised ash

Figure 8.1J ◆ Dumping waste

Case study: The tropical rain forests

Nowhere is the environmental impact of the economic relationship between developing and developed countries better illustrated than in the tropical rain forests. Rain forests girdle the equator (Figure 8.1K) covering 35 million sq km of land in some of the world's poorest countries.

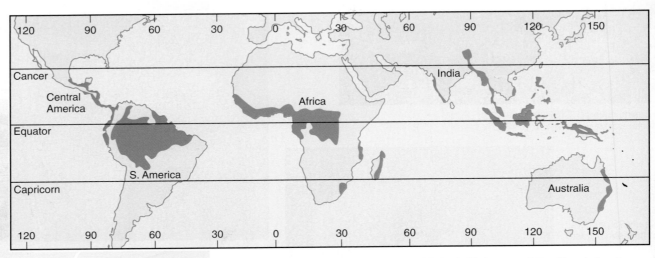

Figure 8.1K The world's tropical rain forests. Rain forests cover about 8 per cent of the world's land with dense vegetation. The rain forests contain about 50 per cent of all growing wood and the greatest variety of life of any ecosystem on earth. It is estimated that in the rain forest of Peru more than 30 million species of insect alone live in the forest canopy (see Topic 1.2)

Rain forests affect the world's climate.

Rain forests are often described as the world's lungs. The forests' giant trees recycle carbon dioxide and oxygen through photosynthesis. Moisture absorbed by the forest evaporates back into the atmosphere, crossing oceans and continents to fall as rain thousands of miles away (Figure 8.1L).

An area of rain forest the size of Belgium is now being lost each year. Cutting down and burning the trees takes away the soil's protective covering of plants, leaving it open to fierce tropical rain storms. Erosion soon sets in as soil is washed into the rivers and carried out to sea (Figure 8.1M). In the dry season the Sun's heat soon bakes the bare soil dry. Chemical reactions in the drying soil produce a hard, brick-like layer called **laterite**.

Why are we clearing the rain forests? 'We' is used deliberately because it is *our* demand for food and goods which use rain forest resources that is partly the reason for clearing. Other reasons are caught up in the social, political and economic development of the rain forest countries. The effects of this mix of national and international pressures on the rain forest environment are highlighted in Figure 8.1N.

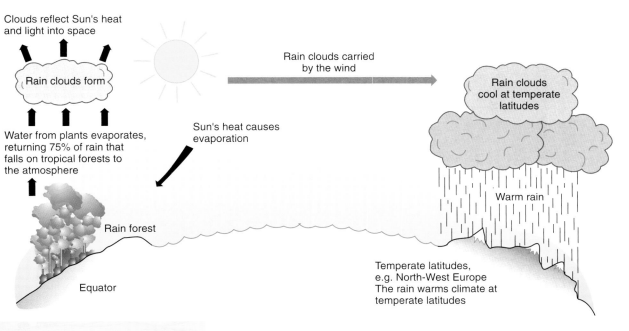

Clouds reflect Sun's heat and light into space

Rain clouds form

Water from plants evaporates, returning 75% of rain that falls on tropical forests to the atmosphere

Rain clouds carried by the wind

Sun's heat causes evaporation

Rain clouds cool at temperate latitudes

Warm rain

Rain forest

Equator

Temperate latitudes, e.g. North-West Europe The rain warms climate at temperate latitudes

Figure 8.1L 🔺 How rain forests affect the world's climate. Without them, the Equator would be warmer and temperate latitudes cooler. The pattern of rainfall would change worldwide, with widespread effects on agriculture

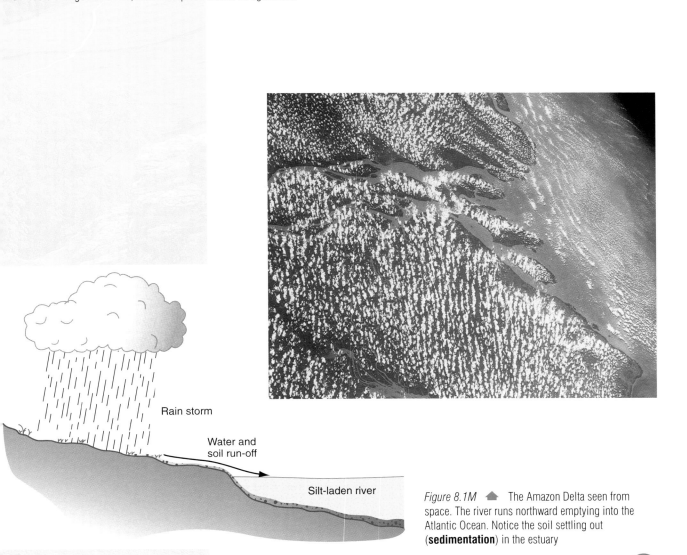

Rain storm

Water and soil run-off

Silt-laden river

Figure 8.1M 🔺 The Amazon Delta seen from space. The river runs northward emptying into the Atlantic Ocean. Notice the soil settling out (**sedimentation**) in the estuary

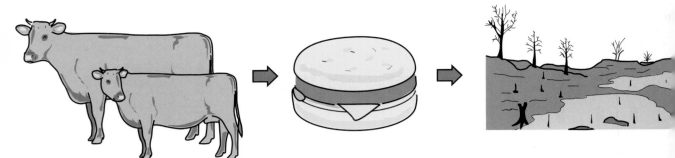

Beef
About 20 000 km² of Brazilian forest are cleared each year to make way for cattle ranches

Cheap beef is exported to North America and made into hamburgers

Forest soil is poor; with clearing, nutrients disappear. Soil is soon exhausted. Semi-desert develops: the ranchers move on to clear a new area

Logging
● Only 4 per cent of trees are felled for timber, but another 40 per cent are damaged or destroyed in the process
● Valuable hardwoods like teak and mahogany are taken to provide luxury goods and expensive furniture for the developed world. Malaya earns about £1 billion per year from timber exports

Hydro-electric power (HEP) and mining
● A huge dam and HEP station have been built on the Toncantins river, Brazil. It provides the power for developing industry, *but* the lake behind the dam swamped 2200 km² of forest
● The Carajas mountains contain rich deposits of iron ore, copper, bauxite, nickel and manganese. The cheapest way of obtaining these is by open-cast mining. *But* this causes much damage to the forest

Figure 8.1N ⬆ Exploiting tropical rain forest

Pollution

Figure 8.10 🔺 Blackpool crowded with holidaymakers

The manufacture of goods maintains our standard of living. In Britain, nearly every household has a colour television, washing machine and telephone. However, manufacturing goods and producing food result in pollution which is disposed of in the environment:

- **Air** is polluted by gases, dust and smoke from vehicles and industry.
- **Water** is polluted by wastes from factories, oil and the run-off of surplus agrochemicals from the land.
- **Land** is polluted by agrochemicals and the dumping of rubbish and waste.

The problems are discussed in Topic 6 and Topic 7. Here three case studies deal with particular examples.

Case study: leisure

Coastal resorts are popular with holiday makers and for seaside leisure activities. Blackpool's pleasure beach, for example, is visited by an estimated 7.3 million people each year (see Figure 8.1O).

The influx of enormous numbers of people to popular bathing beaches puts pressure on sewerage systems most of which were built over 100 years ago. Then the contents of sinks and lavatories emptied into pipes that took the untreated waste straight into the nearest stretch of sea. Here sunlight and salt water destroyed most of the harmful disease-causing microorganisms (see Topic 19.1). The system worked quite well, but it was not designed to cope with the present size of seaside towns, especially when summer visitors swell the population.

Seawater contaminated with sewage is unfit for bathing.

Figure 8.1P 🔺 Sewage outlet off the coast of Britain. Sea water is fit for bathing only if 100 cm³ contains less than 2000 bacteria found in faeces. Bacteria causing diarrhoea, salmonella, hepatitis, cystitis, infections of the ear, nose and throat, typhoid, polio are found in sea water polluted with sewage

New sewage works (see Topic 24.6) and replacement of antiquated piping are helping to make beaches fit for bathing. More than 90% of the 463 beaches monitored in 1999 met the required water standards (see Figure 8.1P).

127

Case study: oil spills at sea

Prince William Sound was a beautiful unspoiled bay in Alaska. It was home to a huge variety of sea animals. Death struck over an area of 1300 square kilometres in 1989. The supertanker *Exxon Valdez* left Valdez with a cargo of oil from Alaska. Only 40 km out of port, the tanker hit a submerged reef and 60 million dm^3 of crude oil leaked from her tanks. Fish, sea mammals and migrating birds from all parts of the American continent perished in the giant oil slick.

Figure 8.1Q a) The sea currents at the time of the *Jessica* incident

b) Marine iguana (a species unique to the Galapagos Islands) basks in the sun on the rocky shore of San Cristóbal – one of the Galapagos Islands. The oil tanker *Jessica* is in the background, run onto rocks offshore

Sea birds dive to obtain food. If the sea is polluted by a discharge of oil, the birds may find themselves in an oil slick when they surface. Then oil sticks to their feathers and they cannot fly. They drift on the surface, becoming more and more waterlogged until they die of hunger and exhaustion. Thirty-five thousand sea birds died in the *Exxon Valdez* disaster.

Environmentally, the Galapagos Islands are perhaps an even more sensitive environment than Prince William Sound. Charles Darwin developed his theory of evolution through natural selection following his visit to the islands in 1835 (see Topic 26). Thousands of species, including the Galapagos penguin and flightless cormorant (see Figure 26.1F(a)) are unique to the islands. Any oil spills are a potential disaster to the ecology of the area – and disaster has threatened: on January 17th 2001, the oil tanker *Jessica* ran aground just off the coast of one of the islands, San Cristobal, spilling $655\,000\,dm^3$ of oil into the sea. Although less than the 60 million dm^3 dumped by the *Exxon Valdez*, the spill could have been devastating for the wildlife and the wildlife tourism on which the economy of the islands depends. Figure 8.1Q illustrates the threat. Fortunately, winds and currents carried the oil away from the islands out to sea.

The accident has led to calls for greater protection of so-called 'particularly sensitive sea areas' (PSSA) by the World Wide Fund for Nature (see p. 134). Australia's Great Barrier Reef and the coast off Cuba are designated PSSAs. Making a PSSA work needs international cooperation to ensure that shipping traffic obeys the rules which aim to protect the environment from oil and other pollutants.

Oil spills at sea are also the results of capsizings, collisions and accidental spills during loading and unloading at oil terminals. There is another source. After a large tanker has unloaded, it may have around 200 tonnes of oil left in its tanks. While in port, the tanker is flushed out with water sprays, and the cleaning water is collected in a special tank, where the oil separates. Some captains save time by flushing out their tanks at sea and

Science at work

Bacteria can be used to clean out a tanker's storage compartment. The empty tank is filled with sea water, nutrient, air and bacteria. When the tanker reaches its destination, the tank contains clean water, a small amount of recoverable oil and an increased number of bacteria. The bacteria can be used as animal feed.

SUMMARY

Spillage of oil from large tankers is a source of pollution at sea. It kills marine animals and washes ashore to pollute beaches.

pumping the wash water overboard. This is illegal. Maritime nations have tried to set up standards to stop pollution of the seas, but several nations have not signed the agreements. Enforcing agreements is very difficult as it is impossible to detect everything that happens at sea. The Extension File Assignment in *Key Science Biology* 'Cleaning up an oil spill' illustrates various methods used to remove oil from the surface of the sea.

CHECKPOINT

▶ **1** (a) What are the causes of oil spills at sea?
　　(b) What damage do they do?
　　(c) Who pays to clean up the mess?
　　(d) Suggest what can be done to stop pollution of the sea by oil.

People fear radioactive nuclear waste because:
● it is the cause of some types of cancer
● it increases the risk of women giving birth to deformed babies.

Case study: disposing of radioactive nuclear waste

Radiation is probably the most feared type of pollutant. Non-natural sources come from the testing and use of nuclear weapons and from leakages from nuclear reactors which also produce radioactive nuclear waste. The safe disposal of radioactive nuclear waste is a major problem for the environment. Radiation is feared because it can alter the proper growth of cells. Exposure to radioactivity impairs our ability to have children and increases the risk of having deformed babies. Radiation also causes sickness, skin burns and loss of hair. Some cancers have been shown to be due to radiation. Table 8.2 shows the sources and the estimated percentages of the radiation each of us receives in the United Kingdom.

Table 8.2 ▼　Sources of radiation in the UK

Source	Percentage
Natural radioactivity	86.8
Medical	11.6
Weapons testing	0.5
Exposure at work	0.5
Aircraft journeys and luminous watches	0.5
Disposal of radioactive waste	0.1

What is radioactivity?

Certain atoms are unstable and their nuclei break down releasing small particles or rays. These atoms are said to be radioactive. The particles are **alpha** and **beta** particles and the rays are **gamma rays**. The atoms that give out (radiate) the particles or rays are called **radioactive isotopes**. Some of these isotopes give out their radiation for thousands of years and are known as long-lived isotopes. Others give out their radiation in a few seconds or days, and are called short-lived isotopes.

Where does the waste come from?

Radioactive waste comes from different sources:
● contaminated water and equipment from nuclear power stations

- solids and liquids from the reprocessing of spent nuclear fuel rods
- protective clothing, laboratory equipment and waste gases and liquids from nuclear research establishments and hospitals.

There are many nuclear research establishments in the UK. All of these produce radioactive waste.

Categories of radioactive waste

Radioactive isotopes vary greatly in the danger of their radiation and the rate at which they give it out. It is safe to be near some of the isotopes after a short time and some should be kept well away for thousands of years. This is why there are three different categories of radioactive waste and different places for them to be dumped or stored.

1 *Low-level waste* holds little radioactive danger. Huge volumes of lightly contaminated water are piped into the sea, rivers and sometimes lakes. Radioactive gases are filtered to remove solids and then passed into the air via chimney stacks. Solid waste such as laboratory equipment and protective clothing is buried in trenches at Drigg, next to the Sellafield site, in Cumbria.

2 *Intermediate-level waste* is a little more radioactive. It is made up of fuel-element cladding (casing) and material contaminated with plutonium. Plutonium is formed from the uranium fuel. A large volume of this type of waste is produced and it cannot simply be released into the sea and air. Plans to bury it deeply below ground at sites chosen for their thick clay deposits were cancelled owing to public pressure. None of us seems keen to have a radioactive dump on our doorsteps! This category of waste is being stored at nuclear establishments. Most of it will be trapped in cement and then kept in stainless steel drums. Perhaps it will be disposed of at sea or deep below ground in the future.

3 *High-level waste* is so radioactive that it generates great heat. It is the material separated from uranium and plutonium during reprocessing at Sellafield. It is remarkably dangerous and is stored in special tanks on the Drigg site. Only small volumes are produced. It has been estimated that the amount generated in supplying electricity from nuclear energy for a lifetime for one person would be the size of a tennis ball. The waste was to be stored very deeply below ground but plans have yet to be finalised.

Accident-prone Sellafield

Over a period of thirty years at Sellafield (see Figure 8.1R) there have been many accidents. Some have been at the Calder Hall nuclear reactor on the site, and some at the reprocessing plant. Release of two isotopes, strontium-90 and caesium-137 has caused particular concern. They are both absorbed by grass, eaten by cows, and find their way into milk. If we drink the milk the radioactive strontium goes to our bone marrow and can cause leukaemia (see Figure 8.1S). The caesium concentrates in muscle tissue. Critics say that the reprocessing plant has been badly managed. The plant was called Windscale for many years before being re-christened Sellafield.

Most of the problems of reprocessing have involved the levels of radioactivity of waste passed into the Irish Sea. Plutonium is the world's most poisonous substance. The sea around Sellafield contains a quarter of a million tonnes of plutonium due solely to the reprocessing operation. It was thought that the plutonium would attach to the silt at the bottom of the Irish Sea. It is now known that currents carry the contaminated silt to Scotland and Norway. The plutonium is long-lived

High-level radioactive waste is a serious hazard to health.

Figure 8.1R ⬆ Sellafield reprocessing complex

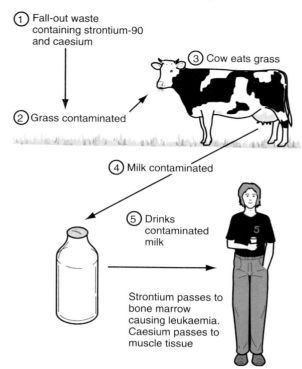

① Fall-out waste containing strontium-90 and caesium

② Grass contaminated

③ Cow eats grass

④ Milk contaminated

⑤ Drinks contaminated milk

Strontium passes to bone marrow causing leukaemia. Caesium passes to muscle tissue

Figure 8.1S ▲ Follow 1–5 for the route of radioactive waste from the environment into the tissues of the body

and will contaminate the sea for thousands of years. Criticism about Sellafield has grown as data suggests a greater incidence of cancers in the area than in the rest of Britain. The radiation could be causing some cancers.

What damage is the radiation doing to marine life? The Irish Government called for the piping of the waste to be stopped. The environmental group Greenpeace tried to block the waste pipeline using a team of divers in 1983. They failed and were fined for the illegal attempt. Their attempts did attract publicity and the pressure helped to clean up Sellafield.

Sellafield has gradually reduced the radiation in its liquid discharges. Further improvements are promised. In 1985 an Ion Exchange Effluent Plant was opened at Sellafield. It removes the radioactive isotopes strontium-90 and caesium-137 from the waste before it goes into the Irish Sea.

The environment tested

Various authorities regularly test radiation levels to see that they are below accepted limits. Seafood, seaweed, milk, air, rivers, lakes and seas are sampled in areas which could be contaminated. The modern unit of measuring radiation is the Becquerel but this takes no account of the body's reaction to different types of radiation. Medical people like to use the Sievert as a unit of radiation because it does take account of this difference. A radiation dose of 1 microSievert a year (apart from natural radiation) should be a person's limit.

Discharges of radioactive waste are controlled by the Government. The Secretary of State can give authority for these discharges. Regular checks are made to see that agreed levels of waste are not being exceeded.

SUMMARY

Resources are needed for the manufacture of goods which in turn results in the pollution of the environment. Land is used to produce food and as a convenient rubbish dump, and is covered with buildings. Increasing leisure puts further pressure on the environment. Radioactive nuclear waste is a long-term threat to our health.

CHECKPOINT

▶ **1** Suggest reasons for the fall in fish catches in the North Sea since 1950. Referring to Topic 3.6 will help you with the answer.

▶ **2** Summarise the impact on coastal waters of popular seaside resorts like Blackpool.

▶ **3** What is nuclear waste? Why are there different categories of nuclear waste?

▶ **4** Why can nuclear waste be very dangerous?

▶ **5** Refer to Table 8.2 and then answer the following questions:

  a) What is the greatest unavoidable source of our exposure to radiation?

  b) Although radioactive waste contributes only 0.1% to our total exposure to radiation, why do you think strict safety precautions for the disposal of radioactive waste are very important?

  c) Do you support the banning of testing nuclear weapons? Give reasons for your answer.

8.2 ▶   # What has it got to do with us?

It's a fact!

The Amazon rain forest is thought to supply one-third of the world's oxygen. Its destruction means a build-up of carbon dioxide enhancing the greenhouse effect.

It's a fact!

The international trade in wildlife is worth billions of pounds every year. It is a factor that is partly responsible for the decline in numbers of many species of animals and plants.

Remember that it is our demand for resources that damages the environment and drives species to extinction.

It's a fact!

Wildlife has inspired poets, writers and artists through the ages. Perhaps the aesthetic value of plants and animals is another important reason for conserving the environment.

Figure 8.2A ◀ The Kenyan authorities burned hundreds of tonnes of poached ivory to prevent it reaching the market-places of the world

Future generations will benefit by our conserving the environment. We have a duty of care to manage wild species which can be of direct use to us in terms of food and other materials. We must realise that over-exploitation of non-renewable resources will leave us poorer in the future. It is also in our interests to safeguard the existence of species that are a part of the delicate balance of nature in our world. There are many animals that we shall never see. The giant otter, wood bison, Parma wallaby, spectacled bear and Atlantic walrus are thought to be extinct or close to extinction. The destruction of habitats to enable agriculture to feed ever more human mouths is one cause. The over-exploitation of species for commercial use is another. In a world where species are disappearing so fast, perhaps we should look again at the sort of legacy we wish our children to inherit.

Elephant populations in some African countries have recovered. In some areas there are too many elephants! Recently the CITES committee has supported the return of hunting elephants and the sale of ivory taken from the elephants killed. Botswana, Namibia and Zimbabwe are able to sell ivory to Japan legally. CITES have defended the move, claiming that the sales will support conservation and community projects in the different countries.

Figure 8.2B ⬆ Some people think that furs are fashionable

Figure 8.2C ⬆ Birdcages and birds for sale. Many of the birds sold in such markets are endangered species

Many of the species that are approaching a premature end, could in fact become food for us if they were managed. The Saiga antelope in Russia has been saved from the brink of extinction and has now increased to numbers that allow some animals to be killed for food. The same is true of the American bison, eland and antelope of the African Savanna.

Plant and animal breeders combine desirable characteristics to produce improved strains, e.g. crops with high yields and farm animals with more meat. To produce the new varieties, breeders often use the original wild ancestors of modern stock. If we fail to preserve rare breeds their gene banks will not be available for use at some future date (see p. 438).

We have to take action if fish and whale stocks are to be conserved (see p. 70). There has been a serious decline in the numbers of many marine molluscs from tropical waters, such as cowries, scallops, cone shells and clams. This is because too many are harvested and exported to the USA and Europe, leaving too few to breed.

Despite a ban on hunting elephants there was a large increase in the illegal poaching of ivory in the 1970s. During the decade half of the elephants in Kenya were lost to poaching and Uganda lost 90% (see Figure 8.2A). Approximately 50 tonnes of poached ivory is used in Europe each year, equivalent to 10 000 elephants. Illegal poaching also concentrates the profits in the hands of a few people. Proper management with the sale of ivory from animals that have died naturally and from those killed to control numbers is thought by some people to be the only way to make progress.

Often the conservation of wildlife is a question of personal responsibility. The trade in furs and other animal skins is an obvious example (see Figure 8.2B). If we stopped buying furs, then the slaughter would stop, and after all there are plenty of imitation furs available. Prosperity increased in the 1960s and more people were able to afford furs. Shooting and trapping took their toll on the wild fur-bearing species. The Convention on International Trade in Endangered Species (CITES) asked governments to pass laws to protect threatened species. But some governments have yet to recognise CITES and widespread poaching and illegal fur trading still occur.

The trade in exotic birds takes 10 million birds from the wild every year; about half of these die even before they reach their destination. High prices are paid for parrots, waxbills, lovebirds and other exotic species. Also birds of prey, such as the peregrine falcon are exported to be used in falconry. Despite the protection of CITES illegal trafficking continues, with the loss of many birds in transit (see Figure 8.2C).

People have introduced new species of animals into many countries (see Topic 3.1). Some of these have been beneficial; others have upset the natural balance of the ecosystems into which they were introduced. Cattle have been introduced into the Americas, Australia and Africa, where there has been a large increase in meat production. On the other hand, rabbits were introduced into Australia only to become a pest and to threaten native species (see p. 59). Rats have spread all round the world on ships and have carried their diseases with them.

Some good news

The news has not all been bleak. In the early 1970s, the tiger was on the verge of extinction, with just 1800 left. Hunting, trapping and the destruction of its natural habitat were the main causes. In 1973, the World Wildlife Fund (now called the World Wide Fund for Nature),

SUMMARY

Conservation of species and protection of their habitats has had a low priority in the eyes of the nations of the world in the past. Constantly bringing wildlife issues to the attention of politicians has led to some progress.

with the co-operation of the Indian government, launched 'Operation Tiger'. They attempted not only to save the tiger but also to conserve a whole ecosystem. Laws were passed to protect the species and tiger reserves were set up. Poachers were given very stiff sentences. The operation resulted in an increase in the tiger population. However, poaching to satisfy the market for 'alternative medicines' based on tiger products continues to threaten the tiger's survival.

Captive breeding programmes have also saved some species from extinction. The Arabian oryx has been bred in captivity in Phoenix Zoo, Arizona. Herds have since been set up in other parts of the world and some released into the wild in Oman and other countries in the Middle East (see Figure 8.2D). Other species saved by captive breeding include Przewalski's horse, the European bison and Pere David's deer. Captive breeding may also be the saviour of tigers. Around 1000 individuals are held in zoos.

a  b

Figure 8.2D 🔺 Arabian oryx (a) in a zoo (around 400 animals are held in captivity) (b) after release into the wild (around 250 individuals roam free in Oman)

8.3 ▶ Conservation in action

FIRST THOUGHTS

Conservation aims to balance exploitation of resources with protection of the environment. The World Conservation Strategy is a blueprint for action.

Our well-being depends on keeping a balance between using and protecting environments and resources. Conservation involves working to achieve this balance and care for the environment. Conservation means different things to different people, as Figure 8.3A shows.

Many people feel very strongly about conservation. Organisations like Friends of the Earth and Greenpeace inform people about environmental issues and try to persuade governments to adopt conservationist policies. But environmental issues are not just the concern of governments – after all, they are rooted in the long-term processes of ecology and outlive successive governments.

The world conservation strategy

In 1980 a plan called the **World Conservation Strategy** was prepared by the United Nations Environment Programme, World Wildlife Fund (now called the World Wide Fund for Nature) and the International Union for the Conservation of Nature and Natural Resources. The plan set out an agenda for development that met human needs but which also protected the environment for future use. Figure 8.3B summarises its programme.

Different initiatives worldwide aim to protect wildlife in danger of extinction (**endangered**) because of human activities. The:

- **Convention on International Trade in Endangered Species** (CITES – see p.133) aims to control the trade in, and products from, endangered wildlife.

- **World Heritage Convention** identifies areas of international importance for science and/or conservation.

- **Ramsar Convention** lists wetlands which are important internationally for the conservation of endangered wildlife.

- **Convention on Biological Diversity** aims to encourage biological diversity (see p. 4) while seeking ways to exploit wildlife in a sustainable way (see p. 138).

- **International Convention for the Regulation of Whaling** aims to protect whales while seeking ways to exploit them in a sustainable way (see p. 138)

The World Conservation Strategy is an international plan for protecting the environment and its resources.

Figure 8.3A ▲ What is conservation? (a) Preserving wildlife and places of natural beauty (b) Maintaining the balance between human needs and resources (c) Providing attractive places to live and opportunities for leisure (d) Returning to the simple life

For the first time the World Conservation Strategy showed that development and conservation went hand-in-hand. In fact, in the long run one cannot succeed without the other. The updated version, World Conservation Strategy 2, is a plan for international action into the twenty-first century. It is aimed at politicians, whose decisions affect the environment and its natural resources, as well as at people concerned with conservation. It aims to encourage governments:

- to realise that ecology, the economy and equal sharing of resources must be partners if nations are to develop successfully in the long term

- to accept that people and not just governments should have a say in managing the development and resources of their own country

- to improve the economic analysis of long-term development

- to check regularly that development is not at the long-term expense of the environment

- to recognise that we share the planet Earth with wildlife which has a right to exist

- to control the harmful effects of industrial development, over-use of energy resources and growing populations increasingly concentrated in big cities

- to recognise the harmful effects of economic insecurity and warfare (including nuclear war) on the environment

- to put right ecosystems damaged or destroyed by human activities and upon which we depend for food, space and indeed life itself.

Erosion

Soil is the lifeblood of farming. But more and more soil is lost each year through erosion caused in part by clearing forests and poor farming methods. Each year erosion removes around 25 billion tonnes of topsoil worldwide. As a result, crop yields decline. Evaporation accelerates the loss of soil water and land gradually turns to semi-desert. A cycle of drought and flood is established.

Ways of conserving soil

1 Reforestation – replanting trees and hedgerows. Plant roots hold soil in place; leaf fall adds humus (see Topic 2.1) which improves soil structure and water-holding properties. However, not all trees are suitable: quick-growing species like eucalyptus and conifers (used for making paper and furniture) may cause further damage to the soil.

2 Improved farming methods – increasing the variety of crops grown, rather than monocultures (see Topic 5.1).

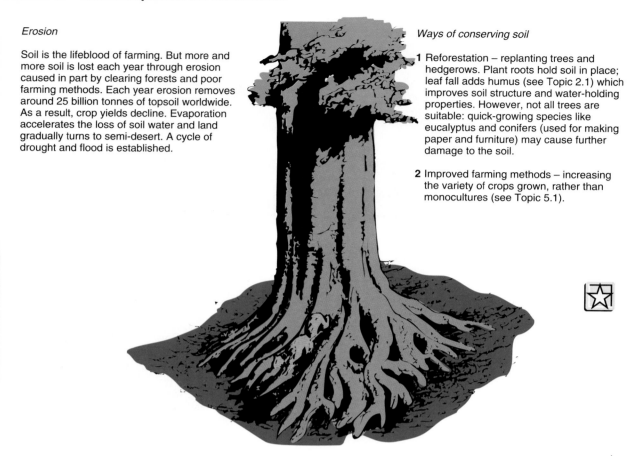

CONSERVING WILDLIFE AND ECOSYSTEMS

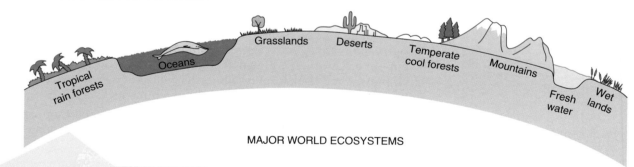

MAJOR WORLD ECOSYSTEMS

Extinction

More and more species are becoming extinct as a result of human activities. About 100 mammal species and 162 bird species have been lost in the past 400 years.

6000 animal species and 25 000 plant species are now threatened with extinction. These are the known losses. It is likely that rain forest clearance wipes out many species which have not yet been identified.

Ways of conserving wildlife and ecosystems

1 Wildlife and ecosystems need to be managed in a way which avoids damage but which also benefits humans. This may mean looking for alternative ways of tackling problems. For example, the African savanna (grassland) is infested with tsetse fly (see Topic 3.7). The tsetse fly transmits a parasite which causes sleeping sickness in cattle and humans. If farmers wish to raise cattle, they have to wipe out the tsetse. This often means clearing the savanna of trees and scrub which are the tsetse's habitat, and killing wildlife which harbours the parasite. This clearance is costly, it damages the ecosystem, and it is not 100% successful – so that repeated clearance is necessary.

However, the wild antelope of the savanna are immune to tsetse fly. Properly managed, they provide more meat per hectare than cattle. The antelope offer the possibility of long-term development of food resources together with conservation of the savanna ecosystem.

2 Managing ecosystems for human benefit is difficult. However, rain forest products such as fruit, nuts, rubber latex and medicinal plants can be exploited while leaving the ecosystem intact. For example, vinblastine and vincristine are powerful anti-cancer drugs. They come from the rosy periwinkle plant that grows in the rain forests of Madagascar. Many more beneficial plants may remain undiscovered – it is vital to preserve areas of undisturbed rain forest if people are to benefit in the future.

CONSERVING RESOURCES

Renewable resources
Without human interference, renewable resources are replaced as fast as plants and animals can reproduce and grow in their particular environments. But if a resource is over-used it will decline (see Topic 3.6). Damage to the environment also limits its ability to produce renewable resources.

Non-renewable resources
There are only limited amounts of non-renewable resources such as fossil fuels and metals in the world (see Topic 8.1). Economic development depends on them. Once used up, they cannot be replaced – substitutes must be found, or new deposits discovered (perhaps in the future we may find deposits on the moon or nearby planets). Careful use of resources helps conserve stocks until acceptable solutions to the problem of dwindling supplies are found

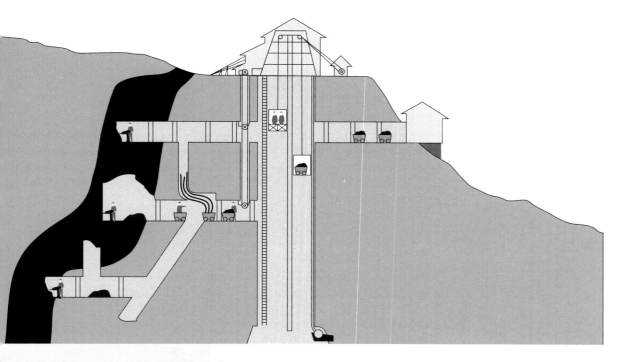

Figure 8.3B ▲ Aims of the World Conservation Strategy

What we do now will affect the future of generations to come. We owe it to them to leave them an environment that meets *their* needs as it meets *ours* today and a world where resources are shared equally by rich and poor.

Recycling makes sense

Using materials again (recycling) is an important way of conserving and sharing resources. Too often we just throw away things when they go out of fashion or are no longer useful to us. We must not forget that they may be useful to someone else. Many common substances can be recycled rather than dumped. Glass, metals, paper, plastics, oil, rags, heat and vegetable waste can all be recycled, *but why bother?* There are many reasons why we should bother with recycling:

raw materials are saved because the substance does not have to be made in such large quantities. Some raw materials are in short supply and may cease to exist if present demand continues.

Less energy is used to recycle than to make the substance from 'scratch'. Energy is always costly and reserves of some fuels, such as oil, are limited. The table shows the energy required to recycle as a percentage of the energy to make the substance from scratch:

Substance	Steel	Aluminium	Copper	Glass	Plastics	Newspapers
% energy for recycling	53	5	10	92	3	77

There is always a saving of energy when recycling.

Costs of disposing of the waste material to landfill sites are saved if the substance is recycled. Landfill sites (see Topic 8.1) are in short supply and space is being saved for waste that cannot be recycled.

The rag and bone man was the originator of recycling. The modern equivalents to the rag and bone man are charity shops, car-boot sales, jumble sales and garage sales. Today people are asked to take their substances for recycling to central points where special containers are provided. Recycling parks are popular throughout Europe. Glass, paper, cardboard, motor-oil, plastics, textiles, batteries and aluminium and steel cans may be deposited separately.

■ Packaging crazy

You can hardly leave a shop without your purchased item being double-wrapped. Many items are displayed in *'blister packs'* of plastic on card to attract the eye. A disposable paper or plastic bag will also be supplied on purchasing the item. In the UK 20 million paper bags are used each day. In the USA, 35% of municipal waste is composed of packaging. Much of the packaging is wasteful. In the UK 75% of glass, 40% of paper, 29% of plastic and 14% of aluminium is used solely for packaging. We have gone packaging mad, encouraging a litter problem.

■ Paper recycling

Each year every person in the UK uses up one tree's worth of paper and board. The trees are not being replanted at the high rate of use. Yet we rely on trees to replenish oxygen in the air. To make just 1 kilogram of paper, 2.2 kg of timber, 445 litres of water and 7 kilowatt-hours of energy are required. Recycling paper requires timber and less energy and water. Recycling of paper makes extra sense in reducing imports of wood-pulp and preserving the environment.

Sustainable development aims to use renewable resources only as fast as they are naturally replaced by reproduction and growth.

See www.keyscience.co.uk for more about sustainability.

Recycling rubbish saves on:
- raw materials used to manufacture goods,
- energy needed to manufacture goods from 'scratch',
- landfill sites into which waste material is dumped.

See www.keyscience.co.uk for more about decay.

Waste paper is collected, sorted and baled. It is pulped, de-inked, cleaned and dried. Some mills use 100% waste paper to make products, other mix the pulp with new wood pulp. The recycled paper is used for newspaper, cardboard, envelopes, writing-paper and tissues. Some special papers such as waxed paper and carbon paper cannot be recycled. It would be best if we kept to the sensible saying *'please don't choose what you can't re-use'*.

Highways for wildlife

Many people see the motorway as an enemy of wildlife. However, the grassy verge of a motorway is one of the few places where people are not allowed to go. Wildlife therefore enjoys a private, if rather narrow, sanctuary hundreds of kilometres long (see Figure 8.3C).

Road verges and railway embankments are a refuge for wildlife.

Figure 8.3C ▲ Kestrels often hunt by motorways for rodents (voles and mice) which flourish undisturbed by people

Parks, gardens and farms also have conservational value even if they are not primarily intended for wildlife. Their populations of living things are reservoirs which boost numbers overall. Road verges and railway embankments are vital links between these island populations connecting habitats and ecosystems. In this way members of a population keep 'in touch', are able to breed and therefore maintain their numbers.

Land use conserves particular habitats

The wide grassy spaces and heather-covered hillsides of upland Britain are moorlands. Notice in Figure 8.3D that the moorland landscape is practically treeless although the climate and soil will allow trees to grow. Why is there no woodland?

* **Deforestation** (cutting down trees) started as human settlement began. Clearings around villages joined up creating an open landscape with separated pockets of forest. Increasing agricultural pressures removed hedges and woodlands in the more fertile lowland areas.
* **Grazing** by the introduction of large numbers of sheep on the hillsides completed the process of deforestation. Sheep can be raised on the poor soil where crops fail to grow. They nibble almost anything, eliminating plants unable to survive the frequent close cropping. Tree seedlings especially are prevented from growing into young trees. Sheep (and also rabbits) are therefore responsible for the characteristics of the moorland environment.

Figure 8.3D ▲ Grazing sheep help to maintain the moorland environment

Zoos, botanic gardens and seed banks

The reasons for breeding wild species of animal and plant in captivity are to:

- **increase** the number of individuals of particular species
- **educate** people by showing them that wildlife is worth preserving
- **research** the needs of species so that the knowledge gained can be applied to conservation in the wild.

Increasing numbers of a species depends on a successful breeding programme. The Arabian Oryx programme at Phoenix Zoo, Arizona described in Topic 8.2 and pictured on p. 134 is a case in point. Following its extinction in the wild in 1972, successful captive breeding and subsequent re-introduction into Oman and other countries in the Middle East seem to have secured the oryx's future. However, the conservation work does not stop with release of animals into the wild:

- Involving the local people in the security of the released animals is essential if the destruction which caused the extinction in the first place is to be avoided.
- Finding the right habitat which is reasonably protected and self-contained is another requirement.
- Watching over the newly introduced animals to ensure they settle down is essential.

Conservation in action is clearly a long-term commitment, and the programme you have just read about is for one species with a successful conclusion. Think of all of the other endangered species, some of which are proving more difficult to conserve (the cheetah is an example). For some people involvement in conservation is a lifetime's work, but with it comes the reward of knowing action helps save some part of life on Earth.

Seed banks are cold stores of seeds. The facility at the Royal Botanic Gardens, Kew, London keeps the seeds of wild species threatened with extinction. The aim is to safeguard over 24 000 plant species worldwide from extinction and secure the future of the flowering plants native to the UK (see Figure 8.3E).

Captive breeding is one way of securing the future of species close to extinction.

There are 250 000 known species of flowering plant worldwide.

Figure 8.3E A repository of samples in the Royal Botanic Gardens at Kew

Millions of hectares of potatoes are grown throughout the world. Even so, the industry depends on a restricted number of varieties. In order to retain and improve the genetic basis for new varieties, gene banks of potato plants are maintained. The most extensive collection in the world is held in Peru (the home of wild potatoes). Many hundreds of different varieties of potato are duplicated at two separate centres (see Figure 8.3F). Should one of the centres be destroyed, the potato genes are still protected at the other. Different countries including the Irish government help finance these potato gene banks in Peru.

Figure 8.3F ▲ Samples of potato species ready for storage in a bank in Huancayo, Peru

Why conservation?

Throughout the Universe we know for certain of only one planet where there is life – our own. The quality of its environment is our concern and responsibility. We have only one planet Earth (see Figure 8.3G). All of us must make sure that it remains a beautiful, varied and worthwhile place in which to live.

Figure 8.3G ▲ Our home in space

SUMMARY

Conservation means different things to different people: preserving the environment, maintaining a balance between human needs and resources, providing attractive places to live, living a simple life. The World Conservation Strategy sets out a plan of action for conservation at government level. Examples of conservation in action include human environments which have wildlife value, zoos, botanic gardens and seed banks.

CHECKPOINT

▶ **1** What do you think 'conservation of the environment' means?

▶ **2** Why should we be concerned about the destruction of tropical rain forests?

▶ **3** Briefly describe how you would manage a **named** ecosystem for commercial gain while conserving its wildlife and environment.

▶ **4** 'Motorways are a curse or a blessing?' On which side of the argument are you? Briefly justify your position.

▶ **5** Do you think keeping animals in zoos is justified? Give reasons for your answer.

Theme Questions

1. In the early years of the 20th century most farms in the eastern region of England produced a variety of crops. Since the late 1940s many farms have specialised in the production of only one type of crop. Farmers have managed to increase yields by increasing the amount of land for agriculture and by the use of chemical fertilisers and pesticides.

   (a) Many hedgerows have been removed to increase the size of fields.

      (i) Farmers often claim that removing hedgerows helps to keep their costs down. Suggest possible reasons for this.

      (ii) Explain how the loss of hedgerows may have reduced the numbers of birds and small mammals in the countryside.

   (b) The table shows the average yield of wheat in England and Wales between 1940 and 1980.

Year	Yield (tonnes per hectare)
1940	2.4
1950	2.6
1960	3.6
1970	4.4
1980	4.7

      (i) Some of the increase in yield is due to the introduction of new varieties. Explain how these varieties could have been developed.

      (ii) Explain how the use of insecticides and selective herbicides has helped to increase the yields.

   (c) Explain how the increased use of organic fertilisers may have added to environmental problems.

   (d) Scientists have been trying to transfer genes from nitrogen fixing bacteria into cereal plants. Suggest how this might reduce the need to use inorganic fertilisers. (OCR)

2. Read the following information carefully.
   In the USA there is a large lake called Clear Lake that is used for recreational purposes such as fishing. In the early 1940s 'run off' of fertiliser from surrounding farmland caused eutrophication which disturbed the natural ecosystem. This led to an increased population of small insects called midges. The midge larvae were sprayed with DDD (an insecticide like DDT) in 1949, 1954 and 1957. The first two sprays killed 99% of the midges but the population quickly recovered. The third spray, in 1957, had no effect. Small fish ate the midge larvae and when some were caught they were found to have up to 200 parts per million (ppm) of DDD in their flesh. This disturbance of the ecosystem lasted for many years. It even resulted in the death of a population of over 1000 grebes – birds that ate the small fish in the lake. These dead grebes were found to have up to 1600 ppm of DDD in their flesh.

   (a) (i) Explain how eutrophication was caused in Clear Lake.

      (ii) Apart from the midges, how would the recreational use of the lake be affected by eutrophication?

   (b) Explain the observed differences in the levels of DDD in the small fish and the grebes.

   (c) Instead of using a spray, the midge larvae could be controlled biologically. Explain what is meant by biological control?

   (d) The widespread use of DDD around the lake probably killed other insects. State another problem that this would cause. (AQA)

3. Lichens are organisms formed from an association between fungi and algae. They are easily damaged by air pollution. A large number of different types of lichen is a good indicator of clean air. Lichens therefore may be used as indicators of pollution levels. The table shows how many different types of lichen were recorded at set distances from a city centre.

Distance from city centre (km)	Number of types of lichen found in a given area
0	4
2	7
3	10
5	20
6	25
7	40

   (a) Draw a graph of these results, plotting Number of types of lichen found in a given area/Distance from city centre (km).

   (b) Use your graph to estimate the number of types of lichen at 4 km from the city centre.

   (c) Use your graph to state a pattern that links the number of types of lichen with the distance from the city centre.

   (d) Since these data were collected, pollution in cities has decreased. Suggest **two** ways that the pollution in city centres has been reduced.

   (e) Burning some fossil fuels produces acid rain. Explain how acid rain is formed and state **one** of its effects.

   (f) Read the introduction to the question once again. Suggest what is meant by 'indicators of pollution levels'. (SEG)

4 Soil contains aluminium compounds. Acid rain washes these aluminium compounds out of the soil and into rivers and lakes.

(a) Explain how acid rain is formed.

(b) Graphs **X** and **Y** show the survival of fish in water at different pH values, with and without aluminium compounds.

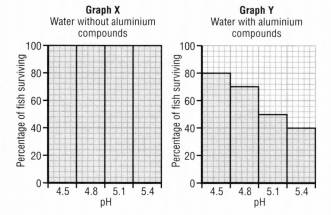

(i) What percentage of fish survive at pH 4.8 in water containing aluminium compounds?

(ii) What percentage of fish survive in the most acidic water containing aluminium compounds?

(iii) What evidence from the graphs shows that aluminium compounds have a greater effect than pH on fish survival?

(c) Describe an experiment you could do to find out if aluminium compounds cause fish to grow slowly.

5 (a) What is the difference between oxygen and ozone?

(b) What converts oxygen into ozone?

(c) What converts ozone into oxygen?

(d) What is the ozone layer? Where is it?

(e) Why is the ozone layer becoming thinner?

(f) Why does the decrease in the ozone layer make people worry?

6 The diagram shows how mercury, released into the sea in chemical waste from a plastics factory, results in the poisoning of local fishermen.

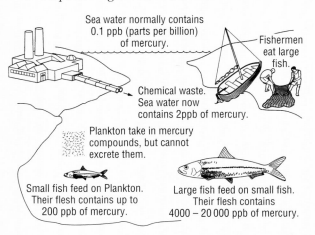

Sea water normally contains 0.1 ppb (parts per billion) of mercury.

Fishermen eat large fish.

Chemical waste. Sea water now contains 2ppb of mercury.

Plankton take in mercury compounds, but cannot excrete them.

Small fish feed on Plankton. Their flesh contains up to 200 ppb of mercury.

Large fish feed on small fish. Their flesh contains 4000 – 20 000 ppb of mercury.

(a) Draw a food chain to show how the fishermen become poisoned with mercury.

(b) (i) What information, missing from the diagram, is needed in order to construct a bar graph to show the levels of mercury in each organism?

(ii) Suggest why the fishermen are poisoned by the mercury, but the other organisms in the food chain are not affected.

(c) As a result of the mercury poisoning, the fishermen suffer from uncontrollable shaking. Suggest which organ is likely to have been affected.

(d) (i) Mercury compounds are non-biodegradable. Explain the term *non-biodegradable*.

(ii) The factory was making plastics. Suggest **two** ways by which these plastics could pollute the environment. (UCLES)

7 The diagram shows the water cycle.

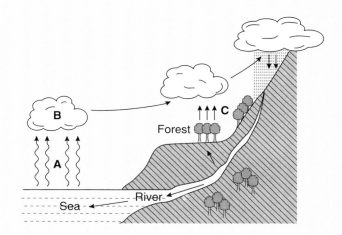

(a) The arrows **A** represent evaporation.

(i) What is the main source of energy for this process?

(ii) 1. What change occurs to the water vapour in region **B**?
2. What causes this change?

(iii) What process do arrows **C** represent?

(iv) State **three** factors which could affect the rate at which process **C** occurs.

(b) A timber company wishes to cut down the forest.

(i) Suggest how cutting down the forest is likely to affect the climate inland (away from the sea).
Explain your answer.

(ii) State **three** other undesirable effects of deforestation. (UCLES)

Cell biology

Carbohydrates, fats, proteins and nucleic acids are organic compounds and are compounds of living matter. Cells are to living matter as bricks are to houses. Cells build organisms. Different types of cell are specialised to perform particular tasks.

Topic 9 The cell

9.1 ▶ # Seeing cells

Cells were first seen in 1665 when the English scientist Robert Hooke looked at cork under his microscope. His discovery is one example of the way in which an advance in technology – in this case the microscope – leads to scientific discoveries. Modern microscopes reveal the structure of cells in great detail.

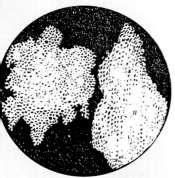

Figure 9.1A 🔺 The microscope which Robert Hooke used to draw the outline of cork cells (from his book *Micrographia*, 1665)

The **plasma membrane** which surrounds the cell is sometimes called the **cell surface membrane**.

When in 1665 the English scientist Robert Hooke looked at thin strips of cork through his microscope, he saw the little box-like outlines illustrated in Figure 9.1A. He called the outlines **cells** because they reminded him of the small rooms (called cells) in which monks live.

The importance of Hooke's discovery passed unnoticed until 1838 when two German biologists, Mathias Schleiden and Thoedor Schwann put forward a **cell theory of life**. Their theory, updated in the light of later discoveries, states that:

- all living things are made of cells
- new cells are formed when old cells divide into two
- all cells are similar in structure and function (the way they work), but not identical
- the structure of an organism depends on the way in which the cells are organised
- the functions of an organism (the way it works) depend on the functions of its cells.

The more we know about the structure and function of cells, the better able we are to understand how a whole organism works.

Microscopes

Most cells are too small to be seen with the naked eye. Without microscopes, we should know very little about them. Figure 9.1B shows a modern **optical** (light) **microscope**. The photographs (called **photomicrographs**) are images of plant cells and animal cells seen through the microscope. A lamp lights the specimen of cells which the observer views through two **magnifying lenses**. The magnification of the cells is worked out as:

$$\text{Total magnification of cells} = \text{Magnifying power of the eyepiece lens} \times \text{Magnifying power of the objective lens}$$

The magnification of the cells shown in Figure 9.1B is given by:

$$\times 10 \text{ (eyepiece)} \times \times 40 \text{ (objective)} = \times 400$$

Good quality optical microscopes can magnify cells up to 2000 times (×2000) their original size. At higher magnification, the image is enlarged but less clear because together the eyepiece lens and objective lens of the microscope cannot distinguish between cell structures lying side-by-side – the structures blur together appearing as one fuzzy image. The ability of a microscope to distinguish between structures lying close together is called **resolving power**.

Figure 9.1C shows a different type of microscope called the **transmission electron microscope**. Instead of a beam of light, it passes a beam of electrons through the specimen of the cells. It can magnify up to 1 million times (×1 000 000), producing images which show the structures of cells in minute detail. Such detail is possible because the resolving power of the transmission electron microscope is much greater than that of the optical microscope.

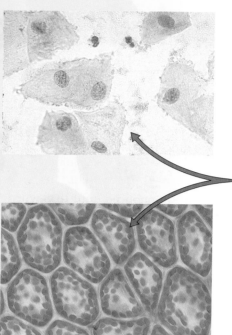

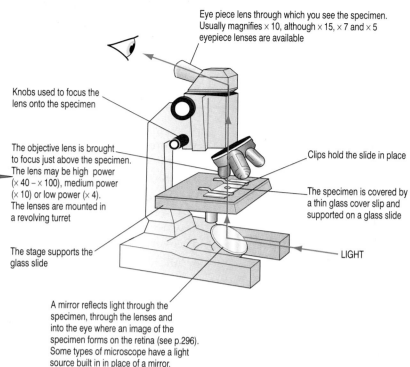

Eye piece lens through which you see the specimen. Usually magnifies × 10, although × 15, × 7 and × 5 eyepiece lenses are available

Knobs used to focus the lens onto the specimen

The objective lens is brought to focus just above the specimen. The lens may be high power (× 40 – × 100), medium power (× 10) or low power (× 4). The lenses are mounted in a revolving turret

The stage supports the glass slide

Clips hold the slide in place

The specimen is covered by a thin glass cover slip and supported on a glass slide

LIGHT

A mirror reflects light through the specimen, through the lenses and into the eye where an image of the specimen forms on the retina (see p.296). Some types of microscope have a light source built in in place of a mirror.

Figure 9.1B ⬆ An optical microscope

It's a fact!

In 1933, the German physicist Ernst Ruska made an electron microscope capable of producing a clear image at ×1200 magnification.. The first commercial instrument was built in 1938. It was used to study metals. Later, biologists found to their surprise that the electron beam did not destroy biological specimens. The powerful new microscope opened up a new world. It revealed cell structures which no-one had ever seen before. The science of cell biology took off.

SUMMARY

The invention of the light microscope led to the discovery of cells. The electron microscope has revealed the fascinating world inside the cell and opened up the study of cell biology.

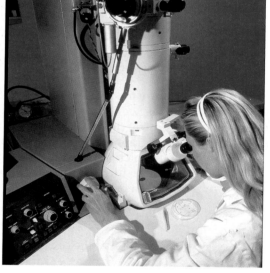

ribosomes   mitochondrion

Figure 9.1C ⬆ A transmission electron microscope (TEM)

As we cannot see electrons, the electron beam is directed through the specimen onto a fluorescent screen to give an image which can be photographed (**electron photomicrograph**). Notice the electron photomicrograph of an animal cell shown in Figure 9.1C. Many more structures are visible compared with the photomicrographs of the animal cells illustrated in Figure 9.1B. Also notice that structures called **mitochondria** (plural – see p. 245) are just visible under the optical microscope. In Figure 9.1C look at the appearance of a mitochondrion (singular) under the transmission electron microscope. Many more of its details are visible – an example of resolving power in action! Resolving power also makes visible tiny bead-like objects strung along the stacks of membranes which are an important part of the structure of cells. The 'beads' are **ribosomes**. Here proteins are made by the cell (see p. 153).

■ Preparing specimens

Cells are fragile and most of their structures are transparent. Different methods are used to prepare cells so that clear images can be seen with the microscope. In outline, the methods of preparing cells for viewing through an optical microscope and an electron microscope are similar, but with important differences. Here we shall just concern ourselves with learning how specimens are prepared for viewing through an optical microscope.

The specimen is cut into very **thin** slices (ideally only the thickness of a single layer of cells), so that light passes through the cells easily. Then the cells are coloured (**stained**) with dyes (**stains**). Figure 9.1B gives you the idea. Some parts of each cell have taken up (absorbed) stain better than other parts. As a result the stained parts show up against unstained areas.

A specimen is a sample of tissue or other material to be looked at or investigated in some way.

 See www.keyscience.co.uk for more about cell biology.

Cell structure

Some structures are found in plant cells, animal cells and the cells of bacteria. The structures are needed for survival, regardless of the type of cell. Figure 9.1D illustrates the similarities between plant cells and animal cells; but also notice the differences. Some structures are found only in plant cells.

Structures found *only* in plant cells

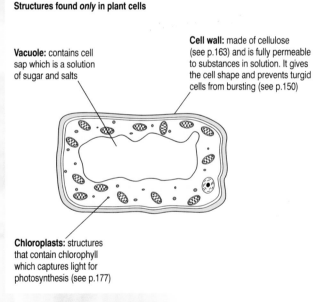

Vacuole: contains cell sap which is a solution of sugar and salts

Cell wall: made of cellulose (see p.163) and is fully permeable to substances in solution. It gives the cell shape and prevents turgid cells from bursting (see p.150)

Chloroplasts: structures that contain chlorophyll which captures light for photosynthesis (see p.177)

Structures found in animal cells *and* plant cells

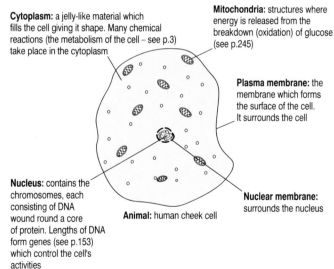

Cytoplasm: a jelly-like material which fills the cell giving it shape. Many chemical reactions (the metabolism of the cell – see p.3) take place in the cytoplasm

Mitochondria: structures where energy is released from the breakdown (oxidation) of glucose (see p.245)

Plasma membrane: the membrane which forms the surface of the cell. It surrounds the cell

Nucleus: contains the chromosomes, each consisting of DNA wound round a core of protein. Lengths of DNA form genes (see p.153) which control the cell's activities

Animal: human cheek cell

Nuclear membrane: surrounds the nucleus

Figure 9.1D ◆ Comparing the structures of plant cells and animal cells visible through the light microscope ×1000

Plant cells and animal cells have mitochondria and a distinct nucleus embedded in cytoplasm which is surrounded by a plasma membrane. Only plant cells have a cell wall, chloroplasts and a large sap-filled vacuole.

Bacterial cells do not have mitochondria or a distinct nucleus.

Figure 9.1E shows that bacterial cells contain fewer structures than the cells of either plants or animals. Bacterial cells are so small (see Figure 1.5B) that to see the differences a transmission electron microscope must be used. Figure 9.1E is a diagram of an electron photomicrograph of a bacterium. Notice that the bacterium does not contain a distinct nucleus. Nor does it have mitochondria or many of the other structures visible in Figure 9.1C.

SUMMARY

Plant cells, animal cells and the cells of bacteria have many features in common. There are also differences between the three types of cell. Each type comes in different varieties, each variety with special features which enable it to carry out its specific tasks (see Topic 9.5).

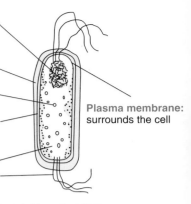

Chromosome: a coiled strand of DNA in the cytoplasm of the cell. It is *not* surrounded by a nuclear membrane (see Figure 9.1D) but forms a **nuclear zone**. At cell division several nuclear zones may be visible. Loops of DNA called **plasmids** are often found in bacterial cells (see p. 404 and p. 411)

Slime capsule: not present in all bacteria

Ribosomes: the structures where proteins are made. They are *not* supported on membranes (compare with Figure 9.1C)

Cell wall: gives bacteria their distinct shape (see pp. 14–15). It is *not* made of cellulose (see Figure 9.1D), but other types of carbohydrate (see p. 163) as well as lipids and proteins

Flagellum: propels the bacterial cell through liquid. It is *not* found in all bacteria

Plasma membrane: surrounds the cell

Cytoplasm

Figure 9.1E 🔺 A bacterium. The cells of most bacteria are only just visible under a light microscope, but the transmission electron microscope shows their structure in detail

CHECKPOINT

▶ **1** Compare the structures of the three types of cell in Figure 9.1D–F. Then copy and complete the table below. Put a tick in each space if the structure is present. Add any comments needed to distinguish between the structures. As an example chromatin has been filled in for you.

Structure in cell	Bacterium	Plant cell	Animal cell
Cell membrane			
Cell wall			
Cytoplasm			
Nucleus			
Nuclear membrane			
Chromatin			
Chromosome	✓ one chromosome	✓ many chromosomes	✓ many chromosomes
Endoplasmic reticulum			
Mitochondria			
Ribosomes			
Chloroplast			
Vacuole			

9.2 ▶ ## Movement into and out of cells

FIRST THOUGHTS

Cells need a non-stop supply of water and the substances dissolved in it to stay alive. This is why there is constant movement of solutions inside cells and also into and out of cells. After reading this section you will be able to explain what is happening to the plant in Figure 9.2A.

Diffusion

The molecules of a gas or liquid move at random. However, there is a better than even chance that some molecules will spread from where they are highly concentrated to where they are fewer in number. As a result there is a net movement of molecules of a substance along a **concentration gradient** from where the substance is in high concentration to where it is in low concentration. The movement of a substance through a solution or through a gas is called **diffusion**. The greater the difference between the regions of high and low concentration, the steeper the concentration gradient and the faster is the substance's rate of diffusion. Diffusion of a substance continues until the concentration of the substance is the same throughout the gas or the solution. Substances move into and out of cells by diffusion.

Figure 9.2A 🔺 (a) A plant wilting through lack of water

(b) The same plant recovering after watering

It's a fact!

Different ways of thinking about osmosis

Water molecules are in constant motion. Osmosis occurs when there is a net movement of water molecules through a partially permeable membrane from a region where the water molecules are in higher concentration to a region where they are in lower concentration. As water molecules move they hit the membrane and generate a pressure which is called the **water potential**. Regions with a high concentration of water molecules (dilute solutions) have a high water potential; regions with a low concentration of water molecules (concentrated solutions) have a low water potential. Water potential, therefore, is a measure of the tendency of water to leave a solution.

Osmosis in plant cells

Movement of substances takes place through the plasma membrane which separates the cell contents from the surroundings. Plasma membranes (and the other membranes inside cells) allow some substances to pass through and stop other substances: they are **partially permeable**. In general, substances pass through a membrane if their particles are smaller than the pores in the membrane (see Figure 9.2B).

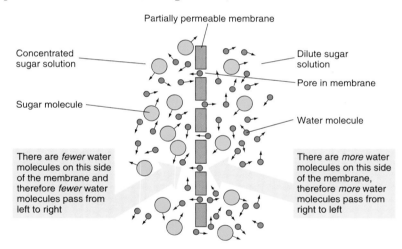

Figure 9.2B 🔺 Osmosis

When two solutions are separated by a partially permeable membrane, water passes through the membrane in both directions. The net flow of water is from the more dilute solution to the more concentrated solution. The flow continues until the two concentrations are equal. The flow of water from a more dilute solution to a more concentrated solution is called **osmosis**.

▓ Turgor

Osmosis takes place across the plasma membrane separating solutions inside the cell from the solutions outside the cell. The changes happening inside plant cells due to osmosis bring about visible changes in the plant. When plant cells are placed in a solution *less* concentrated than the solution inside them, water passes through the cell wall and plasma membrane, through the cytoplasm and into the vacuole. The cell fills with water and becomes **turgid** (see Figure 9.2C).

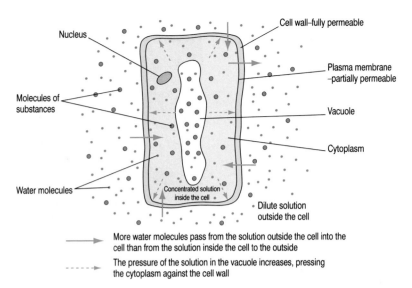

Water flows from the dilute solution outside the cell, through the cell wall and plasma membrane into the cytoplasm and into the vacuole. The increased pressure of the solution in the vacuole (**turgor pressure**) presses the cytoplasm against the cell wall. When the cell contains as much water as it can hold, the cell is fully turgid

Figure 9.2C 🔺 A turgid cell

▪ Plasmolysis

Sometimes plant cells may be placed in a solution *more* concentrated than the solution inside them, although this does not often happen in nature. Then water passes out of the vacuole, out of the cytoplasm, out through the plasma membrane and cell wall, and into the solution outside the cell. The pressure of the vacuole on the cytoplasm decreases until the cytoplasm pulls away from the cell wall. The cell becomes **flaccid** (limp), rather like a partly blown-up balloon. Cells in this condition are said to be **plasmolysed** (see Figure 9.2D).

Osmosis refers to the diffusion of water through a partially permeable membrane.

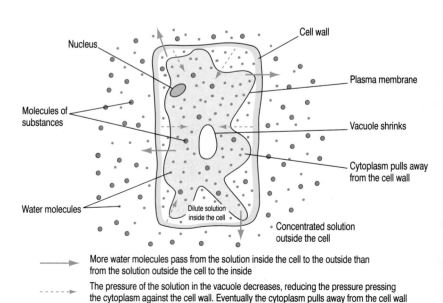

Figure 9.2D 🔺 A plasmolysed cell

Osmosis in animal cells

If human cheek cells are placed in a solution of substances that is *less* concentrated than the solution inside them, then the cells swell up because of the inflow of water. Eventually they burst (**lysis**) because the delicate plasma membranes break. Cheek cells do not have a cell wall; in plants the cell wall holds in the cell contents, preventing lysis.

Placed in a solution of substances that is *more* concentrated than the solution inside them, cheek cells shrink and crinkle (**crenate**) because of the outflow of water. For animals, maintaining an osmotic balance between cells and body fluids is very important, otherwise lysis or crenation occurs with disastrous results. Maintaining the osmotic balance is called **osmoregulation** (see p. 307)

■ Active transport

Sometimes substances move across the plasma membrane and other cell membranes from a region where they are present in low concentration to a region where they are present in high concentration. That is, they move against a concentration gradient in the reverse direction to normal diffusion. Such movement is called **active transport**. By this means, cells may build up stores of substances which would otherwise be spread out by diffusion (see Figure 9.2E). Active transport requires *more* energy than normal diffusion.

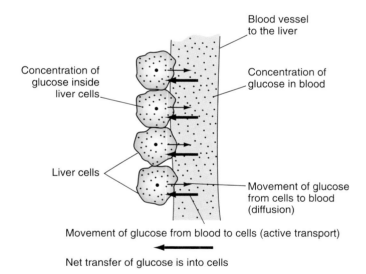

Figure 9.2E ▲ Active transport

CHECKPOINT

▶ **1** Refer to Figure 9.2A. Explain fully, with the aid of a diagram, what has happened in the plant cells to produce the change from photograph (a) to photograph (b).

▶ **2** The following three conversations take place in a kitchen. Give scientific explanations for all three observations.

   (a) *Michael*   Why are you putting that celery in a jug of water?

      Kate   It will keep crisper in water.

   (b) *Jonathan*   I've sprinkled sugar on my strawberries and it's gone all pink.

      Beth   That's because it makes the juice come out.

   (c) *Jasper*   I cut some chips earlier and put them in salt water to keep.

      Miranda   They seem to have shrunk!

3 (a) Explain what is meant by a partially permeable membrane.

(b) The figure opposite shows a bag made from visking tubing (a partially permeable membrane) filled with a 15% salt solution (15 g of salt per 100 g of water). Draw sketches to show the changes that will happen if the beaker contains (i) distilled water (ii) 7% salt solution (iii) 15% salt solution (iv) 30% salt solution.

4 Peas are sometimes preserved by drying them.

(a) How much water do dried peas take up when they are soaked in water? Plan an experiment to find out. You should plan a quantitative experiment – that is to measure the exact quantity (e.g. mass) of water absorbed by a known mass of dried peas.

(b) What makes the cells of a pea absorb water when the pea is placed in water? Describe what happens inside the cells. Use scientific words to describe the difference between the cells in the dried pea and the cells in the soaked pea.

(c) Water does not flow into a dried pea indefinitely. What makes the flow of water stop?

5 Explain the difference between osmosis and active transport. Give one example of the importance of active transport.

Glass tube

Level of salt solution

Beaker containing water

Visking tubing tied at the bottom to form a bag, filled with 15% salt solution, and tied tightly at the top around the glass tube

---

Making proteins

Figure 9.3A ⬥ Enzyme treatment which weakened the cell wall was followed by immersion in water which burst the bacterium, spilling its DNA

It's a fact!

The chemical reactions which take place in a cell are catalysed by **enzymes** (see p. 168), and enzymes are proteins. In controlling which proteins are made in a cell, DNA controls enzyme production and therefore the structure and all the functions of the cell as a whole.

Cells make **proteins** (see Topic 10.3) – the building material of cells and bodies. To make one molecule of a certain protein, hundreds or thousands of amino acid molecules must combine in the right order. The substance responsible for getting the order right is deoxyribonucleic acid (**DNA** – see p. 170). It carries instructions – called a **code** – for combining amino acids in the right order. Figure 9.3A shows the DNA of a bacterium. In bacteria, the DNA is not separate from other structures in the cytoplasm of the cell. In most other types of cells, the DNA is surrounded by a membrane (the **nuclear membrane** – see p. 147) within a distinct **nucleus**.

DNA contains the bases **adenosine, thymine, guanine** and **cytosine (A, T, G** and **C**). **Ribonucleic acid (RNA)** is a substance similar to DNA but the base **uracil (U)** replaces thymine (see p. 171). There are twenty amino acids. The instructions needed to assemble one amino acid in its correct place in the protein molecule are contained in a row of three bases: a **codon** (see Figure 9.3B).

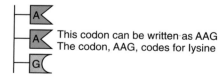

This codon can be written as AAG
The codon, AAG, codes for lysine

Figure 9.3B ⬥ A row of three bases forms a codon

Each amino acid has a different codon. For example, GGC codes (gives instructions for) the amino acid glycine; AAG codes for lysine.

DNA is present in the nucleus. Proteins are synthesised on the ribosomes in the cytoplasm. *How do the instructions for making proteins travel from DNA to the ribosomes?* DNA employs a messenger to take instructions to where they are needed. The messenger is RNA – so-called **messenger RNA (m-RNA)** so as to distinguish it from other types of RNA in the cytoplasm of the cell.

SUMMARY

The nucleus of a cell contains DNA. The sequence of bases in DNA is a 'code' of instructions for synthesising proteins. Protein synthesis takes place in the cytoplasm. DNA employs m-RNA to carry the code out of the nucleus to the ribosomes in the cytoplasm. Amino acids brought to the m-RNA–ribosome complex by t-RNA are sorted into the correct order by the m-RNA on the ribosomes. The amino acids combine to form a protein. By controlling protein synthesis, DNA controls the whole life of the cell.

Messenger-RNA forms in the nucleus after a DNA double helix unwinds. Each single strand of DNA attracts bases according to the rules of **base pairing** described on p. 171. The bases combine to form a strand of m-RNA. Because of the rules of base pairing, m-RNA is a **complement** of one strand of DNA. It detaches from the strand of DNA and passes through minute gaps (nuclear pores) in the nuclear membrane into the cytoplasm where it bonds to a ribosome (see p. 146). Another type of RNA is involved at this stage. Transfer-RNA (t-RNA) brings amino acids to the m-RNA–ribosome complex. The amino acids become attached to the m-RNA–ribosome complex and then combine with one another to form a long chain. The chain of amino acids is a protein molecule. The order in which amino acids combine forming the chain is the result of the code on m-RNA which, because of the rules of base pairing, is a copy of the code on the strand of DNA for which it is the complement (see Figure 9.3C). In this way a particular type of protein is made according to the particular sequence (**order**) of codons of the DNA coding ('instructing') amino acids to assemble in a particular order. The sequence of codons forming a length of DNA which codes for the whole of one protein is called a **gene**. This is why the DNA code is called the **genetic code**.

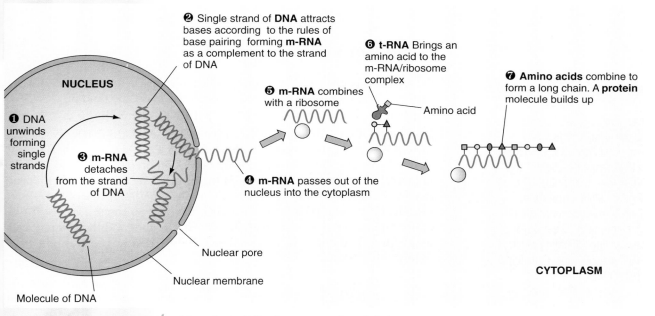

Figure 9.3C 🔺 How DNA controls protein synthesis. Follow the sequence of events 1–7

CHECKPOINT

▶ 1 Explain the difference between a codon and a gene.

▶ 2 Copy the following sequence of bases.

A A T C C T G A C T A G

How many codons are contained in this section of DNA? Assume the first codon begins at the left hand end and that codons do not overlap. Mark off the codons on the sequence. Assuming that each codon codes for one amino acid, how many amino acids are represented in this section of DNA?

▶ 3 (a) Where in the cell is protein made?

(b) Why is making new protein essential for the cell?

9.4 ▶ Cell division

New cells (daughter cells) are formed when old cells (parent cells) divide. During cell division the cytoplasm and nucleus divide. Division of the nucleus passes on the chromosomes of the parent cell to each of the daughter cells. Mitosis passes on the full number of chromosomes to each daughter cell; meiosis halves the number of chromosomes in each daughter cell.

New cells are formed when old cells divide (see Figure 9.4A). The cells which divide are called **parent cells**, and the new cells are called **daughter cells**. Two overlapping processes take place:

● The nucleus divides. There are two ways of dividing: **mitosis** and **meiosis**. The cells of the body divide by mitosis. The cells of the sex organs that give rise to the sex cells (sperms and eggs) divide by meiosis.

● The cytoplasm divides.

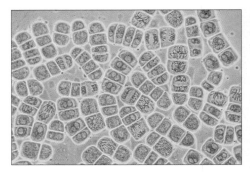

Figure 9.4A ⬆ Cells dividing

DNA replication

Before mitosis takes place, new DNA is made inside the nucleus. The process is called **replication** because each chromosome (and its component DNA) makes an exact copy – a replica – of itself.

Figure 9.4B shows the sequence of events. The DNA double helix **unwinds** forming two single strands of DNA. Each strand attracts bases according to the rules of base pairing described on p. 171. A new strand of DNA builds up alongside each of the original strands of DNA to form two new molecules of DNA. Each new molecule is a **replica** of the original because of the way bases pair up. Now you can see why daughter cells are genetically identical to each other and their parent cell which divides by mitosis. Each cell receives an identical copy of the parent cell's DNA.

Occasionally the wrong base adds by mistake during the formation of new DNA. Then the new DNA formed is slightly different from the original. The change is called a **mutation** and alters the code for making a particular protein (see p. 152) so that another different protein is made instead. Most mutations are harmful because the altered protein does not work as well as the original protein (see p. 416 for the example of sickle-cell anaemia), or may not work at all. If mutations occur in the sex cells (sperms and eggs), after fertisilation of the egg, the embryo (see p. 366) may not develop properly or may die at an early stage. Mutations in body cells may lead to uncontrolled mitosis (see the page opposite) and the development of cancer.

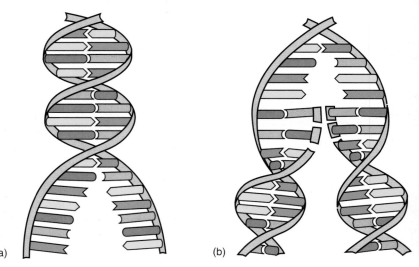

(a)　　　　　　　　　　　(b)

Figure 9.4B ⬆ (a) The DNA double helix unwinding (b) Two new molecules of DNA form as a new strand of DNA builds up along each of the original strands. The horizontal shapes joining two strands represent the bases A, T, G or C (see p. 171).

Remember the differences between mitosis and meiosis. A cell dividing by mitosis produces 2 diploid daughter cells. A cell dividing by meiosis produces 4 haploid cells. New body cells are produced by mitosis. New sex cells (sperms and eggs) are produced by meiosis.

Remember that each chromosome appears under the microscope as a pair of chromatids, each one joined to the other by a centromere.

Remember that matching chromosomes form homologous pairs (see p. 156) only during meiosis. The chromosomes of each pair each carry the same genes (see p. 379) as its partner but not necessarily the same alleles (see p. 379) for a particular gene.

What is a chromosome?

Figure 9.4C shows that a chromosome consists of a folded strand of DNA coiled round a core of protein which itself loops, twists and coils, forming an intricate and complicated structure. It may contain up to 10 000 times its own length of DNA. The folding and coiling allows the DNA to cram into the confined space of a cell nucleus. The DNA component of the chromosome controls the **inheritance** of characteristics (see Topic 23.1).

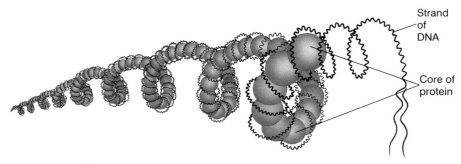

Strand of DNA

Core of protein

Figure 9.4C ⬆ The structure of a chromosome much simplified. Further coiling forms each of the structures seen in Figure 23.1E on page 384

What are mitosis and meiosis?

Mitosis divides the chromosomes of the parent cell into two identical groups. Cells which through mitosis receive the full number of chromosomes from their parent cells and in turn hand them on to their daughter cells are described as **diploid** (or **2n**).

Meiosis halves the number of chromosomes in the parent cell. The half number of chromosomes in the daughter cells is described as **haploid** (or **n**).

The processes of mitosis and meiosis are compared in Figure 9.4D. Notice that each parent cell dividing by mitosis produces *two* diploid daughter cells, but that each parent cell dividing by meiosis produces *four* haploid daughter cells. In each case the first sign of cell division is when the chromosomes can be seen more easily under the microscope. This happens because the genetic material coils up and thickens to form visible chromosomes. At this stage each chromosome then divides into a pair of identical **chromatids** attached to each other by a structure called the **centromere**.

The importance of mitosis

Because daughter cells each receive an identical full (diploid) set of chromosomes from the parent cell, daughter cells and parent cell are genetically identical as we have seen. This is why mitosis is the way in which living things repair damage to the body, grow and reproduce asexually. For example:

- parent cheek cells divide by mitosis into identical daughter cheek cells
- parent skin cells divide by mitosis into identical daughter skin cells
- parent root cells divide by mitosis into identical daughter root cells.

Mitosis replaces old body cells with new ones, with the result that the individual stays the same. This is why our appearance is unchanged this week compared with last week, even though the surface cells of the skin will have been replaced in the intervening period.

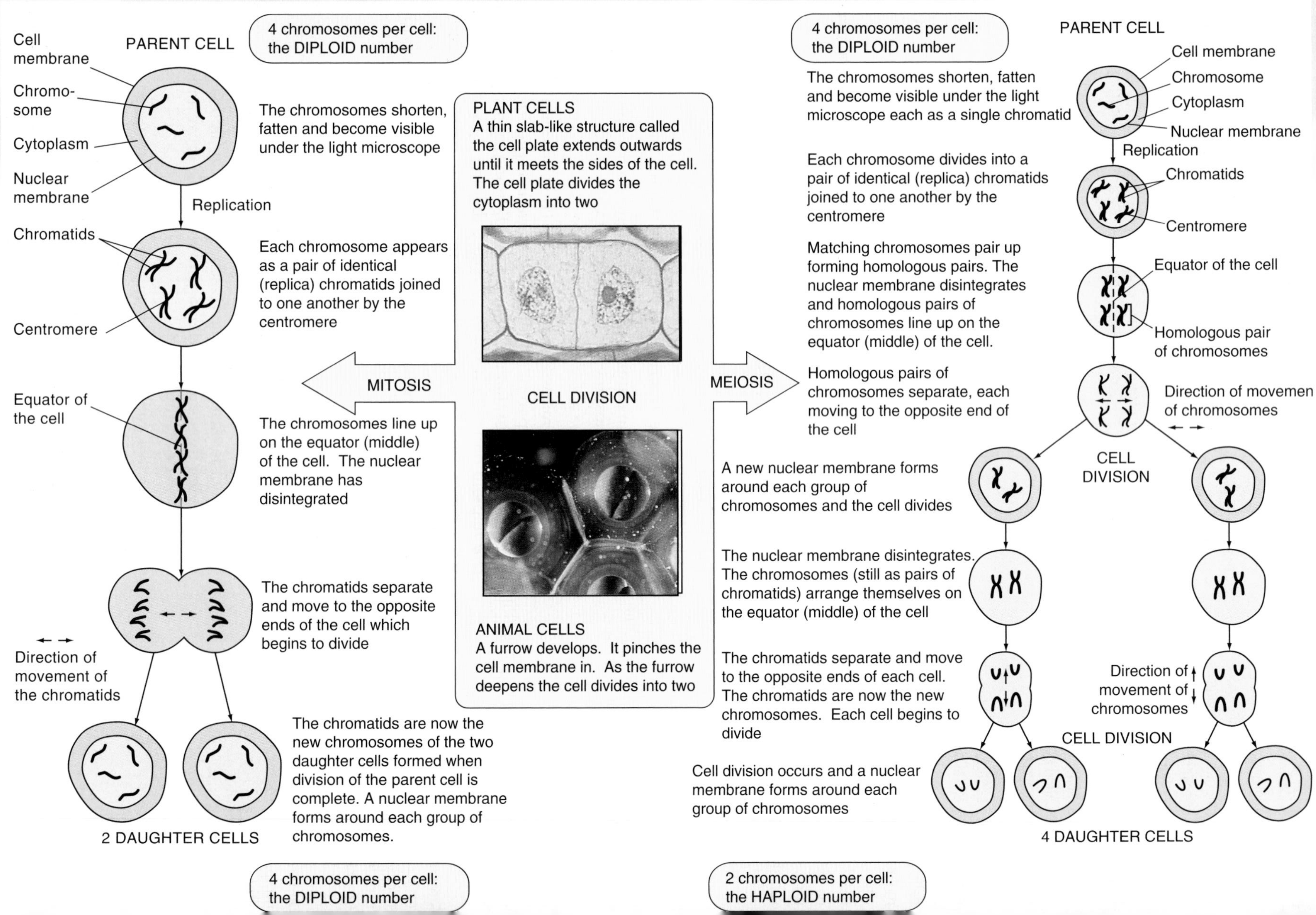

Figure 9.4D ⬆ Cell division

PARENT CELL

4 chromosomes per cell:
the DIPLOID number

Cell membrane
Chromosome
Cytoplasm
Nuclear membrane

Replication

Chromatids
Centromere

Equator of the cell

Homologous pair
of chromosomes

Direction of movemen
of chromosomes

CELL DIVISION

Direction of
movement of
chromosomes

CELL DIVISION

4 DAUGHTER CELLS

2 chromosomes per cell:
the HAPLOID number

The chromosomes shorten, fatten
and become visible under the light
microscope each as a single chromatid

Each chromosome divides into a
pair of identical (replica) chromatids
joined to one another by the
centromere

Matching chromosomes pair up
forming homologous pairs. The
nuclear membrane disintegrates
and homologous pairs of
chromosomes line up on the
equator (middle) of the cell.

Homologous pairs of
chromosomes separate, each
moving to the opposite end of
the cell

A new nuclear membrane forms
around each group of
chromosomes and the cell divides

The nuclear membrane disintegrates.
The chromosomes (still as pairs of
chromatids) arrange themselves on
the equator (middle) of the cell

The chromatids separate and move
to the opposite ends of each cell.
The chromatids are now the new
chromosomes. Each cell begins to
divide

Cell division occurs and a nuclear
membrane forms around each
group of chromosomes

MEIOSIS

PLANT CELLS
A thin slab-like structure called
the cell plate extends outwards
until it meets the sides of the cell.
The cell plate divides the
cytoplasm into two

CELL DIVISION

ANIMAL CELLS
A furrow develops. It pinches the
cell membrane in. As the furrow
deepens the cell divides into two

MITOSIS

PARENT CELL

4 chromosomes per cell:
the DIPLOID number

Cell
membrane
Chromo-
some
Cytoplasm
Nuclear
membrane

Replication

Chromatids

Centromere

Equator of
the cell

Direction of
movement of
the chromatids

The chromosomes shorten,
fatten and become visible
under the light microscope

Each chromosome appears
as a pair of identical
(replica) chromatids joined
to one another by the
centromere

The chromosomes line up
on the equator (middle)
of the cell. The nuclear
membrane has
disintegrated

The chromatids separate
and move to the opposite
ends of the cell which
begins to divide

The chromatids are now the
new chromosomes of the two
daughter cells formed when
division of the parent cell is
complete. A nuclear membrane
forms around each group of
chromosomes.

2 DAUGHTER CELLS

4 chromosomes per cell:
the DIPLOID number

156

SUMMARY

In mitosis, the parent cell divides into two. A full set of chromosomes is passed to each daughter cell. The chromosomes in each daughter cell are identical to those of the parent cell. In meiosis the parent cell divides into four. A half set of chromosomes is passed to each daughter cell.

The importance of meiosis

Daughter cells (sex cells) each receive a half (haploid) set of chromosomes from the parent. During **fertilisation** (when sperm and egg join together – see Topic 20.2), the chromosomes from each cell combine. As a result the fertilised egg (**zygote**) is diploid but inherits a new combination of genes (contributed 50 : 50) from the parents. The new individual which develops from the zygote inherits characteristics from both parents, not just from one parent as in asexual reproduction.

CHECKPOINT

▶ **1** (a) Why do the cells of a tissue need to undergo mitosis?
   (b) What effect does mitosis have on the chromosomes of the parent cell?
   (c) What is the relationship between the chromosomes of the parent cell and the chromosomes of the daughter cells?
   (d) What is the importance of this relationship for the health of the tissue?

▶ **2** (a) In everyday language, what is a 'replica'?
   (b) What is formed by the replication of DNA?
   (c) Refer to Figure 9.4B, which shows the replication of DNA. Explain briefly what is happening.

9.5 ▶ Cells, tissues and organs

FIRST THOUGHTS

All cells have certain features in common (see Topic 9.1). This section deals with the differences in shape and structure between different types of cell.

Do you remember that some cells exist as independent organisms? You can see that the organisms on pp. 16 and 17 are made of single cells. Other cells cannot survive on their own; they exist as members of **multicellular** organisms. The bodies of multicellular organisms are made of different types of cell. Each type of cell is specialised to perform a particular biological task.

Tissues and organs

A group of similar cells makes a **tissue**. Different tissues together make up an **organ**, and different organs are combined in an **organ system**. Figure 9.5A shows how one organ system, the human heart and its blood vessels, is organised.

Types of cell

The human body is constructed from more than two hundred different types of cell. Fewer types of cell form the bodies of flowering plants (e.g. foxglove), fungi (e.g. field mushroom) and simple animals. The fresh water animal *Hydra* (see p. 24) has only seven different types of cell. The appearance of cells often helps us understand what they do. Study the

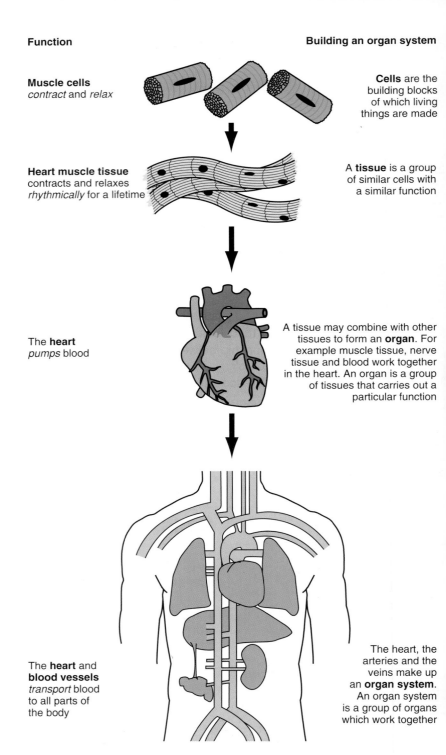

Function

Building an organ system

Muscle cells
contract and *relax*

Cells are the building blocks of which living things are made

Heart muscle tissue
contracts and relaxes *rhythmically* for a lifetime

A **tissue** is a group of similar cells with a similar function

The **heart**
pumps blood

A tissue may combine with other tissues to form an **organ**. For example muscle tissue, nerve tissue and blood work together in the heart. An organ is a group of tissues that carries out a particular function

The **heart** and **blood vessels**
transport blood to all parts of the body

The heart, the arteries and the veins make up an **organ system**. An organ system is a group of organs which work together

Figure 9.5A ⬆ The human heart and blood vessels: an organ system

red blood cell and sperm cell in Figure 9.5B and the xylem cells and root hair cell in Figure 9.5C. *How does the shape of each type of cell help you to understand what it does?*

Figures 9.5B, C and D show cell types from different kingdoms. Notice the features that all the cells have in common, such as cytoplasm, cell membranes and nuclei. Notice also how specialisation for particular biological tasks makes each type of cell different.

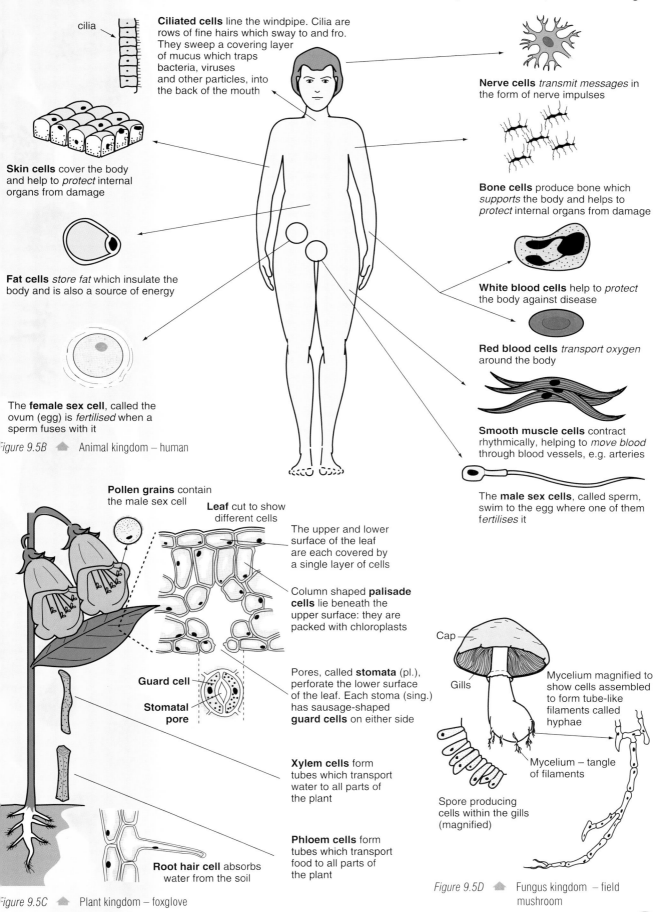

cilia

Ciliated cells line the windpipe. Cilia are rows of fine hairs which sway to and fro. They sweep a covering layer of mucus which traps bacteria, viruses and other particles, into the back of the mouth

Skin cells cover the body and help to *protect* internal organs from damage

Fat cells *store fat* which insulate the body and is also a source of energy

The **female sex cell**, called the ovum (egg) is *fertilised* when a sperm fuses with it

Figure 9.5B ⬆ Animal kingdom – human

Nerve cells *transmit messages* in the form of nerve impulses

Bone cells produce bone which *supports* the body and helps to *protect* internal organs from damage

White blood cells help to *protect* the body against disease

Red blood cells *transport oxygen* around the body

Smooth muscle cells contract rhythmically, helping to *move blood* through blood vessels, e.g. arteries

The **male sex cells**, called sperm, swim to the egg where one of them *fertilises* it

Pollen grains contain the male sex cell

Leaf cut to show different cells

The upper and lower surface of the leaf are each covered by a single layer of cells

Column shaped **palisade cells** lie beneath the upper surface: they are packed with chloroplasts

Guard cell

Stomatal pore

Pores, called **stomata** (pl.), perforate the lower surface of the leaf. Each stoma (sing.) has sausage-shaped **guard cells** on either side

Xylem cells form tubes which transport water to all parts of the plant

Phloem cells form tubes which transport food to all parts of the plant

Root hair cell absorbs water from the soil

Figure 9.5C ⬆ Plant kingdom – foxglove

Cap

Gills

Mycelium magnified to show cells assembled to form tube-like filaments called hyphae

Mycelium – tangle of filaments

Spore producing cells within the gills (magnified)

Figure 9.5D ⬆ Fungus kingdom – field mushroom

Remember, the larger the cell the smaller is its surface area to volume ratio.

Cell size and surface area to volume ratio

All cells (tissues, organs, organisms) exchange gases, food and other materials between themselves and their environment. The exchanges occur mostly by diffusion across surfaces. Look at the calculations for surface area (SA), volume (V) and surface area to volume ratio SA/V in Figure 9.5E.

Notice that:

- the SA/V of cube B is half that of cube A
- the SA/V of cube C is two-thirds that of cube B and one third that of cube A

In other words the larger the cube becomes, the smaller its SA/V.

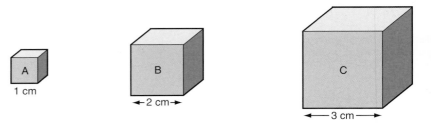

Figure 9.5E

	Cube A	**Cube B**	**Cube C**
Surface area of one face	1 cm × 1 cm = 1 cm²	2 cm × 2 cm = 4 cm²	3 cm × 3 cm = 9 cm²
Surface area of cube	1 × 6 cm² = 6 cm²	4 cm² × 6 = 24 cm²	9 cm² × 6 = 54 cm²
Volume of cube	1 cm × 1 cm × 1 cm = 1 cm³	2 cm × 2 cm × 2 cm = 8 cm³	3 cm × 3 cm × 3cm = 27 cm³
Ratio: surface area/volume	$\frac{6\ cm^2}{1\ cm^3}$ = 6 : 1	$\frac{24\ cm^2}{8\ cm^3}$ = 3 : 1	$\frac{54\ cm^2}{27\ cm^3}$ = 2 : 1

Figure 9.5E shows three cubes of different sizes. A cube has six faces. Cells are not cube-shaped, but the calculations on cubes apply to any shape of cell. What they show is that when a cell grows the surface area of the cell membrane increases more slowly than the volume of the cell (because surface area increases with the **square** of the side, and volume increases with the **cube** of the side). As the cell grows, it needs to take in more food and gases. But the area of the cell membrane is not increasing at the same rate as the volume. After the cell reaches a certain size, the surface area becomes insufficient to meet the needs of the larger volume. At this point, the cell either divides into daughter cells or dies.

Cell size depends on the rate at which it uses materials. Cells with a fast turnover of materials are usually smaller than cells with a low turnover.

Figure 9.5F gives an idea of the range of size of cells. Comparing the different sizes is rather like comparing the difference in size between a mouse and a whale. The range is enormous! However, the size of most cells falls somewhere in the middle: invisible to the naked eye but easily seen under a light microscope.

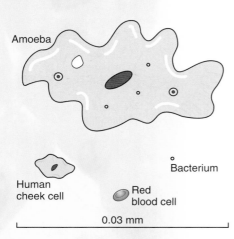

Figure 9.5F A range of cells drawn to scale

The problems of exchanging materials are the same for cell surfaces and body surfaces. Figure 9.5G shows that different organs and organ systems are specialised to increase the available surface area for the exchange of materials with their surroundings.

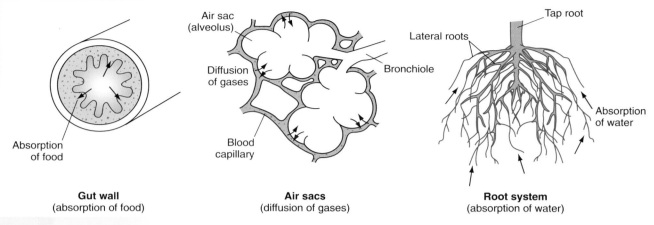

Gut wall
(absorption of food)

Air sacs
(diffusion of gases)

Root system
(absorption of water)

Figure 9.5G 🔺 Increasing surface area

CHECKPOINT

▶ **1** Copy and complete the following paragraph. You may use the words in the list once, more than once or not at all.

(a) **an organ   organs   cells   types   tissues**

Living things are made of _____. Groups of similar _____ with similar functions form _____ that can work together as _____. A group of _____ working together form _____ system.

▶ **2** (a) What effect does the division of a cell into small daughter cells have on the surface area/volume ratio?

(b) Why are cells with a high material turnover usually smaller than those with a low material turnover?

Topic 10   The chemicals of life

10.1 ▶   Carbohydrates

FIRST THOUGHTS

What do sugar and starch, the cell walls of plants and the exoskeletons of insects have in common? This section will tell you!

Sugars

Carbohydrates are compounds containing carbon, hydrogen and oxygen only. Carbohydrates do a vital job in all living organisms: they provide energy. When carbohydrates react with oxygen, carbon dioxide and water are formed and energy is released. This reaction takes place slowly inside cells. Energy is released in a controlled way and used by the cells for all their activities. The process by which living cells release energy is called **cellular respiration** (see p. 243). Most of our energy is obtained from the respiration of carbohydrates (usually glucose), although living organisms also respire fats, oils and proteins.

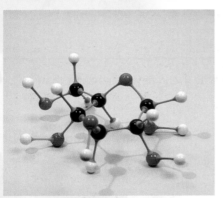

(a) Model of a glucose molecule

Figure 10.1A 🔺   Glucose

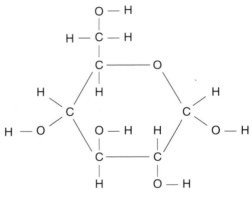

(b) The formula for glucose in full

(c) The formula in a shorthand form

Carbohydrates include sugars, starches and cellulose. Two of the simplest carbohydrates are the sweet-tasting sugars **glucose** and **fructose**. They both have the formula $C_6H_{12}O_6$, but the structures (the arrangement of atoms) of their molecules are different. The structure of a molecule of glucose is shown in Figure 10.1A. As you can see, six of the atoms are joined to form a ring. Sugars with one ring of atoms are called **monosaccharides** (mono = one). There are two kinds of monosaccharides: **hexoses**, which contain 6 carbon atoms, e.g. glucose and fructose, and **pentoses**, which contain 5 carbon atoms, e.g. ribose, $C_5H_{10}O_5$.

Two molecules of glucose can combine to form one molecule of the sugar maltose, $C_{12}H_{22}O_{11}$ (see Figure 10.1B).

$$\text{Glucose} \quad \rightarrow \quad \text{Maltose} \;\; + \text{Water}$$
$$2C_6H_{12}O_6(aq) \quad \rightarrow \quad C_{12}H_{22}O_{11}(aq) \; + \; H_2O(l)$$

Maltose is a **disaccharide**: its molecules contain two sugar rings. Sucrose (table sugar) is another disaccharide. It is formed from glucose and fructose.

$$\text{Glucose} + \text{Fructose} \quad \rightarrow \quad \text{Sucrose} + \text{Water}$$

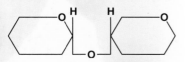

Figure 10.1B 🔺   The formula for maltose in shorthand form

Starch, glycogen and cellulose

Starch (a stored food substance in plant cells), **glycogen** (a stored food substance in animal cells) and **cellulose** (a component of plant cell walls) are all carbohydrates, although they do not taste sweet. Starch and glycogen are slightly soluble in water; cellulose is insoluble. These compounds are **polysaccharides**. Polysaccharides are carbohydrates whose molecules contain a large number (hundreds) of sugar rings (see Figure 10.1C).

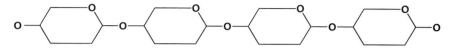

Figure 10.1C 🔺 Part of a starch molecule

In starch, glycogen and cellulose, the molecules are long chains of glucose rings. Polysaccharides differ in the length and structure of their chains (see Figure 10.1D).

As starch and glycogen are only slightly soluble they can remain in the cells of an organism without being dissolved out and therefore make good food stores (see Figures 10.1E and F). Cells convert starch and glycogen into glucose, which is soluble. Glucose is oxidised in cells to release energy (see Topic 15.1).

Some polysaccharides are building materials. Cellulose fibres make the walls of plant cells (see Figure 10.1G). The tough framework of fibres makes the plant cell wall rigid. Another polysaccharide that provides a strong structure is **chitin**. It forms the **exoskeleton** round insects' bodies which protects their insides (see Figure 10.1H).

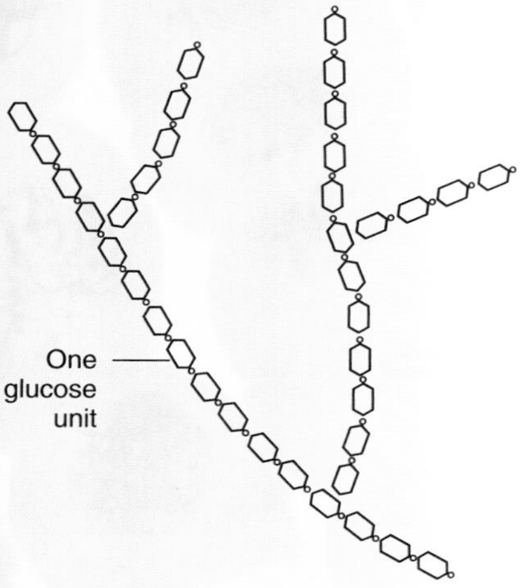

Figure 10.1D 🔺 A model of part of a starch molecule

One glucose unit

Figure 10.1E 🔺 Starch grains (blue) in the cells of a potato. They are larger than those found in the cells of most other types of plant

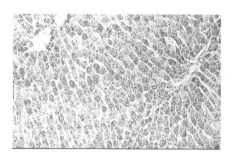

Figure 10.1F 🔺 A thin section of liver. Note the glycogen stained pink

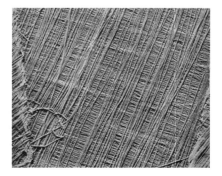

Figure 10.1G 🔺 Plant cell wall (×10 000). Note the frame work of cellulose fibres

Figure 10.1H 🔺 The chitin-impregnated exoskeleton of a rhinoceros beetle

SUMMARY

Carbohydrates:

- Monosaccharides
  - sugars, e.g. glucose and fructose with a six-carbon ring and ribose with a five-carbon ring
- Disaccharides
  - made up of two monosaccharides
  - sugars such as sucrose and maltose
- Polysaccharides
  - made up of many monosaccharides
  - starch and glycogen, are stores of food energy in plants and animals respectively
  - cellulose, strengthens plant cell walls
  - chitin, strengthens insect exoskeletons

CHECKPOINT

> **1** Runners in the London marathon ate pasta the evening before. Why did they think this would give them energy?

> **2** Explain why rice, bread and potatoes are 'high-energy foods'.

> **3** Give two examples of the importance of cellulose and chitin as building materials in living things.

> **4** Glucose, maltose and starch are carbohydrates. Answer the following questions about them.
>   (a) Which one tastes sweet?
>   (b) Which two are very soluble in water?
>   (c) List the elements that make up all three.
>   (d) The formula of glucose is $C_6H_{12}O_6$. What is the ratio Number of atoms of hydrogen : Number of atoms of oxygen?
>   (e) The formula of maltose is $C_{12}H_{22}O_{11}$. What is the ratio Number of atoms of hydrogen : Number of atoms of oxygen?
>   (f) Using your answers to (d) and (e), try to explain where the name 'carbohydrate' comes from.
>   (g) The 'shorthand' formula for glucose is
>
>   Draw the shorthand formulas for maltose and starch.

10.2 ▶ Lipids

Have you seen brands of margarine and cooking oil described as 'high in polyunsaturates' and wondered what it means? This section will tell you.

Figure 10.2A ⬆ Cutting up whale blubber

What are lipids?

Lipids is a name for **fats** and **oils**. Lipids are made of carbon, hydrogen and oxygen. A difference between fats and oils is that most fats are solid at room temperature and most oils are liquid at room temperature. Fats and oils have a 'greasy' feel and are insoluble in water. Fats and oils are important as:

- **Sources of energy**. Fats and oils are stores of energy. In cellular respiration, lipids provide about twice as much energy per gram as carbohydrates.

- **Insulation**. Mammals have a layer of fat under the skin. This layer helps to keep the animals warm. Whales and seals which live in Antarctic waters have an especially thick layer of fat, called blubber, to protect them from extreme cold (see Figure 10.2A).

- **Protection**. Delicate organs, such as the kidneys, are protected by a layer of hard fat.

- **Food**. Vitamins A, D and E are soluble in fats and oils. Foods containing lipids provide animals with these essential vitamins.

Fats and oils are also used to build cells. Look at the magnified picture of the **plasma membrane** of a cell (see p. 145) in Figure 10.2B. You can see two 'tramlines', which scientists think are composed of molecules of a **phospholipid** – a fat containing phosphorus.

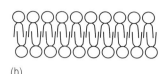

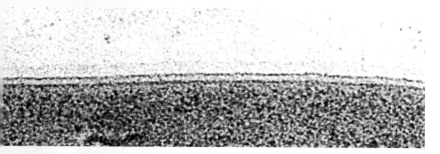

(a) A photograph of the plasma membrane of a cell taken through an electron microscope. The membrane shows as two 'tramlines' (×250 000)

Figure 10.2B ◀ Plasma membrane

(b)

A magnified view of the 'tramlines' shows a double layer of phospholipid molecules

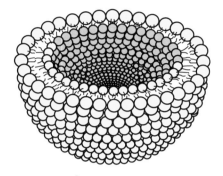

SUMMARY

Fats and oils are together called lipids. They are sources of energy. A layer of fat insulates the bodies of mammals from the cold. Fats form part of the structures of cells.

Ŏ One phospholipid molecule

(c) The diagram shows how scientists interpret the photograph

Saturated and unsaturated fats and oils

Do you watch your diet? Many people are concerned about having too much fat in their diet, particularly **saturated fats**. Let us consider what the term saturated means here.

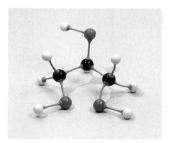

(a) Model of a molecule of glycerol

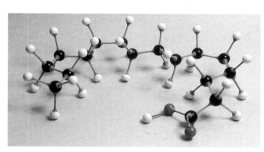

(b) Model of a molecule of a fatty acid (hexadecanoic acid)

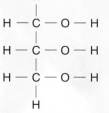

(c) The formula of glycerol

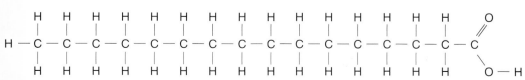

(d) The formula of hexadecanoic acid

Figure 10.2C ◀ Glycerol and hexadecanoic acid

165

Fats and oils are compounds of fatty acids and glycerol (see Figure 10.3C). One molecule of glycerol can combine with three molecules of a fatty acid to form a molecule of a triglyceride and three molecules of water. The shapes represent the different molecules reacting to form a triglyceride. **Lipids** are mixtures of triglycerides.

Glycerol + Fatty acid ➔ Triglyceride + Water

$$\begin{array}{l} \text{—OH} \\ \text{—OH} \\ \text{—OH} \end{array} + 3\text{HA} = \begin{array}{l} \text{—A} \\ \text{—A} \\ \text{—A} \end{array} + 3\text{H}_2\text{O}$$

Like other organic compounds, fatty acids may be saturated or unsaturated. If there is one double bond between carbon atoms, the compound is described as **monounsaturated**. If there is more than one double bond between carbon atoms, the acid is described as **polyunsaturated**. Below is part of the carbon chain in a saturated compound. This compound is described as **saturated** because it cannot combine with any more hydrogen atoms.

Make sure that you can distinguish between saturated fats and unsaturated fats.

Below is part of the carbon chain in a monounsaturated compound. This compound is described as unsaturated because it can add more hydrogen atoms across the double bond.

Below is part of the carbon chain in a polyunsaturated compound

SUMMARY

Lipids (fats and oils) are mixtures of compounds of glycerol with fatty acids. These compounds can be saturated (with only single bonds between carbon atoms) or monounsaturated (with a double bond between carbon atoms) or polyunsaturated (with more than one carbon–carbon double bond per molecule).

The fats which contain glycerol combined with saturated fatty acids are called saturated fats. Fats which contain glycerol combined with unsaturated fatty acids are called monounsaturated fats or polyunsaturated fats, depending on the number of double bonds in a molecule. Animal fats contain a large proportion of saturated compounds and are solid. Plant oils contain a large proportion of unsaturated compounds and have lower melting points. Many scientists believe that eating a lot of saturated fat increases the risk of **heart disease**.

CHECKPOINT

▶ **1** (a) Which of the following items are fats and which are oils?
   soft margarine, butter, lard, hard margarine, cooking oil
   (b) What is the main physical difference between fats and oils?

▶ **2** Refer to Figure 10.2A. What feature helps the seals to keep warm?

▶ **3** Why are fats and oils better stores of energy than carbohydrates?

▶ **4** Briefly explain the difference between saturated and unsaturated fats.

▶ **5** The manufacturers of *Flower* margarine claim that their product encourages healthy eating because it is 'low in saturated fats'.
   (a) What do they mean by 'low in saturated fats'?
   (b) Name a product which contains much saturated fat.
   (c) What particular benefit to health is claimed for products such as *Flower*?

▶ **6** The following equation represents the formation of a fat.

$$
B\!\left[\begin{array}{l} -\,OH \\ -\,OH \\ -\,OH \end{array}\right] + 3HA \;=\; C\!\left[\begin{array}{l} -\,A \\ -\,A \\ -\,A \end{array}\right] + 3H_2O
$$

   (a) Name the substance B.
   (b) Name one substance which could be HA.
   (c) Name one fat and one oil which could be C.

10.3 ▶ Proteins

FIRST THOUGHTS

Muscle, skin, hair and haemoglobin: what do they have in common? You can find out in this section.

Proteins are compounds which contain carbon, hydrogen, oxygen, nitrogen and sometimes sulphur. Proteins are vitally important because:

- proteins are the materials from which new tissues are made. If organisms are to grow and if they are to repair damaged tissues, they need proteins

- enzymes are proteins. None of the reactions which take place in living things would take place rapidly enough without enzymes

- hormones are proteins which control the activities of organisms

- proteins can be oxidised in respiration to provide energy.

Every protein molecule is made from a large number of molecules of **amino acids**. The simplest amino acid is **glycine** (see Figure 10.3A).

$$
\begin{array}{ccc}
H & H & O \\
\backslash & | & \parallel \\
N - & C - & C \\
/ & | & \backslash \\
H & H & O - H
\end{array}
$$

Figure 10.3A 🔺 Model of a glycine molecule

Hair, nails, claws and feathers consist chiefly of the protein **keratin**

There are 20 amino acids. They have the general formula

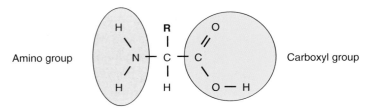

R is a different group of atoms in each amino acid; it can be —H, —CH$_3$, —CH$_2$OH and many other groups. In the following equations, each shape represents a different amino acid.

Two amino acids can combine to form a **peptide** and water.

$$\square + \bigcirc \;\;\rightarrow\;\; \square\!-\!\bigcirc + H_2O$$

The peptide which is formed can combine with more amino acids. Many different amino acids can combine to form a long chain.

$$\square + \bigcirc + \triangle + \mathbb{0} \;\;\rightarrow\;\; \square\!-\!\bigcirc\!-\!\triangle\!-\!\mathbb{0} + 3H_2O$$

Peptides are molecules with up to 15 amino acids.
Polypeptides are molecules with 15–100 amino acids.
Proteins have still larger molecules. Part of a protein molecule is shown below.

$$\bigcirc\!-\!\triangle\!-\!\mathbb{0}\!-\!\square\!-\!\square\!-\!\mathbb{0}\!-\!\triangle\!-\!\bigcirc\!-\!\square\!-\!\triangle$$

It's a fact!

The role of a protein in a living organism depends on the shape of its molecule. The shape is decided by the:

● **sequence** (ordering) of amino acids in the molecule
● **folding and twisting** of the chain(s) of amino acids.

Essential and non-essential amino acids

Animals are able to make some amino acids in their bodies, but they cannot make all the amino acids they need. They must obtain the ones they cannot make from their diet. The amino acids that animals need to obtain from their diet are called **essential amino acids**. The amino acids which animals can make themselves are called **non-essential amino acids**.

A protein which contains all the essential amino acids is called a **first-class protein**. Such proteins are found in lean meat, fish, cheese and soya beans. A protein which lacks essential amino acids is called a **second-class protein**. Such proteins are found in flour, rice and oatmeal. In parts of the world where rice is the staple diet, **kwashiorkor** is common.

It's a fact!

Fifteen million children die each year of starvation and disease. You will have seen pictures of children with swollen abdomens. They are suffering from a disease called **kwashiorkor**. They are getting some food but are starved of protein.

Enzymes, an important group of proteins

Catalysts help chemical reactions to take place but are not used up in the reaction itself (see *Key Science Chemistry* page 274). They are effective in *small* amounts, they remain *unchanged* at the end of the reaction and usually *speed up* the rates of chemical reactions.

Enzymes are *organic* catalysts made by living cells. Most of them are proteins. They control the rates of chemical reactions in cells and speed up the digestion of food in the gut (see Topic 14.5). All of the features of catalysts stated above are also features of enzymes. Also, enzymes are **specific** in their action which means that each enzyme catalyses a particular chemical reaction or type of chemical reaction. The substance that the enzyme helps to react is called the **substrate**. The substance(s) formed in the reaction are called the **product(s)**. Enzymes are sensitive to changes in **pH** (acidity/alkalinity) and **temperature**.

Figure 10.3B highlights the activity of the digestive enzymes (see p. 235) **amylase** and **pepsin** to illustrate the different features of enzymes.

Thousands of different enzymes control the many chemical reactions (metabolism) taking place inside a cell. Different enzymes also help digest food in the gut (see Topic 14.5).

The term **hydrolysis** literally means 'splitting by water'. It refers to a type of chemical reaction in which water splits a large molecule into smaller component molecules (see p. 235)

(a) How the enzyme amylase catalyses the breakdown of starch

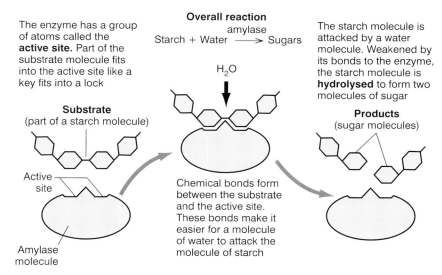

The enzyme has a group of atoms called the **active site.** Part of the substrate molecule fits into the active site like a key fits into a lock

Overall reaction

Starch + Water $\xrightarrow{\text{amylase}}$ Sugars

The starch molecule is attacked by a water molecule. Weakened by its bonds to the enzyme, the starch molecule is **hydrolysed** to form two molecules of sugar

H_2O

Substrate
(part of a starch molecule)

Products
(sugar molecules)

Active site

Amylase molecule

Chemical bonds form between the substrate and the active site. These bonds make it easier for a molecule of water to attack the molecule of starch

(b) The effects of temperature and pH on the rates of reactions catalysed by the enzymes amylase and pepsin. The rate of reaction is a measure of a particular enzyme's **activity**.

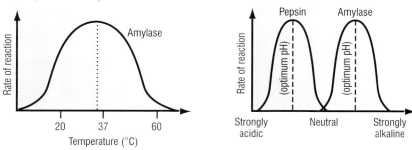

Rate of reaction

Amylase

Temperature (°C)

20   37   60

The activity of most enzymes at first increases as temperature increases, reaching a maximum. Activity then decreases if the temperature increases further, as more and more enzyme is denatured

Rate of reaction

Pepsin   Amylase

(optimum pH)

(optimum pH)

Strongly acidic   Neutral   Strongly alkaline

Activity is greatest at the optimum pH for a particular enzyme. Strong acid/ alkali denatures most enzymes

Figure 10.3B ◀ How the enzyme amylase catalyses the breakdown of starch. Notice that the enzyme is unchanged at the end of the reaction and able to catalyse the breakdown of more starch

■ Questions and answers about enzymes

Why enzymes?
The higher the temperature, the faster the rates of chemical reactions (see *Key Science Chemistry* p. 45). However, temperatures above 60 °C kill most cells. Enzymes help chemical reactions in cells to occur quickly even at the relatively low temperatures (around 37 °C for mammals – see p. 311) found in living things.

Why is an enzymes specific for a chemical reaction or type of reaction?
Only the shape of the particular substrate molecule fits the active site of the enzyme.

What does 'optimum' mean when we talk about the effect of pH and temperature on enzyme activity?
The 'optimum' is the value of pH or temperature at which the rate of chemical reaction is at a maximum.

Why at pH/temperature values only slightly higher or lower than the optimum, does the rate of reaction fall off?

The enzyme is less active and therefore less efficient as a catalyst. If the temperature approaches 0 °C, most enzymes stop working. We say that they are **deactivated**. At temperatures above 60 °C most enzymes are destroyed. We say that they are **denatured** (see p. 181).

Why is an enzyme less active and therefore less efficient above/below its 'optimum'?

A change of pH or temperature from optimum alters the shape of the active site of the enzyme, preventing the substrate molecule from making a close fit. The enzyme-catalysed reaction may not occur.

Breaking and joining molecules

Some enzymes break down large molecules into smaller molecules. Other enzymes join up small molecules into larger molecules.

Amylase is an example of a 'breaking down' enzyme. It breaks down the large molecules of starch into smaller molecules of maltose as Figure 10.3B shows. **Phosphorylase** is an example of a 'joining up' enzyme. It joins glucose molecules together to make larger starch molecules. The reaction is very important in plants. Sugar made in leaves by photosynthesis (see Topic 11.1) is turned into starch which is stored in roots (e.g. carrots) or tubers (e.g. potatoes – see p. 163). **DNA polymerase** is another example of a joining up enzyme. It joins nucleotides to form a strand of DNA (see below).

SUMMARY

Many parts of living organisms are made of proteins. The molecules of proteins are long chains of amino acids. The molecules of peptides and polypeptides are shorter chains of amino acids.

Enzymes are an important group of proteins. They catalyse the reactions which take place in living cells.

CHECKPOINT

▶ 1 What is the difference between (a) a peptide and a polypeptide (b) a polypeptide and a protein?

▶ 2 (a) To what group of chemical compounds do enzymes belong?

   (b) What vital job do enzymes do?

   (c) Give the names of three substrates on which enzymes work and the names of the products that are formed.

10.4 ▶ Nucleic acids

FIRST THOUGHTS

When the scientists James Watson and Francis Crick found out the structure of DNA, there was great excitement in the scientific world, and they won Nobel prizes. Why is DNA so important? What is so special about its structure? You can find out in this section.

RNA and DNA

The nucleic acids are **ribonucleic acid (RNA)** and **deoxyribonucleic acid (DNA)**. DNA occurs in cell nuclei. The genes that living things inherit from their parents are lengths of DNA. The information that a cell needs to assemble molecules of amino acids in the correct order to make protein molecules is present in its DNA. RNA occurs in nuclei and cytoplasm. It transfers the information contained in DNA to the places in the cell where proteins are made.

RNA and DNA have large complex molecules made up from smaller molecules of compounds called **nucleotides** (see Figure 10.4A).

There are two sorts of nucleic acid: ribonucleic acid (RNA) and deoxyribonucleic acid (DNA). Each sort of nucleic acid is made up (synthesised) from nucleotides.

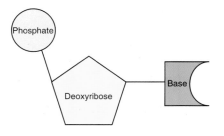

Figure 10.4A ◀ A nucleotide in DNA. The five-carbon sugar deoxyribose occurs in DNA. The base is one of four: adenine, cytosine, guanine or thymine. (In RNA the sugar ribose occurs, and the base uracil is present instead of thymine)

Erwin Chargaff, a biochemist, discovered that the amounts of adenine (A) and thymine (T) in DNA are equal and the amounts of guanine (G) and cytosine (C) are equal. This led to the idea of the one-to-one pairings of adenine–thymine and cytosine–guanine. Chargaff's rule, as it came to be known, helped Francis Crick and James Watson to construct their model of DNA.

Remember that the bases in a strand of DNA are adenine (A), thymine (T), guanine (G) and cytosine (C). The combination of two strands of DNA to make a double helix molecule depends on A combining with T (or T with A) and G combining with C (or C with G). The combinations are called 'base pairs'.

Nucleotides combine to form nucleic acids. The **phosphate** group of one nucleotide molecule combines with the **sugar** group of another. A strand of alternate sugar groups and phosphate groups is formed. The **bases** attached to the sugar groups stick out from the strand (see Figure 10.4B).

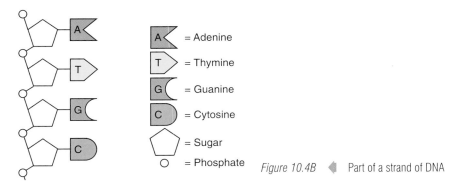

Figure 10.4B ◀ Part of a strand of DNA

A DNA molecule consists of two such strands bonded together. The bases sticking out from one sugar–phosphate strand bond to bases sticking out from a second sugar–phosphate strand. In Figure 10.4C you will notice that C bonds to G and A bonds to T – the base pairing rules. Because the rules always apply, one strand of DNA is a complement of its partner (see p. 152). (Remember, in RNA, the base uracil is present instead of thymine.)

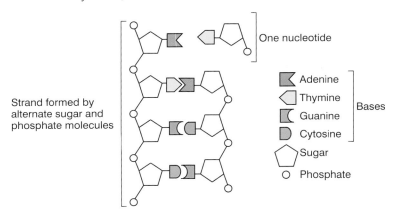

Figure 10.4C ◀ Part of a molecule of DNA. Notice how two sugar-phosphate strands are joined along their lengths as their bases combine

As two strands of sugar–phosphate groups combine to form a molecule of DNA, the result resembles a ladder: the sugar–phosphate strands form the sides and base pairs form the rungs. The 'ladder' is in fact twisted into a spiral shape (see Figure 10.4D). The shape is called a **double helix** – two intertwined spiral strands.

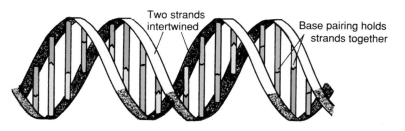

Figure 10.4D ◀ The double helix: two connected spiral strands

The DNA story

Scientists recognised the vital importance of DNA in cells, and were therefore keen to discover its structure. Many scientists worked on the problem, using different techniques. Success came in 1953 when James Watson and Francis Crick combined all the evidence from the different research workers. They succeeded in building a model of DNA which fitted all the facts (see Figure 10.4E). This model was the double helix (see Figure 10.4F). So excited were Watson and Crick that they dashed out of the Cavendish Laboratory of Cambridge University, England, and ran along the street, looking for people to tell about their discovery.

The technique of **X-ray crystallography** was of prime importance. It involves passing a beam of X-rays through a crystal on to a photographic plate. X-rays affect photographic plates, and a pattern can be seen when the plate is developed. The pattern depends on the arrangement of the atoms in the crystal.

Rosalind Franklin of King's College, London, together with Maurice Wilkins, obtained the X-ray crystallography pictures of DNA crystals that gave Watson and Crick the vital clues they needed to build the DNA model. Her work was essential to the discovery. She died in 1958. Whether, if she had lived longer, she would have shared the Nobel Prize that Watson, Crick and Wilkins later won, we cannot say.

Figure 10.4F 🔺 Rosalind Franklin

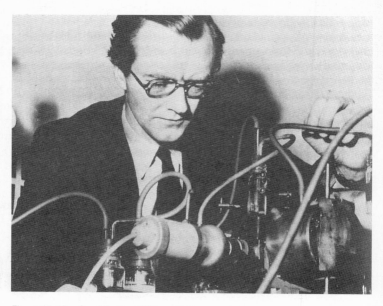

Figure 10.4G 🔺 Maurice Wilkins with X-ray equipment similar to that used by his co-worker, Rosalind Franklin

Figure 10.4E 🔺 A computer graphics image of the DNA double helix looking from the side

The famous double helix, the structure of DNA, was worked out by James Watson and Francis Crick on the basis of the X-ray crystallography of DNA by Rosalind Franklin.

Figure 10.4H ▲ James Watson, an American biologist (left) and Francis Crick, a British physicist turned biologist (right) with their model of DNA at the Cavendish Laboratory, Cambridge. You can see the spiral pattern of the double helix. Notice the model sugar ring at Watson's left shoulder

Francis Crick, James Watson and Maurice Wilkins received the Nobel Prize from the King of Sweden for finding the structure of DNA in 1962. Also receiving prizes that year were Max Perutz for finding the structure of haemoglobin and John Kendrew for his work on the structure myoglobin (the first protein to have its three dimensional structure worked out).

Figure 10.4I ▲ Six of the 1962 Nobel Prize winners on the eve of the award ceremony. From left to right: Maurice Wilkins, John Steinbeck (winner of the literature prize), John Kendrew, Max Perutz, Francis Crick and James Watson

CHECKPOINT

▸ **1** (a) Name three types of group present in a nucleotide.
   (b) Which sugar gives DNA its name?
   (c) Briefly explain why DNA got the name 'the double helix'.
   (d) The letters ATCG stand for bases. Name the four bases.
   (e) Briefly explain the meaning of 'base pairs'.

Theme Questions

1 Match each of the biological terms in Column A with the descriptions in Column B.

Column A	Column B
Ribosome	Contains genetic material
Cell wall	Contains cell sap
Vacuole	Site of aerobic respiration
Mitochondrion	A food reserve in animal cells
Nucleus	Site of photosynthesis
Starch grain	Site of protein synthesis
Cell membrane	Made of cellulose
Chloroplast	A food reserve in plant cells
Endoplasmic reticulum	Controls the passage of substances into and out of the cell
Glycogen	Network of channels running through the cytoplasm

2 The diagram below shows a cell of yeast.

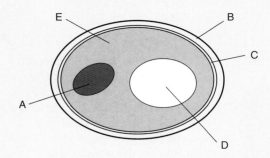

(a) Use the label letters **A**, **B**, **C**, **D** and **E** to identify the cell wall, the cell membrane, the nucleus, the vacuole and the cytoplasm.

(b) Give **two** ways in which this cell is different from a leaf mesophyll cell.

(c) Give **two** ways in which this cell is different from a cheek cell.

3 The diagram shows the bacterium *Salmonella*.

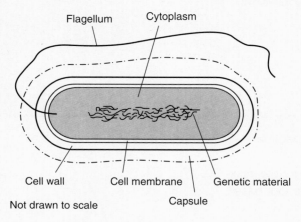

Flagellum   Cytoplasm

Cell wall   Cell membrane   Genetic material

Not drawn to scale   Capsule

(a) Give **one** similarity between the bacterial cell structure and the structure of a palisade cell in a leaf.

(b) How is the genetic material in bacterial cells organised?

(c) What method of asexual reproduction occurs in bacteria?

4 Diagram **A** shows the appearance of a plant which had not been watered for two weeks. Diagram **B** shows the same plant a few hours after it had been watered.

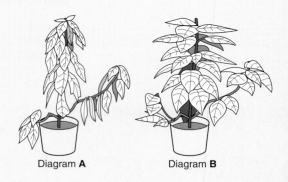

Diagram **A**   Diagram **B**

Explain what has happened in the plant to cause the change in the position of the leaves between diagram **A** and diagram **B**. (MEG)

5 The sentences below describe the process of protein synthesis. Write out the sentences and use words from the list to complete them.

amino acids	DNA	ribosomes
bases	nucleus	sugars
cytoplasm	polypeptides	tRNA

(a) The process of protein synthesis begins in the ....

(b) Strands of .... begin to unwind.

(c) .... in groups of three along the DNA form triplets.

(d) mRNA forms along the DNA strand and moves to the ....

(e) Triplets along mRNA code for ....

(f) Amino acids join to form ....

(Edexcel

6 A student looked at various cells using a microscope.

Red blood cell
large surface area

Motor neurone
long, thin, flexible

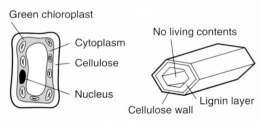

Palisade cell

Xylem cell
long, thin, strong

(a) Which are animal cells and which are plant cells?
(b) What job does each cell carry out?
(c) Describe how each cell is adapted for its particular job. (MEG)

7 Graph 1 shows how temperature affects an enzyme-controlled reaction.

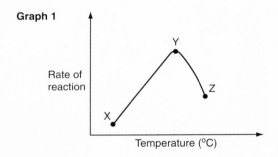

(a) Explain what is happening (i) between **X** and **Y** (ii) between **Y** and **Z**.
(b) At what temperature would you expect the reaction to be quickest?
(c) Suggest how high temperature might affect the active site of the enzyme.
Graph 2 shows how pH affects the rate of two different enzyme-controlled reactions.

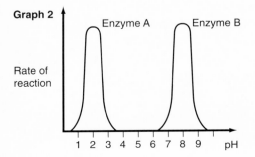

(d) Which enzyme is likely to be most active in (i) the stomach (ii) the duodenum?
(e) What does the graph tell you about the range of pH over which an enzyme is active?

8 Match the chemicals in Column A with their correct chemical composition in Column B and function in Column C.

Column A	Column B	Column C
Lipids	CHON	Growth and repair of cells
Nucleic acids	CHO	Solvent
Carbohydrates	HO	Store of energy
Water	CHON(SP)	Material of heredity
Proteins	CHO	Main source of energy

9 The diagram represents part of a model of a DNA molecule which has a 'ladder-like' structure.

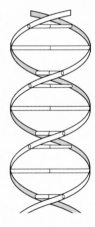

(a) Name the chemicals which make up the 'sides' of the ladder.
(b) The 'rungs' of the ladder are made of chemicals called bases. Describe how these bases are arranged.
(c) A mutation occurs when there is a change in the structure of the DNA molecule.
  (i) State **two** factors which can increase the risk of a mutation occurring.
  (ii) Explain briefly how such a change could bring about variation in a subsequent generation. (MEG)

Green plants as organisms

P lants make food by photosynthesis and obtain water, nutrients and carbon dioxide. Materials are transported to where they are needed. Plants respond to their surroundings by growth movements. All of these different processes play important roles in the life of plants.

Topic 11

Photosynthesis and mineral nutrition

11.1 ▶ Leaves are adapted for photosynthesis

FIRST THOUGHTS

Life on Earth depends on light. Why? Because light energy is converted into the energy of the chemical bonds in food substances by photosynthesis, and all living things need food.

The sun floods the Earth with light. Plant cells use light to help them to make food by **photosynthesis**. They trap the energy in sunlight and use it to convert carbon dioxide and water into sugars. A summary of the chemical reactions that take place is

catalysed by chlorophyll

Carbon dioxide + Water → Glucose + Oxygen

$$6CO_2(g) + 6H_2O(l) \rightarrow C_6H_{12}O_6(aq) + 6O_2(g)$$

How does photosynthesis get its name? 'Photo' means light (in Greek) and 'synthesis' means putting together. Photosynthesis takes place in the **chloroplasts** found in the cells of green plants. Chloroplasts contain the green pigment **chlorophyll**, which acts as a **catalyst** in the chemical reactions of photosynthesis.

Leaves are adapted for photosynthesis:

* The leaf mosaic maximises the number of leaves directly exposed to sunlight.
* Each leaf has a large surface area for the absorption of sunlight.
* Inside the leaf, palisade cells, each packed with chloroplasts, lie just under the upper surface of the leaf exposed to sunlight.
* Inside each palisade cell, chloroplasts concentrate where light is most intense.
* Each chloroplast is filled with membranes layered with chlorophyll, thereby maximising the surface area of chlorophyll exposed to sunlight.

Leaves, light, water and gases

Leaf cells are 'factories' which collect sunlight, water and carbon dioxide, and manufacture food. Figure 11.1A traces the path of water and carbon dioxide molecules from the air into a leaf, a leaf cell and a chloroplast. Notice the letters on Figure 11.1A. They are a checklist of the ways in which leaves are adapted for photosynthesis.

A Leaves are arranged so that the ones above do not overshadow those below. The arrangement is called the **leaf mosaic**. More leaves therefore are exposed to direct sunlight. The leaf blade itself is *flat* which means that a large surface area is exposed for absorption of light. It is also *thin*. As a result light reaches the lower layers of cells which contain chloroplasts. The cells are called **spongy mesophyll**.

B Water reaches leaves in the **transpiration stream** through the xylem (see p. 190). Carbon dioxide diffuses into the leaf through the small gaps called **stomata** that pierce the undersurface. Oxygen produced as a by-product of the chemical reactions of photosynthesis diffuses out of the leaf, also through the stomata. Leaves therefore not only manufacture food but are also surfaces for **gaseous exchange** (see p. 247).

C The cells of the upper leaf-surface (**epidermis**) do not contain chloroplasts and are transparent, so that light reaches the palisade cells beneath. **Palisade cells** are column-shaped, tightly packed and filled with chloroplasts. Many chloroplasts, therefore, are exposed to bright light, maximising the rate of photosynthesis. Spongy mesophyll cells are less tightly packed. As a result there are air spaces between them. The spaces allow gases and water vapour to circulate freely within the leaf, bringing the raw materials for photosynthesis to the leaf cells. The cells of the lower leaf surface do not have chloroplasts except the **guard cells** which flank the **stoma**. The shape of the guard cells depends on photosynthesis and controls the size of the opening of the stoma (see p. 191). Size of opening controls the rate at which gases and water vapour diffuse into and from the leaf.

See www.keyscience.co.uk for more about photosynthesis.

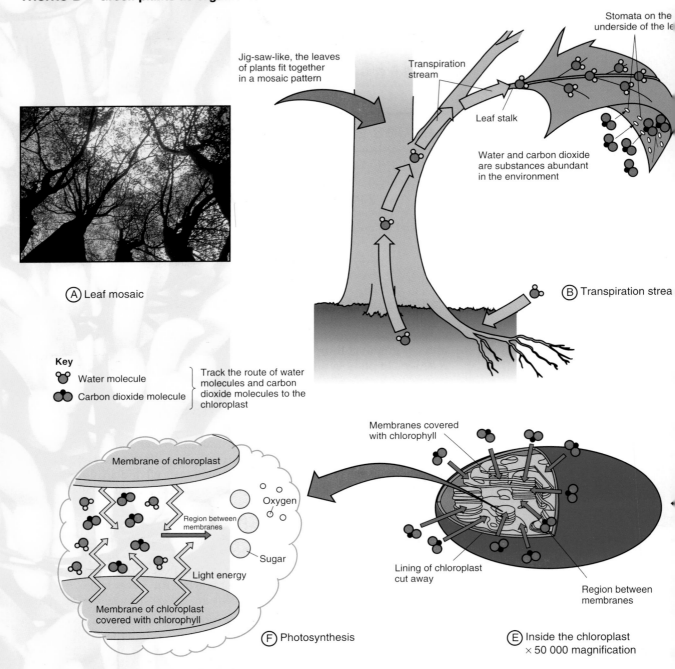

Jig-saw-like, the leaves
of plants fit together
in a mosaic pattern

Transpiration
stream

Stomata on the
underside of the le

Leaf stalk

Water and carbon dioxide
are substances abundant
in the environment

(A) Leaf mosaic

(B) Transpiration strea

Key

Water molecule

Carbon dioxide molecule

Track the route of water
molecules and carbon
dioxide molecules to the
chloroplast

Membrane of chloroplast

Oxygen

Region between
membranes

Sugar

Light energy

Membrane of chloroplast
covered with chlorophyll

Membranes covered
with chlorophyll

Lining of chloroplast
cut away

Region between
membranes

(F) Photosynthesis

(E) Inside the chloroplast
× 50 000 magnification

Figure 11.1A 🔺 The leaf and photosynthesis

D Chloroplasts pack the inside of each palisade cell. They stream in the cytoplasm (**cyclosis**) to the region of the palisade cell where light is brightest. As a result the rate of photosynthesis is maximised.

E Chlorophyll covers the membranes inside chloroplasts. The membranes are arranged like stacks of pancakes, maximising the surface area of chlorophyll exposed to light.

F Photosynthesis is a two-stage process:
Stage 1 chlorophyll, activated through absorbing light, makes light energy available for the synthesis (making) of sugar;
Stage 2 enzymes catalyse the conversion of water and carbon dioxide into sugar. These enzymes are located in the region of the chloroplast between the membranes.

The sugars formed by photosynthesis are the source of energy for the plant. When animals eat plants, the food substances built up in the plant become available to them.

**EXTENSION FILE
ACTIVITY**

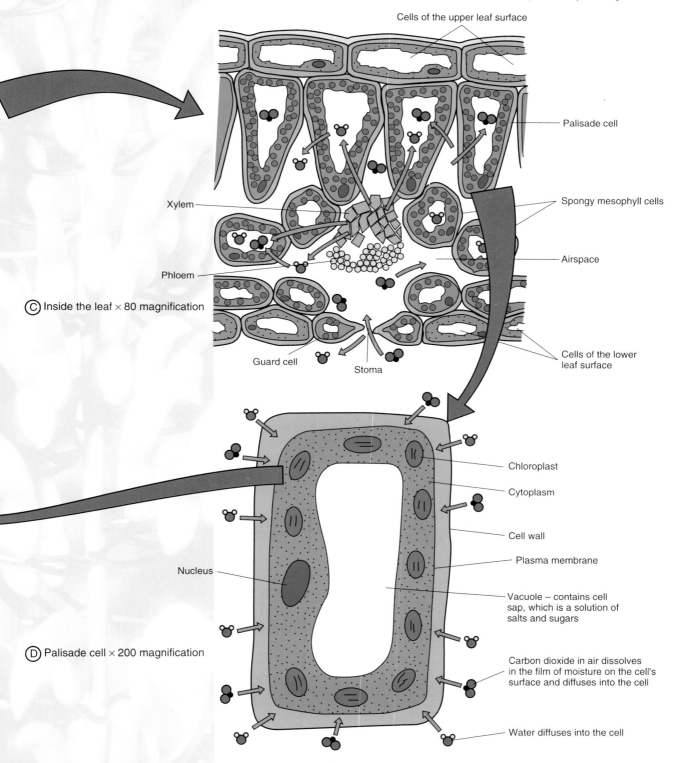

Cells of the upper leaf surface

Palisade cell

Xylem

Spongy mesophyll cells

Airspace

Phloem

Ⓒ Inside the leaf × 80 magnification

Guard cell

Stoma

Cells of the lower leaf surface

Chloroplast

Cytoplasm

Cell wall

Plasma membrane

Nucleus

Vacuole – contains cell sap, which is a solution of salts and sugars

Carbon dioxide in air dissolves in the film of moisture on the cell's surface and diffuses into the cell

Ⓓ Palisade cell × 200 magnification

Water diffuses into the cell

Different pigments in photosynthesis

Chlorophyll absorbs light of all colours except green. It therefore appears green. In fact there are several kinds of chlorophyll. The green **pigments**, chlorophyll A and chlorophyll B, are found in most plants. They strongly absorb red and blue light. Other pigments, called carotenoids, are also

Chlorophyll absorbs most light at the red and blue ends of the spectrum.

present. They are red, yellow and orange and absorb light of other colours. Together, the assortment of pigments enables a plant to absorb sunlight over a wide spectrum (Figure 11.1B).

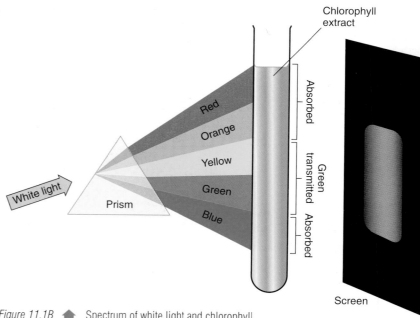

Figure 11.1B Spectrum of white light and chlorophyll

SUMMARY

Photosynthesis is the process by which plants make food. Carbon dioxide and water combine in the presence of the catalyst, chlorophyll, to form glucose and oxygen. The reaction takes place in sunlight. The energy of sunlight is converted into the energy of the chemical bonds in glucose.

In summer nearly all leaves are green, because the colours of the carotenoids are masked by the chlorophylls. In autumn, however, when the production of chlorophylls tails off, the colours of the carotenoids blaze through. Tomatoes, carrots and ripe fruits of different kinds get their colours from carotenoids.

What conditions are best for photosynthesis?

Light intensity, **temperature** and supplies of **carbon dioxide** and **water** together affect the rate at which plants make sugars by photosynthesis. They are called **limiting factors** because if any one of them falls to a low level, photosynthesis slows down or stops.

A greenhouse provides everything plants need for photosynthesis. The Assignment 'Growing crops under glass' in *Key Science: Biology Extension File* gives more information. Conditions are controlled so that photosynthesis occurs at maximum efficiency and plants grow quickly. In other words, limiting factors are eliminated.

Buttercups in the sun

Dog's mercury in the shade

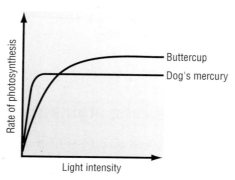

Figure 11.1C Light intensity as a limiting factor

■ Light

Look at Figure 11.1C showing plants growing in different light conditions. The graph shows that in both cases the rate of photosynthesis increases as the light intensity increases. However, once a particular light intensity is reached the rate of photosynthesis stays constant even if the light intensity increases further.

Different limiting factors affect the rate of photosynthesis.

Notice that the rate of photosynthesis in dog's mercury is maximal at a lower light intensity than for buttercups, but that the maximum rate of photosynthesis for buttercups is greater than the maximum rate for dog's mercury. *How has dog's mercury adapted to shady conditions? Why do you think buttercups are not usually found in shaded woodland and dog's mercury in sunny meadows?* Try to answer the questions using the words *competition* and *competitive exclusion* (see p. 65 and p. 66).

■ Carbon dioxide

Figure 11.1D shows two crops of lettuces grown in a glasshouse. One crop has had carbon dioxide added to the atmosphere; the other has not. Increasing the amount of atmospheric carbon dioxide is only practical in a glasshouse. Outdoors, the level of carbon dioxide in the air is about 0.03%. The graph shows how the rate of photosynthesis increases with increasing carbon dioxide concentration.

The rate of photosynthesis stays constant after a certain concentration of carbon dioxide is reached. In the short term this limit point is about 0.5%, but over long periods the best level is about 0.1%. Above this level the amount of light often limits the plant's ability to use the extra carbon dioxide.

Glasshouse crops like lettuces and tomatoes are commonly grown in an atmosphere of 0.1% carbon dioxide. The extra cost to the grower is more than offset by the increased crop yield.

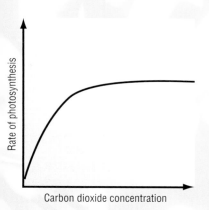

Figure 11.1D ⬆ Carbon dioxide concentration as a limiting factor: notice the larger size of the lettuces grown in a carbon dioxide enriched atmosphere

■ Warmth

Most plants grow best in warm conditions. The higher the temperature, the faster the chemical reactions of photosynthesis – within limits! **Enzymes** are proteins that control chemical reactions (including photosynthesis) in living things (see p. 168). Temperatures approaching 0 °C deactivate and above 60 °C destroy (**denature**) proteins and reduce enzyme activity (see p. 170).

Plants which grow well at 20–5 °C grow in the UK; plants which thrive at 35–40 °C are found in tropical regions.

■ Water

Water is one of the raw materials for photosynthesis. However, lack of water affects so many cell processes that it is impossible to single out its direct effect on photosynthesis. Figure 11.1E highlights the importance of water for plant growth. It makes farming possible – even in hot deserts!

Figure 11.1E ⬆ Turning the desert green

CHECKPOINT

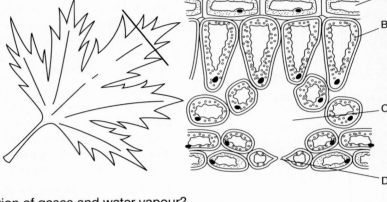

▶ **1** The diagram opposite shows a leaf and a small portion of it cut through lengthways.

(a) How does the structure of the leaf enable it to collect light?

(b) Name the parts labelled A, B, C and D.

(c) Where in the leaf is most of the plant's food made by photosynthesis?

(d) Where do gases pass into and out of the leaf?

(e) How does C help the circulation of gases and water vapour?

▶ **2** The diagram opposite represents the inside of a chloroplast.

(a) What happens at A during photosynthesis?

(b) How does the structure of A enable it to play a part in photosynthesis?

(c) What happens at B during photosynthesis?

(d) Briefly explain the relationship between what happens at A and what happens at B during photosynthesis.

(e) Which gas is released into the environment during photosynthesis?

▶ **3** The table shows the number of bubbles of gas given off in one minute by water weed illuminated with light of different colours.

Colour of light	Number of bubbles given off in one minute
Blue	85
Green	10
Red	68

What relationship between colour of light and rate of photosynthesis do the results indicate? Suggest how results like this might help a gardener to grow bigger plants in a shorter time in a greenhouse.

▶ **4** The diagram opposite shows an experiment that was carried out to measure the rate of photosynthesis in water weed at 20 °C.

The rate of photosynthesis was assessed at different light intensities by counting the number of bubbles of gas given off by the plant in a given time. The results are shown in the table.

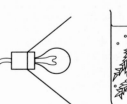

Number of bubbles per minute	6	15	21	25	27	28	28
Light intensity (arbitrary units)	1	2	3	4	5	6	7

(a) Plot the results on a graph, putting 'Light intensity' on the horizontal axis and 'Number of bubbles' on the vertical axis.
   Now answer the following questions, using the diagram and your graph to help you.

(b) At what light intensity did the plant produce 25 bubbles per minute?

(c) Suggest a better way of measuring the rate of photosynthesis than counting the bubbles.

(d) If the experiment was done at the following temperatures, what would be the effect on the rate of photosynthesis?
   (i) 2 °C   (ii) 33 °C   (iii) 65 °C

(e) What other factors affect the rate of photosynthesis?

11.2 ▶ Mineral nutrition

FIRST THOUGHTS

To survive and grow, plants need chemical elements. Photosynthesis provides carbon, hydrogen and oxygen in the form of sugars (see Topic 10.1). Land plants obtain the other elements from the soil. Major elements are needed in large amounts; trace elements are needed in small amounts.

It's a fact!

Bottles of 'plant food' sold at garden centres contain solutions of some of the important minerals needed for healthy plant growth.

Nitrogen, phosphorus and potassium are major elements. Magnesium, iron and calcium are trace elements. Major elements and trace elements are necessary for healthy plant growth.

 See www.keyscience.co.uk for more about minerals.

Healthy plant growth depends on the minerals (**nutrients**) which are absorbed through the roots as solutions of salts in the form of ions. The concentration of mineral ions is usually higher in root tissue than in the soil. Normally, therefore, mineral ions would diffuse down their concentration gradient (see p. 148) from the roots to the soil. However, roots take up mineral ions against their concentration gradient. The process is an example of **active transport** (see p. 151). It uses a lot of energy which is supplied by mitochondria (see pp. 147 and 245) in the hair cells of the root (see p. 159).

The ions of **major elements** are needed in quite large amounts (on a kilogram per hectare scale):

- **Nitrogen (N)** (in the form of nitrates) is used by plants to make protein.
- **Phosphorus (P)** (in the form of phosphates) is used by plants to make cell membranes, for the development of roots and for the synthesis of nucleic acids (see Topic 10.4).
- **Potassium (K)** (as potassium salts) is a component of the processes of photosynthesis.

Nitrogen, phosphorus and potassium are the elements which are most frequently in short supply in soils. Nitrogen is especially important because it combines with sugars produced during photosynthesis to form amino acids. The amino acids join together to form large protein molecules (see Topic 10.3). Plants which are short of nitrates are stunted in growth and have yellow leaves. Shortage of phosphates causes poor root growth and purple discolouration of leaves. Yellow leaves with dead spots occur when potassium is in short supply.

Sugar + Nitrogen → Amino acids → Protein
(from photosynthesis) (in the form of nitrates from the soil – see Topic 2.7)

Most artificial fertilisers are NPK fertilisers. Huge quantities are used worldwide to improve crop growth (see Topic 5.3).

The ions of **trace elements** (micronutrients) are needed in much smaller amounts than major elements:

- **Magnesium (Mg)** is a component of the chlorophyll molecule.
- **Iron (Fe)** is also a component of the chlorophyll molecule.
- **Calcium (Ca)** is needed for the formation of cellulose cell walls.

If magnesium and/or iron are in short supply then leaves become mottled or pale (Figure 11.2A). Copper (Cu), sodium (Na) and manganese (Mn) are also trace elements. Shortage of any one of them causes poor plant growth.

Figure 11.2A ▶ Poor chlorophyll formation because of magnesium/iron shortage causing mottled leaves

Hydroponics and nutrient film technique

You can grow plants without soil providing the plants are supplied with a solution of all of the substances they need for a healthy existence (nutrient solution). The technique is called **hydroponics** (or nutrient culture). Air is bubbled through the solution (aeration) so that the roots have enough oxygen for aerobic respiration (see Topic 15.1).

What are the advantages of hydroponics?

- *Plants grow faster* because their roots are supplied with nutrient solution all of the time. Lots of root growth to find nutrients in short supply is not necessary.

- *Weeds are eliminated*, reducing the workload of farmers and cutting out the need for herbicides (see p. 96).

- *Insect pests which live in the soil are eliminated*, reducing the need for insecticides (see p. 96).

- *Quality and taste are better* because the plants are supplied with all of the nutrients they need.

The more popular systems for hydroponics are the:

- **capillary (wick) system:** the nutrient solution is drawn up the roots by capillary action. The system is used to grow house plants, herbs and other slow-growing plants

- **flood and drain system:** every time the nutrient solution floods from the lower reservoir into the upper growing tray, the roots are bathed in fresh solution. The roots are aerated (supplied with air) as the nutrient solution drains back to the reservoir

- **nutrient film technique:** a film of nutrient solution flows along a channel. Part of the root system grows down into the nutrient solution and part grows above the water level, receiving air. The system is used to grow lettuce, spinach and some herbs

The nutrient film technique is one way of growing crops in desert areas (see Figure 11.1E). However, the expense of the technical facilities and maintenance may make the system uneconomic in practice.

Hydropononic techniques are methods of growing plants without planting them in soil.

Figure 11.2B ⬆ Nutrient film hydroponics is being used to grow a crop of lettace in a large glasshouse. Nutrient solution is fed into the channels, supplying the plants with all of the substances needed for growth.

CHECKPOINT

▶ 1 A farmer brings you plants taken from different fields. From their appearance you suspect that the plants are suffering from shortage of different minerals.
Field A – the plants were stunted in growth and the older leaves on each plant were yellow
Field B – the plants showed poor root growth and the younger leaves on each plant were purple
Field C – the plants had yellow leaves with dead spots.
What advice would you give the farmer to deal with the problem?

11.3 ▶ Growth and support

The cotyledons of plants like the broad bean stay below ground during germination. During germination of plants like the French bean the cotyledons are carried above ground by the developing shoot. The turgor pressure (see p. 149) of their cells and wood support plants.

The embryo plant is protected within the seed (see p. 351):

- The radicle develops into the root.
- The plumule develops into the shoot.
- Cotyledon(s) store food which is a source of energy for germination.

For plants, the sugars produced during photosynthesis are a source of energy and materials for growth. Figure 11.3A gives you the idea.

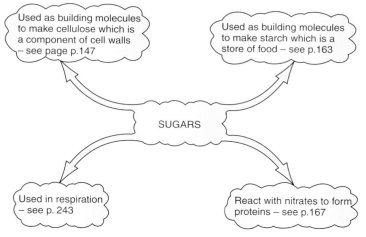

Used as building molecules to make cellulose which is a component of cell walls – see page p.147

Used as building molecules to make starch which is a store of food – see p.163

SUGARS

Used in respiration – see p. 243

React with nitrates to form proteins – see p.167

Figure 11.3A 🔺 The uses of sugar in plants

The growth of plants (and animals and other living things) can be measured as an increase in an individual's mass, length or number of cells and the data can be represented as a growth curve. Figure 11.3B shows the growth curve of a broad bean seed as the embryo inside it develops into a new plant.

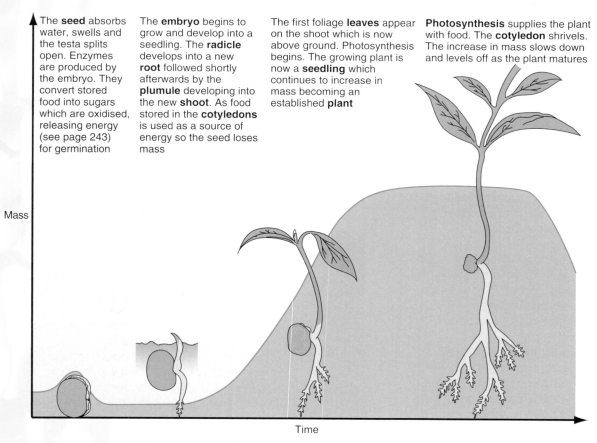

The **seed** absorbs water, swells and the testa splits open. Enzymes are produced by the embryo. They convert stored food into sugars which are oxidised, releasing energy (see page 243) for germination

The **embryo** begins to grow and develop into a seedling. The **radicle** develops into a new **root** followed shortly afterwards by the **plumule** developing into the new **shoot**. As food stored in the **cotyledons** is used as a source of energy so the seed loses mass

The first foliage **leaves** appear on the shoot which is now above ground. Photosynthesis begins. The growing plant is now a **seedling** which continues to increase in mass becoming an established **plant**

Photosynthesis supplies the plant with food. The **cotyledon** shrivels. The increase in mass slows down and levels off as the plant matures

Mass

Time

Figure 11.3B 🔺 Growth curve of a broad bean from seed to mature plant.

The growth curves of nearly all annual plants (see p. 23) look like Figure 11.3B. Flowering time is followed by the formation of fruits and seeds which are then dispersed. The plant then quickly loses mass and eventually dies.

Figure 11.3C shows a root tip. Behind the tip of the developing root of a germinating seed there is a region where cells divide very quickly. Behind the region of dividing cells is another region where cells become longer, growing to ten times or more their original length.

As the cells get longer they become different from one another: we say that they become **differentiated** (see p. 302). Some form the sieve cells of the phloem, others form xylem cells (see Topic 9.5 and Topic 12). Differentiation in the shoot is more complicated than in the root since the developing shoot produces leaves and flowers.

Differentiation therefore, lays down the tissues of the plant body. The word **primary** is used to describe these tissues because they are the first tissues to develop. Plants which have only primary tissues are called **herbaceous** plants (see p. 23).

■ Where does growth occur in plants?

In animals most cells can divide by mitosis (see Topic 9.4) which means that nearly all parts of the body can grow. In plants only some cells divide by mitosis. These cells form the growing points of a plant, mostly at the tips of the shoots and roots.

Trees grow year after year because they contain cells which divide to produce more and more xylem. Such growth is called **secondary growth** because the new xylem develops after differentiation during germination has laid down the primary tissues. This is how **wood** forms. It is impregnated with **lignin** (see Topic 12.2) which gives it extra strength.

In countries like the UK with a temperate climate, wood grows seasonally. The cycle is repeated each year to form cylinders of wood, which in cross-section appear as **annual rings** (see Figure 11.3D).

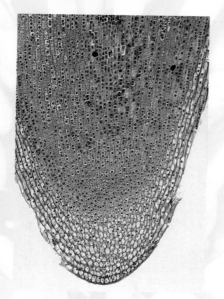

Figure 11.3C 🔺 The growing root tip

It's a fact!

The stages of growth from the embryo inside the seed to the time when the seedling no longer depends on stored food is called **germination**. To germinate successfully most seeds need:

- **warmth**. If the temperature is too low germination is prevented. Prevention safeguards the plant from ice and frost which would otherwise kill the seedling.
- **water** which the seed absorbs. The testa (seedcoat – see p. 351) splits. Growth of the embryo plant begins.
- **oxygen**. The seedling needs more oxygen than when it was an embryo inside the seed. The oxygen is used during aerobic respiration (see Topic 15.1) which releases energy for cell division and the processes of growth.

Figure 11.3D 🔺 The cut end of a Douglas fir tree showing annual growth rings

Support in plants

Plants are supported by the strands of **xylem** and **phloem** that run in vascular bundles from the roots, through the stem to the leaves and flowers (see Figure 12.1A). Non-woody (herbaceous) plants are supported by the firmness of their cells. The pressure of the cell contents against the cell wall makes a plant cell **turgid** (see Topic 9.2). Individual cells press against each other and hold the plant upright. The wood formed from lignified xylem gives even more support. This is why trees can grow to be tall and heavy.

Wood is one of the places in a plant where wastes are stored. Plants dispose of the wastes by combining them with inorganic salts to form insoluble crystals. Figure 11.3E shows where the insoluble crystals of wastes are stored and how the wastes stored in leaves are eventually removed from the plant.

Water moving into plant cells by osmosis (see p. 149) increases the pressure inside the cells, making them turgid.

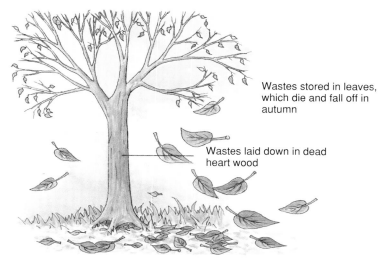

Wastes stored in leaves, which die and fall off in autumn

Wastes laid down in dead heart wood

Figure 11.3E ⬆ Plant waste disposal

CHECKPOINT

▶ **1** Which region of the root tip contributes most to the extension of the root through the soil? Give reasons for your answer.

▶ **2** Where in the root tip are you likely to find:
(a) the smallest cells,
(b) cells containing the highest percentage of water?

▶ **3** Complete the following paragraph using the words provided below. Each word may be used once, more than once or not at all.

| nucleus | expand | mitosis | grow | newly | differentiate | transport | large |
| water | dividing | small | osmosis | elongation | | | |

Roots _____ at their tips. The cells of the region immediately behind the root tip divide by _____ . Repeated divisions mean that there are _____ numbers of _____ cells. Further back these _____ formed cells begin to _____ due to the uptake of _____ by _____ . This expansion produces rapid _____ of the root through the soil. We say that they _____ .

▶ **4** Figure 11.3B shows the growth curve of a broad bean during germination. Study it and answer the following questions.
(a) Define the word 'germination'.
(b) Why does the mass of seed increase as germination begins?
(c) After the initial increase, why does the germinating plant then lose mass?
(d) After losing mass why does the germinating plant then increase in mass?

Topic 12

Transport and water relations

12.1 ▶ ## The transport system of plants

FIRST THOUGHTS

How do adequate supplies of water and food reach all parts of a plant? A transport system is the answer!

Moss plants (see p. 20) do not have xylem and phloem. They rely on capillary action to soak up water.

See www.keyscience.co.uk for more about transport in plants.

Water and food move into and out of plant cells by **osmosis, diffusion** and **active transport** (see Topic 9.2). Flowering plants require special tissues for the transport of these materials to where they are needed. The tissues are **xylem** and **phloem** (see Figure 12.1A). Xylem transports water and phloem transports food. Xylem and phloem tissue form vascular bundles which is why flowering plants, along with conifers and ferns, are sometimes referred to as **vascular** plants.

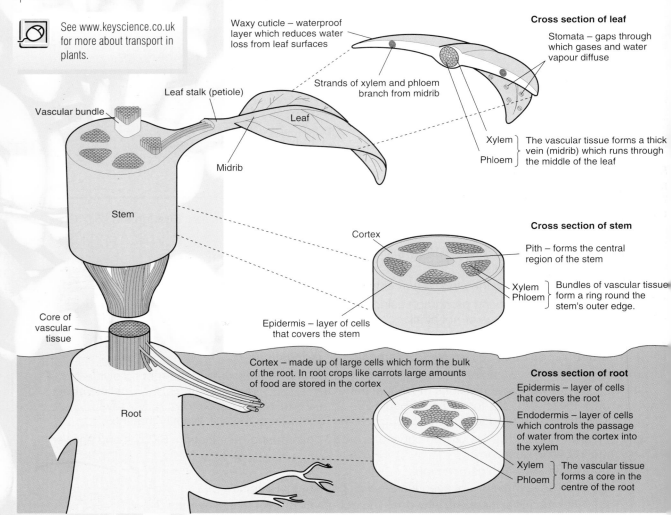

Waxy cuticle – waterproof layer which reduces water loss from leaf surfaces

Cross section of leaf

Stomata – gaps through which gases and water vapour diffuse

Strands of xylem and phloem branch from midrib

Leaf stalk (petiole)

Vascular bundle

Leaf

Midrib

Xylem
Phloem } The vascular tissue forms a thick vein (midrib) which runs through the middle of the leaf

Stem

Cross section of stem

Cortex

Pith – forms the central region of the stem

Xylem
Phloem } Bundles of vascular tissue form a ring round the stem's outer edge.

Epidermis – layer of cells that covers the stem

Core of vascular tissue

Cortex – made up of large cells which form the bulk of the root. In root crops like carrots large amounts of food are stored in the cortex

Cross section of root

Epidermis – layer of cells that covers the root

Endodermis – layer of cells which controls the passage of water from the cortex into the xylem

Root

Xylem
Phloem } The vascular tissue forms a core in the centre of the root

Figure 12.1A ⬆ The arrangement of xylem and phloem (vascular tissue) in a flowering non-woody plant. (Notice that the central core of vascular tissue in the root changes to a ring of bundles in the stem at ground level)

12.2 ▶ Xylem transports water and mineral ions

Xylem tissue consists of dead cells. How then can it transport water to all parts of the plant? Reading this section will help you find out.

Mature xylem cells are dead. They are cylindrical and join end-to-end to form strands of xylem tissue. The **cross-walls** separating adjoining cells break down so that each xylem strand becomes a long, hollow **vessel** through which water can pass freely (see Figure 12.2A). Xylem vessels run from the tips of roots, through the stem and out into every leaf as Figure 12.2B shows.

Mineral ions enter the plant through the root hairs by active transport (see p. 183). Water enters the plant through the root hairs by osmosis. It moves across the root cortex by osmosis and passes into the xylem vessels. Once in the xylem vessels, water forms unbroken columns from the roots, through the stem and into the leaves. Water evaporates from the leaves – mainly through the tiny pores in the underside of the leaf called stomata (see Figure 12.2E). The process is called **transpiration**.

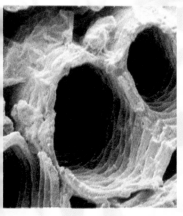

Figure 12.2A ⬆ Xylem in a stem – notice the rings of **lignin**, a substance which strengthens and waterproofs the cell walls of each vessel

Wood forms from cells which divide to produce more and more xylem. Trees grow as they produce wood. Year after year cylinders of wood form appearing as annual rings. Lignin impregnates wood giving it extra strength.

It's a fact!

In 1872 an Australian eucalyptus tree was measured at 100 m – the tallest tree ever recorded. Today the tallest tree is a coastal redwood growing in California. It measures more than 100 m.

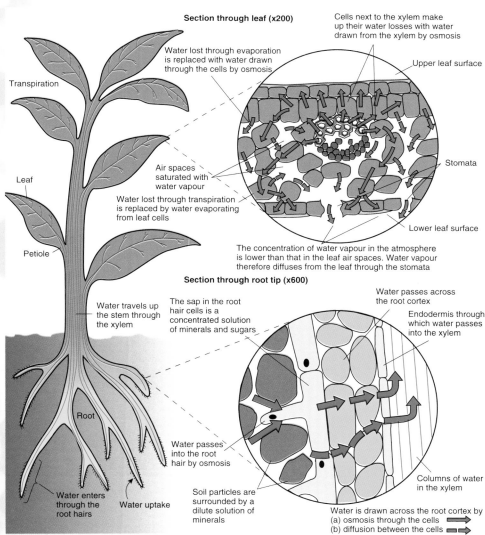

Section through leaf (x200)

Cells next to the xylem make up their water losses with water drawn from the xylem by osmosis

Water lost through evaporation is replaced with water drawn through the cells by osmosis

Upper leaf surface

Transpiration

Leaf

Petiole

Air spaces saturated with water vapour

Water lost through transpiration is replaced by water evaporating from leaf cells

Stomata

Lower leaf surface

The concentration of water vapour in the atmosphere is lower than that in the leaf air spaces. Water vapour therefore diffuses from the leaf through the stomata

Section through root tip (x600)

Water travels up the stem through the xylem

The sap in the root hair cells is a concentrated solution of minerals and sugars

Water passes across the root cortex

Endodermis through which water passes into the xylem

Root

Water passes into the root hair by osmosis

Water enters through the root hairs

Water uptake

Soil particles are surrounded by a dilute solution of minerals

Columns of water in the xylem

Water is drawn across the root cortex by
(a) osmosis through the cells ▭▭▶
(b) diffusion between the cells ▭▭▶

Figure 12.2B ⬆ The transpiration stream

EXTENSION FILE ACTIVITY

Transpiration causes water to be pulled along xylem vessels.

Cactus spines also deter animals from feeding on the plant's fleshy moist tissues.

As water transpires, more is drawn from the xylem. This movement of water exerts 'suction' on the water filling the xylem vessels of the stem. As the water is 'sucked' upwards through the xylem of the stem, more water is supplied to the bottom of the xylem by the roots. There is therefore a continuous moving column of water from the roots to the leaves. This is called the **transpiration stream** (see Figure 12.2B).

The force that drives the transpiration stream is equivalent to as much as thirty atmospheres pressure – sufficient to move water to the top of the tallest tree.

■ Desert survival

Cacti are adapted to survive in the hot, dry environment of deserts (Figure 12.2C). Loss of water from the cactus is reduced by:

- a thick waxy cuticle which covers the plant's surfaces
- leaves which are often little more than spines, reducing the surface area for water loss
- deep hair-fringed pits in the plant's surface surrounding the stomata, which trap water vapour, increasing the humidity of the microenvironment around the stomatal opening
- a shiny surface which reflects heat and light
- very thick stems storing water
- shallow, widespread roots which cover a large area absorbing what little water may be available.

Why do you think each of the adaptations listed helps cacti conserve water?

Figure 12.2C ⬆ A desert cactus

If diffusion of substances through the stomata is restricted and the leaves reduced to spines, *how then do cacti obtain carbon dioxide and absorb light for photosynthesis?* Find out about the unusual method cacti have of obtaining the raw materials for photosynthesis.

It is not a good idea to plant out seedlings on a bright, breezy day. A high rate of transpiration means that the young plants lose a lot of water and may wilt.

Figure 12.2D ▶ Environmental factors affecting the rate of transpiration

Make sure you can answer an exam question which asks about the distribution of chloroplasts in the cells of the leaf. Remember that guard cells (and palisade cells, spongy cells) contain chloroplasts but that the other cells of the lower leaf surface (and upper leaf surface) do not.

It's a fact!

The chloroplasts of guard cells do not seem to produce sugar. Instead they make energy available for the accumulation or removal of potassium.

■ Factors affecting transpiration

If the loss of water through transpiration is not made up by intake of water from the soil, then the stomata close. The closing of the stomata reduces transpiration. However, if the plant still does not get enough water, then its cells begin to lose **turgor** and the plant **wilts** (see p. 149).

Other factors in the environment also affect the rate of transpiration and water intake. For example, plants transpire more quickly (and therefore absorb more water from the soil) when it is light than when it is dark. Light stimulates the stomata to open wide. This means that water vapour can transpire more easily. Figure 12.2D shows the effects of other environmental factors on the rate of transpiration. *How do the different factors have their effects?*

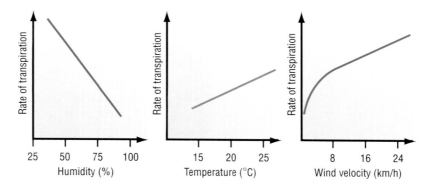

■ The size of the stoma

Control of the size of a stoma (see p. 152) depends on the two sausage-shaped guard cells which surround the opening (**pore**). Unlike the other cells of the lower leaf surface, guard cells contain chloroplasts. In the light, the concentration of potassium in the guard cells increases compared with neighbouring cells. There is a net flow of water by osmosis (see p. 149) into the guard cells. In the dark the concentration of potassium in the guard cells drops. There is a net outflow of water by osmosis.

The thickest part of the wall surrounding each guard cell is the inner part next to the stomatal pore. Thickness means that the inner part is not very 'stretchy'. As water flows in, the guard cells on either side of a stomatal pore swell and become turgid. However, the inner wall of each cell does not stretch as much as the rest of the cell wall. As a result, the guard cells curve outwards and the stomatal pore between them opens. When water flows out of the guard cells, each one becomes less turgid and therefore less curved. As a result the stomatal pore closes.

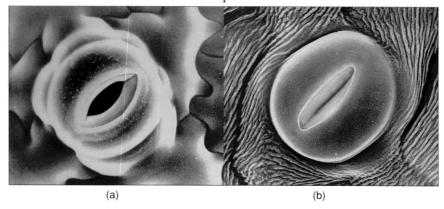

(a)   (b)

Figure 12.2E ▲ A stoma (a) open (b) closed. The pore is never completely closed. Chloroplasts are present in the guard cells but absent in the other cells of the lower leaf surface

12.3 ▶   Phloem transports food

FIRST THOUGHTS

Phloem tissue consists of living cells. The processes which transport food are different from those that transport water.

Phloem tissue consists of different types of cell. **Sieve cells** and **companion cells** are the most important. Figure 12.3A shows that sieve cells are cylindrical and joined at their end walls to form strands of tissue called sieve tubes.

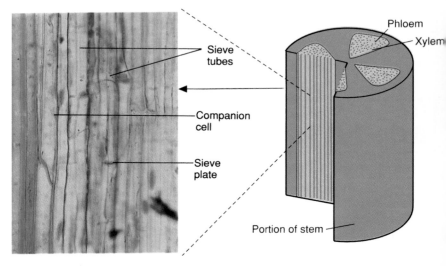

Figure 12.3A   🔺   Sieve tubes in the phloem of the stem (×20)

Mature sieve cells do not have nuclei. However, they are alive: each cell contains cytoplasm. The end walls, called **sieve plates**, are pierced with holes (hence their name). A companion cell lies next to each sieve cell. Companion cells and sieve cells work together to transport sugars to where they are needed in the plant. The movement of sugars and other substances from one region to another through the sieve cells is called **translocation**.

Figure 12.3B shows how translocation takes place. Sugar passes from the leaf cells to the sieve cells by **active transport** (see p. 151). Once in the sieve tubes, the sugar, along with other substances, is carried to where it is needed. Sugar that passes to the roots can move out of the sieve tubes and into the cells of the cortex. The sugar is converted into starch and stored.

Scientists understood 300 years ago that xylem transports water in plants. *How do we know that phloem transports food?* The evidence is more recent. Two of the experiments that provided the answer are summarised opposite and on page 194:

- **Tree ringing experiments** A ring of bark is cut from the stem of a woody plant. The strip of bark removed is just thick enough to contain phloem tissue from the vascular bundles. The xylem remains intact. After a few days there is a bulge of growth above the ring and no further growth below the ring. *Why do you think there is extra growth above the ring? Why does this result suggest that food is transported in the phloem? What will happen to the plant eventually?* (See Figure 12.3C.)

Look out for the question that asks you to distinguish between transpiration and translocation.

Phloem transports sugar up and down the stem to the growing and storage tissues of the plant.

It's a fact!

Do you know why a car parked under a lime tree is soon covered with a stickly substance? Aphids feeding on the tree sprinkle the juice which passes through them onto the car beneath.

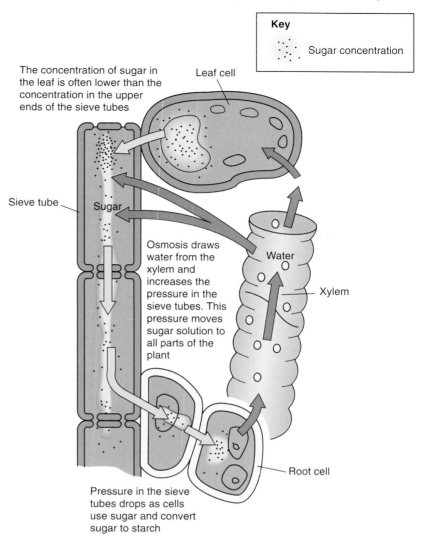

Key

Sugar concentration

The concentration of sugar in the leaf is often lower than the concentration in the upper ends of the sieve tubes

Leaf cell

Sieve tube

Sugar

Osmosis draws water from the xylem and increases the pressure in the sieve tubes. This pressure moves sugar solution to all parts of the plant

Water

Xylem

Root cell

Pressure in the sieve tubes drops as cells use sugar and convert sugar to starch

Figure 12.3B ⬆ Translocation depends on the differences in concentration of sugar in different parts of the plant. Diffusion transports sugar from where it is in high concentration to where it is in low concentration

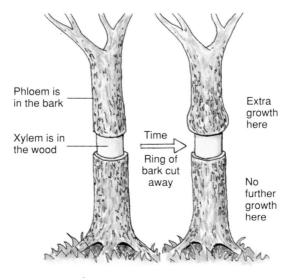

Phloem is in the bark

Xylem is in the wood

Time

Ring of bark cut away

Extra growth here

No further growth here

Figure 12.3C ⬆ Tree ringing

nber that their method
eeding makes aphids
ficient vectors (see p. 74) of
virus diseases. The insects
inject viruses into the sieve
tubes through their needle-like
mouthparts.

Aphids and radioactive tracers The aphid can be used to study the transport of sugar in the phloem sieve tubes. To feed, the aphid inserts its tube-like mouthparts (called a stylet) through the surface of a stem into a sieve tube. The pressure of the liquid in the sieve tube force-feeds the aphid to the point where drops ooze from its anus. When an aphid is cut from its stylet, liquid will continue to ooze from the stylet for several days. Radioactively labelled glucose is injected into a leaf and samples of liquid are collected regularly from a newly placed stylet. *What test would you do to find out if the liquid on which the aphid was feeding contains sugar? If sugar is transported in the phloem what change would you expect in the samples of liquid obtained from the stylet? How would you detect the change?* (See Figure 12.3D.)

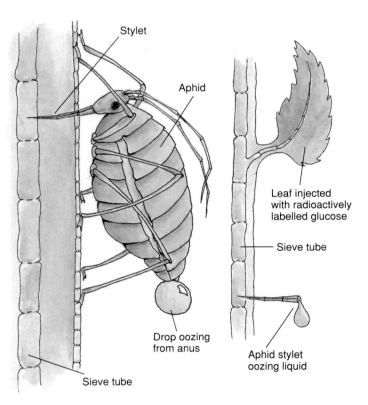

Figure 12.3D Aphids and radioactive tracers

CHECKPOINT

1 With the help of diagrams describe the differences in structure between xylem and phloem.
2 Distinguish between transpiration and translocation.
3 Describe the passage of a water molecule from the soil through a plant and out into the atmosphere.
4 Why does a plant lose more water on a dry, warm, windy day than when it is cool, still and wet?
5 By what route does sugar made in the leaves move to the roots?
6 Atmospheric pressure can support a column of water 11 metres high. How tall could a tree grow and water still reach its topmost branches?
7 Why does pressure in the sieve tubes drop as cells use sugar and convert it to starch?

Topic 13

Hormonal control of plant growth

13.1 ▶ Growth movements

'Responses' and 'stimuli' are everyday words but what do we mean when we talk about living things responding to stimuli? Think of it like this. The environment is changing all the time. Some changes are long-term, others short-term. Because these changes cause plants and animals to take action, the changes are called **stimuli**. The actions which plants and animals take are called **responses**. Being able to respond to stimuli means that living things can alter their activities according to what is going on around them. Figure 13.1A shows plants responding to different stimuli.

Make sure that you know the difference between responses and stimuli. Living things **respond** to **stimuli**.

 See www.keyscience.co.uk for more about hormones.

(a) The leaves of some species of the mimosa plant close up when touched

(b) The stems of plants respond to a source of bright light by growing towards the window

(c) The roots of plants grow to where water is in greatest abundance

Figure 13.1A ▲ Responses to stimuli

Plant growth movements

Do you keep plants in the house? If you do you may have noticed that they bend towards the window. Plants do so because light is a stimulus. Plants respond by growing towards the light (see Figure 13.1A(b)). The benefit to the plant of this response is clear: the leaves receive as much light as possible for photosynthesis. Stems twist and turn and flowers and leaves move in daily rhythms to follow the light.

Plant movements are **growth movements** in response to stimuli. There are two kinds of growth movement:

- **Nastic movements** are responses to stimuli which come from all directions. For example, temperature change is the stimulus for tulip and crocus flowers to open and close. They open when the temperature rises and close when it falls.

- **Tropic movements (tropisms)** are responses to stimuli which come from one direction. Tropisms are **positive** if the plant grows towards the stimulus; **negative if** it grows away (see Figure 13.1B).

Positive phototropism - stems grow towards light

Negative geotropism - stems grow away from the pull of gravity

Positive hydrotropism - roots grow towards water

Positive geotropism - roots grow towards the pull of gravity

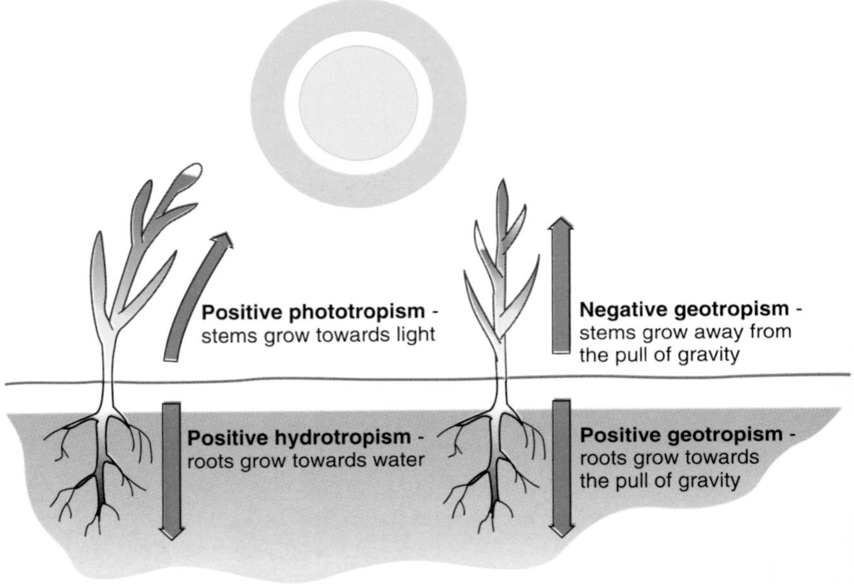

Figure 13.1B Different tropisms

It's a fact!

Does it matter which way up a seed is planted? Positive phototropism means that the shoot will always grow upwards. Positive geotropism means that the roots will always grow downwards. So it doesn't matter which way up a seed is. Its shoot and roots will always grow in the right direction.

It's a fact!

Plant growth substances are sometimes referred to as plant hormones. We shall call them growth substances to avoid confusion with animal hormones – see Topic 17.4.

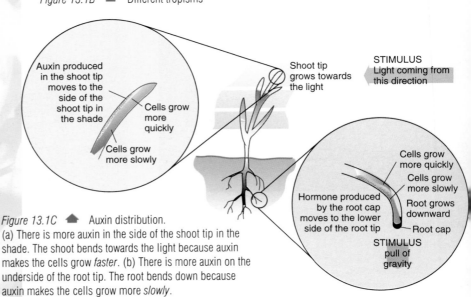

Auxin produced in the shoot tip moves to the side of the shoot tip in the shade

Cells grow more quickly

Cells grow more slowly

Shoot tip grows towards the light

STIMULUS Light coming from this direction

Cells grow more quickly

Cells grow more slowly

Root grows downward

Root cap

Hormone produced by the root cap moves to the lower side of the root tip

STIMULUS pull of gravity

Figure 13.1C Auxin distribution.
(a) There is more auxin in the side of the shoot tip in the shade. The shoot bends towards the light because auxin makes the cells grow *faster*. (b) There is more auxin on the underside of the root tip. The root bends down because auxin makes the cells grow more *slowly*.

Plant growth substances regulate cell growth and development.

■ Control of growth movements

Many factors affect the growth of plants. Important among them are compounds called **plant growth substances** (sometimes called plant hormones – see p. 300). **Auxin** was the first plant growth substance to be discovered. It is produced in shoot tips. Under the influence of auxin, it seems that the cellulose walls of plant cells become more elastic and the cells elongate rapidly. Figure 13.1C on the previous page shows a widely accepted explanation as to why the shoot top grows towards the light. Notice that a different response of cells to auxin seems to control the response of the root tip to gravity.

Experiment 1

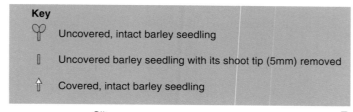

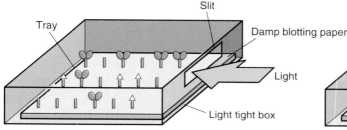

Barley seedlings about 25 mm high are grown in a light tight box with a slit at one end. Foil caps on some seedlings exclude light from the shoot tip

The seedlings with their shoot tips removed or covered by foil caps do not grow towards the light. The uncovered, intact seedlings grow towards the light

(a) Does the whole shoot or just the shoot tip respond to light?

Experiment 2

Key

♀ Intact barley seedling

Barley seedling with tip cut off. Tip placed on a slip of metal foil and replaced on the rest of the shoot. The metal foil prevents diffusion of chemicals from the shoot tip to the rest of the shoot

Barley seedling with tip removed. Tip placed on agar block and replaced on the rest of the shoot. The agar block allows diffusion of chemicals from the shoot tip to the rest of the shoot

Barley seedling with tip cut off. An agar block is placed on the remaining part of the shoot

Barley seedling with tip cut off. An agar block is soaked in a mash made up of the shoot tip and then placed on top of the remaining part of the shoot

Barley seedling with tip cut off. An agar block is soaked in a solution of auxin and then placed on top of the remaining part of the shoot

Two days later

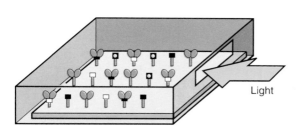

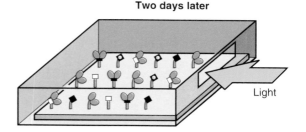

Barley seedlings grown as before

The seedlings with their tips cut off and replaced with plain agar blocks and the seedlings with their tips separated from the rest of the shoot by metal foil do not grow towards the light. The other seedlings grow towards the light

(b) Is it a chemical produced in the shoot tip or just the presence of the shoot tip itself which causes the response to light?
Figure 13.1D ◆

Two experiments designed to investigate the responses of plants to light are illustrated in Figure 13.1D on the previous page. Examine the diagrams carefully. *What conclusions do you draw from the results of the experiments? Is it the shoot tip that responds to light? Is there a chemical in the shoot tip which controls growth?*

Recent work has raised some doubts about these explanations of plant growth. In one experiment (Figure 13.1E) removal of the tips of growing seedlings caused growth to stop. However, if a barrier to auxin was placed behind the tip, so that auxin could not diffuse from the tip, growth was normal. In other words, auxin from the tip of the seedling is not necessary for growth.

As so often happens, scientific explanations are changing as new experiments provide new information. Research continues to help us try to understand how plants respond to a source of bright light.

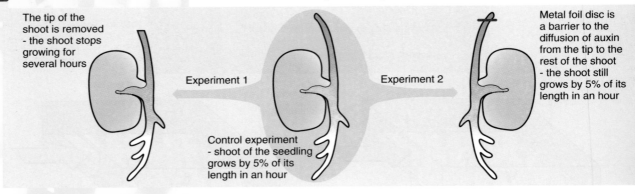

The tip of the shoot is removed - the shoot stops growing for several hours

Experiment 1

Control experiment - shoot of the seedling grows by 5% of its length in an hour

Experiment 2

Metal foil disc is a barrier to the diffusion of auxin from the tip to the rest of the shoot - the shoot still grows by 5% of its length in an hour

Figure 13.1E The results of the experiment seem to contradict the results of experiments reported in Figure 13.1D

CHECKPOINT

▶ 1 Comment critically on the evidence that shoot tips respond to light.

▶ 2 Complete the paragraph below using the words provided. Each word may be used once, more than once or not at all.

tip   phototropism   auxin   geotropism   elongate   nastic   towards   faster

The _____ of the shoot produces a growth substance called _____ which causes cells behind the tip to _____ . When the shoot is lit from one side, _____ accumulates on the shaded side. The cells on the shaded side grow _____ and the shoot bends the light. The response of the plant is called positive _____ .

13.2 ▶ Using growth substances

After the discovery of auxin the search was on for other plant growth substances. Today we know that a range of substances control plant growth. For example, growers control ripening by keeping fruit in sheds in an atmosphere that contains ethene. As little as 1 part of ethene per million parts of air is enough to speed up ripening. Figure 13.2A shows some of the uses of growth substances.

The substance 2,4-D (2,4-dichlorophenoxyethanoic acid) is a synthetic auxin. It kills plants by making them grow too fast. The plants become tall and spindly. However, 'grassy' plants, including the food crops barley, wheat and oats are not affected by 2,4-D at concentrations which destroy broad-leaved plants like docks, daisies and dandelions. Gardeners and farmers put this **selective** effect of 2,4-D to good use as Figure 13.2B shows.

The cutting on the left was placed in a dilute solution of auxin (10 mg/dm³), 14 days before the photograph was taken. At the same time, the cutting on the right was placed in water. Gardeners use rooting preparations which contain auxin to encourage root growth in cuttings. Too much auxin prevents root growth

The plants above have been treated with a substance called gibberellin. The rapid growth of the stem normally occurs only as a result of an environmental stimulus like the onset of winter frosts

EFFECTS OF PLANT HORMONES

Root, stem and fruit development are controlled by plant growth substances

(a)　　　　　　　　　(b)　　　　　　　　　(c)

Seeds as well as shoots produce auxin: (a) shows a normal strawberry, the strawberry shown in (b) remains underdeveloped after all of its seeds have been removed; (c) shows a strawberry which has had some of its seeds removed. Where seeds remain, development is normal. Market gardeners treat the unpollinated carpels (see p. 345) of some crop species with auxin paste to produce fruit without fertilisation. Examples are seedless cucumbers and tomatoes

Figure 13.2A ⬆ Using growth substances

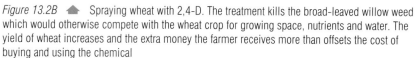

Figure 13.2B ⬆ Spraying wheat with 2,4-D. The treatment kills the broad-leaved willow weed which would otherwise compete with the wheat crop for growing space, nutrients and water. The yield of wheat increases and the extra money the farmer receives more than offsets the cost of buying and using the chemical

Always use herbicides safely by following the instructions for their application.

Substances that kill unwanted plants (weeds) are called herbicides (see Topic 5.4). Dicamba is another herbicide which behaves like auxin. It can be used in the form of a 'weed pencil' (see Figure 13.2C) to dab a spot on single deep-rooted plants such as dandelion and bindweed.

Figure 13.2C ⬆ Using a weed pencil to kill a dandelion plant growing in a lawn

SUMMARY

Plant growth substances have a variety of commercial applications. They help us to produce more food and food 'out of season'.

CHECKPOINT

▶ **1** Complete the following paragraph using the words below. Each word may be used once, more than once or not at all.

weedkiller   ripens   growth substance   unfertilised   tip   seedless   slowly

Auxin is a _____ produced in the _____ of the shoot. Synthetic auxin is used as a _____ to kill unwanted plants. Auxin paste applied to the carpels of some crop species produces _____ fruits. Fruit stored in an atmosphere containing ethene _____ more quickly.

▶ **2** What is a herbicide? Explain the meaning of 'selective herbicide'.

Theme Questions

1 The diagram shows a section through a leaf.

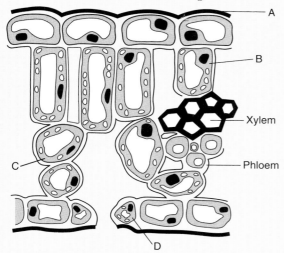

(a) Name the layer labelled **A** and the cells labelled **B**, **C** and **D**.

(b) Which cells of the leaf contain chloroplasts?

(c) Describe how water reaches a leaf and enters the cell labelled **B**.

(d) Describe how sugar produced in a palisade cell reaches the roots.

(e) The sugar produced by photosynthesis may be converted into other substances that the plant needs. Name two of the substances.

(f) Some students investigated gas exchange between living organisms and their environment. They used four pieces of water weed of equal mass, six pond snails of equal mass and equal volumes of indicator.
The table below shows how the colour of the indicator changes in different concentrations of carbon dioxide.

Colour of indicator	Concentration of carbon dioxide
Red	As in the atmosphere
Purple	Very low
Yellow	Very high

The diagram shows how the students set up four test tubes, sealed them and left them for two hours. State and explain the concentration of carbon dioxide in each test tube after two hours.

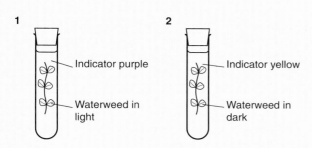

3

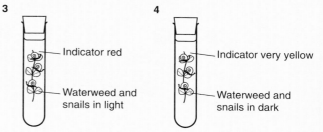

2 A scientist carried out an experiment growing cereal plants in a field. He took a sample of the plants every four hours and measured how much sugar had been made in their leaves. The sugar concentrations, expressed as a percentage of the dry mass of the leaves, are given in the following table.

Time of day	Sugar concentration
4.00 a.m.	0.45
8.00 a.m.	0.60
12 noon	1.75
4.00 p.m.	2.00
8.00 p.m.	1.45
12 midnight	0
4.00 a.m.	0.45

(a) Plot the data on graph paper, with time of day on the horizontal axis.

(b) From the graph suggest the likely sugar concentration at (i) 10 a.m. and (ii) 2 a.m.

(c) At what time of day is the sugar concentration likely to be at a maximum?

(d) Look at the graph and try to explain the variations that occur in the sugar concentration over 24 hours.

3 The data in the table represent the uptake of sulphate ions by barley plants under three different conditions. The plants were given sulphate labelled with a radioactive isotope of sulphur (^{35}S) and the amount taken up was measured using a Geiger–Muller tube (counts per minute).

Sulphate uptake (counts per min) Conditions	Time (minutes)								
	0	30	60	90	120	150	180	210	240
Aerobic	0	210	300	355	400	450	490	510	540
Anaerobic	0	100	195	205	210	230	260	280	295
The presence of a metabolic poison	0	100	170	190	200	205	215	230	250

(a) Plot **three** curves to represent the data in the table above. Label each of the curves.

(b) A scientist suggested that uptake of sulphate ions by barley plants involved the process of **active transport**. Using the information in the graph, describe the evidence that would support this suggestion. Explain why the evidence is supportive.

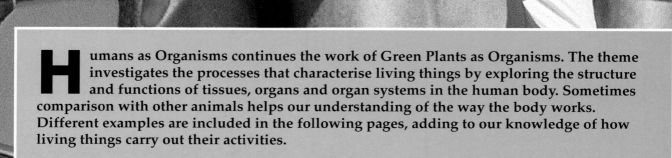

Theme E

Humans (and other animals) as organisms

Humans as Organisms continues the work of Green Plants as Organisms. The theme investigates the processes that characterise living things by exploring the structure and functions of tissues, organs and organ systems in the human body. Sometimes comparison with other animals helps our understanding of the way the body works. Different examples are included in the following pages, adding to our knowledge of how living things carry out their activities.

Topic 14   Nutrition

14.1 ▶ Food – why do we need it?

Cars need good maintenance and fuel to keep them going. Without petrol they wouldn't get very far! *Have you ever thought how we keep going?* Food is the fuel that our bodies need if we are to keep active. But food is more than just a fuel. It also provides the raw materials (**nutrients**) that we need to build up our bodies and to keep them working properly. It is vital for:

- **Energy** – it is the fuel that keeps us going.

- **Growth** and repair of cells and tissue.

- **Metabolism** – all the complex chemical reactions that take place in the cells of our bodies (see p. 3).

Food contains **water**, and food from plants contains **fibre**. However, water and fibre are not usually thought of as nutrients, although fibre is an important part of our food and water is essential for life. Adults can survive for many weeks without nutrients but only for a few days without water.

As well as nutrients, water and fibre, food also has flavour, colour and texture. Small quantities of a range of substances give food these qualities. Cooks and food technologists aim to enhance them to make food more attractive.

It's a fact!

Water makes up about two-thirds of your body weight.

Carbohydrates, fats and proteins are sources of energy which power life's activities (see p. 3).

Energy nutrients

All the processes of life need **energy**. Every time you move, think, speak, read or blink, you use energy. Your body needs energy to grow, to resist disease, and to heal itself, and for all the other processes that make it work.

Energy for living comes from the **carbohydrates, fats** and **proteins** (see Topic 10) in food. It is stored in the chemical bonds of their molecules, and released in the oxidation reactions of **cellular respiration** (see p. 243). Carbohydrates, fats and proteins are known as the **energy nutrients** and give foods their energy content.

The energy value of food is measured in the laboratory with an instrument called a **bomb calorimeter** (Figure 14.1A). Food is burnt and releases food energy as heat. The heat given off by the burning food is transferred to the surrounding water through the heat exchange coil. The change in water temperature is measured with the thermometer and used to work out the energy value of the food. *How is the heat transferred from the heated wire to the water?* Food labels usually carry information on the energy content of the food. Next time you go shopping look at the 'Nutrition information' or 'Nutrition – typical values' on a selection of food labels. The food's energy content is often called its **energy value**, and is usually given in kilojoules (kJ) per 100 g of food (Figure 14.1B).

Food burns more easily in an oxygen-rich atmosphere.

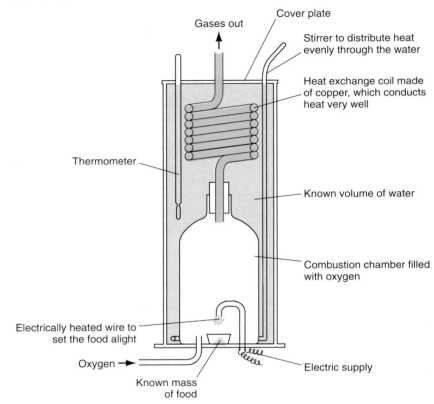

Figure 14.1A ▲ A bomb calorimeter

Calorimetry is used to measure the energy content of food.

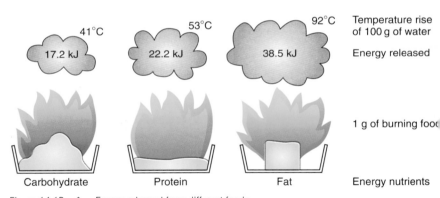

Figure 14.1B ▲ Energy released from different foods

The rate at which the person shown running in Figure 14.1D consumes oxygen is used to estimate his metabolic rate.

Our energy needs

The rate at which the body uses energy is called the **metabolic rate**. It is lowest (called the **basal metabolic rate**) when the body is at rest (sleeping). Breathing, the heartbeat, maintenance of body temperature, repair and replacement of cells and growth are some of the body activities that contribute to the basal metabolic rate. Any kind of activity increases the metabolic rate. Figure 14.1C shows the energy needed each day by people doing different things.

Measuring energy needs

The man in Figure 14.1D is running on a treadmill. The scientists are measuring the amount of energy he is using. The runner has not eaten and has rested for at least 12 hours before this test. *Why has he taken these*

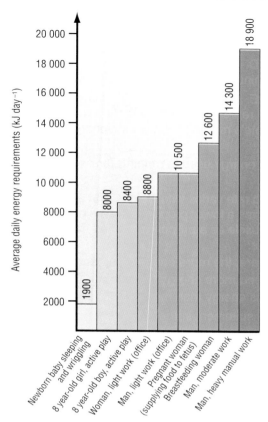

Figure 14.1C ▲ Average daily energy requirement of different people undertaking different activities

precautions? The amount of energy he uses in running is found by measuring how much oxygen he consumes. His nose is plugged so he breathes through the tubes in his mouth.

Figure 14.1E shows the balance between energy used and food energy needed. We are in 'energy balance' when the amount of energy we get from food equals the amount of energy our bodies use. Measure the height of each block on the right-hand side of the see-saw. The total height represents the total amount of energy the man in Figure 14.1D used. *How much energy did he use for running? What is the answer as a fraction of the total energy he used?*

What will happen to the runner's weight if the energy see-saw drops to the left? The right?

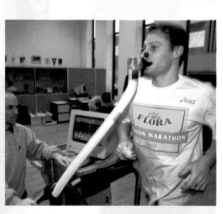

Figure 14.1D ▲ Using energy

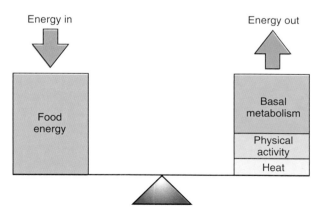

Figure 14.1E ▲ The energy balance

205

CHECKPOINT

▶ 1 Look at Figure 14.1D. The runner uses 5 litres of oxygen to provide 84 kilojoules of energy. In running 1 km in 6 minutes, he uses 25 litres of oxygen. How much energy does he use?

▶ 2 Look at Figure 14.1E. Measure the height (in mm) of each part of the right-hand block. Let the total height (in mm) represent your answer (in kilojoules) to question 1. What height (in mm) represents the athlete's running? How much energy (in kilojoules) is represented by the height of this block?

▶ 3 What fraction is the energy used in running (answer to 2) of the total energy used by the athlete (answer to 1)?

▶ 4 The basal metabolic rate varies with weight, age and sex. Women are generally lighter than men because muscle forms a lower proportion of their body weight, and they usually have a higher proportion of fat. Muscle as a proportion of body weight decreases with age. Think about the following statements:

- The heavier you are, the more energy you need.
- Young people need more energy than adults.
- Women need less energy than men.

Briefly explain each statement in the light of what you have read in this section and the background information given. Why do you think young people need more energy than adults? In what circumstances do you think the energy needs of a woman would increase sharply?

It's a fact!

Astronauts need more energy in the weightless conditions of space than when on Earth. They take in 12 600 kJ of food energy daily, yet still lose weight on space flights lasting more than two weeks. To help to prevent this, space-travellers are provided with high-energy foods. A typical meal prepared by NASA for astronauts aboard the space shuttle is cream of mushroom soup, mixed vegetables, smoked turkey and strawberries.

Remember that the body's primary use of protein is for growth and repair.

Nutrients for growth, repair and the control of metabolism

■ Protein

Although protein is an 'energy nutrient' this is not its primary role in the body. Its most important use is for **growth** and **repair**. Muscle, blood and other body tissues are made of protein.

Amino acids are the 'building blocks' of protein. Of the 20 naturally occurring amino acids, nine cannot be made by the body and must be supplied in food. They are called **essential amino acids**. The other eleven can be made by the body and are called **non-essential amino acids**.

Some proteins in food contain all of the essential amino acids the body needs for growth and repair. Other proteins lack one or two essential amino acids. Figure 14.1F shows the percentage and type of protein in different foods.

■ Minerals

Some minerals are also important for growth and repair of the body. Others control metabolism (see p. 3). We need small amounts of them in our food for good health. Minerals such as calcium sulphate and sodium chloride are present in the blood and tissue fluids.

Calcium is needed for making strong **bones** and **teeth** and for clotting blood. An adult needs about 1.1 g of calcium each day. Calcium deficiency can cause **rickets** in children. The bones are soft and bend easily (Figure 14.1G).

Iron is needed to make the blood protein **haemoglobin**. Insufficient iron in the diet causes **anaemia**. An adult needs about 16 mg of iron each day.

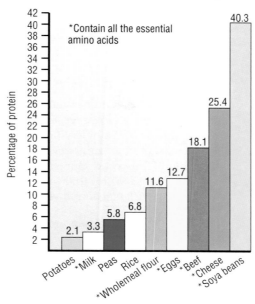

Figure 14.1F ▲ The protein content of different foods

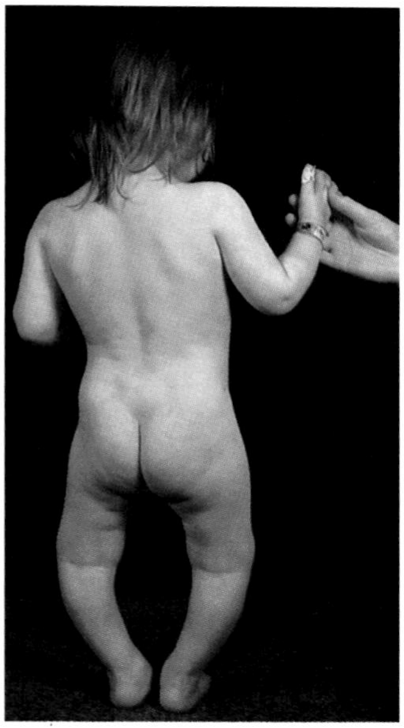

Figure 14.1G ▲ The badly bowed legs of a child with rickets

Different minerals are important for growth and repair of the body, and for the control of metabolism.

Since only a small amount of iron is needed compared with calcium, iron is called a **trace element**. Minerals that are needed in larger amounts are called **major elements**. Diseases caused by mineral deficiency are called **deficiency diseases**. Table 14.1 summarises the information on the minerals we need for good health.

Table 14.1 ▼ Summary of minerals needed by humans – *What are beri-beri and goitre?*

Mineral	Some sources	Importance in body	Deficiency disease
MAJOR ELEMENTS			
Calcium	Milk, cheese and other dairy products, bread	Making bones and teeth Blood clotting	Rickets (soft bones)
Sodium and Chlorine	Table salt, cheese, green vegetables	Keeping level and make-up of body fluids correct Transmission of nerve impulses	Cramp
Phosphorus	Most foods	Making bones and teeth Important in nucleic acids and energy release in cells	Rarely deficient
Sulphur	Dairy products, beans and peas	Part of vitamin B_1	Beri-beri
Potassium	Meat, potatoes, most fruit and green vegetables	Keeping level and make-up of body fluids correct Transmission of nerve impulses	Rarely deficient
Magnesium	Cheese, green vegetables, oats, nuts	Energy metabolism Calcium metabolism	Rarely deficient
TRACE ELEMENTS			
Iron	Liver, egg, meat, cocoa	Making haemoglobin	Anaemia
Fluorine	Water, tea, sea-food	Helps tooth enamel to resist decay	
Iodine	Fish, iodised table salt, water	Making the hormone thyroxin (which controls growth) in the thyroid gland	Goitre
Zinc	Meat, peas and beans	Protein metabolism Enzymes	Poor healing, skin complaints
Copper	Liver, peas and beans	Making haemoglobin Energy release	Rarely deficient
Cobalt	Meat, yeast, comfrey (a herb)	Part of vitamin B_{12}	Pernicious anaemia (failure to produce haemoglobin)

■ Vitamins

We also need small amounts of vitamins for good health. A few vitamins are made in the body. The rest come from food or are made by the bacteria that live in our intestines. They are organic substances, which play an important role in the control of metabolism (see p. 3).

As different vitamins were discovered they were labelled alphabetically (A, B, C, etc.). However, in some cases a substance which was first thought to be a single vitamin later turned out to be several related substances, and numbers were added to the letter label (B_1, B_2, etc.). Table 14.2 lists the sources and functions of some of the vitamins needed by humans. Notice that some vitamins are soluble in fat and some in water. *Which vitamins are more likely to be stored in the body?*

Table 14.2 ▼ Summary of vitamins needed by humans. (The B vitamins help to release energy from food and to prevent anaemia. They promote healthy skin and muscle)

Vitamin	Some sources	Importance in body	Deficiency disease
FAT-SOLUBLE			
A	Liver, fish-liver oil, milk, dairy produce, green vegetables, carrots	Gives resistance to disease Protects eyes Helps you to see in the dark	Infections Poor vision in dim light
D	Fish-liver oil, milk, butter, eggs Made by the body in sunlight	Helps the body to absorb calcium from food	Rickets in children Brittle bones in adults
E	Milk, egg yolk, wheatgerm, green vegetables	Antioxidant, protects vitamins A, C, D, K and polyunsaturated fatty acids	Poorly understood in humans Causes sterility in rats
K	Green vegetables, pig's liver, egg yolk Produced by bacteria in gut	Helps to make blood clot	Spontaneous bleeding Long clotting time
WATER-SOLUBLE			
B_1	Whole cereals, wheatgerm, yeast, milk, meat	Helps body to oxidise food to release energy	Beri-beri Nervous disorders
B_2	Fish, eggs, milk, liver, meat, yeast, green vegetables	Helps body to oxidise food to release energy	Dry skin, mouth sores, poor growth
B_6	Eggs, meat, potatoes, cabbage	Helps to digest protein	Anaemia
B_{12}	Meat, milk, yeast, comfrey (herb)	Helps in the formation of red blood cells	Pernicious anaemia
C	Oranges, lemons and other citrus fruits, green vegetables, potatoes, tomatoes	Helps to bond cells together Helps in the use of calcium by bones and teeth	Scurvy (bleeding gums, and internal organs)

EXTENSION FILE ACTIVITY

Vitamins are important for the control of metabolism.

Deficiency diseases occur when the body lacks particular vitamins. They are easily cured by supplying the missing vitamins. Many sailors died of **scurvy**, a deficiency disease, on long sea voyages in the sixteenth and seventeenth centuries (see opposite above). It causes bleeding in various parts of the body, particularly the gums (Figure 14.1H).

Vitamin D promotes absorption of calcium by the small intestine and incorporation of calcium into bones. A severe shortage of vitamin D reduces the calcium availability and may cause **rickets** in children (Figure 14.1G). Bones fail to grow properly and become soft, so when children with rickets start walking the bones bend with the weight of the body. Eating vitamin D-rich foods like fish-liver oil, butter, eggs and milk, prevents rickets. Bright sunlight also changes a chemical in the skin into vitamin D. In hot countries, where there is a lot of sunshine, people

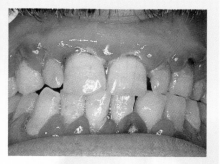

Figure 14.1H ⬆ Scurvy: notice the swollen and bleeding gums

Science at work

Deficiency diseases such as rickets are rare in countries where vitamins are added to foods like margarine.

It's a fact!

Some desert animals such as the oryx take in all the water they need from their food and never drink.

EXTENSION FILE
ACTIVITY

It's a fact!

Fog rolling in from the coast of south-west Africa is virtually the only source of water in the Namib desert. When it is foggy the 'head standing beetle' creeps to the top of a sand dune and stretches its back legs, tilting its body forward, head down. The beetle drinks as fog condenses on to its body and runs down to its mouth.

KEY SCIENTIST

Scurvy

In 1747 the naval doctor James Lind made the following report.

'On the 20th May I took twelve patients in the scurvy, on board the *Salisbury* at sea.... They all in general had putrid gums.... Two had each two oranges and one lemon given them every day.... The consequence was, that the most sudden and visible good effects were perceived from the use of the oranges and the lemons; one of those who had taken them being at the end of six days fit for duty.' *(From James Lind's account of how he treated scurvy aboard the HMS Salisbury.)*

The other ten sailors received different treatments and did not recover.

By the early 1780s a daily ration of lemon juice was a compulsory part of a sailor's rations and scurvy was no longer a problem on board ship. Look at Table 14.2 to find out which vitamin is responsible for preventing scurvy. *Why do you think sailors on long sea voyages were particularly prone to scurvy?*

make most of the vitamin D they need in this way. If they move to cooler, cloudier climates, they need extra vitamin D in their food to prevent rickets. Children under five years old need about 0.1 mg of vitamin D daily; over-fives need about 0.0025 mg daily. In adults the bones have stopped growing and lack of vitamin D causes a disease called **osteomalacia**. The bones lose calcium, become brittle and snap.

Water

Water makes up about two-thirds of your body weight. It is taken in either directly by drinking or indirectly as part of food.

Chemical reactions in the body also produce water. For example, the oxidation of carbohydrates and fats to release energy produces water.

Water is used in the body:

- as a solvent in which chemical reactions take place,
- as a solvent for waste matter which passes out of the body in solution,
- for transporting substances round the body (water is a major part of **blood** and **lymph**),
- as a means of keeping cool.

An adult needs about 2500 cm³ of water each day. The body loses and gains water through its different activities. Losses and gains roughly balance as Table 14.3 shows.

Table 14.3 ⬇ Daily water balance sheet for an adult

Daily gains (cm³ day⁻¹)		Daily losses (cm³ day⁻¹)	
Drinks	1400	Urine	1500
Food	800	Faeces	100
Cellular respiration	300	Evaporation from lungs	350
		Sweat	550
	Total 2500		Total 2500

Dietary fibre

Distinguish between the two sorts of fibre:

- Soluble fibre decreases the absorption of different substances from food e.g. cholesterol.
- Insoluble fibre promotes the passage of food through the intestine.

Dietary fibre comes from plant foods. There are two types:

- **Soluble fibre** from fruit pulp, vegetables, oat bran and dried beans.

- **Insoluble fibre** from the cellulose of plant cell walls and the bran husk that covers wheat, rice and other cereal grains (Figure 14.1I). Wholemeal bread contains much more fibre than white bread because it is made from wholemeal flour – flour made from the whole grain.

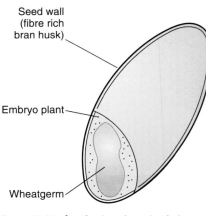

Figure 14.1I 🔺 Section of a grain of wheat

In white flour the outer husks of the grain have been removed. The wheatgerm and the bran also contain most of the vitamins.

Soluble fibre dissolves in water to produce a gel. Insoluble fibre does not dissolve, but it 'holds' water and swells up if mixed with water.

The two types of dietary fibre have different effects. Insoluble fibre adds bulk to food. The muscles of the intestine can work against it and help food to pass through quickly. As a result, disease-causing substances produced by bacteria in the intestine and in the food do not remain in the intestine for very long.

Soluble fibre has the opposite effect. It slows down the passage of food through the intestine. It also seems to decrease the absorption of some minerals, and of glucose and **cholesterol**, into the body. Many people want to lower their cholesterol level in order to reduce the risk of heart disease developing (see Topic 16.4). Perhaps this is why porridge is popular. With its high content of soluble fibre, porridge may reduce the absorption of cholesterol.

CHECKPOINT

▶ 1   Why do we need water?

▶ 2   Look at the amount of water needed daily by an adult (Table 14.3). What percentage of it is produced by 'cellular respiration'?

▶ 3   (a) What is meant by 'fat-soluble vitamin' and 'water-soluble vitamin'?
         (b) Which type is difficult to store in the body?

▶ 4   What are the sources and functions of (a) vitamin C and (b) vitamin D?

▶ 5   What can happen when we do not have enough (a) vitamin C and (b) vitamin D?

▶ 6   Which foods are sources of calcium and iron?

▶ 7   What can happen when a person does not have (a) enough calcium and (b) enough iron in his/her diet?

▶ 8   Why is fibre important in the diet?

▶ **9** Susan and Mary want to find out the energy values of sugar and butter.
Their apparatus is shown in the diagram.

Susan's apparatus

Beaker 1 — Thermometer
— Water
— Flames
— Some sugar
— Stand for food container

Mary's apparatus

Beaker 1 — Mary stirring
— 200 cm³ of water
— 5 g of sugar

Beaker 2 — Thermometer
— Water
— Flames
— Some butter
— Stand for food container

Beaker 2 — Mary stirring
— 200 cm³ of water
— 5 g of butter

Susan set the sugar and butter alight and measured the rise in temperature of the water in each beaker after three minutes. All the sugar had burnt away but the butter was still alight.
Her results were: Rise in water temperature in beaker 1 = 3 °C
Rise in water temperature in beaker 2 = 5 °C
Mary also set alight the sugar and butter but she let all the food burn away before measuring the rise in temperature of the water in each beaker.
Her results were: Rise in water temperature in beaker 1 = 6 °C
Rise in water temperature in beaker 2 = 10 °C

(a) Susan and Mary both believe that butter contains more energy than sugar. Why do you think that Mary's experimental evidence is better than Susan's? You should find four reasons.

(b) What could Mary do to improve the accuracy of her experiment?

(c) Since butter is mainly fat, Susan's and Mary's results might give them the idea that all fats contain more energy than sugar. Briefly explain how they could test this idea.

14.2 ▶   Diet and food

FIRST THOUGHTS

What is your diet? Is it healthy? This section will help you assess the food you eat and understand what makes a healthy diet.

Your **diet** is the food you eat and drink. It should contain nutrients, water and fibre in the correct amounts and proportions for good **health**. If it does, then your diet is said to be **balanced** or complete.

A healthy diet

By the time you are 70 years old you will have eaten about 30 tonnes of food. Advice about what you eat is big business. 'Experts' tell you that some foods are healthy and others are not. *Who is right? Who should you believe?*

EXTENSION FILE
ASSIGNMENT

Eating a variety of items from each of the four different food groups promotes a healthy and balanced diet.

See www.keyscience.co.uk for more about diet.

If a diet consists of a single food then the food will be 'unhealthy' no matter what the food is, because no single food contains all the nutrients in the proportions we need for healthy living. For example, both beef and wholemeal bread lack vitamins A, C and D and are low in calcium. Beef also lacks dietary fibre, which wheat provides. Wheat lacks vitamin B_{12}, which beef provides. Together beef and bread provide more nutrients than either on its own, but between them vitamins A, C and D and calcium are still missing. If salad, fruit and vegetables are added, then vitamins A and C are brought into the diet. Milk and cheese add the missing calcium and vitamin D.

A healthy diet, therefore, is a mixture of foods which together provide sufficient nutrients. Notice that each of the foods above lacks some nutrient which the other foods make up between them. They each represent one of the group of foods shown in Figure 14.2A.

Nutritionists have developed the idea of the 'basic four' **food groups** to help us choose a balanced diet. You should aim to eat at least one helping of food from each group daily.

You should also aim for variety. Daily helpings of the same food from each group may contribute to a balanced diet but not necessarily a healthy one. For example, the vitamin C content of fruits may range from next-to-nothing in raw pears to 150 mg per 100 g in stewed blackcurrants. Also, eating the same foods all the time is not only boring but may provide too much fat, sugar and salt. In excess these foods, together with too much alcohol, are a major cause of disease in developed countries.

Diet and age

Children are not miniature adults and their diet should match their needs for energy and nutrients. The needs are greatest in very young children because they are growing rapidly and have a high basal

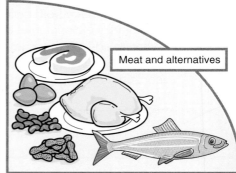

Milk and milk products

Meat and alternatives

Bread and cereals

Fruits and vegetables

Figure 14.2A ⬆ The four basic food groups

A pregnant woman needs extra nutrients for the developing fetus and placenta (see p. 367). Increases in the efficiency of her metabolism and absorption of nutrients from the intestine into the blood meet most of the extra demand. However, intake for a few nutrients is more critical:

- zinc and folate (a B-group vitamin: see p. 208) are important for growth and development
- calcium is needed for making strong bones and teeth
- iron helps make fetal haemoglobin

Adding foods and/or tablets containing these substances to the pregnant woman's diet helps to make sure that she gives birth to a healthy baby.

The diet of children, adolescents and adults meets different needs.

- Children and adolescents need a diet that meets their requirements for rapid growth and development.
- Adults need a diet that maintains good health.

metabolic rate (see p. 204). Each tissue and organ of the body has its own **critical period** for growth. For example, the critical periods for the growth of muscles and bones are during infancy and adolescence (the teenage years). An adequate supply of nutrients needed for growth during the critical periods helps avoid setbacks in development (see Topic 17.4).

Soft solid food may be added to an infant's diet of milk between the ages of four to six months. At around the age of twelve months an infant can move on to cut-up and mashed adult food. The growth rate slows over the next three-or-so years and children may be less enthusiastic about food than during their first year. However, providing a child's diet consists of different items chosen from each of the four basic food groups (Figure 14.2A) in sufficient quantities, and if he or she is growing normally and remains healthy then there is little cause for concern.

Adolescence may be a time of dietary chaos! Increased appetites, especially during growth 'spurts' may be satisfied by 'snacking' through the day. Although many of the foods may be nutritionally poor, the sheer volume eaten helps most adolescents meet their energy and nutrient needs (Figure 14.2B).

Figure 14.2B ⬆ Many fast foods fit into the basic food groups, but a fast-food diet is likely to be unbalanced – high in fat, low in fibre and deficient in minerals and vitamins (especially calcium and vitamins A and C)

Diet during childhood and adolescence affects present and future health. Insufficient intake of energy nutrients, protein, iron or zinc slows growth and results in underweight individuals vulnerable to infectious diseases. Sufficient calcium for bone formation is particularly important. During growth 'spurts' as much as 10 g of calcium per year accumulates in the skeleton. A poor diet may also lead to iron deficiency in adolescent girls who are menstruating (see p. 363).

In adults, reduced physical activity, muscle mass and metabolic rate result in decreased energy needs, but the requirements for protein, calcium and vitamins A and D may increase. During adulthood, therefore, the focus should shift from diets that promote growth to ones that maintain health (Table 14.4).

Table 14.4 ⬇ Recommendations for the adult diet

Consume less	Consume more
Alcohol	Wholegrain cereals
Sugar	Vegetables
Fat	Fruit
Salt	

Elderly people should follow the dietary recommendations for younger adults.

In addition:

- Small daily doses of vitamin D are recommended for those who spend most of their time indoors. *Why do you think the recommendation important?*

- Women may help reduce the risk of calcium loss from bones (**osteoporosis**) by maintaining a good daily intake of calcium-rich food (milk, cheese and other milk products).

- Regular intake of fatty fish (or fish oil) may reduce the risk of heart disease developing.

Many elderly people have altered their diet to take account of nutritional advice – it is never too late to make changes for the better.

CHECKPOINT

The table compares the amount of energy and some nutrients that an adult human needs each day with the energy/nutrients in milk.

Energy/ foodstuff	Daily needs of adult	Content of 100 g of cow's milk
Energy	12 000 kJ	272 kJ
Protein	72 g	3.2 g
Calcium	0.5 g	0.10 g
Vitamin A	0.75 mg	0.06 mg
Vitamin C	30 mg	1.5 mg

Study the table and answer the following questions:

▶ 1 On a diet of milk only, how much would an adult have to drink to satisfy daily energy needs?

▶ 2 Milk is a 'balanced diet' for a baby. What is meant by a 'balanced diet'?

▶ 3 Give one reason why milk is particularly important for the development of bones and teeth in babies.

▶ 4 How much milk would an adult need to satisfy his/her daily need for calcium?

**EXTENSION FILE
ASSIGNMENT**

Alcohol

Beers, wines and spirits contain **ethanol** (alcohol in everyday language). Drinking alcohol is part of many people's social life. Figure 14.2C shows the amount of alcohol consumed each week by men and women. *Which age group drinks the most alcohol each week?* Units of alcohol are shown in Figure 14.2D.

Drinking too much alcohol is one cause of diseases such as **cirrhosis** of the liver, heart disease and damage to the nervous system. The liver breaks down (metabolises) alcohol, and the link between heavy drinking and cirrhosis is well-established. Figure 14.2E shows the number of deaths from all causes and the number of deaths from cirrhosis in Paris between 1935 and 1965. Notice that deaths from cirrhosis fell by 80% when wine was rationed during World War II (1939–1945). The number of cirrhosis-related deaths rose rapidly to pre-war levels when wine rationing stopped. More recent figures from France suggest wine consumption has fallen. Estimates predict that a 50% decrease would cut cirrhosis by about 58%.

= 1 measure of whisky

= 1 glass of sherry

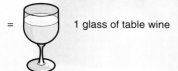

= 1 glass of table wine

= half a pint of beer

Figure 14.2D ▲ Units of alcohol

It's a fact!

One reason women are more affected by alcohol than men is because of the water content of their bodies. In men 55–65% of the body weight is water; in women it is 45–55%. Alcohol is distributed by the body fluids, so in men it is more dilute.

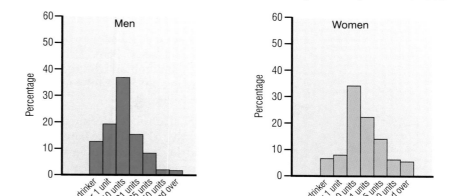

Source: Statistical Bulletin (1999)

Figure 14.2C ▲ The amount (in units, see Figure 14.2D) of alcohol drunk by men and women each week in the UK

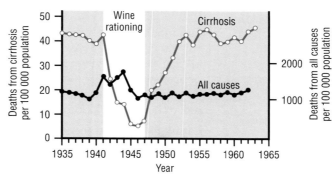

Figure 14.2E ▲ Deaths from all causes and deaths from cirrhosis in Paris, 1935–65

How much alcohol is too much? This depends on a person's sex, age, size and metabolic rate (see p. 204). For example, the 'safe' level of alcohol for a woman is only about two-thirds as much as for a man of the same weight. Also, different drinks contain different concentrations of alcohol (Figure 14.2D).

It is difficult to give 'safe' limits for drinking alcohol because the level varies so much from person to person. The Royal College of Physicians suggests that a man should not drink more than the equivalent of four pints of beer a day. Other experts think this is too much and suggest two pints of beer a day is enough. All agree that drinking alcohol affects your behaviour and heavy drinking harms your health (Figure 14.2F). One of the first things to be affected by drinking alcohol is your driving ability. The message is clear –
DO NOT DRINK AND DRIVE.

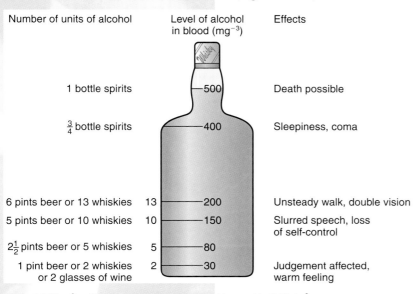

Number of units of alcohol		Level of alcohol in blood (mg⁻³)	Effects
1 bottle spirits		500	Death possible
$\frac{3}{4}$ bottle spirits		400	Sleepiness, coma
6 pints beer or 13 whiskies	13	200	Unsteady walk, double vision
5 pints beer or 10 whiskies	10	150	Slurred speech, loss of self-control
$2\frac{1}{2}$ pints beer or 5 whiskies	5	80	
1 pint beer or 2 whiskies or 2 glasses of wine	2	30	Judgement affected, warm feeling

Figure 14.2F ▲ The effects of alcohol: one pint is equal to 0.58 dm³

■ Alcohol and pregnancy

Exchange of substances between mother and fetus occurs across the placenta (see p. 366). Alcohol passes across the placenta very easily. A pregnant woman who regularly drinks beer, wine and/or spirits increases the risk of the fetus developing abnormally. Fetal growth is also reduced.

Fetal alcohol syndrome may occur if the pregnant woman drinks heavily. At birth, the baby is smaller than average, may be mentally retarded and may suffer heart problems. Also the facial bones are incorrectly developed giving an abnormal look to the face.

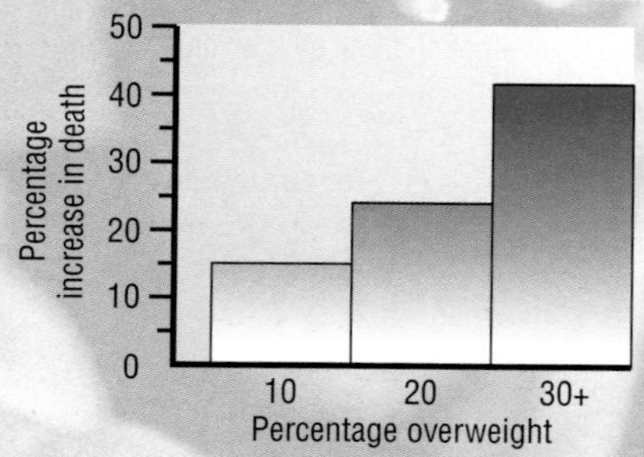

Sugar and fat

The amounts of sugar and fat we eat also affect our risk of developing heart disease (p. 279). If people eat too much sugar and fat they tend to put on weight. Overweight people have a higher risk of heart disease (Figure 14.2G).

Too much of the wrong sort of fatty food also increases the level of a substance called **cholesterol** in the blood. Cholesterol is found in nearly all body tissues. Large amounts of cholesterol are found in **atheroma** deposits (see p. 278), which block arteries and restrict the flow of blood through them. The more cholesterol there is in the blood, the greater the risk of heart attack (Figure 14.2H).

Eating food containing a lot of **saturated fats** (see p. 165) seems to raise the level of cholesterol in the blood and therefore increase the risk of heart attack.

Look at Figures 14.2I, 14.2J and 14.2K. We need some fat in our diet for good health. It comes from different foods. Different fats contain different proportions of saturated and unsaturated fatty acids. *Which foods are high in saturated fats? You should eat less of these. Which foods can be substituted because they contain more unsaturated fats? What are the trends in consumption of saturated and unsaturated fats?*

The P/S ratio, which is the ratio of polyunsaturated fats to saturated fats in the diet, helps us calculate what proportion of saturated and unsaturated fats to eat. It can be worked out as follows

$$\frac{P}{S} = \frac{\text{Mass of polyunsaturated fat in diet}}{\text{Mass of saturated fat in diet}}$$

The national average for P/S is around 0.25. However a P/S value of 0.45 is recommended to reduce the risk of heart disease. In other words for every gram of fat eaten, 0.45 g should be polyunsaturated.

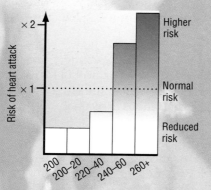

Figure 14.2G ◆ Increase in deaths from heart disease due to overweight

Figure 14.2H ◆ Cholesterol and the risk of heart disease

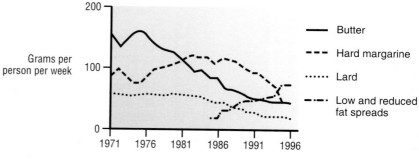

Figure 14.2I ◆ Consumption of saturated and unsaturated fat. Butter, hard margarine and lard contain a high proportion of saturated fat. Low and reduced fat spreads contain less saturated fat

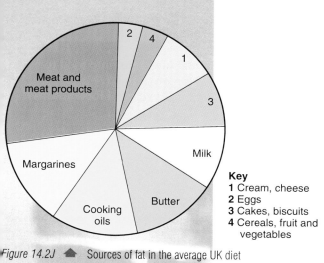

Figure 14.2J ▲ Sources of fat in the average UK diet

Key
1 Cream, cheese
2 Eggs
3 Cakes, biscuits
4 Cereals, fruit and vegetables

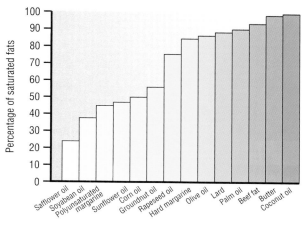

Figure 14.2K ▲ Saturated fat as a proportion of total fat in different foods

So far the evidence seems straightforward: the higher your cholesterol level the greater your risk of developing heart disease. In fact the story is more complicated than this and more research is needed. However, it is much better that you act now to improve your diet than do nothing while waiting for more information.

Salt

When we talk about salt in food we mean **sodium chloride** (see *Key Science Chemistry* p. 72).

The muscles, nervous system and kidneys need salt to work properly. Salt also helps to maintain the correct **osmotic** (see Topic 9.2) balance between blood and tissues.

Too much salt in the diet can raise **blood pressure** (see p. 276) and a person with blood pressure higher than normal is more likely to suffer from heart disease (see Figure 14.2L).

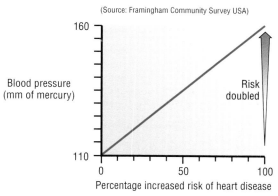

Figure 14.2L ▲ Blood pressure and the risk of heart disease

High blood pressure can also damage the kidneys and eyes, and increase the risk of an artery tearing open. When an artery tears which supplies blood to the brain it is called a **cerebral haemorrhage** or 'stroke'.

Very few young people have high blood pressure, but after the age of 35 it becomes more common. The reason is not clear, but different studies show that lifestyle can have an important effect. Eating less salt helps to keep blood pressure within normal levels.

It's a fact!

Your body loses salt when you sweat, which is why sweat tastes salty. People who work in hot places sweat a lot and may suffer from muscle cramps because of the salt they lose. Taking salt tablets helps replace the salt lost.

Food additives

Food additives are substances that manufacturers put into food to make it tastier, improve its texture, make it look more attractive and prevent it from spoiling. Table 14.5 summarises the main types of food additive and what they do.

Table 14.5 ▼ Types of food additive and what they do

Types of food additive	What they do
Preservatives	Stop microorganisms such as bacteria and fungi from spoiling food
Antioxidants	Prevent oxygen in the atmosphere oxidising fats and oils, turning them rancid
Emulsifiers	Keep oil and water in sauces mixed together
Drying agents	Prevent foods like flour from caking
Thickeners	Make foods like soups less 'runny'
Supplements	Nutrients added to help prevent deficiency diseases, e.g. vitamins A and D added to margarine
Colourings	Enhance natural colours to make foods look more attractive
Flavourings	Make foods tastier by bringing out their flavours
Moisteners (humectants)	Prevent food from drying out

When you look at the labels on food containers for information about energy values, you may notice the letter E printed with a number after it. This is an 'E Number', given to additives that are recognised as safe for use in food by the European Community. Each additive has its own E Number. Some examples are given in Table 14.6.

Table 14.6 ▼ The identity of some common E numbers

E Number	Additive	E Number	Additive
	Antioxidants		**Emulsifiers**
E300	l-Ascorbic acid	E400	Alginic acid
E320	Butylated hydroxyanisole (BHA)	E406	Agar
	Colours		**Preservatives**
E102	Tartrazine	E210	Benzoic acid
E110	Sunset yellow FCF	E220	Sulphur dioxide
E120	Cochineal		**Sweeteners**
E162	Beetroot red (betanin)	E421	Mannitol
		E420	Sorbitol

Food additives are carefully tested to make sure they are safe to eat before they are given an E Number (see Figure 14.2M). However some approved additives can make some people ill. For example, some people are sensitive to the yellow colouring tartrazine (E102). If they eat food containing E102 they may have an asthmatic attack or develop a skin rash. More examples of the side-effects of some additives are listed in Table 14.7.

Doubts about the long-term effects of additives have led to some of them being withdrawn from use, even though they have been declared 'safe'. You must remember that additives help to stop food from spoiling and make it more convenient to use. However, public worries have led food manufacturers to cut down on their use, which in turn may create new

It's a fact!

Hygienic food handling helps prevent food poisoning. People working in the food industry (processors, packers, shop workers) must make sure that their hands and the equipment they use are clean and not contaminated with harmful bacteria. The precautions should extend to the home kitchen! In food shops and the home, cooked meat should be kept separately from raw meat; cooked vegetables from fresh vegetables. Food should be protected from flies which spread disease (see Topic 4.3).

Figure 14.2M ⬆ The ingredients of a vegetable soup mix. How many have side-effects?

What are the reasons for the recent rapid increase in the reported cases of food poisoning? Is it:

- an increased frequency of reporting
- an increased use of convenience foods
- an increased consumption of 'fast' foods
- because more people are 'eating out'
- due to a decrease in the use of food preservatives because of public fears about their side-effects?

Perhaps the cause is a combination of reasons. *What do you think?*

Table 14.7 ▼ Additives with side-effects

E Number		Effects on some people
E102	Tartrazine	Skin rashes, blurred vision, breathing problems, hyperactivity in children
E122	Carmoisine or Azorubine	Skin rashes, swellings
E150	Caramel	None proven although suspect for many years
E220	Sulphur dioxide	Irritation of gut
E320*	Butylated hydroxyanisole	Raises lipid and cholesterol levels in blood
E321*	Butylated hydroxytoluene	Skin rashes, behavioural effects, blood cell changes

*Not allowed in baby foods

problems. For example, bacteria and fungi multiply much more quickly in food without preservatives (in everyday language, the food goes 'bad'). Eating 'bad' food causes food poisoning.

Look at Figure 14.2N. The rapid increase in the number of cases of food poisoning since 1980 coincides with the growing public pressure to cut down additives in food. *Do you think the two are linked?* (Scientists call this 'cause and effect'.) *Can you interpret the data in a different way?* For example, there has been an increase in the consumption of ready-made meals and takeaways. Also people go out for meals more often than they used to. *Do you think these factors influence the data in the graph?* Briefly summarise your opinion on food additives in the light of the evidence. **Remember** that the number of cases of food poisoning is probably an underestimate. One study in 1999 claims that cases of a particular type of poisoning are more than seven times higher than reported.

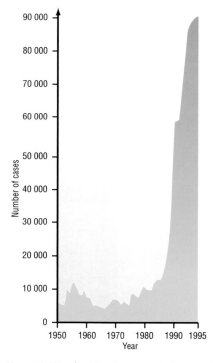

Figure 14.2N ▲ The rise in reported cases of food poisoning

CHECKPOINT

- ▶ **1** What is meant by the word 'preservative'?
- ▶ **2** Why is a diet rich in plant fat probably more healthy than one rich in animal fat?
- ▶ **3** Why do people who work in very hot countries sometimes take salt tablets?
- ▶ **4** List some diseases caused by drinking too much alcohol.
- ▶ **5** How many glasses of table wine contain the same amount of alcohol as $1\frac{1}{2}$ pints of beer?
- ▶ **6** List some good effects and some bad effects of food additives.

Vegetarianism – fad or healthy alternative?

Vegetarians do not eat meat: some because they think it cruel to kill animals for food; others for religious and cultural reasons. A growing number of people in Britain claim to be vegetarian, but what does it mean to be one? Table 14.8 compares the different types of vegetarian diet and the diets of carnivores and omnivores.

Table 14.8 ▼ Different types of diet. Lacto refers to milk, ovo to eggs

Type of diet	Foods eaten				
	Beef and other 'red' meats	Pork, poultry, fish, seafood	Eggs	Milk, cheese and other milk products	Vegetables, fruits, grains, and grain products, legumes, nuts, seeds, oils, sugars
Omnivore	✓	✓	✓	✓	✓
Semivegetarian		✓*	✓	✓	✓
Lacto-ovovegetarian			✓	✓	✓
Lactovegetarian				✓	✓
Vegan					✓
Carnivore	✓	✓			

*May not include all types of food in this group

Many semivegetarians say that they 'feel better' for not eating 'red' meat, so practical reasons govern their choice of diet. Vegans choose their diet because of a principle, believing it not only cruel to kill animals for food but also to use them to produce dairy products.

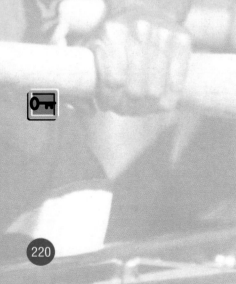

It's a fact!

Since the mid-1990s, fear of contracting the human form of 'mad cow disease' (BSE) from eating the meat of infected cattle has reduced the consumption of beef in the UK and other member countries of the EU.

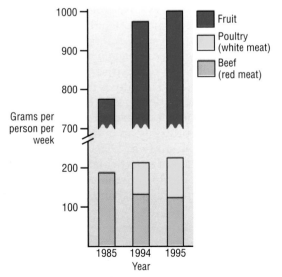

Source: National Food Survey

Figure 14.20 ▲ Changes in food consumption on England and Wales

Look at Figure 14.20 and notice how consumption of red meat has declined and consumption of white meat and fruit has increased. People are more aware that food affects health and are changing their diets accordingly. *Do you think these changes are for the better?* Give reasons for your answer.

If foods are selected carefully then it is possible to obtain a complete range of nutrients from a vegetarian diet, especially if eggs, milk and cheese are included. In fact, properly balanced vegetarian diets have advantages over some non-vegetarian diets. They contain:

- more dietary fibre,
- less saturated fat and cholesterol,
- less high-energy food.

Vegetarians are less likely to be overweight than people who regularly eat meat. They are also less likely to develop heart disease, diabetes and some types of cancer. *Why should this be so?*

CHECKPOINT

▸ 1   Which nutrient found in meat could be in short supply in a vegetarian diet?

▸ 2   How do eggs, milk and cheese help vegetarians to obtain a complete range of nutrients?

▸ 3   Why do vegans have more difficulty than other groups in obtaining all the nutrients they need?

EXTENSION FILE
ASSIGNMENT

People's 'ideal' weight depends on their age, gender, height and size of skeleton.

Thin or fat – what are the facts?

Magazines, newspapers, television and advertisements bombard us with body images. The message is: thin is beautiful, fat is ugly. This has not always been the case. In the late sixteenth century a much plumper body image was fashionable. Comparison between the past and the present shows how fashion trends shape our image of thinness and fatness.

Life insurance companies have calculated healthy weights for people of different heights (Figure 14.2P). Overweight people are more likely to be ill (Figure 14.2G) and are therefore a greater risk to insure.

Notice that Figure 14.2P shows a range of weights for each height. The range allows for differences in the size of the skeleton which forms the body frame. A small-framed person will tend to weigh in at the lower limit of the weight range for his/her particular height; a large-framed person of the same height will tend to be at the upper limit of the weight range.

Sports people and others with a muscular physique may weigh in as overweight for their height. Muscle is more dense than fat, so although the graph seems to be saying that they are overweight, they probably have less body fat than people of the same weight who do not take regular exercise.

Weight-for-height figures for children are more difficult to calculate than for adults. As you grow up the proportions of water, muscle and fat in your body alter as much as the lengths of your arms, legs and trunk. However, childhood **obesity** is a growing problem in the UK and other developed countries. As well as the increased risk to health, obese children may become targets for other children's teasing and ridicule. Although some obesity has medical causes, lack of exercise seems to be a major factor. In the USA there is a decline in levels of physical activity among children and obese children tend to be much less physically active than children of normal weight. Watching too much television could be part of the problem (Figure 14.2Q). Next time you watch television, count how many advertisements you see in a given time and how many are for food. *What percentage of advertisements are for food? What types of foods are being advertised?*

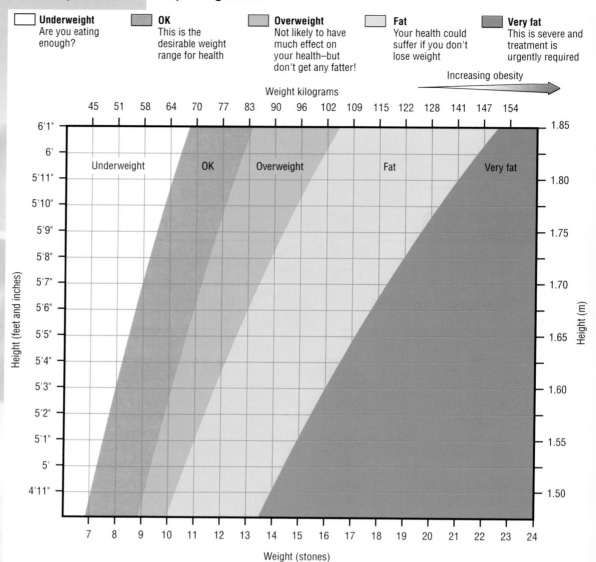

Figure 14.2P 🔺 Healthy weights

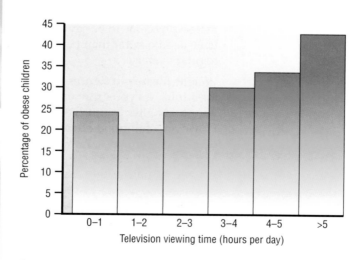

Figure 14.2Q 🔺 The relationship between obesity in children and the time spent watching television

Slimming

Congratulations, Natalie Whitelaw, for slimming from 115 kg to 70 kg in less than a year – a loss of 45 kg (Figure 14.2R). How did you do it?

If a person eats more food than is necessary for his or her energy needs, then the excess is turned into fat and stored in fat cells under the skin and he or she puts on weight (see Figure 14.1E)

The only way to lose weight and slim is to make sure that the energy input (food) is less than the energy output (metabolism and physical activity). The options are:

● take more exercise, which increases energy output,

● eat less high-energy food, which decreases energy input.

It's a fact!

A weight loss of 1.00 kg per week requires a reduction in food energy intake of 32 300 J over that time (4600 J day^{-1}).

See www.keyscience.co.uk for more about malnutrition.

Figure 14.2R ▲ Slimming (before and after)

The first option is not very effective on its own. For example, a man trying to lose weight uses about 420 kJ hour^{-1} more taking a brisk walk than sitting down. If the walk makes him thirsty and he drinks a pint of beer at the end of it, he will take in more energy than he used up. The result: his weight increases.

Fortunately the second option is effective if carried out properly. The sensible approach to slimming includes:

● eating smaller amounts of food,

● eating fewer high-energy foods (Figure 14.2S),

● more physical activity.

To be effective a slimming diet should allow the person to maintain the loss of weight long-term.

A programme for slimming designed with these points in mind not only results in weight loss but also helps to maintain it. The aim is to alter gradually a person's exercise and eating habits. It is much easier to adjust to modest changes – smaller amounts of food, using stairs instead of lifts, for example – than to make sudden drastic changes. Weight is reduced quickly at the start of a weight control diet. Most of the popular diets promoted by the multi-million pound 'slimming' industry owe their success to sudden weight loss. However, few people stick to a diet that demands major upheavals in their eating habits and the weight soon goes back on.

Figure 14.2S ▲ Many low-energy foods have a high fibre content so eating lots of bread, fruit and vegetables can help you slim. Substituting artificial sweeteners for sugar in tea and coffee and low-fat cottage cheese for full-fat cheese could also help you lose weight

Anorexia

Some people take slimming too far, by following a strict low-energy diet to lose weight. They lose their appetite, eat little food and become dangerously thin. This disease is called **anorexia nervosa**. It is most common in teenage girls and young women from middle- to high-income families.

The effects of anorexia nervosa:

- Muscle tissue is used as a source of energy once the body's fat reserves are exhausted.
- Body temperature, metabolism and heart rate decrease.
- Depression sets in.
- Growth and sexual development in teenagers stop.
- Thoughts become dominated by food and eating.

People with anorexia nervosa do not recognise that they are in effect starving themselves (see Figure 14.2T). They often have a low opinion of themselves. Treatment focuses on building up their self-image. If these problems are overcome, then normal eating patterns and weight gain often follow. Many people recover from anorexia nervosa, and the earlier it is discovered, the more likely it is that treatment will be successful.

> ### It's a fact!
>
> Rapid weight loss at the start of a weight control diet is mainly due to losses in body water. After the first week further weight loss is from the body's fat stores.

Anorexia is an eating disorder which in effect starves the individual. **Bulimia** is another eating disorder characterised by 'bingeing' followed by vomiting the food consumed.

Figure 14.2T ▶
The effects of anorexia nervosa can be devastating

CHECKPOINT

▶ **1** The table shows the percentage of British people in different age groups who are overweight.

Age group	Percentage overweight	
	Men	Women
20–24	22	23
25–29	29	20
30–39	40	25
40–49	52	38
50–59	49	47
60–65	54	50

(from: Report of Royal College of Physicians,)

(a) Draw two bar charts to show the percentage of men and women overweight on the vertical axis and age group on the horizontal axis.

(b) How many times more overweight men are there in the 60–65 age group than in the 25–29 age group? Suggest reasons for the increase in weight.

(c) Why do you think that more men than women in all age groups except the first are overweight?

(d) Plan two menus – one that could lead to people becoming overweight, and one that overweight people could use to reach normal weight. Describe briefly the differences between the menus which bring out the gain and loss in weight. Do both of your menus give a balanced diet?

(e) 'Overweight people are more likely to suffer from ill health than normal weight people.' Briefly give reasons why you think this statement is true or false.

▶ **2** 'You are what you eat.' Discuss the meaning of this popular saying by comparing a meal of hamburgers, egg and chips and steamed jam pudding with a meal of fish, salad and fruit.

14.3 ▶ Keeping food fresh

FIRST THOUGHTS

Have you ever felt ill soon after eating a meal? If so, you may have experienced food poisoning. This section tells you how keeping food fresh helps to prevent the build up of microorganisms which cause food poisoning and other diseases.

Keeping food fresh means keeping food free from the moulds and bacteria which make food 'bad' and cause **food poisoning** and other diseases.

Fungi and bacteria are nature's 'refuse collectors'. They cause the decay and decomposition which clears away dead organic matter (see Topics 1.5 and 2.6). Their activities release chemical elements from dead organisms into the environment making the elements available for absorption and the growth of new plants. Animals obtain the elements essential for their growth, development and healthy living through feeding on plants and/or other animals.

The problem

Moulds and bacteria decompose food, releasing substances which cause food poisoning.

Food contains salts, which are inorganic; also food is dead organic matter. Fungi and bacteria can come into contact with our food (Figure 14.3A). Some of them cause diseases. They begin to decompose the food (Figure 14.3B) and release substances which make us unwell.

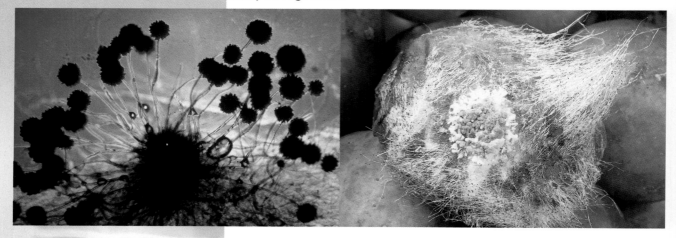

Figure 14.3A 🔺 Moulds are a type of fluffy fungus – the black shiny capsules contain thousands of spores which are wafted away on air currents when the capsules burst. If the spores settle on food they grow into new moulds

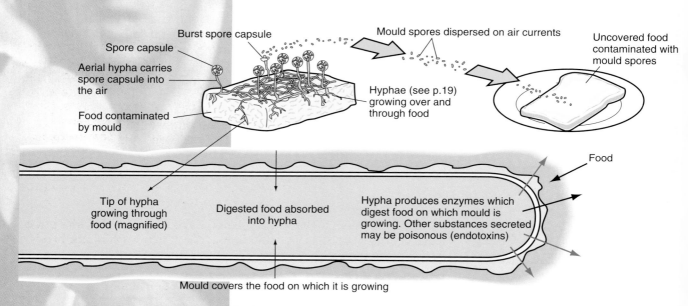

Spore capsule

Burst spore capsule

Mould spores dispersed on air currents

Uncovered food contaminated with mould spores

Aerial hypha carries spore capsule into the air

Hyphae (see p.19) growing over and through food

Food contaminated by mould

Food

Tip of hypha growing through food (magnified)

Digested food absorbed into hypha

Hypha produces enzymes which digest food on which mould is growing. Other substances secreted may be poisonous (endotoxins)

Mould covers the food on which it is growing

Figure 14.3B 🔺 Mould covers the food on which it is growing

Rearing chickens intensively provides ideal conditions for the spread of *Salmonella* bacteria which cause food poisoning.

Case study: *Salmonella* food poisoning

Currently most chickens are kept intensively (see Figure 14.3C) although a recent change in the law means that chickens in the future will be able to live more 'natural' lives (see p. 88). The crowding in today's poultry houses provides an ideal environment for spreading disease-causing bacteria from bird to bird. The bacterium *Salmonella enteriditis* is found in the gut of most chickens without causing ill effects. However, after slaughter *Salmonella* bacteria from the gut and soiled skin contaminate the carcasses before they are sent to shops for sale. *Salmonella*-infected chickens almost certainly means *Salmonella*-contaminated eggs.

Have you seen the Lion Quality logo stamped on egg shells (see Figure 14.3D)? The logo is a symbol for rearing chickens and producing eggs according

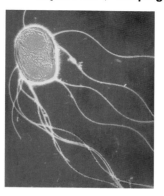

Figure 14.3C ◀ (a) Intensively reared chickens – their droppings contain *Salmonella* bacteria   (b) *Salmonella enteriditis* is a rod-shaped bacterium

Figure 14.3D ◀ Eggs stamped with the *Lion Quality* logo

Chicken must be cooked thoroughly to at least 68 °C to kill *Salmonella* bacteria.

It's a fact!

Different bacteria can cause food poisoning, including *Escherichia coli* (see p. 15). *Clostridium botulinum* is a particular danger to health. It produces a poison called **botulin** which may cause paralysis and even death.

 See www.keyscience.co.uk for more about food preservation.

to a code of practice that reduces the risk of infecting consumers with *Salmonella*. Currently about 74% of eggs sold in the UK are *Lion Quality*. However, the risk of *Salmonella* infection still remains, despite development of the *Lion Quality Code of Practice* and the changes for keeping hens described in Topic 5.1.

Someone who eats food contaminated with *Salmonella* soon develops the symptoms of food poisoning. The bacteria invade cells lining the small intestine. They multiply and produce a poison (**endotoxin**) which inflames the tissue and causes fever and acute pain. Symptoms also include vomiting and diarrhoea which results in an enormous loss of water. The body quickly dehydrates and the victim feels tired and unwell. Young children and elderly people are particularly vulnerable.

Cooking chicken thoroughly to at least 68°C kills *Salmonella* bacteria and helps ensure it is 'safe' to eat. Extra care is needed when cooking chicken from frozen. The bird should be completely thawed, otherwise the centre of the carcass may not reach the 'safe' temperature.

Particular attention to personal hygiene is also important. Handling raw chicken and then preparing other food without first washing the hands increases the danger of contamination and *Salmonella* food poisoning. Improving the standards of hygiene at farms by 'mucking out' poultry houses more frequently also helps to control the disease.

Most victims of *Salmonella* food poisoning recover within a week. Antibiotics are not very helpful since an insufficient amount of the drug gets into the gut cells where the bacterium is causing the problem. Treatment aims to replace water, glucose and salts lost through the severe diarrhoea (see Topic 4.3).

Preserving food

Preventing food-borne diseases means keeping food fit for people to eat. 'Fresh' in this sense means preserving food in different ways.

- **Sterilisation** kills bacteria. Food is heated to a high temperature and then sealed in cans or other air-tight containers. Food is preserved for a long time but its flavour is affected.

- **Pasteurisation** of milk and cheese is a partial sterilisation. **Flash** heating milk to 72 °C for 15 seconds kills most bacteria but does not affect flavour. Some bacteria survive, so pasteurised milk should be kept in the refrigerator below 5 °C.

EXTENSION FILE ASSIGNMENT

EXTENSION FILE ASSIGNMENT

SUMMARY

Bacteria and fungi contaminate food, spoiling it. Some bacteria and fungi make us unwell. Different methods of preserving food either kill the microorganisms or make them inactive.

- **Refrigeration** below 5 °C stops bacteria from reproducing and slows their other activities. However, when the food warms up, the bacteria begin to reproduce once more and decomposition sets in. The food quickly spoils.

- **Freezing** between −18 °C and −24 °C stops all bacterial activity. As with refrigeration, bacteria become active again as the food thaws.

- **Drying** is used to preserve vegetables, fruits and some meats. Bacteria deprived of water cannot reproduce. Preserving food by leaving it in the sun to dry is a method that has been in use for thousands of years.

- **Freeze drying** is used to preserve foods such as custard powder, coffee and soups. The food is first frozen and the ice is then drawn off in a vacuum. The dried food is stored in sealed containers.

- **Ohmic heating** cooks and sterilises food by passing an electric current through it. The food quickly heats up because of its resistance to the current.

- **Chemical preservatives** are substances which either stop the growth of bacteria or kill them. Important preservatives are sulphur dioxide, nitrites and nitrates. Each one is given an 'E' number (see Topic 14.2).

- **Pickling** food in **vinegar** produces an acid environment which prevents bacterial growth. *What is the acid in vinegar?* Pickling in **brine** (a concentrated solution of sodium chloride) draws water from the food by osmosis (see Topic 9.2). Bacteria are prevented from growing on the food. Also they lose water by osmosis and are killed. Pickled food has a distinctive taste.

- **Jam making** preserves food in a concentrated sugar solution. Again the food and bacteria lose water through osmosis. Prompt storage of treated food in clean, sterilised containers is necessary because some moulds can grow on jam.

- **Smoking** food over burning wood or peat deposits a thin coating of nitrites and nitrates on the food's surface. This kills bacteria and mould.

- **Irradiation** exposes the food to γ-**radiation** from a radioactive source. This kills bacteria and moulds and so prevents food spoilage. However, any poisons released by the microorganisms in the food before irradiation are not affected by the treatment. Irradiation does not affect enzymes in the food so that the ripening and texture of fruit, for example, are not affected. Food must not be put on sale for twenty-four hours after irradiation. The amount of radiation used is tightly controlled.

CHECKPOINT

▶ 1 What is the link between fungi and bacteria causing decomposition and food poisoning?

▶ 2 Why does rearing chickens intensively increase the risk of *Salmonella* food poisoning?

▶ 3 List the different methods used to preserve food. Discuss three of the methods in more detail.

14.4 ▶ Teeth

When animals take in (**ingest**) food we say they are feeding. Different animals have different structures for feeding (see Topic 1.5). **Teeth** are the feeding structures in most vertebrates (except birds).

Tooth structure

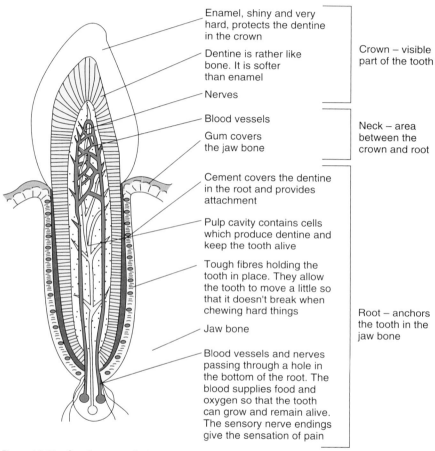

Enamel, shiny and very hard, protects the dentine in the crown

Dentine is rather like bone. It is softer than enamel

Nerves

Blood vessels

Gum covers the jaw bone

Cement covers the dentine in the root and provides attachment

Pulp cavity contains cells which produce dentine and keep the tooth alive

Tough fibres holding the tooth in place. They allow the tooth to move a little so that it doesn't break when chewing hard things

Jaw bone

Blood vessels and nerves passing through a hole in the bottom of the root. The blood supplies food and oxygen so that the tooth can grow and remain alive. The sensory nerve endings give the sensation of pain

Crown – visible part of the tooth

Neck – area between the crown and root

Root – anchors the tooth in the jaw bone

Figure 14.4A ▲ Structure of a human tooth

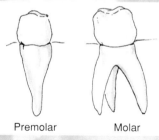

Premolar   Molar

Broad surface made uneven by 'bumps' called cusps for crushing and grinding. A molar has three 'prongs' to its root; some premolars have two

Figure 14.4A shows the internal structure of a human tooth. Notice that blood vessels and nerves enter the pulp cavity of the tooth through a hole at the bottom of the root. In carnivores and omnivores (see Topic 2.4) this hole becomes smaller when the tooth is fully grown, reducing the supply of blood. The tooth is then said to have a **closed root**; dentine is no longer produced, and the tooth stops growing. The roots of herbivores' teeth stay **open**, so that the supply of blood is enough for dentine production to continue. Their teeth keep growing throughout life. These animals feed on rough material like grass. Continuous growth of the teeth makes sure that the abrasive food does not wear them away.

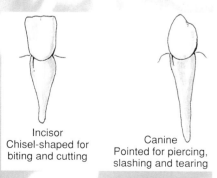

Incisor
Chisel-shaped for biting and cutting

Canine
Pointed for piercing, slashing and tearing

Figure 14.4B ▲ The four basic types of human teeth

Type of teeth

The teeth of fish, amphibia and reptiles are usually cone-shaped. The teeth of mammals are different shapes and sizes (Figure 14.4B).

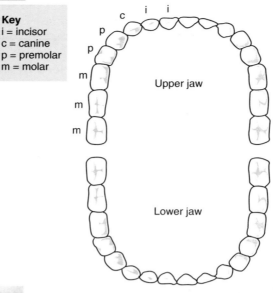

Key
i = incisor
c = canine
p = premolar
m = molar

Upper jaw

Lower jaw

Figure 14.4C 🔺 The arrangement of teeth in an adult human jaw. There are 32 teeth in total; eight on each side of the upper and lower jaw

EXTENSION FILE
ACTIVITY

'Milk' teeth are said to be deciduous because they fall out and are replaced by the permanent teeth between the ages of six and twelve.

It's a fact!

What do teeth have in common with hair, nails, claws and feathers? They all contain keratin. See Topic 10.3.

The different types of teeth are positioned in the mouth according to their functions. Figure 14.4C shows the arrangement of teeth in an adult human jaw.

Humans have two sets of teeth. The first set of 'milk' (or **deciduous**) teeth form in the jaw before birth and begin to appear (or **erupt**) about three to six months after birth. There are 24 teeth in the first set, 20 of which are gradually replaced by the permanent teeth between the ages of six and twelve. The third molars (the 'wisdom' teeth) do not appear until the age of about 18 years, if at all.

The word **dentition** is used to describe the number and arrangement of teeth in an animal. Humans and other omnivores have all four basic types of teeth to deal with a mixed diet of plants and meat. The dentition of adult humans is described in a **dental formula** using the key letters in Figure 14.4C.

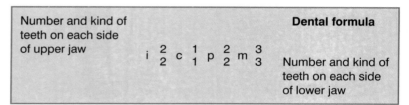

Number and kind of teeth on each side of upper jaw	Dental formula
	$i\ \frac{2}{2}\ c\ \frac{1}{1}\ p\ \frac{2}{2}\ m\ \frac{3}{3}$ Number and kind of teeth on each side of lower jaw

Herbivores and carnivores have different dentition, to enable them to deal with their particular diets.

■ Herbivore dentition

Sheep and cattle are herbivores. They eat tough plants and grasses. Instead of having incisors in the front of the upper jaw, sheep and cattle have a tough, horny pad which the incisors of the lower jaw bite against. There are no canines in the upper jaw and the canines in the lower jaw look like incisors. There is a space in both jaws in front of the premolars. This space is called the **diastema** (Figure 14.4D). The animal can push its long, muscular tongue through the diastema to sweep grasses into its mouth, where they are cut off by the action of the lower incisors against the pad in the upper jaw (Figure 14.4E).

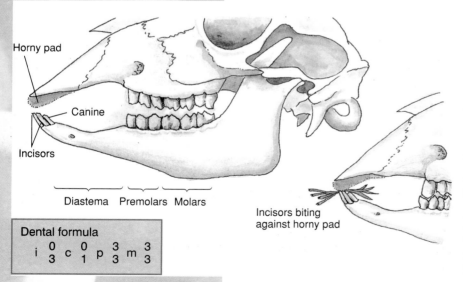

Horny pad

Canine

Incisors

Diastema   Premolars   Molars

Incisors biting against horny pad

Dental formula
$i\ \frac{0}{3}\ c\ \frac{0}{1}\ p\ \frac{3}{3}\ m\ \frac{3}{3}$

Figure 14.4D 🔺 The dentition of a sheep and its dental formula

Figure 14.4E ▲ The cow uses its tongue to sweep hay (dried grass) into its mouth

Make sure that you can distinguish between the dentition of herbivores and the dentition of carnivores.

Figure 14.4G ▲ Carnivore dentition: notice the long canine teeth. The carnassials are visible on either side of the upper jaw

The premolars and molars have layers of cement, enamel and dentine which wear away at different rates. This causes **ridges** of enamel to form, which make a good surface for grinding the food (Figure 14.4F). The joint of the jaw and skull can move from side to side as well as up and down, which also makes grinding food easier.

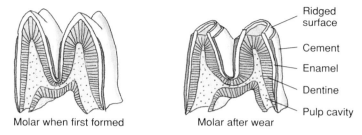

Figure 14.4F ▲ The crown of a sheep's tooth showing how a herbivore's teeth wear into ridges

■ Carnivore dentition

Carnivores' teeth are adapted for catching struggling prey and cutting through soft flesh and hard bones (Figure 14.4G).

Dogs and cats are carnivores. Their canines are long and well-developed for grasping and tearing, and they have powerful jaw muscles. Their incisors, premolars and molars are used for cutting. The last premolar on each side of the upper jaw and the first molar on each side of the lower jaw are very large and are called the **carnassial teeth**. They are especially suitable for cutting through flesh and bone (Figure 14.4H). The jaw joint only allows up and down movements.

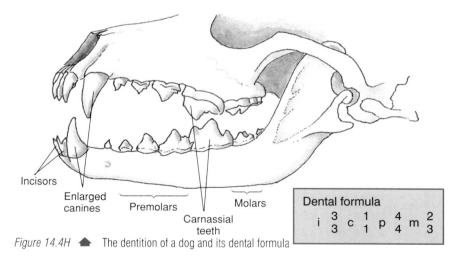

Dental formula

$$i\ \frac{3}{3}\ c\ \frac{1}{1}\ p\ \frac{4}{4}\ m\ \frac{2}{3}$$

Figure 14.4H ▲ The dentition of a dog and its dental formula

CHECKPOINT

▶ 1 Explain the words carnivore, herbivore and omnivore.

▶ 2 Look at Figure 14.4A. Name the innermost part of the tooth? What does it contain?

▶ 3 (a) Name the four basic types of teeth.
   (b) Describe the functions of two of the basic types of teeth in humans.

▶ 4 (a) How are the teeth of a sheep adapted for grinding food?
   (b) How are the teeth of a dog adapted for grasping and cutting food?
   (c) Compare Figures 14.4D and 14.4H and list the differences between sheep's teeth (herbivore) and dogs' teeth (carnivore).

It's a fact!

Even after you have cleaned your teeth, bacteria soon collect on them and multiply. They form **plaque** on the rough surfaces of teeth, in the areas next to the gums and in between teeth. Plaque is almost invisible. It accumulates and hardens, forming **tartar**.

Looking after your teeth

With proper care, a set of human teeth can last a lifetime. However, it is a sad fact that in the UK a child of twelve years has on average eight decayed teeth. What can we do to improve this?

Evidence shows that sugary foods in particular are bad for teeth. Figure 14.4I shows what happens. Reducing the amount of sugary food we eat helps to prevent tooth decay and gum disease. Look at Figure 14.4J. *Do you think cutting down on the amount of sweets you eat will help protect your teeth*? Give reasons for your answer.

Cleaning your teeth properly and regularly – at least after breakfast and last thing at night – is also very important. A disclosing tablet contains a dye, which colours plaque and shows you where extra cleaning is needed (Figure 14.4K).

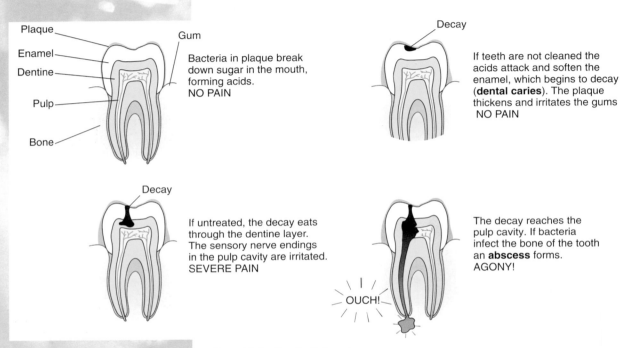

Bacteria in plaque break down sugar in the mouth, forming acids.
NO PAIN

If teeth are not cleaned the acids attack and soften the enamel, which begins to decay (**dental caries**). The plaque thickens and irritates the gums
NO PAIN

If untreated, the decay eats through the dentine layer. The sensory nerve endings in the pulp cavity are irritated.
SEVERE PAIN

The decay reaches the pulp cavity. If bacteria infect the bone of the tooth an **abscess** forms.
AGONY!

Figure 14.4I 🔺 Tooth decay in progress

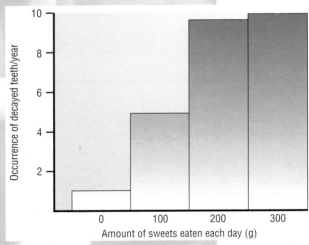

Figure 14.4J 🔺 Tooth decay in children

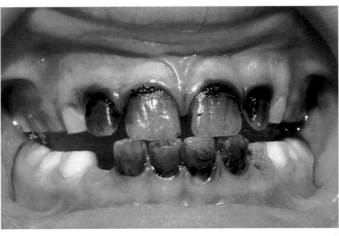

Figure 14.4K 🔺 Dental plaque revealed using a disclosing tablet which stains plaque red

Prevent tooth decay by:

- reducing the consumption of sugary foods,
- regularly cleaning your teeth,
- visiting the dentist every six months for a check-up.

You need a good toothbrush. It should have a fairly small head, dense bristles and a firm handle. You also need to use it correctly (see Figure 14.4L). Your dentist can give you more advice on cleaning your teeth. You will need a new toothbrush about every three months. After brushing, you can use a soft thread called **dental floss** to clean between your teeth where the brush cannot reach (Figure 14.4M).

Having your teeth checked every six months by a dentist helps to make sure that your teeth and gums stay healthy. Dentists like to concentrate on preventing tooth decay and gum disease.

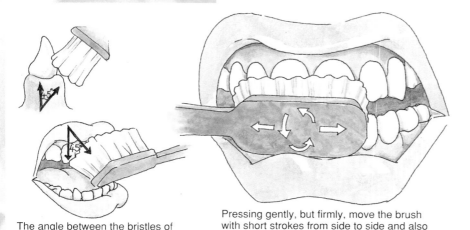

The angle between the bristles of the toothbrush and the neck of the tooth should be about 45°

Pressing gently, but firmly, move the brush with short strokes from side to side and also with an up–down circular movement to help clean in the spaces

Work round the mouth on only a few teeth at a time, brushing all the front, back and biting surfaces

Figure 14.4L ▲ One way of cleaning teeth properly with a toothbrush

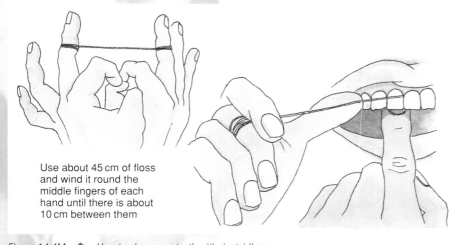

Use about 45 cm of floss and wind it round the middle fingers of each hand until there is about 10 cm between them

Starting with the upper teeth, use your fingers to guide the floss gently between two teeth until it reaches the gum line. Slowly slide it up and down the sides of both teeth. Repeat between all teeth, unwinding the floss from one hand to the other occasionally, for a fresh length

Figure 14.4M ▲ How to clean your teeth with dental floss

EXTENSION FILE
ASSIGNMENT

■ Fluorides for healthy teeth

Why do dentists recommend that people use toothpastes containing the salt calcium fluoride? Tooth enamel reacts with calcium fluoride to form a harder enamel which is better at resisting attack by mouth acids. To make sure that everyone gets protection from tooth decay, many water companies add a small amount of calcium fluoride to drinking water. The concentration of fluoride ions must not rise above 1 p.p.m. of fluoride ions (parts per million parts of water – see p. 98). Spending pennies per person each year on fluoridation saves the National Health Service pounds per person each year in dentistry. Some people are opposed to the fluoridation of water supplies. This is because they are worried about the effects of drinking too much fluoride.

KEY SCIENTIST

The fluoride story started in 1901 when Fred McKay began work as a dentist in Colorado, USA. Many of his patients had mottled teeth or dark brown stains on their teeth. He could find no information about this condition and called it 'Colorado Brown Stain'. He became convinced that the stain was caused by something in the water. After the town of Oakley in Idaho changed its source of drinking water, the children there no longer developed mottled or stained teeth. The drinking water in a number of towns was analysed and showed that teeth were affected when there were over 2 p.p.m (parts per million) of fluoride in the drinking water. It affected children's teeth more than adults' teeth. McKay also noticed that people with 'Colorado Brown Stain' had less dental decay than other patients!

CHECKPOINT

▶ 1 (a) Make a list of sugary foods. Tick the four that you eat most often.
      (b) How does eating sugary foods cause tooth decay and gum disease?

▶ 2 (a) How often should you visit the dentist?
      (b) What is dental floss?

▶ 3 Chlorine is added to drinking water to kill bacteria that would otherwise cause diseases like cholera and typhoid. Are you in favour of treating drinking water with chlorine? Give reasons for your answer.

▶ 4 In 1945, a British dentist called Weaver inspected the teeth of children from North Shields and South Shields. These two towns are on opposite banks of the River Tyne. Tables 1 and 2 show the results (rounded off) of Weaver's inspection of 1000 children aged 5 and 1000 children aged 12. The figures give the number of DMF (decayed, missing or filled) teeth per 1000 teeth. The numbers of teeth in each position in both upper and lower jaws and on both left and right sides have been added together. At the time, the water supplies contained:
North Shields 0.25 p.p.m fluoride, South Shields 1.40 p.p.m fluoride.

Table 1 ▼ Survey of 1000 children aged 5 years

Position of tooth	Number of DMF teeth per 1000 teeth	
	North Shields	South Shields
1	265	155
2	200	100
3	135	55
4	675	440
5	725	485

Table 2 ▼ Survey of 1000 children aged 12 years

Position of tooth	Number of DMF teeth per 1000 teeth	
	North Shields	South Shields
1	45	20
2	60	20
3	15	10
4	70	25
5	75	30
6	725	490
7	160	75

(a) On graph paper, draw a bar graph of the number of DMF teeth in each position for 5 year-olds in (i) North Shields and (ii) South Shields. Shade the bars for the two towns differently.

(b) Draw similar bar graphs for 12 year-olds in (i) North Shields and (ii) South Shields.

(c) What do you observe from your bar graphs about the teeth of children in North Shields compared with South Shields (i) at 5 years and (ii) at 12 years?

(d) What is the fundamental difference between the teeth of a 5 year-old and those of a 12 year-old?

(e) How did the fluoride content of the water in North Shields compare with that in South Shields?

(f) What can you deduce from these figures about the effect of fluoride in drinking water on the health of children's teeth?

(g) Remember 'Colorado Brown Stain'? What is the maximum safe fluoride level?

14.5 ▶ Digestion and absorption

FIRST THOUGHTS

How are nutrients in the food you eat converted into substances your body can absorb and use? The answer is – by digestion. This section tells you all about it.

Carbohydrates, lipids and **proteins** are the main components of food (see Topics 10.1, 10.2, 10.3). They are complex *insoluble* molecules which the body cannot absorb directly. To be useful they must be broken down into their *soluble*, basic constituents; substances which the body can absorb.

Nutrient		Absorbed as . . .
Carbohydrate	→	Simple sugars
Protein	→	Amino acids
Fat	→	Fatty acids and glycerol

The gut is a muscular tube through which food moves. It is where food is broken down. The process is called **digestion**.

Digestive enzymes

Digestion uses nearly one hundred different **enzymes** (see Topic 10.3) to speed up its chemical reactions. Without them the components of the food you eat would be broken down so slowly that you would starve to death.

Digestive enzymes catalyse the breakdown of food by **hydrolysis** (see p. 169). Water splits molecules of food components into smaller molecules.

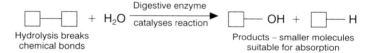

The absence of even one enzyme can cause illness. For example, people who do not produce the enzyme **lactase** cannot digest the sugar lactose ('milk' sugar). They develop cramps and diarrhoea if they consume lactose in milk and milk products, because the sugar builds up in the gut.

Different digestive enzymes are listed in Table 14.9. They are grouped according to the reactions they catalyse. For example, carbohydrate-digesting enzymes are grouped as carbohydrases. Table 14.9 identifies where different digestive enzymes are found in the gut. Look at Figure 14.5A and find their location.

See www.keyscience.co.uk for more about digestion.

EXTENSION FILE
ACTIVITY

14.9 ▼ Enzymes that digest carbohydrates, proteins and lipids

...me group	Example	Where found	Food component digested	Products of digestion
...phydrases (...yse the digestion of carbohydrates)	Amylase Maltase	Mouth Small intestine	Starch Maltose	Maltose Glucose
...ases (...yse the digestion of proteins)	Pepsin Chymotrypsin and dipeptidase	Stomach Small intestine	Protein Polypeptides and dipeptides	Polypeptides Dipeptides and amino acids
...es (...yse the digestion of fats and oils)	Lipase	Small intestine	Fat	Fatty acids and glycerol

The gut is a tube through which food is processed. The sequence reads:

Ingestion
↓
Digestion
↓
Absorption
↓
Egestion

The gut and how it works

Figure 14.5A shows the human gut and its position in the body. At one end food is put into the mouth (**ingested**). At the other end the undigested remains of a meal are removed through the anus (**egested**). In between mouth and anus the **mechanical** and **chemical** processes of **digestion** break down food into substances suitable for **absorption**. These processes are explained in Figure 14.5A. *Why do you think the action of rennin is particularly important in babies? How do the actions of teeth and bile make it easier for enzymes to digest food?*

■ Chemical processes

■ Mechanical processes

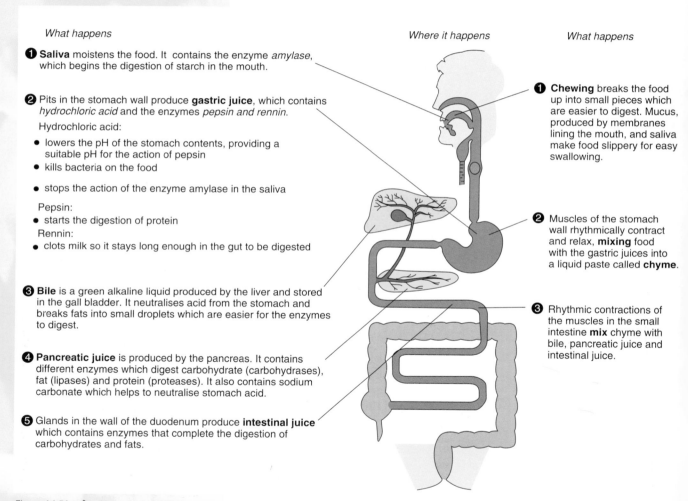

What happens

❶ **Saliva** moistens the food. It contains the enzyme *amylase*, which begins the digestion of starch in the mouth.

❷ Pits in the stomach wall produce **gastric juice**, which contains *hydrochloric acid* and the enzymes *pepsin and rennin*.

Hydrochloric acid:
- lowers the pH of the stomach contents, providing a suitable pH for the action of pepsin
- kills bacteria on the food
- stops the action of the enzyme amylase in the saliva

Pepsin:
- starts the digestion of protein

Rennin:
- clots milk so it stays long enough in the gut to be digested

❸ **Bile** is a green alkaline liquid produced by the liver and stored in the gall bladder. It neutralises acid from the stomach and breaks fats into small droplets which are easier for the enzymes to digest.

❹ **Pancreatic juice** is produced by the pancreas. It contains different enzymes which digest carbohydrate (carbohydrases), fat (lipases) and protein (proteases). It also contains sodium carbonate which helps to neutralise stomach acid.

❺ Glands in the wall of the duodenum produce **intestinal juice** which contains enzymes that complete the digestion of carbohydrates and fats.

Where it happens

What happens

❶ **Chewing** breaks the food up into small pieces which are easier to digest. Mucus, produced by membranes lining the mouth, and saliva make food slippery for easy swallowing.

❷ Muscles of the stomach wall rhythmically contract and relax, **mixing** food with the gastric juices into a liquid paste called **chyme**.

❸ Rhythmic contractions of the muscles in the small intestine **mix** chyme with bile, pancreatic juice and intestinal juice.

Figure 14.5A ⬆ How we digest food

The words gut, intestine and alimentary canal all refer to the same structure – the tube which extends from the mouth to the anus and which processes the food we eat.

Different parts of the gut perform different tasks in processing food as it passes through. The human gut is 7–9 m long. The longest part of it is composed of the **small intestine** and the **large intestine**. These lie folded and packed in the space of the **abdominal cavity**. The **liver** and **pancreas** are connected by ducts to the gut and play an important part in the digestion of food. They also have a major role in the metabolism of food substances once these have been absorbed into the body.

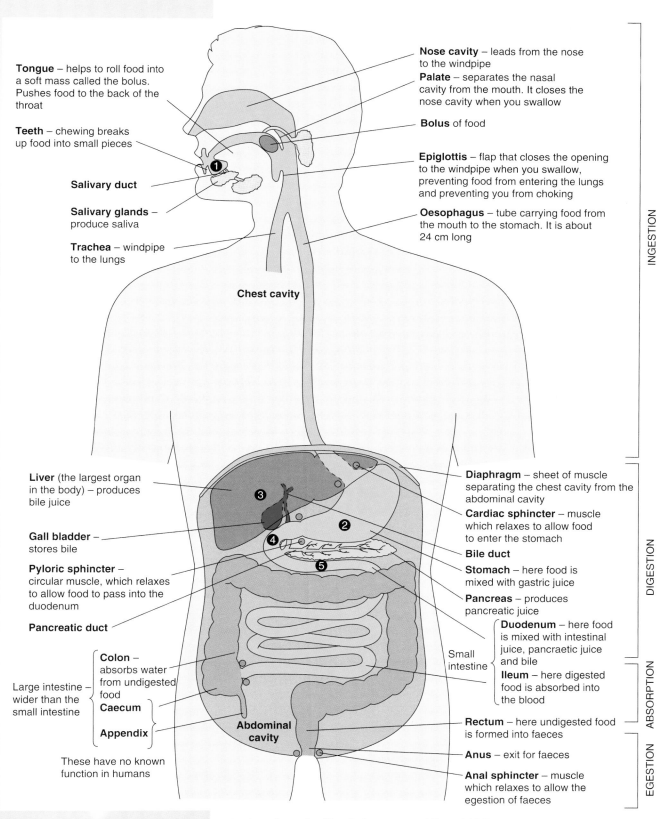

Tongue – helps to roll food into a soft mass called the bolus. Pushes food to the back of the throat

Teeth – chewing breaks up food into small pieces

Salivary duct

Salivary glands – produce saliva

Trachea – windpipe to the lungs

Chest cavity

Nose cavity – leads from the nose to the windpipe

Palate – separates the nasal cavity from the mouth. It closes the nose cavity when you swallow

Bolus of food

Epiglottis – flap that closes the opening to the windpipe when you swallow, preventing food from entering the lungs and preventing you from choking

Oesophagus – tube carrying food from the mouth to the stomach. It is about 24 cm long

INGESTION

Liver (the largest organ in the body) – produces bile juice

Gall bladder – stores bile

Pyloric sphincter – circular muscle, which relaxes to allow food to pass into the duodenum

Pancreatic duct

Large intestine – wider than the small intestine

Colon – absorbs water from undigested food

Caecum

Appendix

These have no known function in humans

Abdominal cavity

Diaphragm – sheet of muscle separating the chest cavity from the abdominal cavity

Cardiac sphincter – muscle which relaxes to allow food to enter the stomach

Bile duct

Stomach – here food is mixed with gastric juice

Pancreas – produces pancreatic juice

Duodenum – here food is mixed with intestinal juice, pancraetic juice and bile

Small intestine

Ileum – here digested food is absorbed into the blood

Rectum – here undigested food is formed into faeces

Anus – exit for faeces

Anal sphincter – muscle which relaxes to allow the egestion of faeces

DIGESTION

ABSORPTION

EGESTION

Figure 14.5B ◆ The human gut (the numbers **1** to **5** refer to the section **Chemical processes** of Figure 14.5A)

It's a fact!

The **amino acid pool** refers to the concentration of amino acids in the blood. It remains constant because of regulation by the liver (an example of homeostasis – see p. 306). Regulation balances:

- *losses* from the pool through deamination (see p. 308) and the incorporation of amino acids in the synthesis of proteins
- *additions* of amino acids from food and the breakdown of body tissue.

The liver in particular is the chemical-processing factory of the body.

It stores:

- glycogen which can be hydrolysed (see p. 236) to glucose and released back into the blood in response to the body's needs
- iron compounds from destroyed red blood cells.

It converts:

- excess of amino acids into urea through the process of **deamination** (see p. 308). Urea is excreted in urine
- amino acids from one type into another (a process called **transamination**) in response to the body's needs. Non-essential amino acids (see p. 168) are made as a result of transamination.

It makes:

- bile
- plasma proteins.

Liver metabolism (see p. 311) releases a lot of heat energy which is distributed all over the body by the blood. Humans (and other mammals and birds) have a high metabolic rate which releases a large amount of heat. This is why we (and other mammals and birds) are **warm-blooded**.

Assimilation follows the absorption of digested food. It describes the processes in cells which convert digested food materials into living matter.

Absorption of digested food

Figure 14.5C summarises what has happened to the food so far. *Which nutrients are not digested? Is fat digested in the stomach?*

The next stage – absorption – takes place mostly in the **ileum**, although alcohol and small amounts of simple sugars and water are absorbed by the lining of the **stomach**. Water is also absorbed by the **colon**.

Figure 14.5B shows that folding and coiling packs as much gut as possible into the restricted space of the **abdominal cavity**. The extra length means that food not only travels greater distances through the gut so that enzymes have more time to digest the food, but also that the surface area for absorption is increased. Closely packed finger-like projections called **villi** line the small intestine and increase its surface area further (Figure 14.5D).

Food at start	In mouth	In stomach	In small intestine	After digestion
Protein				Amino acids
Carbohydrate				Simple sugars
Fats				Fatty acids and glycerol
Minerals				Minerals
Vitamins				Vitamins

Figure 14.5C ⏶ The progress of digestion. Solid colours represent the nutrients in food before digestion

KEY SCIENTIST

In 1822 Alexis St Martin, a Canadian fur trapper, was wounded in his left side by a shotgun blast. He was nursed back to health by an army surgeon, Dr William Beaumont. When his wound healed a small hole remained between the stomach and the outside. Beaumont took samples of food from his stomach through this hole. He looked after St Martin for more than two years and carried out experiments on the role of gastric juice in digestion. His work greatly increased our understanding of how the stomach works.

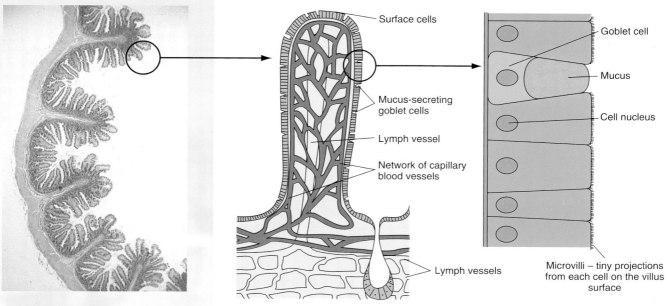

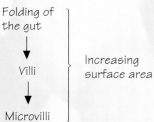

Figure 14.5D 🔺 Section of the small intestine. Five million villi, each about 1 mm long, cover the inside of the small intestine. They provide approximately 10 m² of surface for absorption in your gut!

Figure 14.5E 🔺 Inside a villus (part of its surface is shown at high magnification)

The gut is adapted for the absorption of food. The sequence reads:

Folding of the gut
↓
Villi — Increasing surface area
↓
Microvilli

Figure 14.5E shows inside a villus. Absorption occurs when nutrients pass through its cells and into the blood and lymph vessels. Most fatty acids and fat-soluble vitamins (see Table 14.2) pass into the lymph vessels. Water, glucose, glycerol, amino acids and other substances pass into the network of capillary blood vessels and are transported to the **liver**. Mucus helps to protect the gut wall from its own digestive enzymes. **Microvilli** project from the surface of each cell lining the villus. They increase the surface area for absorption about 20 times more compared with the villi alone! *Altogether, how large a surface area do villi and microvilli make for absorption?* (See Figure 14.5D.)

Undigested food passes out of the small intestine into the colon. Here, water poured on to the food during digestion is absorbed back into the body (Figure 14.5F). The food remnants dry out into a compact mass of **faeces** which is removed from the body through the anus – a process called **defaecation**.

Constipation occurs when the body has difficulty in removing faeces through the anus. Food containing a lot of insoluble fibre (see p. 210) helps prevent constipation.

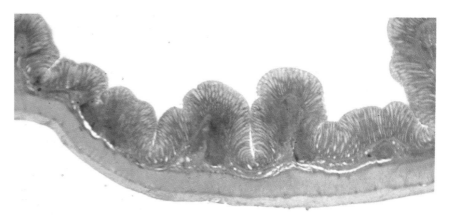

Figure 14.5F 🔺 Section of large intestine. Compare with the section of small intestine shown in Figure 14.5D. *What differences can you see?*

Moving food through the gut

Repeated contraction and relaxation of **muscle** layers in the gut wall move food through the gut. This muscular action is called **peristalsis** (Figure 14.5G).

Peristalsis and segmentation are muscular contractions of the gut wall which move food through the gut.

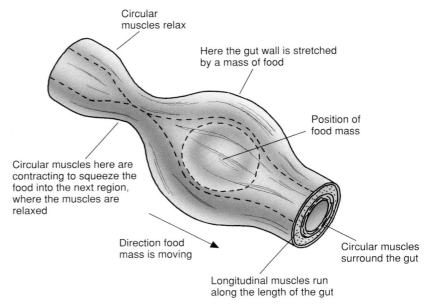

Circular muscles relax

Here the gut wall is stretched by a mass of food

Position of food mass

Circular muscles here are contracting to squeeze the food into the next region, where the muscles are relaxed

Direction food mass is moving

Circular muscles surround the gut

Longitudinal muscles run along the length of the gut

Figure 14.5G ⬆ Peristalsis – the gut narrows when the circular muscles contract, and shortens when the longitudinal muscles contract

Another type of gut movement happens when the circular muscles alone contract. This muscular action is called **segmentation** (Figure 14.5H). Together peristalsis and segmentation move a meal through the gut in about 24 hours.

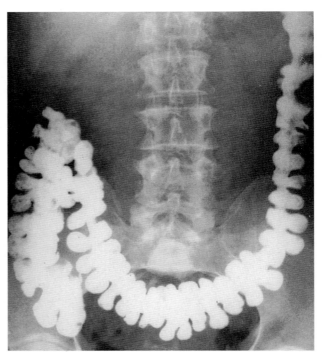

Figure 14.5H ⬆ Segmentation. The X-ray picture shows that contraction of the circular muscle temporarily divides the gut into a series of segments

The gut suits the diet

Different guts are adapted to deal with different types of diet. Most humans, for example, are omnivores and the human gut is adapted to deal with a mixture of plant and animal foods. Rabbits and cows are herbivores, with guts adapted for plant foods, and cats are carnivores, with guts adapted for meat. Figure 14.5I compares them. List the differences between them. *How is each gut adapted to deal with a particular diet?*

Most animals do not possess enzymes to digest the cellulose in the walls of plant cells. However, some species of **bacteria** and **protists** (Topic 1.5) do. These microorganisms live in different parts of the gut of an animal and digest the cellulose in food. In rabbits, for example, the **caecum** and **appendix** are the home of these microorganisms. In sheep and cows they live in the stomach, which consists of four parts. The **rumen** and **reticulum** receive the food first of all. The cellulose-digesting microorganisms live in the rumen. Here the food forms into balls of **cud** before return to the mouth where they are thoroughly re-chewed (animals that chew cud are called **ruminants**). Swallowing takes the cud to the remaining two parts of the stomach, where protein digestion takes place.

Rabbits do not chew cud. Instead they eat the **pellets** of faeces produced from the food's first journey through the gut. The pellets are soft and contain a lot of undigested food. As a result the cellulose-digesting microorganisms have another chance of dealing with the pellets as they pass through the gut a second time. Faecal pellets produced after the food's second journey through the gut are hard.

Cats do not chew their food – they swallow it whole or in large chunks. The stomach is large so that it can store the meal while it is digested. Cats (and other carnivores) do not need cellulose-digesting bacteria. As a result the carnivore gut does not have special parts to house them.

Most animals do not produce enzymes which digest the cellulose in plant food. Different types of microorganism do produce enzymes which digest cellulose. The herbivore gut is a home for cellulose-digesting microorganisms. The relationship between cellulose-digesting microorganisms and herbivores is an example of mutualism (see p. 74).

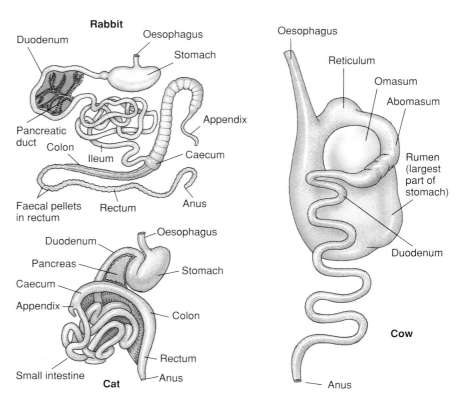

Figure 14.5I ⬆ The guts of the cow, rabbit and cat

CHECKPOINT

▶ **1** Mechanical digestion breaks up food, and chemical digestion converts its insoluble nutrients into soluble substances which the gut can absorb.

(a) Rewrite the following list under two headings – mechanical digestion and chemical digestion – placing each in the correct column.

action of teeth   action of bile
action of amylase   action of saliva
action of peristalsis

(b) Why do you think that mechanical digestion usually occurs before chemical digestion?

▶ **2** Experiments to investigate the effects of temperature and pH on the activity of the protein-digesting enzyme pepsin produced the following results.

Temperature (°C)	Time taken for digestion to be completed (s)	
	pH5.8	pH8.5
20	275	
25	221	
30	169	
35	114	Digestion not
40	137	completed during
50	196	the experiment
60	Digestion not completed during the experiment	

Plot the results on graph paper and then comment on the results.

▶ **3** Relative to the size of the rest of the body, frog tadpoles have a long, coiled gut and adult frogs have a short one. Briefly explain why you suspect frog tadpoles are herbivores and adult frogs are carnivores.

Topic 15 Respiration

What is respiration?

Figure 15.1A ▲ Breathing hard

Figure 15.1A shows a man and his best friend doing something that we all do – breathing: in their case, breathing hard!

Breathing describes **inhaling** (taking in) and **exhaling** (giving out) air. Look at Table 15.1. It shows that the proportions of gases in inhaled and exhaled air are different suggesting that oxygen is used by the body and carbon dioxide is produced. Oxygen and carbon dioxide are exchanged between the inhaled air and the blood across the inner surface of the lungs (**gaseous exchange**). *What is oxygen used for in the body?*

Table 15.1 ▼ Differences between inhaled and exhaled air

Gas	Inhaled air (%)	Exhaled air (%)
Nitrogen	78	78
Oxygen	21	16
Noble gases	1	1
Carbon dioxide	0.03	4
Water vapour	0	1

Cellular respiration

Digested food substances are **oxidised** in cells to release energy. These oxidations are the reactions of **cellular respiration**.

Look at Table 15.1. *Why is there less oxygen in exhaled air than in inhaled air?* The reason is that some of the oxygen inhaled is used by the cells to oxidise food. Cellular respiration that uses oxygen is called **aerobic respiration**.

Aerobic respiration is **exothermic** – it gives out energy.

$$\text{Glucose} + \text{Oxygen} \rightarrow \text{Carbon dioxide} + \text{Water}$$
$$C_6H_{12}O_6(aq) + 6O_2(g) \rightarrow 6CO_2(g) + 6H_2O(l)$$
$$\text{Energy released} = 16.1 \text{ kJ g}^{-1} \text{ glucose}$$

Make sure that you can answer an exam question which asks you to distinguish between gaseous exchange and cellular respiration.

Make sure that you can distinguish between aerobic respiration and anaerobic respiration. In each case how much energy is released from the oxidation of glucose?

An oxygen debt accumulates when muscles which are contracting vigorously switch from aerobic respiration to anaerobic respiration.

In Figure 15.1A aerobic respiration in the cells of the man's leg muscles gives him a flying start. Soon, however, in spite of rapid breathing and strenuous pumping by the heart, oxygen cannot reach the muscles fast enough to supply their needs. The muscles then switch from aerobic respiration to **anaerobic respiration**, which does not use oxygen. Lactic acid is produced and collects in the muscles. The reactions are

$$\text{Glucose} \rightarrow \text{Lactic acid}$$
$$C_6H_{12}O_6(aq) \rightarrow 2CH_3CHOHCO_2H(aq)$$
$$\text{Energy released} = 0.83 \text{ kJ g}^{-1} \text{ glucose}$$

The energy released is less than in aerobic respiration. An **oxygen debt** builds up. After the muscles have respired anaerobically for a few minutes, the lactic acid they have built up stops them from working. The dog's leg muscles will also have been respiring aerobically and then anaerobically. Both the man and the dog will be unable to run further until the lactic acid is removed from the muscles. During the **recovery period**, which lasts several minutes, man and dog pant vigorously (Figure 15.1B). The rush of oxygen to the muscles promotes aerobic respiration. Some lactic acid is oxidised to carbon dioxide and water; the rest is converted into glucose. In this way the oxygen debt is **repaid**.

Figure 15.1B ⬆ Recovery period

Bacteria, yeasts and root cells of plants can also change from aerobic respiration to anaerobic respiration if they are short of oxygen. Certain bacteria live permanently without oxygen: in fact oxygen is poisonous to some of them!

Yeast cells use anaerobic respiration to convert glucose into ethanol and carbon dioxide, with the release of energy. The reaction, called **fermentation**, is used commercially for the production of ethanol (alcohol) by yeast (see Topic 24.1).

Anaerobic respiration by yeast cells produces ethanol (alcohol) and carbon dioxide. The reaction, a fermentation, is used comercially to make bread and alcoholic drinks (see Topic 24).

$$\text{Glucose} \rightarrow \text{Ethanol} + \text{Carbon dioxide}$$
$$C_6H_{12}O_6(aq) \rightarrow 2C_2H_5OH(aq) + 2CO_2(g)$$
$$\text{Energy released} = 1.17 \text{ kJ g}^{-1} \text{ glucose}$$

The energy released is less than in aerobic respiration. Other differences between aerobic and anaerobic respiration are summarised in Table 15.2.

Why is there a difference in energy output? Aerobic respiration completely oxidises glucose to carbon dioxide and water and releases all the available energy from each glucose molecule. Anaerobic respiration converts glucose into ethanol or lactic acid. The oxidation of glucose is incomplete. More energy can be obtained by oxidising the ethanol or lactic acid aerobically.

Cellular respiration is sometimes compared with the combustion of petrol in vehicle engines, but there is a vital difference. When fuel is burnt in an engine, energy is released suddenly in an explosive reaction.

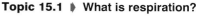

If cells were to release energy from food suddenly, the sharp rise in temperature would kill them. Cellular respiration is not a one-step chemical change, but a series of chemical changes that release energy from food gradually.

Table 15.2 ▼ A comparison of aerobic and anaerobic respiration

Process	Oxygen used or not used	Products of process	Energy released (kJ g^{-1} glucose)
Aerobic respiration	Used	Carbon dioxide and water	16.1
Anaerobic respiration in yeast	Not used	Ethanol and carbon dioxide	1.17
Anaerobic respiration in muscle cells	Not used	Lactic acid	0.83

Stages of cellular respiration

The first stage in both aerobic and anaerobic respiration is the conversion of glucose into a substance called pyruvic acid. The reactions occur in the cell's cytoplasm. Then, if oxygen is present, pyruvic acid is oxidised in the **mitochondria** of cells (see p. 147) to carbon dioxide and water (Figure 15.1C).

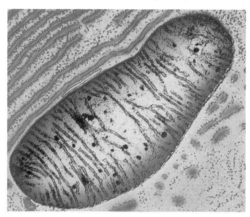

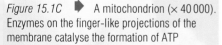
Figure 15.1C ▶ A mitochondrion (× 40 000). Enzymes on the finger-like projections of the membrane catalyse the formation of ATP

If no oxygen is present, anaerobic respiration takes place: in the case of muscle tissue lactic acid is formed; yeast forms ethanol.

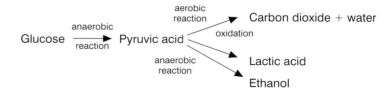

Aerobic respiration occurs in the mitochondria of cells.

Cellular respiration and photosynthesis

Cellular respiration and photosynthesis form a **cycle**. The products of one are the starting materials of the other (Figure 15.1D).

During the day photosynthesis produces more oxygen than is used in aerobic respiration. The surplus oxygen diffuses through the stomata out of the leaf (see Topic 11.1). At night photosynthesis stops but plant cells still need oxygen for aerobic respiration. Oxygen therefore diffuses through the stomata *into* the leaf. The carbon dioxide produced by aerobic respiration diffuses through the stomata *out of* the leaf.

EXTENSION FILE
ACTIVITY

SUMMARY

Cellular respiration is the name given to the processes in cells that release energy from food. Aerobic respiration in the presence of oxygen oxidises glucose to carbon dioxide and water with the release of energy. Anaerobic respiration in the absence of oxygen converts glucose into ethanol and carbon dioxide, or into lactic acid. Less energy is released in anaerobic respiration than in aerobic respiration.

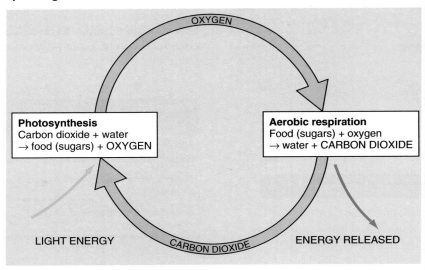

Photosynthesis
Carbon dioxide + water
→ food (sugars) + OXYGEN

Aerobic respiration
Food (sugars) + oxygen
→ water + CARBON DIOXIDE

OXYGEN

CARBON DIOXIDE

LIGHT ENERGY

ENERGY RELEASED

Figure 15.1D 🔼 Respiration and photosynthesis form a cycle

CHECKPOINT

▶ 1 Draw a table with two columns, one headed 'Photosynthesis', the other 'Aerobic respiration'. Study the descriptions below and decide which ones belong to which column.

(a) occurs only in plant cells
(b) occurs in all cells
(c) takes place only in daylight
(d) produces carbon dioxide
(e) takes place night and day
(f) produces food
(g) produces oxygen
(h) uses oxygen
(i) uses carbon dioxide
(j) releases energy
(k) uses water

▶ 2 Complete the following paragraphs using the words provided. Each word may be used once, more than once, or not at all.

(a) **aerobic   aerobically   anaerobically   with   as much   releases   without   respire**

_____ respiration requires oxygen. Anaerobic respiration takes place _____ oxygen. Some cells can _____ either aerobically or anaerobically. If oxygen is available these cells will respire _____ .

(b) **respire   repaid   oxygen   muscles   oxygen debt   liver   lactic acid**

During a sprint your leg _____ will need oxygen faster than it can be supplied by your blood. To compensate, the muscle cells _____ anaerobically. This process forms _____ which stops _____ from working. The oxygen needed to oxidise the lactic acid is called the _____ .

▶ 3 The flow diagram at the bottom of the page is a summary of the chemical changes that take place during cellular respiration.

(a) Which pathway, A, B or C, summarises aerobic respiration?
(b) What is the name of compound X?
(c) Why is there a difference in energy values between pathway A and
(i) pathway B (ii) pathway C?

$$A$$
Glucose → Pyruvic acid → $CO_2 + H_2O + 16.1$ kJ g^{-1} glucose
$$B$$
Glucose → Pyruvic acid → $CO_2 + X + 1.17$ kJ g^{-1} glucose
$$C$$
Glucose → Pyruvic acid → Lactic acid $+ 0.83$ kJ g^{-1} glucose

15.2 ▶ Exchanging gases: lungs

FIRST THOUGHTS

One of the processes vital to living things is the exchange of gases across a surface. Read on to find out why this is so important.

It's a fact!

The ancestors of lungfish were among the first land-living vertebrates 300 million years ago. Breathing air through lungs allowed them to survive times of drought.

The lungs where gaseous exchange occurs are connected to the nostrils and the mouth by the upper respiratory tract.

The upper respiratory tract consists of the trachea and bronchi.

The upper respiratory tract is lined with mucus which traps microorganisms in the air passing from the mouth/ nostrils to the lungs.

The fish in Figure 15.2A breathes air! It has simple sac-like structures, called **lungs**, through which it exchanges oxygen and carbon dioxide in air. It also has **gills** through which it exchanges oxygen and carbon dioxide in water. Breathing air is a safety device for this animal. It enables it to survive periods of drought when the swamps and rivers it lives in dry out temporarily.

Figure 15.2A 🔺 An Australian lungfish

The human lungs and upper respiratory tract

Our lungs are large and honeycombed with passages. Gases are exchanged across the surfaces in the lungs. Figure 15.2B shows the position of the lungs in the human body. They lie inside the **thoracic cavity** (chest cavity). The **upper respiratory tract** is a tube from the nostrils and mouth to the lungs. Figure 15.2B is your guide to the human gas exchange system. Study it carefully.

Notice the opening which leads into the trachea and oesophagas. Here there are mechanisms to prevent food from entering the **larynx** at the top of the trachea from the **pharynx** at the top of the oesophagus. The act of swallowing closes the **glottis** and lifts up the whole larynx so that it is blocked by the base of the **tongue**. At the same time, the flap-like **epiglottis** moves backward and protects the opening of the glottis. If food does lodge in the larynx, it triggers off violent **coughing**, which usually clears the obstruction. *If food were to enter the lungs, why would it be a hazard to health?*

■ The upper respiratory tract at work

The upper respiratory tract is an air-conditioning system. It warms and filters inhaled air. The membrane which lines the upper respiratory tract is well supplied with blood, which warms the air to body temperature. Hairs in the nasal passage filter out large particles of dust. A sheet of mucus lines the upper respiratory tract and traps bacteria, viruses and dust particles (Figure 15.2C). The mucus comes from cells in the membrane lining called **goblet** cells. **Cilia**, rows of fine hairs, sway to and fro and sweep the mucus, and with it trapped bacteria, viruses and dust, into the pharynx. It is either swallowed, sneezed out or coughed up. The air which then passes to the lungs is cleaned and freed from germs. What happens when the body fails to remove all germs is described later in this topic.

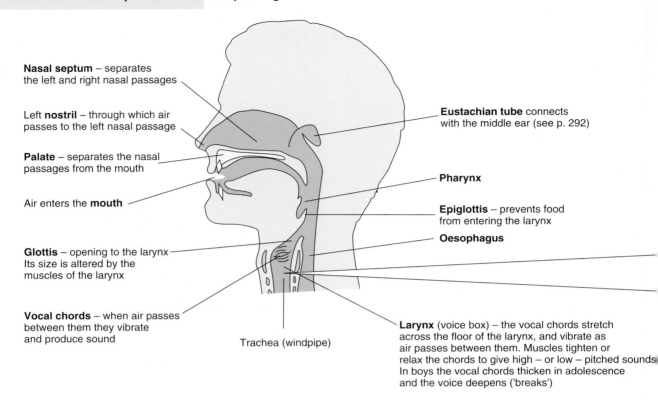

Nasal septum – separates the left and right nasal passages

Left **nostril** – through which air passes to the left nasal passage

Palate – separates the nasal passages from the mouth

Air enters the **mouth**

Glottis – opening to the larynx Its size is altered by the muscles of the larynx

Vocal chords – when air passes between them they vibrate and produce sound

Eustachian tube connects with the middle ear (see p. 292)

Pharynx

Epiglottis – prevents food from entering the larynx

Oesophagus

Trachea (windpipe)

Larynx (voice box) – the vocal chords stretch across the floor of the larynx, and vibrate as air passes between them. Muscles tighten or relax the chords to give high – or low – pitched sounds In boys the vocal chords thicken in adolescence and the voice deepens ('breaks')

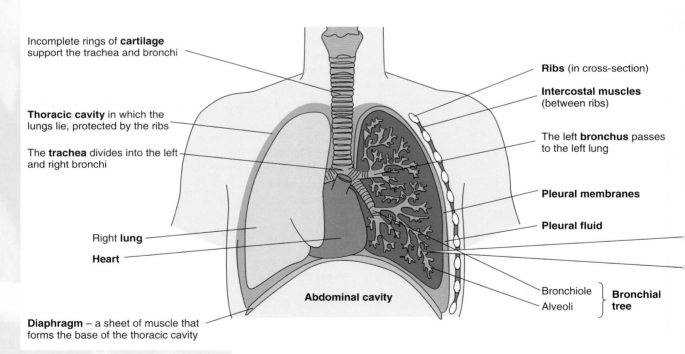

Incomplete rings of **cartilage** support the trachea and bronchi

Thoracic cavity in which the lungs lie, protected by the ribs

The **trachea** divides into the left and right bronchi

Right **lung**

Heart

Diaphragm – a sheet of muscle that forms the base of the thoracic cavity

Ribs (in cross-section)

Intercostal muscles (between ribs)

The left **bronchus** passes to the left lung

Pleural membranes

Pleural fluid

Bronchiole
Alveoli
Bronchial tree

Abdominal cavity

Figure 15.2B ⬆ The lungs (viewed from the front) and the upper respiratory tract (viewed from the left-hand side) in the human body. Two types of intercostal muscle criss-cross between each rib. The **external** intercostals slant forwards and downwards; the **internal** intercostals slant backwards and downwards.

■ The alveoli at work

In the lung, each bronchus branches many times into small tubes called **bronchioles**. These form a network called the **bronchial tree** (see Figure 15.2B). The bronchioles divide and sub-divide into even smaller tubes which end in clusters of small sacks called **alveoli** (singular: alveolus).

The walls of the alveoli are very thin and surrounded by capillary blood vessels (see p. 271). Figure 15.2D shows gas exchange between the walls of the alveoli and the capillary blood vessels.

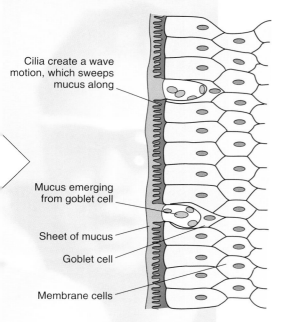

Cilia create a wave motion, which sweeps mucus along

Mucus emerging from goblet cell

Sheet of mucus

Goblet cell

Membrane cells

Figure 15.2C ◆ The membrane lining of the upper respiratory tract

It's a fact!

There are millions of alveoli in a pair of human lungs. Together they give a surface area of approximately 90 m^2!

It's a fact!

On average air exerts a pressure of 100 000 Nm^{-2}. Air moves from where pressure is high to where it is low until the pressures are equal.

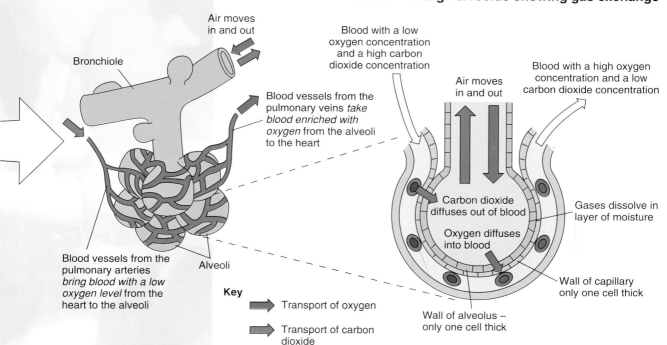

Cluster of alveoli

Air moves in and out

Bronchiole

Blood vessels from the pulmonary veins *take blood enriched with oxygen* from the alveoli to the heart

Blood vessels from the pulmonary arteries *bring blood with a low oxygen level* from the heart to the alveoli

Alveoli

Key

⇨ Transport of oxygen

⇨ Transport of carbon dioxide

Section through alveolus showing gas exchange

Blood with a low oxygen concentration and a high carbon dioxide concentration

Air moves in and out

Blood with a high oxygen concentration and a low carbon dioxide concentration

Carbon dioxide diffuses out of blood

Oxygen diffuses into blood

Gases dissolve in layer of moisture

Wall of capillary only one cell thick

Wall of alveolus – only one cell thick

Figure 15.2D ◆ The alveoli at work (notice that oxygen and carbon dioxide diffuse in solution between the alveolus and blood in the capillary vessel)

See www.keyscience.co.uk for more about breathing.

■ Breathing movements

The cage formed by the ribs and diaphragm is elastic. As it moves the pressure in the lungs changes. It is the change in pressure that causes **inhaling** (breathing in) and **exhaling** (breathing out) (Figure 15.2E).

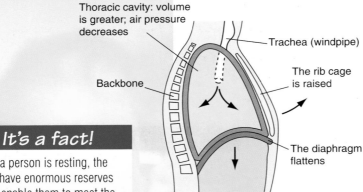
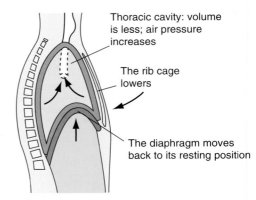

Thoracic cavity: volume is greater; air pressure decreases

Trachea (windpipe)

The rib cage is raised

Backbone

The diaphragm flattens

Thoracic cavity: volume is less; air pressure increases

The rib cage lowers

The diaphragm moves back to its resting position

Inhaling
a) The intercostal muscles contract, lifting the ribs upwards and outwards
b) The diaphragm muscles contract and the diaphragm flattens
The movements *increase* the volume of the thoracic cavity. As a result the pressure of air inside the thoracic cavity decreases and becomes less than atmospheric pressure. Air, therefore, is drawn into the lungs

Exhaling
a) The intercostal muscles relax, letting the ribs drop downwards and inwards
b) The diaphragm muscles relax and the diaphragm returns to its resting position
The movements *decrease* the volume of the thoracic cavity, helped by the **elastic recoil** of the ribs and diaphragm returning to their resting positions. As a result the pressure of air inside the thoracic cavity increases and becomes greater than atmospheric pressure. Air, therefore, is forced out of the lungs into the trachea

Figure 15.2E Inhaling and exhaling

It's a fact!

When a person is resting, the lungs have enormous reserves which enable them to meet the greatly increased demands for gaseous exchange during exercise. For example, oxygen consumption of around 500 cm^3 per minute can increase to about 3000 cm^3 per minute in a moderately fit person who is exercising.

SUMMARY

Gases are exchanged in solution across the gas exchange surfaces of the alveoli. The surfaces are:

- thin
- moist
- large in surface area
- well supplied with blood vessels
- in contact with the atmosphere through the tubes of the trachea, bronchi and bronchioles – all adaptations for efficient diffusion.

FIRST THOUGHTS

What is in the air we breathe? How does it affect our lungs? This section tells you.

Upper respiratory tract and lung diseases

The mechanism for sweeping dust, bacteria and viruses out of the upper respiratory tract and lungs is described in Figure 15.2C. Coughing and sneezing remove the mucus with its load of dust and micro-organisms. Even so, any part of the upper respiratory tract and lungs can become infected by disease-causing microorganisms.

Infection of the:

- throat (pharynx) is called **pharyngitis**,
- voice-box (larynx) is called **laryngitis**,
- windpipe (trachea) is called **tracheitis**,
- bronchus and bronchioles is called **bronchitis**.

Infection of the lungs by a particular type of bacterium causes **pneumonia**. The patient becomes breathless because fluid collects in the alveoli, reducing the surface area available for the absorption of oxygen. Pneumonia is treated with **antibiotic** drugs such as penicillin (see p. 337).

The **pleural membranes** line the rib cage and cover the surface of the lung (Figure 15.2B). Between them the **pleural fluid** stops the lungs sticking to the chest wall. Sometimes bacteria infect the pleural membranes, making them rough and causing pain when they rub together. The infection is called **pleurisy** and it is treated with antibiotics.

People affected by **cystic fibrosis** produce too much mucus in the lungs. The mucus blocks the alveoli and bronchioles. Regular physiotherapy helps improve the person's breathing difficulties. Excess mucus also blocks the pancreatic duct (see p. 388). Reduced release of pancreatic juice with its digestive enzymes leads to inadequate digestion. The trapped enzymes may start to digest the pancreas itself. *Why can people affected in this way develop diabetes* (see p. 301)? Cystic fibrosis is caused by a recessive allele. *Why are only people who are homozygous recessive for cystic fibrosis affected by the condition* (see pp. 379, 388)?

Other diseases include **lung cancer** and **asthma**. Lung cancer can be caused by many different things such as smoking. **Tumours** (see Topic 9.4) form in the lung and if they are not discovered quickly the cancer can spread to other parts of the body.

▪ The air we breathe

Motor vehicle exhaust fumes, smoke and dust from industry, cigarette smoke and all the other substances that human activities put into the air are part of our environment. Breathing this mixture affects our lungs.

Figure 15.2F shows the effect of fine coal dust on a coalminer's lungs. People who work in a dusty atmosphere can be affected in a similar way.

Particles of blue asbestos (in the past much used for insulation against fire) are particularly dangerous. They can lodge in the lungs and cause a type of cancer called **asbestosis**. There is now a ban on the use of the different types of asbestos except white asbestos which can be used for a limited time in situations where safety is essential but where no alternative material is available.

People suffer from **asthma** when the muscles lining the walls of the bronchi and bronchioles contract in spasms, narrowing the airways. The sufferer finds it difficult to breathe, wheezes and may experience a feeling of tightness in the chest. An asthmatic attack can be triggered by an infection, emotional responses or substances in the air which cause an allergic reaction. The occurrence of both lung cancer and asthma can be linked to the quality of the air we breathe.

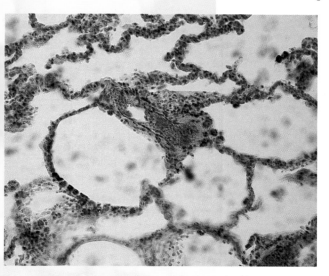

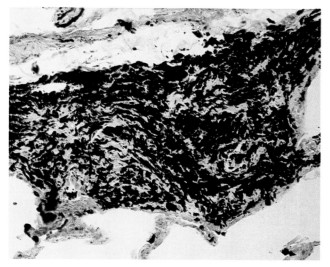

Figure 15.2F (a) Section of lung taken from a healthy person

(b) Section of lung taken from a coal miner with a lung disease called **pneumoconiosis**. Notice the deposits of coal dust in the lung tissue

Smoking is harmful: the evidence and attitudes

Research into the links between smoking and disease was prompted by the sort of data shown in Figure 15.2G.

Scientists soon suspected a correlation (relationship) between smoking cigarettes and **lung cancer** but could not prove the link completely (Figure 15.2H). However, new data strengthened the case against cigarettes.

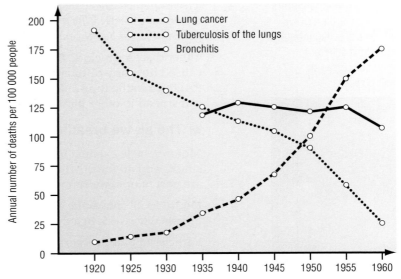

Figure 15.2G ⬆ Deaths from lung disease in England and Wales from 1916–60. Deaths from lung cancer increased sharply when deaths from other forms of lung disease (in this case **tuberculosis**) were falling

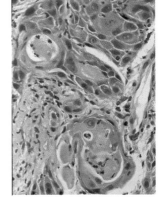

Figure 15.2H ⬆ (a) Healthy lung tissue

(b) Cancerous lung tissue: notice that many of the cell nuclei (dark) are enlarged – a feature of cancer cells (×40)

- When doctors saw the early evidence many of them gave up smoking cigarettes. Deaths from lung cancer among doctors went down compared with the population as a whole (Figure 15.2I).

- Studies clearly established the relationship between the risk of dying from lung cancer and number of cigarettes smoked – the more cigarettes smoked, the greater the risk (Figure 15.2J).

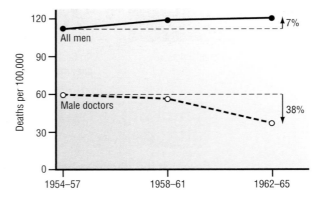

Figure 15.2I ➡ Death rates from lung cancer in male doctors and all men in England and Wales

Substances in cigarette smoke stop the cilia, which move the sheet of mucus lining the upper respiratory tract (see p. 249), from beating. As a result mucus accumulates, causing 'smoker's cough'.

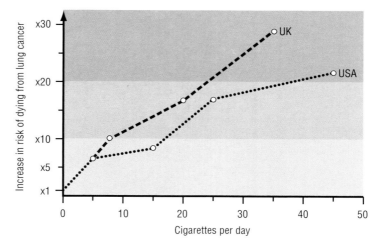

Figure 15.2J ◆ Death rates from lung cancer in men who smoke

Nicotine – one of the most powerful poisons known. It increases heart rate and blood pressure.

Carbon monoxide – a gas that combines with haemogoblin in red blood cells, so reducing the level of oxygen in the blood

Tar – contains over a thousand chemicals. Some of them are **carcinogens** (substances that cause cancer). Tar collects in the lungs as tobacco smoke cools

Figure 15.2K ◆ Dangers to health from the chemicals in cigarette smoke

The chemicals in cigarette smoke affect the fetus. Pregnant women are advised not to smoke.

Why are cigarettes dangerous? A lighted cigarette produces a number of substances. Many of them are harmful (Figure 15.2K). Some irritate the membrane lining the upper respiratory tract. Others stop the cilia from beating (Figure 15.2C). Extra mucus (**phlegm**) forms in the trachea and bronchi, causing 'smoker's cough'; it is the only way to get rid of the build up of phlegm. Smoking also weakens the walls of the alveoli and repeated coughing can destroy some of them. This breakdown of the alveoli is called **emphysema** (Figure 15.2L). *Why does a person with emphysema become breathless and exhausted easily?*

Pregnant women are always advised not to smoke. The chemicals in cigarette smoke enter the mother's bloodstream and reach the developing baby across the placenta (see Topic 22.4).

Babies born to mothers who smoke are generally lighter than babies born to mothers who do not smoke and there is an increased risk of premature birth. It seems that chemicals in cigarette smoke prevent the baby from getting all the nourishment he or she needs from the mother. This is in addition to the other dangers listed in Figure 15.2K.

Children see parents, older brothers and sisters, pop stars and film stars smoking and decide that smoking is the 'grown up' thing to do, so they copy them. Other children copy them, and so on (Figure 15.2M). This is the way smoking usually starts and once 'hooked' it is difficult to give up because the nicotine in tobacco is habit-forming.

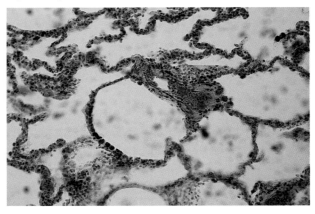

(a) Healthy alveoli

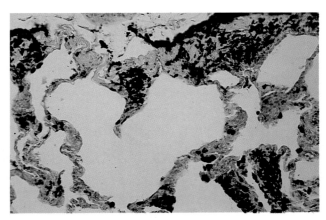

Figure 15.2L ⬆ (b) Alveoli destroyed by emphysema

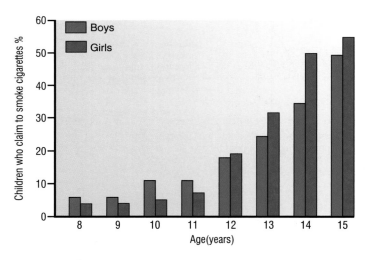

Figure 15.2M ⬆ Percentage of children who say that they have smoked cigarettes (by age and sex). *Can you explain why more girls aged 12 years and older smoke more cigarettes than boys of comparable age?*
Source: Department of Health

■ Living sensibly

In 1971 The Royal College of Physicians said:

'Premature death and disabling illnesses caused by cigarette smoking have now reached epidemic proportions and present the most challenging of all opportunities for preventive medicine...'

Since then the campaign against cigarette smoking has been fought hard. Cigarette sales have gone up and down, but the overall trend is downwards. Smoking is now banned in many public places, because non-smokers also suffer increased risks of ill health when they breathe in smoke from other people's cigarettes. This is called **passive smoking**.

Since 1978 cigarette smoking in Britain has dropped (Figure 15.2N). *By how much?* However, among women deaths from lung cancer continue to increase. *How might Figure 15.2M help to explain the problem?* Cigarette manufacturers are now putting their efforts into exports to the developing countries of the world. Here cigarette smoking is increasing, and so too are deaths from lung cancer. Think about this and look at the evidence once again. The message is clear:
DO NOT START SMOKING.

SUMMARY

Smoking cigarettes is a major cause of lung disease, especially lung cancer. It has taken scientists many years to establish the link. Fewer people smoke today than previously but many people continue to damage their health by smoking cigarettes.

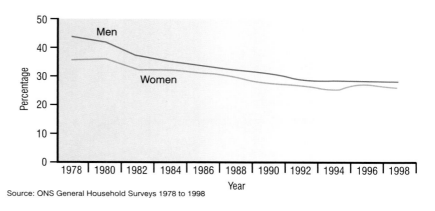

Source: ONS General Household Surveys 1978 to 1998

Figure 15.2N ▲ A Health of the Nation target has been set to reduce the percentage of adults in the UK who smoke to 20 per cent. Health education and heavy taxes on tobacco have persuaded people to give up smoking

CHECKPOINT

▸ **1** What is a gas exchange surface?

▸ **2** Write a brief statement against each of the following items to show how efficient alveoli are as gas exchange surfaces:
   a) large surface area
   b) short diffusion distance
   c) good blood supply
   d) moist
   e) in contact with air

▸ **3** Complete the following paragraph using the words given. Each word may be used more than once or not at all.

   uptake   respire   inhalation   oxygen   alveoli

   energy   exchange   moist   exhalation

   The _____ of oxygen and removal of carbon dioxide occur in the _____ of the lungs. These provide a large surface area (about 90 m²) for efficient gas _____. They are thin-walled, have an excellent blood supply, are _____ and kept well supplied with air by breathing. Air is taken into the lungs by _____ and removed by _____ .

▸ **4** Why do you think people should not smoke in the presence of a pregnant woman?

15.3 ▶ Exchanging gases: gills and other surfaces

FIRST THOUGHTS

Do you remember how the lungs are adapted for gaseous exchange (see p. 250)? Looking at gaseous exchange in plants (see p. 177) and other animals highlights the features characterising surfaces across which oxygen and carbon dioxide diffuse. In this section we look at gaseous exchange across gills and body surfaces of small animals.

It's a fact!

Gas exchange surfaces are:
- large in surface area, thereby maximising the rate of gaseous exchange
- thin, thereby facilitating (making easy) the diffusion (see p. 148) of gases
- moist, because exchange of gases occurs in solution
- well supplied with blood (animals only), which transports the gases to and from the gas exchange surfaces (see p. 249).

EXTENSION FILE
ASSIGNMENT

Figure 15.3A shows some different organisms that live in water exchanging gases because of:
- **photosynthesis** (plants only)
- **cellular respiration** (plants and animals)

Which gas (oxygen or carbon dioxide) does each process (a) use and (b) produce?
Notice that although the organisms are different, their gas exchange surfaces have features in common. The gills of the mayfly nymph (see p. 303) and leaves of the water weed are thin and expose a large surface area to the water in which oxygen and carbon dioxide are dissolved.

Figure 15.3A ◀ A mayfly nymph resting on water weed: the feathery structures projecting from its upper surface and rear are gills

In small animals like *Hydra* and *Planaria* (see Topic 1.5) gas exchange occurs across the body wall. Their surface areas are large enough to supply sufficient oxygen to the inside of the body and let carbon dioxide pass out. Flattening the body of the flatworm *Planaria* increases its surface area allowing gases to diffuse easily through it (Figure 15.3B).

Only small organisms can obtain enough oxygen by diffusion through the body wall. Larger animals usually have special organs for supplying oxygen to tissues in the deepest parts of the body and removing carbon dioxide from them.

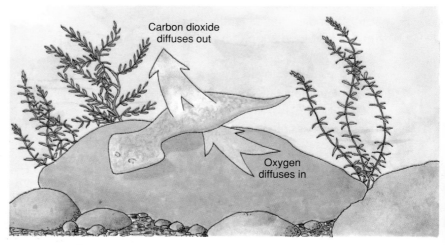

Carbon dioxide diffuses out

Oxygen diffuses in

Figure 15.3B ◀ *Planaria* lives in water where gases are in solution. The gases diffuse into and out of the animal across its body wall

Gills

Figure 15.3C looks inside the mouth of a fish. Notice the series of openings on either side. They are the **gill slits**. In **bony** fish, such as cod, herring and goldfish, each series of gill slits is covered by a flap of bony tissue called the **operculum** (plural: opercula). **Cartilaginous** fish such as shark and skate do not have opercula. The gill slits are easily seen as Figure 18.2A on p. 317 shows.

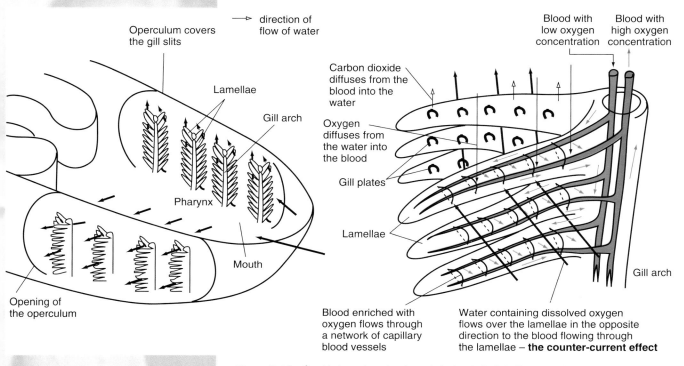

Figure 15.3C ⬆ Horizontal section through the head of a fish. The arrows indicate the direction of water current and blood flow. Part of a gill is shown in more detail

A gill consists of rows of lamellae. The gill plates greatly increase the surface area of each lamelle.

The efficiency of gaseous exchange across the gill plate is increased by a counter-current effect.

Also notice in Figure 15.3C that a partition separates each gill slit from the next in line. Each partition is strengthened by a thin bar of bone called a **gill arch**. A gill consists of rows of leaf-like tissue (called **lamellae**) projecting from either side of the gill arch. Each lamellae is folded into **gill plates**, greatly increasing the **surface area** of the gills overall. Blood vessels at the base of each gill branch into dense networks of capillaries in the lamellae.

■ The filaments at work

As water passes over the lamellae, dissolved oxygen diffuses into the blood flowing through the capillary network. At the same time carbon dioxide diffuses out of the blood and is carried away in the water current. *Where in the water and in the blood flowing through the gills are oxygen and carbon dioxide in greatest concentration?*

Exchange of the gases is helped because the flow of blood through the lamellae and the flow of water over the lamellae are in opposite directions. *How does this **counter-current** effect make the diffusion of oxygen from water to the blood and diffusion of carbon dioxide from blood to the water more efficient?*

SUMMARY

Dissolved oxygen and carbon dioxide in solution are exchanged between water and the blood in the capillary blood vessels which run through the filaments of the gills. A counter-current effect between water and blood enhances exchange of the gases. A large surface area is another important requirement for the efficient exchange of gases.

■ The respiratory current

The flow of water through the mouth and over the gill lamellae is called the **respiratory current**. Figure 15.3D shows what happens. The co-ordinated opening and shutting of the mouth and opercula results in a continuous flow of water over the surfaces of the lamellae. Water cannot enter the gut because muscles at its upper end contract, closing off the opening.

INSPIRATION – lowering the floor of the pharynx *increases* the volume of the mouth cavity and reduces pressure. Water flows in through the open mouth and over the lamellae

EXPIRATION – raising the floor of the pharynx *reduces* the volume of the mouth cavity and increases pressure. The **pressure pump** effect pushes water over the lamellae and against the opercula, pressing them open

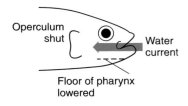

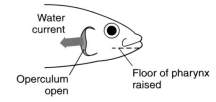

Figure 15.3D ⬆ Respiratory current – inspiration and expiration are continuous processes

■ Gas exchange in the earthworm

The earthworm (see p. 27) exchanges oxygen and carbon dioxide between itself and the atmosphere across its body wall.

Glands in the body wall of the earthworm produce mucus which keeps it moist and therefore provides ideal conditions for the exchanges of gases in solution. The body wall is also well supplied with blood vessels. The blood transports gases to and from the body wall.

Being moist also makes the earthworm vulnerable to loss of water from the body (**desiccation**). This is why earthworms are usually found in soil and not above ground. Soil is a moist habitat where earthworms can survive.

CHECKPOINT

▶ **1** Discuss how gills are adapted for the efficient exchange of gases between water and the blood of a fish.

▶ **2** Describe the pressure changes taking place during inspiration and expiration of the respiratory current in fish.

Topic 16 Circulation and transport

16.1 ▶ ## Blood and its functions

FIRST THOUGHTS

Blood transports gases, food and other vital materials to the tissues of the body. It protects the body from disease by producing antibodies and forming clots. This topic explains how the blood in the human body performs these functions and many others beside.

It's a fact!

Anybody who is healthy, weighs over 50 kg and is between the ages of 17 and 60 (70 for people who are already donors) can give blood. There are about 5 dm³ of blood in the body of an adult human. A donor may give up to 0.5 dm³ of blood at one time. That is 10 per cent of the body's total blood content.

Blood consists of a liquid plasma in which are suspended different types of blood cell.

It's a fact!

1 cm³ of blood contains 3000 white blood cells and 5 000 000 red blood cells.

When a person is seriously injured or undergoes a major operation they often lose a lot of blood. The blood they lose is replaced by **blood transfusion** with blood from blood **donors**. A needle is inserted into a vein in the arm of the donor and blood flows through the needle and into a sterilised bottle via a tube (Figure 16.1A). The blood is mixed with sodium citrate to stop it from clotting and stored at 5°C in regional blood banks. When a hospital needs blood it should be available from its regional blood bank.

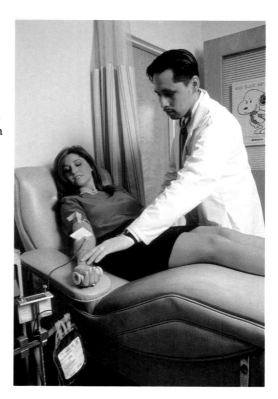

Figure 16.1A ▶ Giving blood

What is blood made of?

Figure 16.1B shows what happens when a sample of blood is spun for a time in a centrifuge. The blood separates into a straw-coloured liquid called **plasma** and a dark red–brown mass of **blood cells**.

The plasma transports **heat** released by metabolism (see p. 3) in the liver, muscles and fat cells to other parts of the body. It consists of:

- **Water** – 90% by volume.
- **Blood proteins** – these include **antibodies** that help to protect the body from disease, and **fibrinogen** one of the proteins that helps blood to clot.
- **Foods, vitamins and minerals** (see Topic 14.1).
- **Urea** (see Topic 17.5).
- **Hormones** – substances which help to co-ordinate different body functions (see Topic 17.4).

The red plug of blood cells at the bottom of the test-tube consists of:

- **Red blood cells** – the cells contain the red pigment **haemoglobin** that gives blood its colour. Red blood cells do not have nuclei.

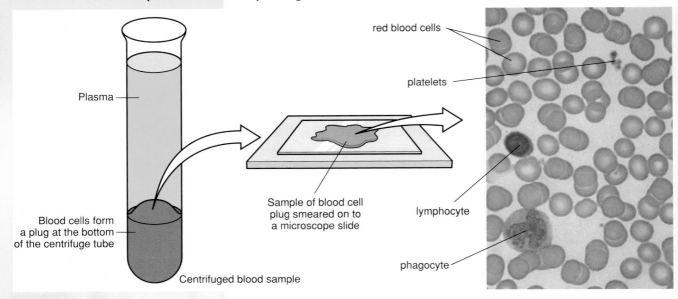

 See www.keyscience.co.uk for more about blood.

Figure 16.1B ⬆ Human blood sample spun in a centrifuge and examined under a microscope. (Notice the characteristic shapes of the nuclei of phagocytes and lymphocytes)

- **White blood cells** – two types, **lymphocytes** and **phagocytes**. The nucleus of each type of white blood cell has a characteristic shape (see Figure 16.1B).
- **Platelets** – these look like fragments of red cells.

Red cells transport oxygen and carbon dioxide

Red cells (and white cells and platelets) are made in the **marrow** of the limb bones (see p. 328), ribs and vertebrae. They contain the protein **haemoglobin**. Haemoglobin readily combines with oxygen in tissues where the concentration of oxygen is high to form **oxyhaemoglobin**. Oxyhaemoglobin breaks down to release oxygen in tissues where the concentration of oxygen is low.

It's a fact!

Sickle-cell anaemia is caused by a mutation in the genes controlling the synthesis of haemoglobin. The amino acid valine replaces glutamic acid (see p. 167). The substitution results in a drastic reduction in the oxygen-carrying capacity of the haemoglobin (see p. 416).

$$\text{Haemoglobin + Oxygen} \underset{\text{other body tissues}}{\overset{\text{lungs}}{\rightleftharpoons}} \text{Oxyhaemoglobin}$$

Blood which contains a lot of oxyhaemoglobin is called **oxygenated blood** and is bright red in colour. Blood with little oxyhaemoglobin is called **deoxygenated blood** and looks a deep red-purple (Figure 16.1C).

Red cells contain the protein ..aemoglobin which absorbs oxygen.

Old red cells are destroyed in the liver which also stores the iron released from the haemoglobin (see p. 238).

Make sure that you can distinguish between oxygenated blood and deoxygenated blood.

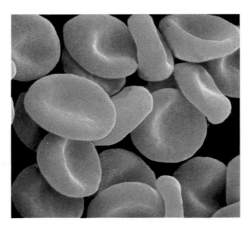

Figure 16.1C ⬆ Oxygen makes the difference

Cellular respiration (see Topic 15.1) releases carbon dioxide which diffuses into the blood. Some of it forms hydrogencarbonate ions (HCO_3^-) in the plasma. Haemoglobin carries the rest of the carbon dioxide in the red cells. When blood reaches the lungs, carbon dioxide diffuses from the plasma and red cells into the alveoli and is exhaled (see Figure 16.1D).

White cells protect the body

Some types of virus and bacteria enter (**infect**) the blood and tissues and cause disease. The two types of white blood cell, **lymphocytes** and **phagocytes**, protect the body by working quickly to destroy viruses and bacteria. Lymphocytes and phagocytes also destroy any other cells or substances which the body does not recognise as its own. Materials 'foreign' to the body are called **antigens**. When antigens come into contact with lymphocytes they stimulate some of the lymphocytes to produce proteins called **antibodies** which begin the process of destruction. The phagocytes finish the job. Figure 16.1E shows what happens.

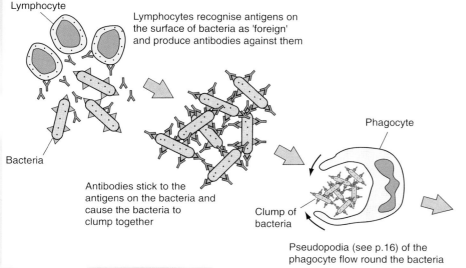

Air containing oxygen and carbon dioxide moves in and out of the alveolus

Transport of carbon dioxide

Alveolus of lung (see p. 249)

Transport of oxygen

Diffusion

Haemoglobin packed in red cells

Oxygen combines with haemoglobin to form oxyhaemoglobin

Capillary

Right side of heart

Left side of heart

Capillary

Red blood cell containing haemoglobin

Red blood cell containing oxyhaemoglobin

Carbon dioxide is carried in the red cells and as hydrogencarbonate ions (HCO_3^-) in the plasma

TISSUE

The breakdown of oxyhaemoglobin releases oxygen to the tissues and the haemoglobin is available to pick up more oxygen from the lungs

Figure 16.1D ⬆ Transport of oxygen and carbon dioxide between the lungs and tissues of mammals

Lymphocyte

Lymphocytes recognise antigens on the surface of bacteria as 'foreign' and produce antibodies against them

Key

 Antibody

 Antigen

Bacteria

Antibodies stick to the antigens on the bacteria and cause the bacteria to clump together

Clump of bacteria

Pseudopodia (see p.16) of the phagocyte flow round the bacteria

Phagocyte

Bacteria are completely enclosed in a vacuole where they are killed

Figure 16.1E ⬆ Lymphocytes and phagocytes at work (bacterial cells are not to scale)

See www.keyscience.co.uk
for more about antibodies.

*Antigens are substances which
stimulate some of the
lymphocytes to produce
antibodies.*

*White cells destroy
microorganisms which invade
the body and which would
otherwise cause disease. Their
action is called an immune
reaction.*

It's a fact!

Some bacteria which invade the
body release toxins (poisons).
Antitoxin antibodies combine
with the toxins, preventing
damage to the tissues.

*Immunisation promotes a
person's immunity to disease-
causing microorganisms.*

*Different methods are used to
produce vaccines.*

■ Antibodies are specific

Antibodies produced against a particular antigen will attack only that
antigen. The antibody is said to be **specific** to that antigen. This means
that antibodies produced against typhoid bacteria will not attack
pneumonia bacteria. It seems that each of us can produce tens of millions
of different antibodies to deal with all the antigens we are ever likely to
meet in a lifetime.

■ Active immunity

The action of lymphocytes and phagocytes against invading microorganisms is called an **immune reaction**. The antibodies produced may
stay in the body for some time ready to attack the same microorganisms
when next they invade the body. Even if the antibodies do not stay in the
body for long, they are soon made again because the first-time battle
between microorganisms and lymphocytes primes the lymphocytes to
recognise the same microorganisms next time. This means that re-
infection is dealt with by immune reactions which are even faster and
more effective than the first reaction. In other words you become
resistant. This is why you rarely catch diseases like chicken pox and
measles more than once.

Immunisation

Immunisation promotes active immunity to disease-causing microorganisms. It involves the doctor or nurse giving you an injection or
asking you to swallow some substance. The substance injected or
swallowed is called a **vaccine** and the process of being immunised is
called immunisation or **vaccination**.

Vaccines are made from one of the following:

- Dead microorganisms, e.g. whooping-cough vaccine is made from
  dead bacteria.
- A weakened form of microorganism which is harmless. Vaccines
  made like this are called **attenuated** vaccines, e.g. the vaccine against
  tuberculosis and Sabin oral vaccine (the vaccine against
  poliomyelitis) are both attenuated vaccines.
- A substance from the disease-causing microorganism but which does
  not itself cause the disease, e.g. diphtheria vaccine.

The processes of **genetic engineering** are also used to produce new
vaccines (see Topic 24.5).

What effect does a vaccine have? Antigens from the dead or attenuated
microorganisms in the vaccine stimulate some of the lymphocytes to
produce antibodies. So, when the same active, harmful microorganisms
invade the body, the antibodies made in response to the vaccine destroy
them.

The active immunity produced by vaccines can protect a person from
disease for a long time, although several more vaccinations called
boosters may be needed after the first one. Boosters keep up the level of
antibodies and so maintain a person's immunity.

Children in the UK are routinely immunised against six diseases which
used to cause many deaths. They are diphtheria, tetanus, whooping
cough, poliomyelitis, tuberculosis and German measles. These diseases
are now rare but those that spread through person-to-person contact
(infectious diseases) would soon increase if the number of people
vaccinated against them fell to levels where infections easily spread
among unprotected individuals.

■ Whooping cough and German measles

Whooping cough is a highly infectious disease caused by the bacterium *Bordetella pertussis*. The patient suffers from a wracking cough with characteristic 'whoops' which may last for 2–3 months. Whooping cough may make the patient vulnerable to **bronchopneumonia** and also increases the risk of **brain damage**. These complications and deaths from the disease are most common in babies under six months old.

Whooping cough vaccine is a suspension of killed *Bordetella pertussis*. It is usually given with diphtheria and tetanus vaccines in a **triple** vaccine.

Vaccination against whooping cough started in 1957. Before then about 100 000 cases of the disease were reported in the UK each year. By 1973 more than 80% of the population had been vaccinated and the number of annual cases fell to around 2400. However, there was a scare over the safety of the vaccine and vaccinations fell to around 30% in 1975. Epidemics (see p. 335) of whooping cough followed: there were nearly 66 000 cases in 1982 and more than 36 000 in 1986. Methods of producing the vaccine were improved. A publicity campaign pointing out the advantages of vaccination helped to restore public confidence. The percentage of people vaccinated increased, halting further epidemics. In 1998 there were around 1500 cases of whooping cough reported in the UK; 95% of children had been vaccinated by the age of two years.

There have been worries over the safety of other vaccines. Recently, concern has focused on the vaccine recommended to protect children against mumps, measles and rubella (German measles). The vaccine for each disease can be combined in one dose (**MMR vaccine**) and given as a single injection at 12–15 months of age. Reports suggested that soon after MMR vaccination some children seemed to experience difficulties such as **autism**. The number of children affected was small but doctors were concerned that parents worried about safety would not have their children vaccinated with MMR vaccine. Despite reassurances, worries remained and another study to examine any possible link between MMR vaccination and autism was undertaken in 1999. No link has been found and most doctors suspect that the earlier reports suggesting a connection were just coincidental. The importance of making sure that people can be confident that vaccines are safe is underlined by the damage that the virus responsible for German measles (rubella) can cause.

German measles in children is not serious. The body, arms, legs and face may be covered with pink spots and at worst there may be a fever and the lymph glands may swell up. These symptoms clear up after two or three days. However, it is much more serious if a woman catches German measles during the first four months of pregnancy. The virus can cross the placenta and affect the developing baby. The baby may be born dead or blind or deaf or with a damaged nervous system. Up to 90 per cent of babies whose mothers catch the disease when pregnant are affected. The component of MMR vaccine which protects against rubella virus avoids the dangers. This is why it is important for girls to be vaccinated at an early age with MMR before they have babies.

If large numbers of children are not vaccinated against different diseases, then the level of protection for the population as a whole falls and outbreaks of disease (with the risk of serious damage to health) increase. The 'scare stories' show how difficult it is to balance individual well-being and freedom of choice against what is good for everybody. *Where do you think the balance lies?*

It's a fact!

Diphtheria and tetanus are serious diseases caused by different types of bacteria. Diphtheria affects the throat. The bacterium causing the disease releases a powerful toxin which damages the heart. As a result the patient may die. Tetanus (lockjaw) causes muscles all over the body to tighten (permanently contract – see p. 315). The muscles of the jaws may 'lock' so the person cannot swallow. Death of the patient often follows.

It's a fact!

Sub-unit vaccines are designed to stimulate immune reactions without the unwanted side-effects of whole vaccines. They are made from particular proteins or small parts of disease-causing microorganisms.

It's a fact!

Autism is an especially distressing condition in children. An autistic child is unresponsive, seeming not to recognise family members or familiar surroundings.

Science at work

Concerns about links between damaging side-effects and vaccines persist. A number of vaccines contain mercury in the form of a preservative called **thiomersal**. The concentration of mercury in the vaccines is very low. However, mercury is a poison that affects the nervous system, and infants and young children are particularly vulnerable to its affects. There are worries that the thiomersal components of vaccines may in some way be responsible for the small (but increasing) number of cases of autism being reported. Some doctors believe that thiomersal should be removed from all vaccines given to children.

■ Passive Immunity

Not all vaccines contain antigens which stimulate the body to produce antibodies. Instead, antibodies can come ready-made from other animals. For example, anti-tetanus vaccine contains anti-tetanus antibodies produced by horses. The bacterium which causes tetanus lives in the soil and multiplies very rapidly in places where there is little air, such as in a deep wound. The bacterium produces a lethal poison which acts so quickly that the body's lymphocytes do not have time to make antibodies against it. This is why if you have a deep dirty cut you should be injected with vaccine containing anti-tetanus antibodies. These can act immediately to stop the disease from developing. Immunity which comes from antibodies made in another animal is called **passive immunity**.

Although it is short-lived, passive immunity is important for babies. While they are breast-feeding, babies receive antibodies from their mother's milk which protect them from disease-causing microorganisms. By the time this protection wears off, the baby is able to make its own antibodies.

Science at work

It is possible to give tetanus vaccine made from attenuated tetanus bacteria.

Antibodies formed in another animal provide passive immunity. They protect against microorganisms whose lethal effect may be quicker than the rate at which the body can produce antibodies against the microorganisms.

A person's blood group is determined by the type of antigen carried on the surface of the cells. In most people there are two types of antigen: A and B.

Blood groups

Although red blood cells all look alike under the microscope, they may carry different antigens called **antigen A** and **antigen B** on the cell surface. Plasma contains antibodies which attack foreign red cell antigens but does not contain antibodies which would attack a person's own red cell antigens. The possible combinations of antibody and antigen commonly found in people are shown in Table 16.1. They determine which blood group a person belongs to.

Table 16.1 ▼ Blood groups

Antigen on red cells	Antibody in plasma	Blood group	Percentage of UK population with blood group
A	Anti-B	A	40
B	Anti-A	B	10
A and B	Neither	AB	5
Neither	Anti-A and anti-B	O	45

The person giving blood is the **donor**. The person receiving blood is the **recipient**.

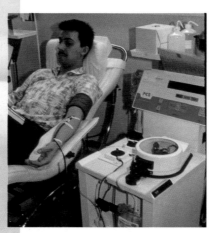

Figure 16.1F ▲ A patient receiving a blood transfusion

Figure 16.1F shows a patient having a transfusion. Blood at the right temperature is fed at the correct rate from the bag through the tube into a vein in the arm.

Before a person receives a blood transfusion it is important to know that the donor's blood group is **compatible** with that of the patient. If it is not then the donor's red blood cells clump in the patient's blood vessels and cause serious harm. Table 16.2 shows which blood groups are compatible. If a donor's blood causes the patients red blood cells to clump, their blood groups are said to be **incompatible**.

Table 16.2 ▼ Blood transfusion: compatibility between donor and recipient

Group	Donate to	Receive from
A	A and AB	A and O
B	B and AB	B and O
AB	AB	All groups
O	All groups	O

Group O people are called **universal donors**. *Why can they give blood to anybody?* Group AB people are **universal recipients**. *Why can they receive blood from anybody?*

Usually whole blood (plasma and cells) is given but sometimes plasma (or **serum**) only is used to restore blood volume, especially in cases of serious burns or blood loss. It takes several weeks for the patient to make new blood cells in the bone marrow to replace the ones lost.

Platelets help to stop bleeding

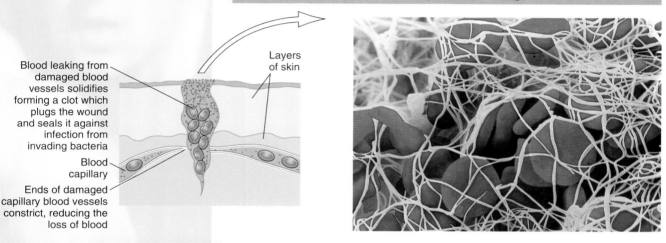

Blood leaking from damaged blood vessels solidifies forming a clot which plugs the wound and seals it against infection from invading bacteria

Layers of skin

Blood capillary

Ends of damaged capillary blood vessels constrict, reducing the loss of blood

Figure 16.1G ▲ A blood clot: red cells trapped by a mesh of fibrin fibres

When you are cut a series of complex events begins which eventually stops the bleeding. A clot begins to form when platelets are damaged by the rough surface caused by a cut or torn tissue. They release a substance which begins a cascade of chemical reactions that ends with the soluble plasma protein (see p. 259) **fibrinogen** changing into insoluble **fibrin**. The fibrin forms a mesh of fibres across the wound and traps red cells forming a clot. A solid plug forms in the blood vessel preventing further loss of blood while repair of the injury takes place (Figure 16.1G).

■ First aid to control bleeding

Blood loss from a deep cut may be too great for a clot to form without help. First aid aims to control bleeding and so help clotting take place:

- **Pressure** on the wound reduces blood loss and gives time for the blood to clot.
- **Elevation** of a cut arm or leg reduces bleeding by lowering blood pressure in the joints.
- **Dressing** with a sterile medicated dressing stops the bleeding.

Most cuts and scratches are not serious and clotting soon stops any bleeding. First aid aims to clean and dress the wound as quickly as possible to reduce the risk of infection.

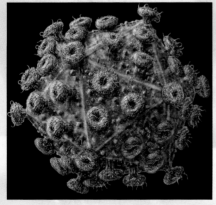

Figure 16.1H ⬆ Human immunodeficiency virus (HIV)

Disorders of the blood

■ Leukaemia

Leukaemia is a condition in which large numbers of immature white blood cells are produced and released into the blood stream. Overproduction of abnormal white cells results in the formation of too few red cells and other blood cells.

The microscopic examination of blood smears and bone marrow smears is an important tool in the diagnosis of leukaemia. The number of white cells and their appearance are helpful indicators of the presence of the disease.

Leukaemia can be treated with drugs and by radiotherapy. For some types of leukaemia the prospects of improvement and recovery in a patient are poor, but other types respond very well to treatment. For example, in more than 95 per cent of cases involving children with certain kinds of leukaemia, treatment stops the disease developing further and improves their chance of a healthy life.

■ Haemophilia

Some people lose a lot of blood if they injure themselves because their blood does not clot properly. The disease is called **haemophilia**. In the most common form of the disease the blood lacks a substance called **factor VIII** which is one of the substances involved in the production of fibrin. Haemophiliacs (people who suffer from haemophilia) are treated with injections of factor VIII. Haemophilia is a **genetic disease** that runs in families (see Topic 23.2). Some of the royal families of Europe carry the genes for haemophilia.

■ AIDS

AIDS (Acquired Immune Deficiency Syndrome) is caused by a virus called the **Human Immunodeficiency Virus (HIV)** (see Figure 16.1H). The virus attacks the lymphocytes which play an important part in the body's defence against disease. This means that the body of a person with HIV is far less well protected than normal. It is not usually HIV itself which causes the suffering of AIDS patients: they may die from another infection, often a type of pneumonia, which develops after HIV has destroyed part of the body's defences. We call AIDS a **syndrome** as there are a variety of symptoms which an HIV-infected person may have. Many of these symptoms are the result of other infections.

Unlike many other viral infections, controlling AIDS presents a particular difficulty. There is a long time interval (from a few months to several years) between the virus getting into a person's body and symptoms developing. During this period the person is infectious and able to pass on the virus to other people without knowing it.

Testing for AIDS depends on detecting antibodies to HIV in blood samples. If testing finds someone has HIV, the person is said to be **'HIV positive'**. If testing shows that the person is not infected with HIV then the person is **'HIV negative'**. Two tests are carried out before a person can be confirmed as 'HIV positive'. **False positive** results, which suggest that a person is infected with HIV when they are not, are therefore very rare.

CHECKPOINT

> **1** List the different types of blood cell.

> **2** What is meant by 'oxygenated blood' and 'deoxygenated blood'?

> **3** What is an antigen?

> **4** Why are diseases like chicken pox and measles rarely caught more than once?

> **5** What is the difference between a vaccine and a vaccination?

> **6** What is an attenuated vaccine?

> **7** How do booster vaccinations help to maintain a person's immunity to disease?

> **8** What is the triple vaccination?

> **9** Briefly explain why it is important for women to be vaccinated against German measles before they have babies.

> **10** What is the difference between active immunity and passive immunity?

> **11** Briefly explain how the presence or absence of antigen A and antigen B determines a person's blood group.

> **12** What is blood serum?

> **13** Before a blood transfusion is given, why is it important to know that the donor's and patient's blood are compatible?

16.2 ▶ Understanding immunology

FIRST THOUGHTS

How does the body establish specific immunity against the millions of antigens it will encounter in a lifetime? Reading this section will help you answer the question, and tell you about the problems of transplant surgery.

Immunology is the science concerned with the processes and mechanisms that establish specific immunity in the body. These processes and mechanisms make up the body's **immune response** which is summarised in Figure 16.2A. Notice that there are two phases: a **primary response** when the antigen first invades the body, and a **secondary response** should the same antigen invade the body again. The secondary response depends upon a specific **immunological memory**.

Action of lymphocytes and phagocytes

Lymphocytes are the white cells in the blood that recognise and react to antigens. They originate in the bone marrow, which is a yellow fatty material that fills the hollow centre of the bone shaft (see Topic 18.3).

Lymphocytes are white blood cells that respond to antigens. There are two types of lymphocyte: B-cells and T-cells.

There are two categories of lymphocyte: **B-cell** lymphocytes and **T-cell** lymphocytes. Contact with an antigen stimulates B-lymphocytes and T-lymphocytes to attack it. The B-lymphocytes divide and produce **antibodies** specific to the antigen that triggers the response. The antibodies circulate in the blood and are some of the **plasma proteins** listed on p. 259. Figure 16.1E on p. 261 shows B-lymphocytes at work.

T-cell lymphocytes do not produce antibodies. They bind with an antigen and divide to form a variety of cells that have different functions. Some of these cells, called **T-helper** cells, control the production of antibodies by B-lymphocytes. Others, called **T-cytotoxic** cells, destroy cells which are infected with a virus.

B-cell lymphocytes produce antibodies, T-cell lymphocytes do not.

Phagocytes are white cells in the blood that behave rather like a 'sanitation squad'. Figure 16.1E on p. 261 shows that some types of phagocyte engulf and destroy bacteria which have been attacked by

Phagocytes are another type of white blood cell. They engulf and destroy bacteria which have been attacked by antibodies.

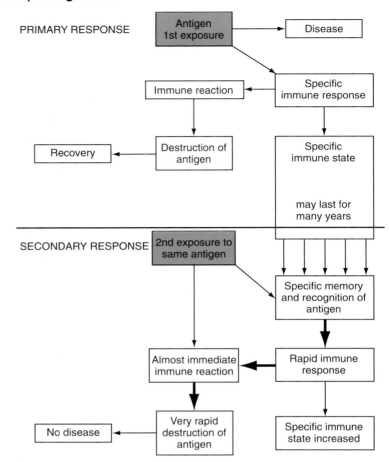

Figure 16.2A The body's immune response

antibodies. Others pass through the walls of the blood vessels and migrate through the tissues of the body to deal with microorganisms which gain entry through cuts, scratches or other openings in the skin. This causes an **inflammatory** action with swelling, redness and heat at the site of infection. Still others damage large parasites like roundworm (see Figure 3.7A on p. 80) which are too large to be engulfed.

Lymphocytes and phagocytes work together to keep most of us healthy for most of our lives. They quickly destroy viruses, bacteria or any other cells or substances which infect the blood or tissues and which the body does not recognise as its own.

Immunological memory

The body takes a few days to produce antibodies against a first-time infection (**primary response**). However, second-time-round the reaction is much quicker (**secondary response**). B-lymphocyte and T-lymphocyte **memory cells** left over from the first-time division of lymphocytes (called **clonal expansion**) are activated. They quickly divide (a second clonal expansion) on re-exposure to the same antigen. The division of lymphocytes is called clonal expansion because the daughter cells are genetically identical to their parents and each other (see p. 158).

Memory cells are specific for a particular antigen. It is due to the action of memory cells that we do not usually catch mumps or chicken pox more than once in a lifetime. The rapid response of immunological memory destroys the viruses before they make us ill.

It's a fact!

'Vaccination' comes from the Latin word *vacca* meaning a cow. The word 'immunisation' is sometimes used instead of 'vaccination'.

It's a fact!

The most frequent types of transplant are:

- skin – treatment for burns
- kidney – alternative treatment to **dialysis** (see p. 309) in cases of kidney failure
- heart – treatment for heart failure
- bone marrow – treatment for children with incurable blood diseases
- liver – treatment for liver failure.

Tissue transplants and rejection

A person who suffers serious burns may need a **skin graft** to prevent infection and loss of water from the exposed areas. If the skin is taken from another part of the victim's body, then the 'new' skin grows and joins up with the skin surrounding the burn. After a time the affected area is as good as new. However, if skin is taken from another person (except the victim's identical twin), tissue rejection sets in after a few days and the 'new' skin dies. Microscopic examination shows the presence of large numbers of T-lymphocytes and phagocytes in the failed graft. These cells seem to be responsible for the rejection. *Why do other transplanted tissues and organs suffer the same fate unless steps are taken to prevent rejection?*

Like viruses and bacteria, mammal cell membranes carry an enormous variety of antigens. In humans these antigens are called **human lymphocyte antigens** abbreviated to **HLA**. Even closely related people rarely have identical types of HLA. If an organ from one person (e.g. the heart) is transplanted to another person, the recipient's T-lymphocytes are activated if the donor's HLA type is different from the recipient's. T-lymphocytes attack the transplanted organ and cause **transplant rejection**.

Preventing rejection

Different methods reduce the chances of rejection:

- **Tissue typing** using **monoclonal antibodies** (see p. 410) identifies the different HLA in the person giving the organ (**donor**) and the person receiving it (**recipient**). Surgeons only undertake transplanation if the HLA of donor and recipient are very similar or match. *Why does tissue typing reduce the chances of rejection?*

SUMMARY

The body produces specific antibodies against the millions of different antigens it meets in a lifetime. This immune response is remembered by memory cells should an antigen invade the body again. The action of T-lymphocytes and phagocytes is probably responsible for the rejection of tissue transplants.

- **Immunosuppressive drugs** are used to prevent an immune response even if the HLA of donor and recipient do not match. **Cyclosporine** from the fungus *Trichoderma polysporum* is less poisonous than other immunosuppressive drugs to other tissues of the body. Given at the time of transplant, it prevents the recipient's T-cell lymphocytes (remember they are culprits in tissue rejection) from acting against the antigens in the transplanted tissue. Because cyclosporine and other immunosuppressive drugs reduce the effectiveness of the immune system, the recipient is more vulnerable to infections. If infection occurs, then the drugs are stopped. However, there is a danger that immune reactions causing rejection of the transplant may then develop. Precautions are taken:

Sterile conditions are provided for the recovery of the recipient for some time after the transplant operation. The risk of infection is reduced while the patient's immune system is supressed and adapting to the presence of the transplanted tissue.

Radiation treatment of the recipient's bone marrow (see p. 328) stops the production of T-cell lymphocytes which would otherwise act against the antigens in the transplanted tissue. During the treatment the recipient's immune system is even more suppressed.

CHECKPOINT

▶ **1** Distinguish between the following pairs:
    antibody   antigen
    primary immune response   secondary immune response
    lymphocyte   phagocyte

▶ **2** What happens when lymphocytes recognise antigens?

▶ **3** Distinguish between the activities of B-cell lymphocytes and T-cell lymphocytes during an immune response.

▶ **4** The diagram shows an immune complex formed when an antibody and antigen stick together. Find out why immune complexes only form between antibodies and the antigen for which the antibodies are specific.

▶ **5** What is inflammatory reaction?

▶ **6** Briefly summarise the basis of immunological memory.

▶ **7** Discuss the biology behind the precautions taken to reduce the chances of rejection of tissue transplants.

16.3 ▶

The blood system

Why do we need a blood system? Small animals like flatworms (see Topic 1.5) do not have one. Osmosis, diffusion and active transport carry gases, minerals and other materials to where they are needed in the flatworm's body. However, large animals like humans require a system specialised to carry materials from one part of the body to another. This is the role of the blood system.

FIRST THOUGHTS

How does the heart work? What is the role of arteries, veins and capillaries? Find out the answer to these questions in this section.

Blood vessels

The blood system consists of a network of tubes called blood vessels through which the heart pumps blood. The major blood vessels are the **arteries** and the **veins**. Table 16.3 compares arteries and veins.

Table 16.3 🔻 Arteries and veins compared. The smooth lining to the vessels helps blood flow and prevents clots from forming

Arteries	Veins

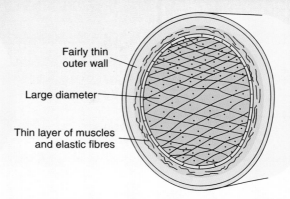

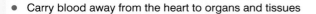

Arteries labels: Thick outer wall, Narrow diameter, Thick layer of muscles and elastic fibres

Veins labels: Fairly thin outer wall, Large diameter, Thin layer of muscles and elastic fibres

Arteries	Veins
• Carry blood away from the heart to organs and tissues	Return blood to the heart from organs and tissues (except the hepatic portal vein)
• Blood at high pressure	Blood at low pressure. Body muscles squeeze the veins to help push the blood to the heart
• Have a pulse because the vessel walls expand and relax as blood spurts from heart	Do not have a pulse since blood flows smoothly at low pressure
• Have thick walls to withstand pressure of blood	Have a thin wall and large diameter reducing resistance to the flow of blood returning to the heart

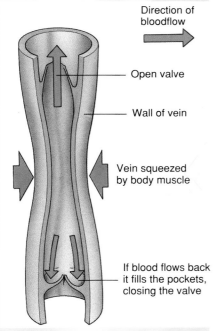

Labels: Direction of bloodflow, Open valve, Wall of vein, Vein squeezed by body muscle, If blood flows back it fills the pockets, closing the valve

Figure 16.3A 🔺 Valves opening and closing inside a vein

Blood in the veins is at a much lower pressure than in the arteries. One-way **valves** inside the veins prevent blood from flowing backwards (see Figure 16.3A). The force of the heart beat keeps blood flowing away from the heart through the arteries, so there is no need for valves inside them.

The arteries and veins in the human body form two circuits: the **lung circuit** and the **head** and **body circuit** (see Figure 16.3B). The veins of the head and body bring deoxygenated blood (see p. 260) to the heart. The heart pumps deoxygenated blood through the pulmonary arteries to the lungs, where it is oxygenated. The oxygenated blood returns to the heart through the pulmonary veins completing the lung circuit. The heart then pumps the oxygenated blood through the arteries of the head and body circuit.

■ Capillaries

Small blood vessels branch from the main arteries and veins. The vessels branching from arteries are called **arterioles**, those branching from veins are called **venules**. Arterioles and venules branch further to form **capillaries**. Capillaries join arterioles to venules and so link arteries and veins. Capillaries form dense networks, called **beds**, in the tissues of the body. This means that no cell is very far away from a capillary (see Figure 16.3C).

Capillaries are tiny vessels only 0.001 mm in diameter with walls one cell thick. The blood in capillaries supplies the nearby cells with the nutrients, oxygen and other materials they need. The blood also carries away urea, carbon dioxide and other wastes produced by the cells' metabolism (see p. 307).

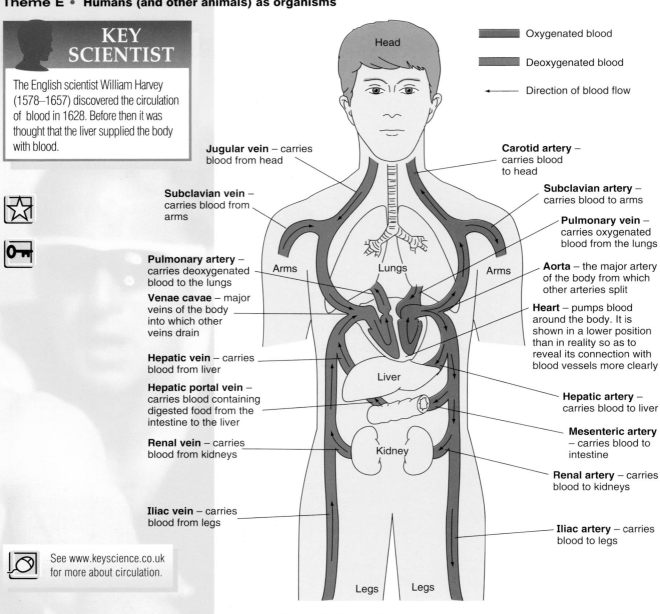

Figure 16.3B ⬆ The lung circuit and the head-and-body circuit. Arteries and veins are illustrated for one side of the body only, making it easier to see the links between blood vessels. Oxygen passes from tissue to blood and carbon dioxide passes from blood to tissue in the lungs. Oxygen passes from blood to tissue and carbon dioxide passes from tissue to blood in all other organs and tissues. *How is the function of the pulmonary artery different from the other arteries of the blood system? How are the functions of the pulmonary vein and hepatic portal vein different from other veins of the blood system?*

The pumping of the heart brings blood at high pressure to the arteriole end of the capillary network. The pressure forces plasma through the thin capillary walls. The liquid is now called **tissue fluid** and carries the nutrients and oxygen to the cells which are not in direct contact with a capillary (see Figure 16.3D).

There is just enough room in the smallest capillaries for red cells to pass in single file. A lot of plasma is forced through the one-cell-thick walls as the red cells squeeze through the capillaries. The pressure drops as blood passes through the capillaries to the venule end of the capillary bed. Tissue fluid can then seep back into the capillaries along with dissolved urea and carbon dioxide. Most tissue fluid returns to the blood by this route. The remaining small amount of fluid drains into the **lymph vessels**.

See www.keyscience.co.uk for more about circulation.

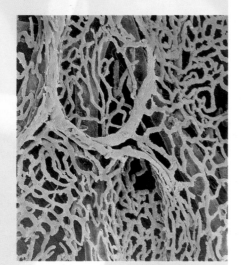

Figure 16.3C ⬆ A capillary network

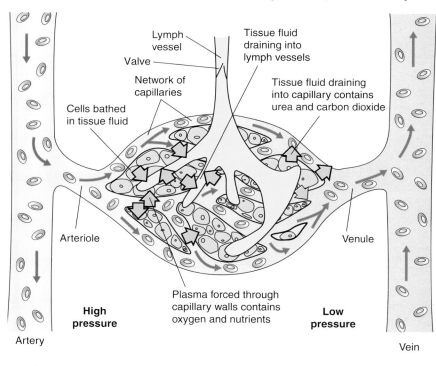

Figure 16.3D ▲ Capillaries at work

The lymph vessels

The lymph vessels return fluids that would otherwise collect in the tissues to the blood system. The smallest lymph vessels are the size of capillary blood vessels; the largest are the size of veins. Capillary lymph vessels are blind-ended tubes which form a network in the body's tissues. Tissue fluid diffuses through their walls and slowly passes into the larger lymph vessels. The fluid, now called **lymph**, is moved by the contraction of the body's muscles during normal daily activity. Valves in the lymph vessels, similar to those found in veins, prevent the backflow of fluid. Figure 16.3E shows how the lymph vessels are arranged in the human body.

The **spleen** is also part of the system of lymph vessels. Lymphocytes (see p. 260) are made in the spleen and in the lymph glands. The spleen also collects damaged and old red blood cells, breaking them down and releasing the haemoglobin in them. Iron in the haemoglobin is removed and re-used by bone marrow to make new haemoglobin.

If the circulation of lymph is upset by disease or poor diet, then fluid may gather in the tissues and cause swelling. This is called **oedema**.

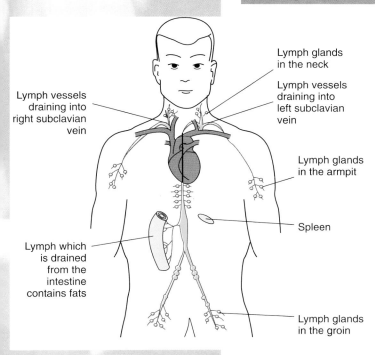

Figure 16.3E ▲ Lymph vessels

273

The human heart

The heart is a pump which propels blood through the arteries and veins. It is made of a type of muscle called **cardiac muscle** which contracts and relaxes rhythmically for a lifetime. The more efficient the heart is, the more efficient are the exchanges of food, oxygen, carbon dioxide and other dissolved materials between the blood and the tissues of the body.

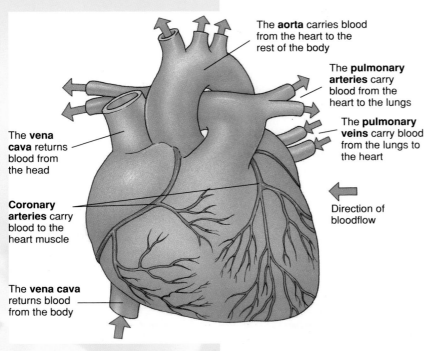

The **aorta** carries blood from the heart to the rest of the body

The **pulmonary arteries** carry blood from the heart to the lungs

The **pulmonary veins** carry blood from the lungs to the heart

The **vena cava** returns blood from the head

Direction of bloodflow

Coronary arteries carry blood to the heart muscle

The **vena cava** returns blood from the body

Figure 16.3F ▲ View of the heart from the front

■ Heart structure

The heart lies in the chest cavity surrounded by a membrane called the **pericardium**. It is protected by the rib cage. Figures 16.3F and G show the heart and the blood vessels leading to and from it.
The heart is full of blood and it seems odd that heart muscle needs an additional blood supply. However, like other tissues, heart muscle needs a steady supply of the food and oxygen dissolved in blood. The walls of heart muscle are so thick that these materials cannot diffuse quickly enough from inside the heart to all of the heart muscle. So, some of the heart muscle is supplied with blood by the **coronary arteries**.

■ Flow of blood through the heart

When heart muscle relaxes (called **diastole**):

● deoxygenated blood from the head and body enters the right auricle (atrium) through each vena cava,

● oxygenated blood from the lungs enters the left auricle (atrium) through the pulmonary veins.

The auricles (atria) fill with blood and then contract (called **auricular systole**). The increase in pressure opens the tricuspid and bicuspid valves and forces blood into the ventricles. When full, the ventricles contract (called **ventricular systole**). The increase in pressure closes the tricuspid and bicuspid valves and forces blood past the semi-lunar valves which guard the openings of the pulmonary artery and aorta. The right ventricle pumps blood into the pulmonary artery on its way to the lungs; the left ventricle pumps blood into the aorta which takes it round the rest of the body. The semi-lunar valves close when the ventricles relax.

In effect the heart is a **double pump**, each ventricle pumping blood along a different route through the body. Look back at Figure 16.3B. The distance travelled by blood through the head and body circuit is greater than the distance it travels through the lung circuit. This difference explains why the wall of the left ventricle is thicker than the wall of the right ventricle. Its contractions must be more powerful to pump blood the greater distance.

EXTENSION FILE
ACTIVITY

The heart is a double pump propelling blood through the blood vessels of the head and body circuit and the lung circuit.

Make sure you can answer an exam question which asks why the wall of the left ventricle is thicker than the wall of the right ventricle.

Answer: Contraction of the left ventricle propels blood through the head and body circuit. The distance the blood travels is greater than the distance through the lung circuit.

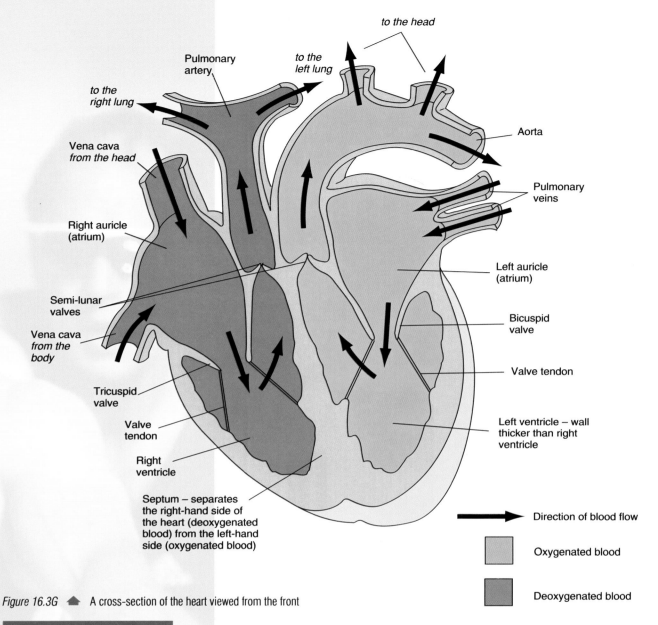

Figure 16.3G ⬆ A cross-section of the heart viewed from the front

■ The heartbeat

Diastole and systole produce an unmistakable two-tone sound which is easily heard through an instrument such as a stethoscope. This sound is the **heartbeat**. In a healthy adult the heart beats on average 72 times a minute but this can vary between 60 beats per minute and 80. Exercise makes the heart beat faster, bringing more blood to the muscles.

The beating of the heart is controlled by the **pacemaker**, which is a group of special muscle cells in the right atrium. Occasionally the natural pacemaker goes wrong. The heart rate slows down causing drowsiness and shortage of breath. When the heart's natural pacemaker does not function properly an electronic pacemaker can be fitted which helps to keep the heart beating at the proper rate.

It's a fact!

Exercise increases the heart rate which means that the pulse rate increases as well. Blood pressure also goes up. When exercise stops, pulse rate and blood pressure return to resting values. The speed of return is a measure of a person's fitness. Resting values are reached more quickly in someone who is fit than in someone who is unfit.

Science at work

Blood pressure is measured in millimetres of mercury (mmHg) rather than the SI unit of pressure, the pascal (Pa). A column of mercury 100 mm high exerts a pressure of 13 300 Pa at its base, so 1 mmHg = 133 Pa.

■ **The pulse**

Each heartbeat sets up a ripple of pressure which passes along the arteries. The ripple can be felt as a **pulse** as the artery's muscular wall expands and relaxes.

Feeling a patient's pulse can help doctors and nurses to tell if the heart is beating properly. The neck pulse and wrist pulse are the most useful pulses (see Figure 16.3H). The pulse is felt with the fingertips pressed lightly over the artery (the thumb is not used since this has its own pulse). The number of beats in a minute are counted against the seconds of a watch.

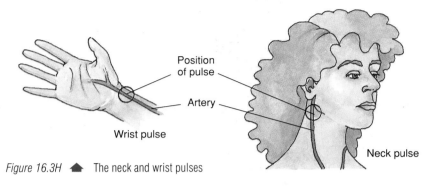

Figure 16.3H 🔺 The neck and wrist pulses

Blood pressure

Figure 16.3I 🔺 Measuring blood pressure with a sphygmomanometer. An electronic version is available

Blood flows along the blood vessels because it is under pressure. We call this **blood pressure**. It is measured with an instrument called a **sphygmomanometer** (see Figure 16.3I). An inflatable armband is wrapped around the patient's upper arm. The armband is inflated until it stops the blood from flowing along the main artery of the arm. The air is then let out of the armband very slowly while the doctor listens for the return of the pulse with a stethoscope placed below the armband. At the point when the blood moves back into the artery below the armband and the pulse returns, the pressure in the armband just equals the pressure of the blood.

Two readings are taken: the pressure of blood when the heart contracts (called **systolic blood pressure**) and the pressure of the blood when the heart relaxes (called **diastolic blood pressure**). The normal systolic pressure of a young adult is about 120 mmHg; normal diastolic pressure is about 75 mmHg.

Constant high blood pressure is harmful. It makes the heart work harder. Eventually the overworked heart may fail altogether. High blood pressure can also damage the kidneys and eyes, and increase the risk of an artery tearing open. A **cerebral haemorrhage**, or stroke, occurs when one of the arteries that supplies blood to the brain ruptures.

The smooth lining of healthy blood vessels allows blood to flow easily

CHECKPOINT

▶ **1** Briefly explain the significance of the difference between arteries and veins.

▶ **2** How does blood, rich in digested food absorbed from the intestine, reach the liver?

▶ **3** Explain the function of the valves in the veins and in the lymph vessels.

▶ **4** What is oedema?

▶ **5** Briefly describe the function of lymph glands.

▶ **6**

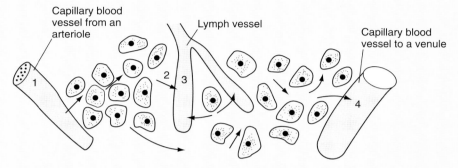

The diagram shows the movement of tissue fluid, plasma and lymph between capillary blood vessels, tissue cells and lymph vessels.

(a) Choose the word which correctly names the fluid in each of the locations numbered 1–4.

water   lymph   salt   solution   plasma   tissue   fluid

(b) Explain why fluid leaves the capillary blood vessel at 1 but enters the capillary blood vessel at 4.

▶ **7** Describe the route that blood takes from the time it enters the heart to the time it reaches the lungs.

▶ **8** What are diastole and systole?

▶ **9** What does 'heart rate' mean?

▶ **10** What is meant by 'blood pressure'?

▶ **11** The diagram shows a section through the human heart viewed from the front.

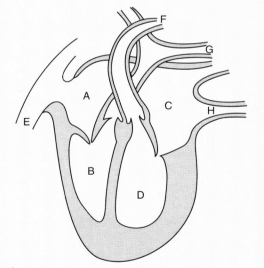

(a) Name the parts of the heart labelled A–H.

(b) Which of the following sequences of letters describes the path taken by blood through the heart?

GBAEHCDF   FDCHEABG

EABGHCDF   HCDFGBAE

(c) Which two parts of the diagram show that chambers A and C are relaxed?

(d) Why is the wall of chamber D thicker than the wall of chamber B?

▶ **12** The diagram shows sections through an artery and a vein.

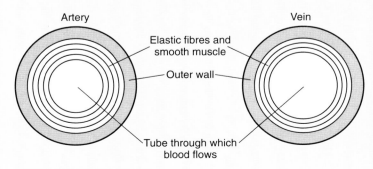

(a) Give two differences in structure between an artery and a vein.

(b) Briefly explain how the differences you have given in (a) are linked to the way in which arteries and veins work.

(c) Complete the following:

An artery carries blood from the _____ at _____ pressure.

A vein carries blood to the _____ at _____ pressure.

Arteries and veins are linked by tiny blood vessels called capillaries which are about _____ mm in diameter and have walls _____ _____ thick. Capillaries form at the end of small blood vessels branching from the arteries and veins. The vessels branching from the arteries are called _____ ; those branching from the veins are called _____ .

16.4 ▶ Understanding heart disease

FIRST THOUGHTS

Heart disease is responsible for about a quarter of all deaths in the UK. More people in the world's developed countries die from heart disease than from any other single cause. Why?

Fatty deposits of atheroma make blood vessels narrower.

Make sure that you can distinguish between a thrombus and a thrombosis.

through them. However, the lining can be damaged and roughened by a fatty deposit called **atheroma**. The build-up of atheroma makes blood vessels narrower and cuts down the flow of blood (see Figure 16.4A). This increases the risk of blood clots forming. A clot can block a blood vessel. The clot is called a **thrombus** and a blockage is called a **thrombosis**.

Atheroma in the coronary arteries is one cause of heart disease. The first signs of trouble may be cramp-like chest pain brought on by quick walking, anger, excitement or any other activity or emotion that makes the heart work harder than usual. The pain is called **angina**.

People live with some types of angina for years but other types get worse and may later result in a heart attack (called a **coronary thrombosis**).

A heart attack happens when the blood supply to the heart is interrupted. The person affected usually experiences a gripping pain in the chest. The pain often spreads to the neck, jaw and arms, and the victim may also sweat and feel faint and sick.

The affected part of the heart is damaged and sometimes a heat attack is so severe that the heart stops beating altogether. This is called **cardiac arrest**. A victim of cardiac arrest will die unless the heart starts beating again within a few minutes.

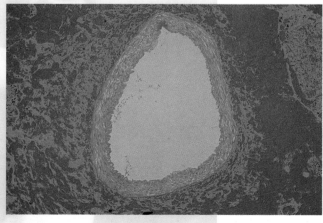

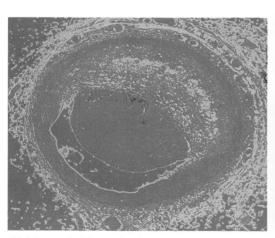

Figure 16.4A ◄ (a) Cross-section through a healthy artery

(b) Atheroma building up inside an artery

Deaths from heart disease

Diseases of the heart and blood vessels kill more men in the UK than any other single cause. They also kill many women. Figure 16.4B compares the percentage of deaths (of men and women aged under 75) due to diseases of the heart and blood vessels with all other causes of death. About 25% of all deaths of men is due to heart disease alone. The problem is not confined to the UK. Heart disease is the biggest single killer of middle-aged men in many of the world's developed countries. However, Figure 16.4C suggests some progress. Death rates from heart disease for men aged 35–74 have fallen by more than 30% in the UK in the last ten years. However, progress has not been as fast as in some other countries: 43% inAustralia, 41% in Sweden and 34% in the USA.

Heart disease is a major cause of death in many of the world's developed countries.

It's a fact!

For women aged under 75, 40% of all deaths is due to different cancers; 33% is due to diseases of the heart and blood vessels.

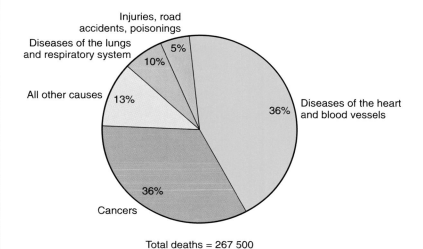

Total deaths = 267 500

Figure 16.4B ▶ Causes of death in people aged under 75 in the UK (Source: Office of Population Censuses and Surveys)

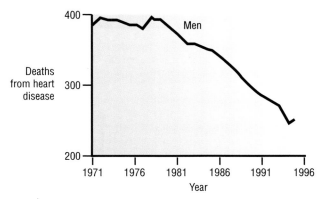

Source: Office for National Statistics

Figure 16.4C ▶ Deaths in the UK of men from heart disease between 1971 and 1995 (Source: Office for National Statistics)

Risk factors

Research has identified some of the causes of heart disease by comparing groups of people who have high rates of heart disease with groups that have low rates. When a difference between the groups is found that might help to explain why some people are more likely to develop heart disease than others, it is called a **risk factor**.

Unavoidable risk factors set the standard against which we can judge how best to take account of avoidable risk factors.

Make sure that you understand the relationship between heart rate, cardiac output and stroke volume, and the heart's effectiveness and fitness.

Three risk factors are unavoidable:

- **Sex** – men are more likely to die of heart disease than women.
- **Age** – the risk of heart disease increases with a person's age.
- **Inherited genes** – the tendency to die from heart disease can run in families.

Being male, old or having a family history of heart disease does not mean that people in these categories will necessarily die from it. If the risk factors are known, people can live sensibly to increase their chances of reaching old age. Living sensibly means avoiding the other, controllable risk factors such as smoking and stress (see Figure 16.4D).

Regular exercise increases heart fitness but is there evidence that exercise directly reduces the risk of heart disease? Many different studies strongly support the idea that regular exercise protects against heart disease and some of the evidence is summarised in Figure 16.4E. *What conclusions can you make from the data?*

The healthy heart at rest beats an average of 60–80 beats per minute (bpm). This is the **heart rate**. The volume of blood pumped from the heart each minute (**cardiac output**) depends on the heart rate and the volume of blood pumped out with each beat (**stroke volume**). Heart rate, cardiac output and stroke volume measure the heart's effectiveness and fitness.

A fit heart pumps more blood to the body's tissues and organs than does an unfit heart. At rest it has 25% more output than an unfit heart, rising

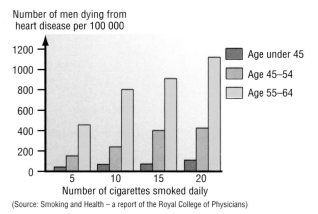

(Source: Smoking and Health – a report of the Royal College of Physicians)

(a) Smoking and the risk of heart disease

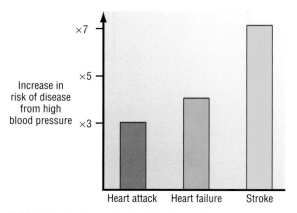

(b) High blood pressure and the risk of heart disease

Figure 16.4D ⬆ The evidence linking the risk of heart disease with avoidable risk factors

SUMMARY

Disease of the heart and blood vessels kills more men in the UK than any other cause. Risk factors increase the likelihood of heart disease. Some risk factors are unavoidable; others can be controlled by diet and exercise. Exercise increases heart fitness.

to 50 per cent or more during vigorous exercise. This meets the increased demand for oxygen from the muscles more efficiently. Stroke volume is also greater and a fit heart beats more slowly. The fit heart therefore does not have to strain to pump blood around the body.

Vigorous exercise increases the rate and depth of breathing (panting) – affects triggered by the response of the parts of the brain controlling breathing movements (see p. 250) to the rising acidity of the blood. The rise in acidity is due to the accumulation of carbon dioxide and increasing levels of lactic acid (oxygen debt – see p. 244) in the muscles. More air is drawn into the lungs and the rush of oxygen to the muscles promotes oxidation of the accumulated lactic acid. Panting also quickly gets rid of excess carbon dioxide. Blood acidity decreases and breathing movements slow to the resting rate.

Exercise and ...

Blood pressure

- Blood pressure is lower in physically fit people
- Regular exercise can help control high blood pressure and reduce the risk of heart disease

Overweight

- Being overweight is often due to lack of exercise and is linked with high blood pressure
- Regular exercise increases fitness, reduces body fat, lowers blood pressure and reduces the risk of heart disease (see Topics 10.3 and 14.2)

Stress

- Stress increases blood pressure and the risk of heart disease as well as making people irritable and depressed
- Regular exercise reduces the risk of heart disease and gives an increased feeling of well being

London Transport Study

- The study found that bus drivers had twice as many fatal heart attacks as the conductors, who climbed the stairs and walked the aisles of the bus all day long

Coronary arteries

- Regular exercise reduces the build-up of atheroma and makes coronary arteries wider reducing the risk of heart disease
- Exercise can promote the growth of new blood vessels after a heart attack

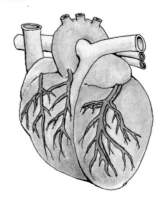

Figure 16.4E ◀ Exercise and heart disease

CHECKPOINT

▶ **1** The diagram shows three places A, B and C where blood vessels supplying blood to the heart could become blocked causing a heart attack.

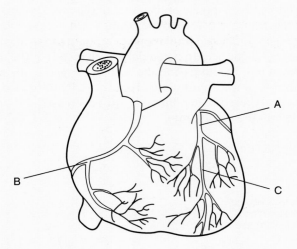

(a) Name the blood vessels supplying the blood to the heart.
(b) What is the blockage of the blood vessel called?
(c) Which of A, B or C would cause the most serious heart attack? Give reasons for your answer.

▶ **2** The diagram shows lengthways sections of a healthy blood vessel and a diseased blood vessel.

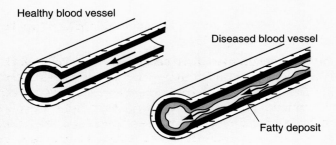

Healthy blood vessel

Diseased blood vessel

Fatty deposit

(a) What is the fatty deposit called?
(b) The artery narrows where the fatty deposit is forming. Briefly explain how the blood vessel could become completely blocked.
(c) What is a heart attack?
(d) Imagine you are walking along a busy street with someone who suddenly has a heart attack. The person remains conscious but needs your help. Explain what you should do.

▶ **3** Briefly explain how a man with a family history of heart disease can cut down the risk of having a heart attack.

▶ **4** 'Exercise is good for you' – so we are told. Using the data available, describe the effect of exercise on heart fitness.

Topic 17

Responses and co-ordination

17.1 ▶ **Senses and the nervous system**

FIRST THOUGHTS

Stimuli are converted by receptors into signals to which the body can respond. The signals are called nerve impulses. Neurones (nerve cells) conduct nerve impulses to muscles which respond by contracting. Muscles are called effectors. Nerves are formed from bundles of neurones and are the link between stimulus and response. The sequence reads: stimulus ➜ receptor ➜ nerves ➜ effector ➜ response. Remember the sequence as you read this section.

Glands are effectors. For example, the smell, taste, and sight of food stimulates the salivary glands to produce saliva (see p. 236).

A **transducer** converts energy from one form to another. Cells are transducers, and sensory receptor cells are examples of these. They convert forms of energy (e.g. light, sound) into the electrical energy of nerve impulses.

 See www.keyscience.co.uk for more about the senses.

Figure 17.1A ▲ Dog and man respond vigorously to stimuli

Look at Figure 17.1A – it illustrates important points about the way we respond to stimuli (see p. 195).

Eyes and ears contain specialised **sensory receptor** cells which convert stimuli into signals the body can respond to. The signals are minute electrical disturbances called **nerve impulses**. They are messages for the man's leg muscles to start working hard. The muscles are called **effectors** because they respond to nerve impulses. Specialised cells called **neurones** (nerve cells) conduct nerve impulses to their destination. Each second, thousands of nerve impulses arrive at the muscle cells making them contract vigorously.

The sensory cells of the man's eyes and ears in Figure 17.1A are linked by neurones to the cells of his leg muscles. Sensory cells and muscle cells are at the beginning and end of the process that allows the man to respond to the fierce dog. The process runs

Stimulus ➜ Receptor ➜ Nerves ➜ Effector ➜ Response

The senses

Our senses keep us in touch with the world around us. Each one consists of sensory cells adapted to detect a particular type of stimulus. The sensory cells of the eye detect light. The sensory cells of the ear detect sound. The sensory cells of the nose and tongue detect different chemicals. These different types of sensory cell are parts of complex organs located in the head. These organs are called organs of **special sense**. The rest of the body has sensory cells too. For example, different sensory cells in the skin detect pressure, pain, cold and heat (see Figure 17.5I).

Motor neurones conduct nerve impulses to muscle.

Neurones and nerves

Neurones link sensory cells with effectors. Figure 17.1B shows a human **motor neurone**. A nerve impulse can travel along an **axon** in milliseconds. It takes more than one neurone to link the cell of a sensory receptor with an effector cell. A number of neurones link up to form a route to take nerve impulses to their destination.

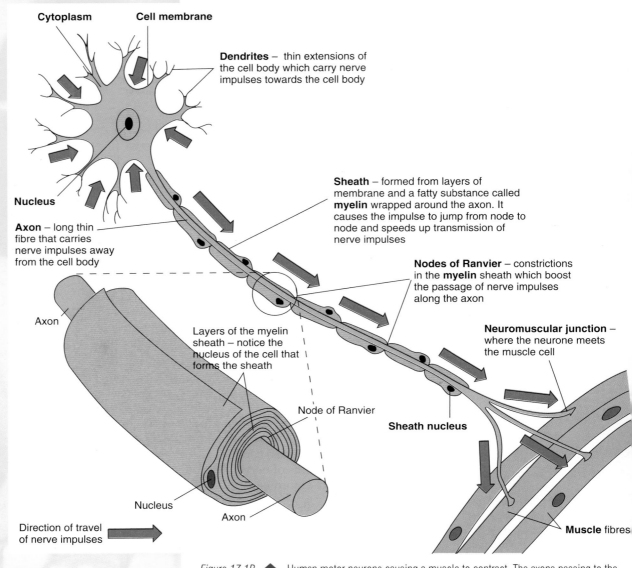

Cytoplasm

Cell membrane

Dendrites – thin extensions of the cell body which carry nerve impulses towards the cell body

Nucleus

Axon – long thin fibre that carries nerve impulses away from the cell body

Sheath – formed from layers of membrane and a fatty substance called **myelin** wrapped around the axon. It causes the impulse to jump from node to node and speeds up transmission of nerve impulses

Nodes of Ranvier – constrictions in the **myelin** sheath which boost the passage of nerve impulses along the axon

Neuromuscular junction – where the neurone meets the muscle cell

Axon

Layers of the myelin sheath – notice the nucleus of the cell that forms the sheath

Node of Ranvier

Sheath nucleus

Nucleus

Axon

Muscle fibres

Direction of travel of nerve impulses

Figure 17.1B ⬆ Human motor neurone causing a muscle to contract. The axons passing to the leg muscles can be up to a metre in length

 See www.keyscience.co.uk for more about neurones.

Minute gaps called **synapses** separate neurones from one another. Each synapse separates the ends of the axon of one neurone from the **dendrites** of the next (see Figure 17.1C). When nerve impulses arrive at the end of the axon they stimulate the production of a special chemical called a **neurotransmitter** which diffuses across the synapse to the dendrites of the neighbouring neurone. The neurotransmitter stimulates the dendrites to fire off new nerve impulses.

The dendrites of different neurones may form synapses with many incoming axons (Figure 17.1D). This allows for an enormous number of linkages. Nerve impulses, therefore, may be switched from one pathway to another within the billions of neurones that make up the nervous system.

It's a fact!

The fastest nerve impulses in humans travel at 8 m s⁻¹.

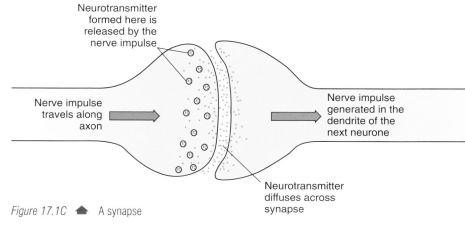

Neurotransmitter formed here is released by the nerve impulse

Nerve impulse travels along axon

Nerve impulse generated in the dendrite of the next neurone

Neurotransmitter diffuses across synapse

Figure 17.1C 🔺 A synapse

■ Nerves and nervous systems

Neurones are grouped together into bundles called **nerves** (see Figure 17.1E) which pass to all parts of the body forming the **nervous system** (see Figure 17.1F). At the head end the nerve cord expands into the **brain**, which makes numerous connections with the sense organs in the head. The sense organs feed information as nerve impulses to the brain. The brain interprets the information and sends nerve impulses through the nerve cords and nerves to effectors. This is how information about the environment travels through the body so that its effectors can respond in a useful way. The process is called **co-ordination**.

The nerve cord and brain form the **central nervous system**. The nerves that join the central nervous system form the **peripheral nervous system**.

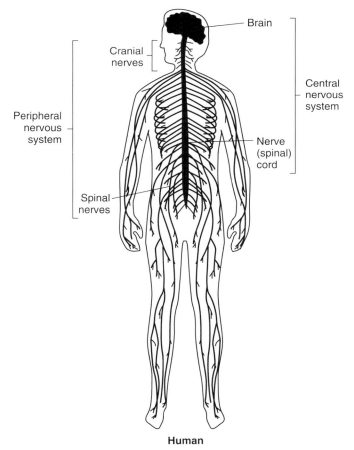

Figure 17.1F 🔺 Plan of the human nervous system

In 1829 Louis Braille, who was blind from the age of three, invented a stystem of writing for the blind. Letters are represented by different combinations of raised dots on paper. The dots are then read by touch.

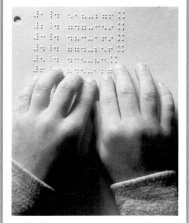

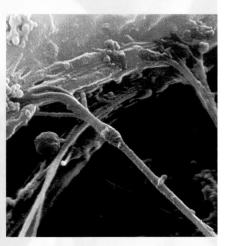

Figure 17.1D 🔺 The nerve cell (yellow) has formed synapses with two incoming axons (purple)

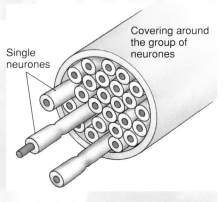

Single neurones

Covering around the group of neurones

Figure 17.1E 🔺 A nerve

■ Reflex responses

If you touch a hot stove you automatically move your hand away. We call the response a **reflex response** and the nerves involved form a **reflex arc** (see Figure 17.1G).

It's a fact!

The central nervous system is surrounded by layers of tissue called **meninges** which cushion it against bumps and jolts. Infection of the meninges causes the disease **meningitis**.

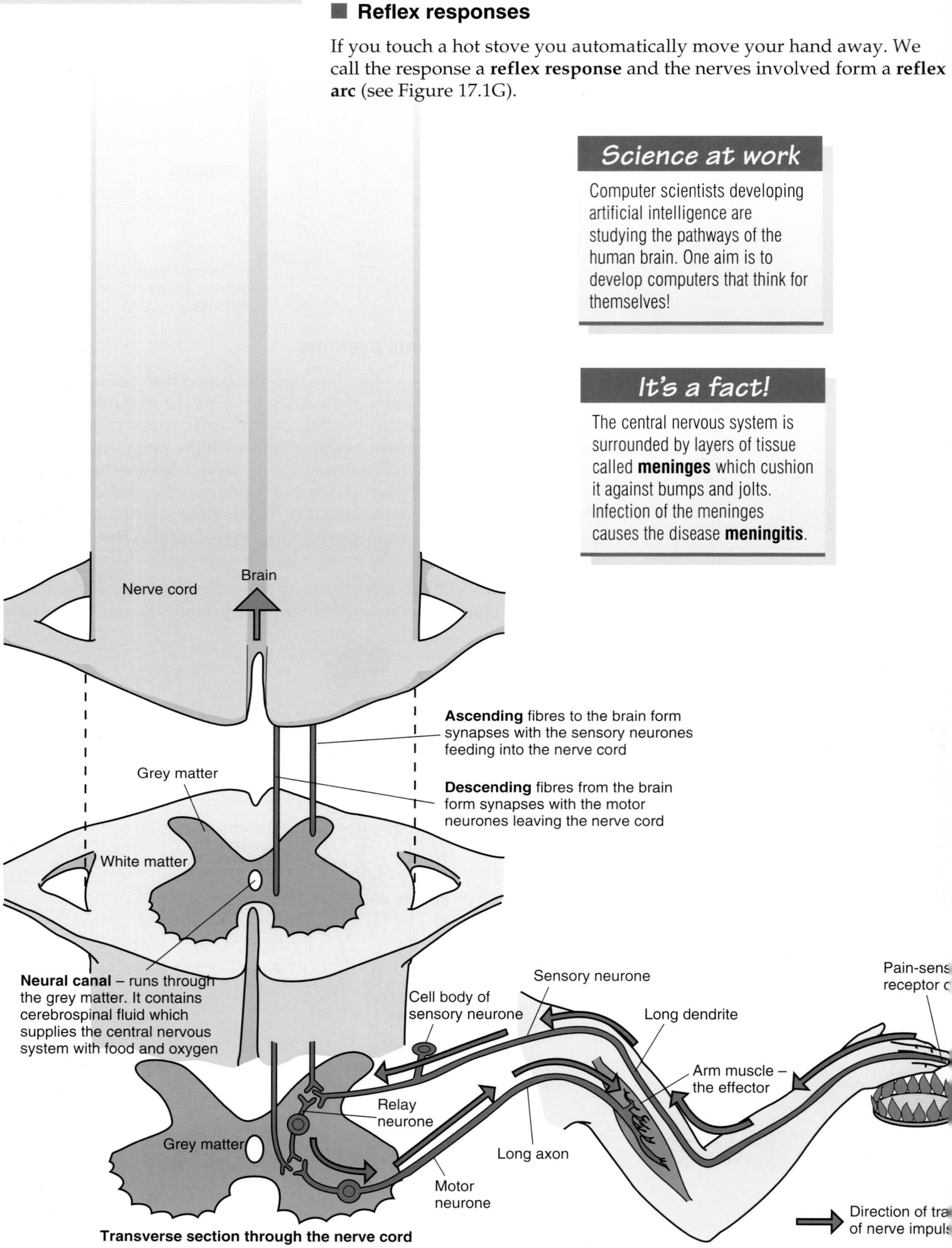

Nerve cord

Brain

Ascending fibres to the brain form synapses with the sensory neurones feeding into the nerve cord

Descending fibres from the brain form synapses with the motor neurones leaving the nerve cord

Grey matter

White matter

Neural canal – runs through the grey matter. It contains cerebrospinal fluid which supplies the central nervous system with food and oxygen

Cell body of sensory neurone

Sensory neurone

Long dendrite

Pain-sens receptor

Arm muscle – the effector

Relay neurone

Grey matter

Long axon

Motor neurone

Direction of tra of nerve impuls

Transverse section through the nerve cord

Figure 17.1G ▲ A reflex arc. To help trace the sequence receptor→effector (see p. 283), each nerve forming the reflex arc is represented by only one neurone.

Reflex actions which occur without you thinking about them are called **involuntary actions**.

Different types of neurone form a reflex arc. A synapse separates each type of neurone from the next neurone in the arc.

- **Sensory neurones** transmit nerve impulses from sensory receptors to the central nervous system (Figure 17.1F). When you touch a hot object a pain-sensitive receptor cell in your finger detects the stimulus – heat – which triggers off nerve impulses. These are transmitted to the nerve cord by sensory neurones.

- **Relay neurones** in the nerve cord receive nerve impulses from the sensory neurones and pass them to the motor neurones.

- **Motor neurones** receive nerve impulses from the relay neurones and transmit them to the effector (Figure 17.1G). In this case your arm muscles, which contract, lifting your finger out of harm's way.

Reflex actions often occur before the brain has had time to process the information. However, when the brain catches up with the events it then takes over and brings about the next set of reactions. These reactions could be a shout of pain or a decision to switch off the stove.

Ascending fibres carry information from the nerve cord to the brain. Descending fibres carry information from the brain to the nerve cord.

The brain catches up with the events because of the bundles of neurones running up and down the nerve cord called **ascending** and **descending** fibres. These form a zone of tissue called **white matter**. The white colour comes from the pale myelin sheaths that cover the axons. In the core of the nerve cord lies an H-shaped mass of **grey matter** that consists mainly of the cell bodies and axons of relay neurones.

KEY SCIENTIST

■ Conditioned reflexes

The influence of the brain on reflex responses was first investigated in a scientific way by the Russian physiologist Ivan Pavlov (1849–1936).

Pavlov noticed that when food was placed in a dog's mouth the flow of saliva increased. He also noticed that the flow of saliva increased as soon as the animal smelt his hand, even before the food was placed in its mouth. The salivary reflex was made stronger by following Pavlov's personal smell with the taste of food. After a period of presenting the dog with both the personal smell and the taste of food, the personal smell alone was enough to make the dog produce as much saliva as if it were given food.

Pavlov used the word 'conditioned' to describe the dog's response because it could be switched on by a non-food stimulus which the dog associated with the meal. Later work showed that in conditioned dogs new nerve pathways had been made. These connected the salivary reflex with other nerve circuits of the nerve cord.

Pavlov also conditioned dogs to salivate in response to other stimuli – the ringing of a bell for example. Conditioning fades unless it is periodically reinforced. Conditioning by personal smell, therefore, must be reinforced from time to time with a meal if the dog's salivary reflex is to remain conditioned. Try out the experiment on your dog or cat if you have one.

Ivan Pavlov

See www.keyscience.co.uk for more about the brain.

■ How the brain works

The brain consists of three regions: **forebrain**, **midbrain** and **hind-brain**. In humans the forebrain forms the **cerebrum** which is so large that it almost covers the rest of the brain (see Figure 17.1H).

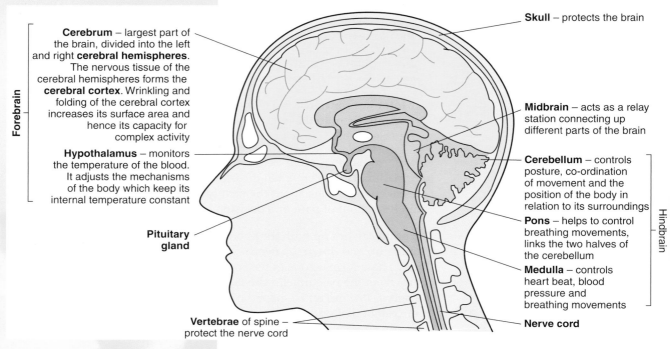

Forebrain

Cerebrum – largest part of the brain, divided into the left and right **cerebral hemispheres**. The nervous tissue of the cerebral hemispheres forms the **cerebral cortex**. Wrinkling and folding of the cerebral cortex increases its surface area and hence its capacity for complex activity

Hypothalamus – monitors the temperature of the blood. It adjusts the mechanisms of the body which keep its internal temperature constant

Pituitary gland

Skull – protects the brain

Midbrain – acts as a relay station connecting up different parts of the brain

Cerebellum – controls posture, co-ordination of movement and the position of the body in relation to its surroundings

Pons – helps to control breathing movements, links the two halves of the cerebellum

Medulla – controls heart beat, blood pressure and breathing movements

Hindbrain

Vertebrae of spine – protect the nerve cord

Nerve cord

Figure 17.1H 🔺 Section through the human head showing the different parts of the brain

Figure 17.1I shows some of the neurones in the brain. They are called **multipolar** neurones because each one has numerous dendrites which can form synapses with incoming axons. Scientists estimate that up to six million cell bodies make up 1 cm^3 of brain matter and that each neurone is connected to as many as 80 000 others.

The human brain weighs approximately 1.3 kg. It is the body's thinking and control centre. Reactions under the brain's control are called **voluntary reactions**. Memory and learning are also under the brain's control (see Figure 17.1I).

The different regions of the brain control a range of voluntary reactions.

Make sure you can answer an exam question which asks you to distinguish between involuntary reactions and voluntary reactions.

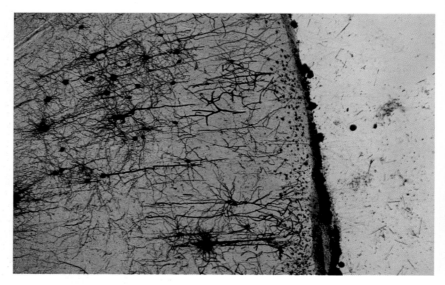

Figure 17.1I 🔺 Multipolar neurones in the brain cortex (× 70)

The different regions of the **cerebral cortex** each have different functions (see Figure 17.1J).

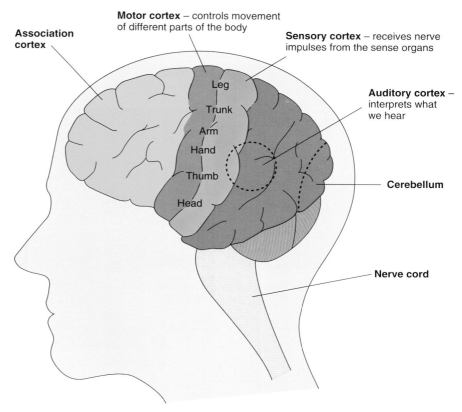

Figure 17.1J Functions of the cerebral cortex

■ Intelligence

What is intelligence? A difficult question. It includes the ability to decide how to tackle a problem and the ability to change your approach if it does not work.

Intelligence is not determined by a fixed centre in the **association cortex**. It depends on the way nerve fibres connect together in the different parts of the cortex and the way they connect the cortex with the rest of the brain. These nerve fibres form **association pathways**.

■ Effect of drugs

Drugs taken under doctor's orders can be of great benefit; they help ill people fight disease and control pain. However, the amount taken and when they are taken must be right or otherwise they may cause harm. For example, if drugs that affect the nervous system are taken under the wrong circumstances people can become **addicted** to them, wanting more and more and not caring about the damage caused to the body. Addiction is expensive because it is illegal to obtain addictive drugs without a doctor's prescription. Addicts, therefore, can only buy their supplies from drug dealers, who therefore make a lot of money. The life of the addict is one of ill health and declining self-respect, and usually short unless the habit is 'kicked'. **Abuse** is the word often used to describe the non-medical use of drugs.

Agonists are drugs that promote the generation of nerve impulses across synapses. They are **stimulants** because they speed up the activity of the nervous system. **Caffeine** in tea, coffee and chocolate is an example of a mild stimulant, **nicotine** in tobacco is another (see p. 254). **Cocaine** is much more powerful. Made from the leaves of the coca shrub from South America, its effects on the nervous system make it a popular drug of abuse. Slang names for cocaine include 'snow', 'coke' and 'C'. The drug may be sniffed ('snorting') (Figure 17.1K) , injected or smoked. 'Crack' is a highly addictive form which is smoked. Cocaine increases confidence and personal energy. A strong **psychological** need for cocaine may also develop and this is one of the drug's main dangers.

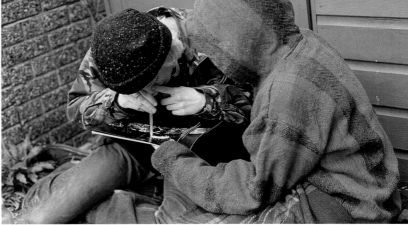

Figure 17.1K Snorting coke

Make sure that you can distinguish between drugs that are *agonists* and drugs that are *antagonists*.

Caffeine, nicotine and the ethanol in alcoholic drinks are not illegal drugs. Cocaine, heroin, cannabis and LSD are.

Antagonists are drugs that block the generation of nerve impulses across the synapses. They are **depressants** because they slow down the activity of the nervous system. **Ethanol** (alcohol in beers, wines and spirits) is an example. Small amounts affect the association cortex of the brain which controls judgement. Larger quantities affect the motor cortex which controls movement. Even more impairs memory. More and more alcohol affects further areas of the brain until vital brain centres that keep us alive are affected. Death may follow (see p. 215). **Heroin** is made from the pain-reducing drug **morphine**, which is produced from opium, a

(a) Smoking heroin is called 'chasing the dragon'

(b) Injecting heroin is called 'main lining'

Figure 17.1L Abusing heroin

chemical substance found in poppy plants. Widely abused, it is also a depressant, although to begin with people take heroin for the 'high' it produces. In its pure form heroin is a white powder which may be smoked or made up in solution and injected into a vein (Figure 17.1L).

Dependence develops rapidly. Addicts continue to use the drug, not so much for the euphoric effect but to avoid the symptoms that follow from giving it up (**withdrawal**). Symptoms vary but are generally unpleasant and can be life threatening. The withdrawal effects are called **cold turkey**. **Methadone** is another heroin-like drug. It is used to treat addicts for symptoms of withdrawal from heroin. Substituted for heroin, withdrawal from methadone is 'tapered off' as the person slowly responds to treatment.

Hallucinogens are another group of drugs that affect the nervous system. **Cannabis** (marijuana) which comes from the leaves of the cannabis plant and **lysergic acid diethylamide** (LSD) are examples. They produce sensations of false reality (hallucinations – hence their name). For example, people may think that they can jump out of high windows without falling. Sniffing the volatile solvents in glues, paints, nail varnish and cleaning fluids (**solvent abuse**) also produces dangerous disorientation. The solvents slow down bodily functions affecting, for example, the nerve centres which control breathing and heart rate. Long-term solvent abuse can lead to damaged liver and kidneys.

CHECKPOINT

▶ **1** Complete the following paragraph using the words provided below. Each word may be used once, more than once or not at all.

largest   nerve cord   axons   skull   vertebrae   cerebral hemispheres

The brain and _____ make up the central nervous system. They control many bodily activities and are well protected. The brain is enclosed in the _____ and the spinal cord runs through a channel in the _____ . The cerebrum is divided into two _____ . It is the _____ part of the human brain.

▶ **2** The diagram opposite shows a motor nerve cell. Name the parts labelled A–F on the diagram and explain their function.

▶ **3** (a) What is a response?

(b) The diagram below shows a reflex arc.

(i) Explain briefly what is happening at the points labelled A–G on the diagram below.

(ii) Name the parts numbered 1–3.

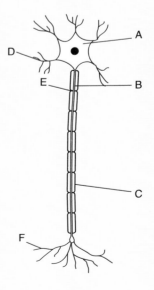

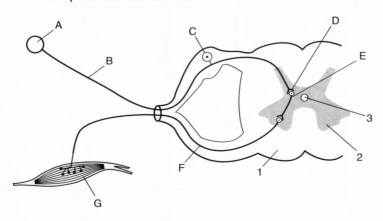

▶ **4** We respond to stimuli all the time. When a door bell rings (stimulus), we answer the door (response): when we feel hot (stimulus), we take jumpers and coats off (response).

    (a) List at least seven stimuli to which you have responded today and state your responses.

    (b) Name the sensory receptors responsible for detecting the stimuli.

    (c) Briefly state the role (i) sensory receptors, (ii) nerves, (iii) muscles or glands, play in the chain of events from stimulus to response.

▶ **5** Match each of the biological terms in Column A with their functions in Column B.

Column A	Column B
Cerebellum	Transmits nerve impulses from the central nervous system to a muscle
Medulla	Transmits nerve impulses from a sense receptor to the central nervous system
Relay neurone	Controls learning and memory
Cerebrum	Controls pulse, breathing movements and other involuntary actions
Sensory neurone	Controls balance
Motor neurone	Links a sensory neurone with a motor neurone or relay neurone

17.2 ▶ The ear

FIRST THOUGHTS

Your ears are vital organs that enable you to receive information from your immediate environment. In this section you will find out how the ear works and how to test your own hearing response.

Have you ever listened to the sound of your own voice played back on a tape recorder? Try it and you will hear yourself as others hear you. To you, your voice will sound different on the recording. When you hear yourself speak, sound waves from your voice travel through your head as well as through the air to reach your ears. Other people only receive the sound of you speaking through the air.

The human ear is a remarkable organ that can detect an enormous range of sound waves. The loudest sounds it can withstand carry over one million million times more energy than the quietest sounds it can hear. It can detect frequencies from about 20 Hz (a bee buzzing) to about 18 000 Hz (a very high-pitched whistle).

The ear sends signals to the brain in response to sound waves arriving at the **eardrum**. Sound waves arriving at the ear make the eardrum vibrate. These vibrations are passed through the middle ear by three tiny bones, the **hammer, anvil** and **stirrup**, to reach the **oval window** of the inner ear. The vibrations of the oval window are transmitted through the fluid of the **cochlea**, making the **basilar membrane** vibrate. Tiny **hair cells**, which are sound-sensitive receptors, are lined up on the basilar membrane. The vibrating membrane activates the hair cells, which fire off nerve impulses to the brain along the **auditory nerve**.

What happens to the ear if very loud sound falls on it? If this happens too often, the ear becomes less and less sensitive and deafness can occur. One reason for this is that the bones of the middle ear vibrate too much and get worn down. They then become less effective at passing the vibrations from the ear drum to the oval window. Operators of noisy machines must wear ear pads or risk suffering permanent loss of hearing. Protect your ears at noisy discos by putting cotton wool in your ears.

Loudness is measured in **decibels** (dB). The faintest sound that the ear can hear is defined as zero decibels (0 dB). Imagine steadily increasing the

It's a fact!

The fleshy lobe of the outer ear is called the **pinna**. It funnels sound waves down the ear tube to the ear drum. Cats, dogs and other mammals can adjust the pinna and cock it towards sources of sound. In most humans it is fixed. The walls of the ear tube produce **wax** which keeps the ear drum soft and supple.

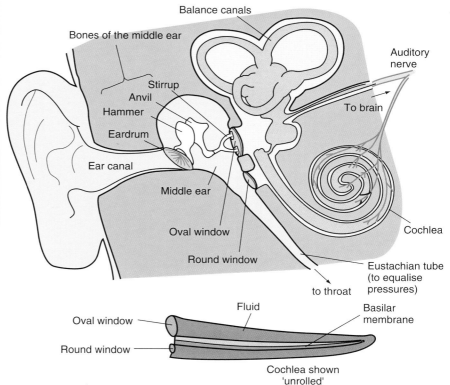

Figure 17.2A 🔺 The human ear

loudness of a radio from zero until it becomes too loud to bear. For every ten decibel (10 dB) increase in loudness, the energy of the sound waves is increased by a factor of 10. The sound would become too loud to bear at about 120 dB. Since this is 12 steps at 10 dB for each step, sound waves at this loudness carry $10\times10\times10\times10\times10\times10\times10\times10\times10\times10\times10\times10 = 10^{12}$ times as much energy as the faintest sound waves. Figure 17.2B shows the decibel levels of some everyday sounds.

The response of the ear to different levels of loudness varies with frequency, as shown in Figure 17.2C. The ear is most sensitive and can detect the softest sounds at about 3000 Hz. It is completely insensitive and cannot detect any sound over 18 000 Hz.

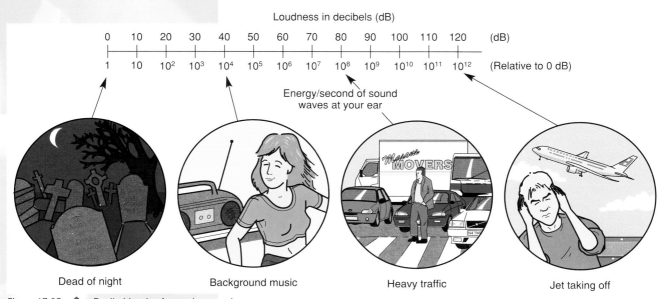

Figure 17.2B 🔺 Decibel levels of everyday sounds

293

SUMMARY

The ear converts signals carried by sound waves into nerve impulses that it sends to the brain. The ear cannot detect frequencies above 18 000 Hz. Loudness levels above 120 decibels can cause deafness if the ear is not protected.

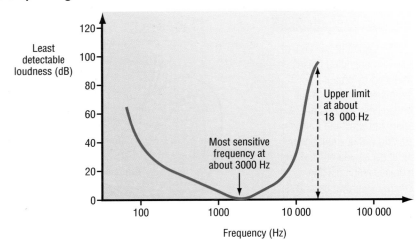

Figure 17.2C ⬥ Frequency response of the ear

CHECKPOINT

▶ 1 (a) One of the first hearing aids was the 'ear trumpet', which was a large hollow horn held to the ear. Why do you think this device improves hearing ability?

  (b) Modern hearing aids are so small that they can be worn behind and even in the ear. Such a device contains an electronic amplifier, a tiny microphone and an earpiece speaker. The amplifier makes electrical signals bigger without changing the frequency of the signal. What is the purpose of the microphone and what is the earpiece speaker for?

▶ 2 What is the function of each of the following parts of the ear:

  (a) the eardrum,   (d) the oval window,
  (b) the pinna,   (e) the hair cells?
  (c) the bones of the middle ear,

▶ 3 Play back your own voice using a tape recorder or mini disc. How does it differ from what you hear when you speak? Compare the recorded voices of your friends. Do their voices seem different from when they speak directly to you?

▶ 4 Here is a passage from Claire's diary describing part of an evening out with her friends.

'It was very noisy and hot in the disco. We could only hear each other when the music stopped. I got a lift home with Michelle and her dad. There was a thunderstorm on the way home. When I got home, the TV was on very loud so I went to my bedroom for some peace and quiet.'

  (a) When was the loudness level greatest and when was it least?

  (b) Estimate the loudness when it was greatest.

  (c) Which do you think was most damaging to the ears: the thunderclap or the disco noise?

▶ 5 If you are in an aeroplane coming in to land, or in a fast train entering a tunnel, you may feel 'popping' sensations in the ears. Look at Figure 17.2A and describe how the Eustachian tubes help overcome this uncomfortable feeling.

▶ 6 Notice in Figure 17.2A that the balance canals are arranged three-dimensionally, each at right-angles to the others. In each canal there is a swelling at one end called an **ampulla**, which contains a structure called the **cupula**. Liquid inside the balance canals pushes the cupula one way or the other depending on the position of the head. The cupula pulls on sensory hairs which send nerve impulses to the brain. Find out and describe how the balance canals work to give you a sense of balance and position. You should include a structure called the **utriculus** as well as the ampulla and cupula in your description. Use diagrams wherever they help to make your answer easier to follow.

▶ 7 Figure 17.2C shows the frequency response of the ear.

  (a) How loud must a sound with a frequency of 18 000 Hz be in order for it to be heard?

  (b) Would you hear a sound with a frequency of (i) 50 Hz (ii) 1000 Hz (iii) 25 000 Hz?

  (c) At around which frequency is the ear most sensitive?

17.3 ▶ The eye

This section concentrates on the eye as an optical instrument and on its biological details. Read on to find out about sight defects and how they are corrected.

Have you ever been told to 'use your eyes' when you complain that you cannot find something or other? Think about what we use our eyes for. They tell us about colour, shape, position and movement. When you look at an object, each eye forms an image of it and sends signals to your brain. Your brain 'reads' these signals and you 'see' the object. Figure 17.3A shows four unusual views of everyday objects. *Can you recognise them?*

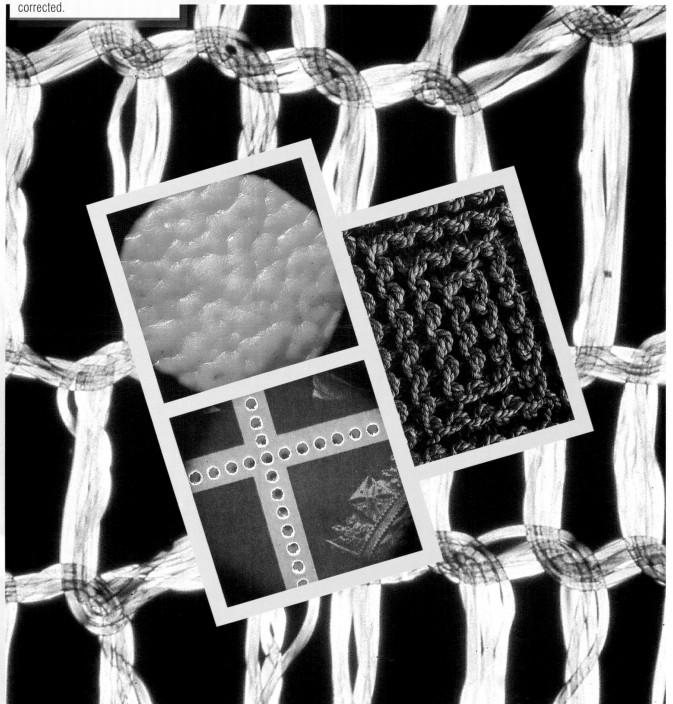

Figure 17.3A ▲ Recognising things

It's a fact!

The front of the eye is covered by a thin transparent membrane called the **conjunctiva**. Dust particles that collect on the conjunctiva are washed away by a watery fluid from the tear glands, which are under the eyelids. This fluid contains **lysosyme** – an enzyme that destroys bacteria. Blinking helps to spread the fluid across the conjunctiva. When the fluid reaches the lower part of the eye, it drains into a tube and goes down into the nose.

 See www.keyscience.co.uk for more about the eye.

Figure 17.3B explains how the parts of the eye work. Light enters the eye through a tough transparent layer called the **cornea**. This protects the eye and it helps to focus the light on to the **retina**, the layer of light-sensitive cells at the back of the inside of the eye. The amount of light entering the eye is controlled by the **iris**, which adjusts the size of the **pupil** – the circular opening in its centre. The eye lens focuses the light to give a sharp image on the retina. Although the image on the retina is inverted, the brain interprets it so you see it the right way up.

Focusing

How does the eye focus on objects at different distances? If you look up from this book and gaze out of the window, your eye lens becomes thinner to keep your vision in focus. The adjustment of lens shape to focus light coming from different distances is called **accommodation**. The **ciliary muscles** alter the thickness of the eye lens. The muscles run round the eye lens and are attached to the lens by **suspensory ligaments**. The tension in the suspensory ligaments and therefore the shape of the eye lens is altered by the contraction and relaxation of the ciliary muscles.

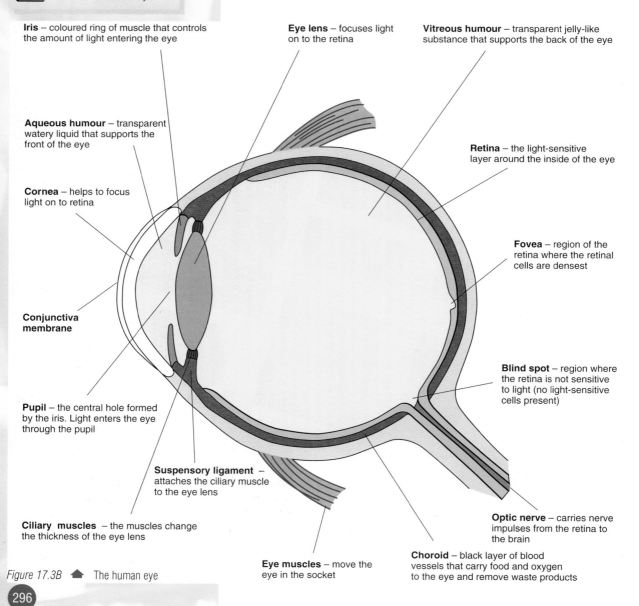

Figure 17.3B The human eye

When the muscle fibres contract, the suspensory ligaments are loosened and the eye lens becomes *fatter*. When the muscle fibres relax, the suspensory ligaments tighten, pulling the lens and making it *thinner*.

What is your range of vision? A normal eye can see clearly any object from far away to 25 cm from the eye.

- The **near point** of the eye is the closest point to the eye at which an object can be seen clearly. The eye lens is then at its fattest (Figure 17.3C).

- The **far point** of the eye is the furthest point from the eye at which an object can be seen clearly. The eye lens is then at its thinnest (Figure 17.3D).

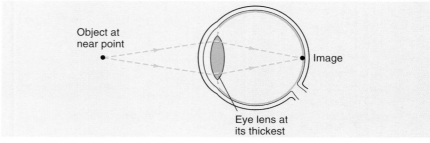

Object at near point

Image

Eye lens at its thickest

Figure 17.3C 🔺 The near point

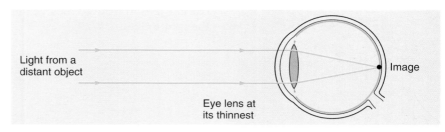

Light from a distant object

Image

Eye lens at its thinnest

Figure 17.3D 🔺 The far point

Seeing shape and colour

How do we recognise the shape of an object? An image of the object is formed on the retina, which consists of lots of light-sensitive cells. When light falls on a cell, the cell sends a nerve impulse as a signal to the brain. The brain recognises the pattern of the signals from the cells covered by the image, and so recognises the object's shape.

How do we tell the colour of an object? There are two types of cells on the retina – **rods** and **cones** (Figure 17.3E). Rods occur mostly near the edges of the retina. They are not sensitive to colour and only respond to the brightness of light. Ask a friend to test you to see if you can tell the colour of something at the edge of your field of vision.

Cones are packed most densely at the middle of the retina. This area is called the **fovea**. Each cone is sensitive to red or blue or green light. For example, when red light falls on the retina, it 'activates' the red-sensitive cones, so you see red. Other colours activate more than one type of cone. For example, yellow light activates the red and the green cones, so they send messages to the brain. When the brain receives signals from adjacent red and green cones, it knows yellow light is on that part of the retina.

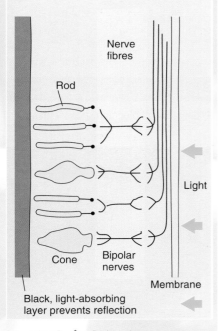

Nerve fibres

Rod

Light

Cone

Bipolar nerves

Membrane

Black, light-absorbing layer prevents reflection

Figure 17.3E 🔺 Rods and cones

A short-sighted person cannot see far away objects clearly. A long-sighted person cannot see near objects clearly.

Sight defects

Sight defects occur when the eye lens cannot form a sharp image on the retina. The lenses of spectacles compensate for sight defects.

Short sight is caused by over-strong eye muscles. A short-sighted eye cannot see far away objects clearly because the eye muscle cannot relax enough to make the eye lens thin enough. A suitable **concave** lens in front of the eye counteracts the effect of the over-strong eye lens (Figure 17.3F).

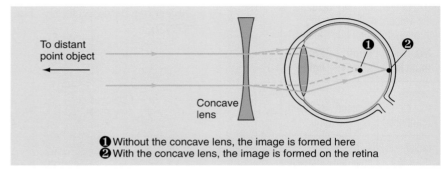

❶ Without the concave lens, the image is formed here
❷ With the concave lens, the image is formed on the retina

Figure 17.3F ▲ Short sight

Long sight is caused by weak eye muscles, and often develops as the muscles weaken with age. The muscles are unable to contract enough around the lens to make it thick enough to focus near objects. A suitable **convex** lens in front of the eye helps the eye lens to form a clear image on the retina (Figure 17.3G).

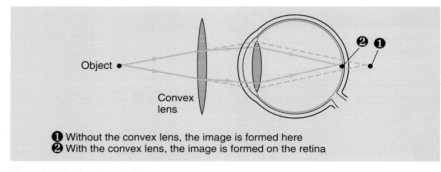

❶ Without the convex lens, the image is formed here
❷ With the convex lens, the image is formed on the retina

Figure 17.3G ▲ Long sight

SUMMARY

The eye lens forms a clear image on the retina. The iris controls the amount of light entering the eye. The thickness of the eye lens changes to accommodate (focus) objects at different distances. Spectacle lenses compensate for defects in the eye muscles controlling the eye lens, to give the wearer normal vision.

CHECKPOINT

▶ **1** (a) Why do your eye muscles relax when you look at a distant object and contract when you look at something close to you?

(b) Why is it easier to study an object in detail if you look straight at it?

(c) Why does the eye pupil dilate (i.e. widen) in dim light?

▶ **2** Explain how

(a) short sight and (b) long sight are caused and how they are corrected.

▶ **3** Try these 'eye tests', which are explained below.

(a) The blind spot test.

(b) The sausage test.

(c) The birdcage test.

(d) The dark room test. Sit in a dark room for twenty minutes or more and you will discover your eyes can see in the dark. The rods become much more sensitive than normal in dark conditions and they make the most of whatever light there is.

The blind spot test

×   ●

1 Position the black spot in front of your left eye. Cover your right eye.

2 Move the book closer and keep staring at the spot.

3 The X disappears when its image falls on the blind spot of your left eye.

The 'sausage' test

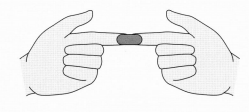

1 Hold your hands in front of your face with the tips of your index fingers touching.

2 Stare past your hands at a distant object and move them towards you.

3 You should see a 'sausage' between the tips of your index fingers caused by overlapping images from each eye.

The birdcage test

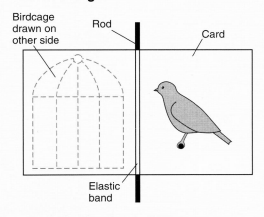

Birdcage drawn on other side

Rod

Card

Elastic band

1 Make the test card as shown.

2 Spin the card about the rod.

3 The bird appears in the cage because the images last for a fraction of a second and hence overlap. This is called 'persistence of vision'.

4 Try spinning the card at different speeds. What difference does it make?.

17.4 ▶ Hormones

Each meal or snack that you eat provides your body with a surge of glucose. Glucose is an important source of energy but high levels seriously disrupt the body's cells and cause them to function very inefficiently.

Your body keeps the level of glucose constant by means of **hormones**. These substances regulate the concentration of glucose in the blood and cope with the surge of glucose at meal-times. The control of glucose levels is one example of **homeostasis** (see Topic 17.5) Hormones are chemicals produced by animals and plants to regulate the organisms' activities. In plants they are **growth substances** (see Topic 13.1) In animals they are produced in the tissues of ductless glands called **endocrine glands** and released directly into the blood system. Hormones circulate in the blood and cause specific effects on the body. (See Figure 17.4A.)

Pituitary gland (connected to the hypothalamus at the base of the brain) – produces nine different hormones which affect:
- *water reabsorption* from kidney tubules
- *growth*
- *sperm and egg* production
- *release of hormones* by other endocrine glands

Thyroid – produces **thyroxin** which affects the *rate of metabolism*

Adrenal gland – produces **adrenalin** which increases heartbeat and the rate of breathing. More blood and the oxygen it contains are therefore delivered to the muscles. Adrenalin also stimulates the release of glucose from the liver (see p.238). As a result the level of glucose circulating in the blood increases. The muscles therefore receive more glucose as well as more oxygen: both necessary for the release of energy which muscle cells require for the rapid contraction needed during sudden action. No wonder adrenalin is sometimes called the fright, flight or fight hormone!

Lungs

Heart

Stomach

Islets of Langerhans – groups of cells in the pancreas which produce **insulin** and **glucagon**. These hormones help to *regulate glucose* levels in the blood

Kidney

Testis (male) – produces **testosterone** which helps to develop and maintain secondary *sexual* characteristics

Ovary (female) – produces **oestrogen** and **progesterone** which *regulate the menstrual cycle* and help to develop and maintain secondary *sexual* characteristics

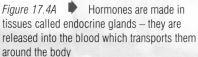
Figure 17.4A ▶ Hormones are made in tissues called endocrine glands – they are released into the blood which transports them around the body

Endocrine glands produce hormones which are usually released into and circulate within the blood system. Each hormone affects a particular target tissue.

The tissue on which a particular hormone (or group of hormones) acts is called a **target tissue**. A hormone affects its target tissue more slowly than a nerve impulse affects a muscle. This is because nerve impulses move rapidly along neurones. Muscles therefore can respond very quickly to changing circumstances. The action of most hormones is longer-term (adrenalin is an exception – see Figure 17.4A).

See www.keyscience.co.uk for more about blood sugar regulation.

Diabetes mellitus occurs when the pancreas does not produce enough insulin.

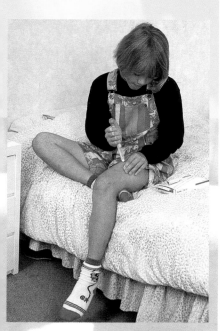

Figure 17.4B ▲ A diabetic child injecting herself with insulin

In teenagers the body matures under the control of hormones over several years. In girls, **oestrogen** controls broadening of the hips and breast development. In boys, **testosterone** controls beard growth, broadening of the shoulders and the deepening of the voice. These developments (called **secondary sexual characteristics**) mark the start of **puberty** and continue through adolescence. Puberty begins in girls at age 11–13 and in boys at age 13–14. Because of the changing balance of hormones in the body, adolescents may experience swings of mood and also skin troubles such as spots and acne. Usually these problems have cleared up by the early twenties.

Hormone regulation of blood glucose

To keep the level of glucose in the blood constant hormones balance the glucose-producing and glucose-using processes of the body.

- **Thyroxine** increases the rate at which glucose is oxidised in **cellular respiration** (see Topic 15.1). This *decreases* the level of glucose in the blood.
- **Insulin** also *decreases* the level of glucose in the blood. It does this by promoting the conversion of glucose into glycogen (see Topic 10.1)
- **Glucagon** *increases* the level of glucose in the blood by promoting the conversion of glycogen into glucose.

If the pancreas does not produce enough insulin a condition called **diabetes mellitus** occurs. The glucose level in the blood becomes dangerously high and can cause blindness or kidney failure. Concentrations of glucose become so high that the kidneys cannot reabsorb all the glucose and glucose is excreted in the urine. A simple test for glucose in the urine of a patient can tell a doctor if a patient is **diabetic**.

Diabetics suffer from thirst and tiredness. If the diabetes is not too severe, a carefully chosen low-sugar diet can control the condition. If the diabetes is severe, diabetics are taught to inject themselves regularly with insulin (see Topic 24.5) to lower their blood glucose levels (see Figure 17.4B). Getting the dose of insulin right is not always easy (see p. 387). If too much insulin is injected, the glucose level in the blood falls too low and diabetics can suffer from unpleasant side effects. Diabetics soon learn to recognise the symptoms and eat a little sugar to boost blood glucose to the right level.

Hormones, growth and muscle building

If a child's pituitary gland does not produce enough **growth hormone** then the child will not grow to a normal height. Providing the condition is diagnosed at an early age, the affected child can be given a growth hormone (see Topic 24.5) to make up for the deficiency. The child puts on a growth spurt and soon 'catches up' with other children of the same age (see Figure 17.4C).

At some time or other you have probably seen athletes straining to lift heavy weights or put the shot long distances. Athletes train very hard to build up the muscles needed to compete in these 'strength' events (see Figure 17.4D). A few, however, cheat and inject themselves with hormones called **anabolic steroids** which develop their muscles even further, giving them an unfair advantage over other competitors.

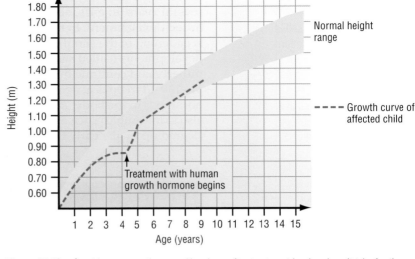

Figure 17.4C Human growth curve. How long after treatment begins does it take for the affected child to 'catch up' with other children?

Anabolic steroids mimic the effect of the hormone testosterone which in men controls the development of secondary sexual characteristics. Testosterone also increases the rate of **protein synthesis** (in women as well as men). Protein makes muscle, which is why athletes who take anabolic steroids develop bigger and stronger muscles. The side-effects of anabolic steroids are very unpleasant. Liver damage is possible and women can develop male secondary sexual characteristics. Men can become sexually impotent.

Anabolic steroids are on the list of drugs banned by the International Olympic Committee. Any athlete caught using them can be prohibited from taking part in future sporting events.

Figure 17.4D The muscles you need for strength events

Hormones control growth and development

A fertilised egg looks very simple under the microscope but it is the starting point for the development of a new individual. The instructions for development are contained in its genes (see Topic 10.4). These are duplicated many times as the embryo develops. Even though all the cells of the multicellular individual have the same set of genes, different types of cell develop (**differentiate**) to do different biological jobs (see p. 186).

How does the same set of genes produce the many different kinds of cell which make up the individual organism? It seems that only some genes are active in each cell type for some of the time.

Development is an exact sequence of events which lays down the different features of the embryo in the right place at the right time. This means that the genes which produce these features must be switched on and off in the right order. Hormones seem to play an important part in switching genes on and off. Thyroxine, for example, controls the growth and development of amphibian tadpoles into adults. The change is called **metamorphosis**. When the production of thyroxine in a tadpole is stopped, the tadpole does not metamorphose into an adult but grows into a giant tadpole.

Hormones also control insect development. The hormone **ecdysone** makes the young insect grow and **moult** its outer body covering, the exoskeleton. It does this in the correct sequence by switching on the genes which control these processes at the right time in development.

Hormones control development by switching genes on and off in the correct sequence.

How insects grow

The growth curve for insects is not like the smooth growth curve for plants shown in Figure 11.3B or that for humans shown in Figure 17.4C. It is stepped (see Figure 17.4E).

Insects grow like this because the hard exoskeleton that surrounds the body cannot stretch. An increase in length occurs only when the old exoskeleton is removed (called moulting or **ecdysis**) and replaced by a new one. The body tissue expands while the exoskeleton is still soft and stretchy.

Insects change a lot as they moult and grow into adults. The changes are another example of metamorphosis. In insects like locusts, metamorphosis is gradual. Young locusts (called hoppers) look rather like miniature adults except that the wings and sex organs are not developed. With each moult (five in all) the hoppers get bigger as Figure 17.4E shows. At the last moult they become adults with fully developed wings and sex organs. This gradual change is called **incomplete metamorphosis** and the young of insects which show incomplete metamorphosis are called **nymphs** (see Figure 17.4F).

Make sure you can answer an exam question which asks you to compare the stepped growth curve for insects with the smooth growth curve for humans.

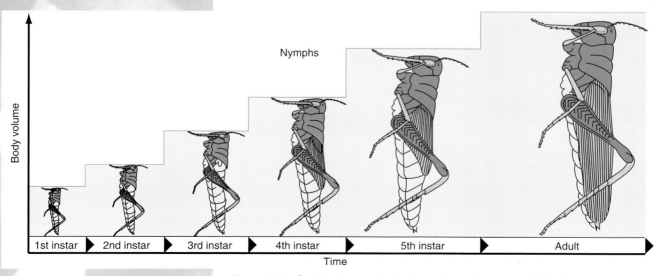

Figure 17.4E ⬆ Growth curve of a locust – each stage between moults is called an instar. A locust nymph passes through five instars before it becomes an adult. It increases in size at every moult

Figure 17.4F 🔺 Nymph and adult locust – incomplete metamorphosis

Make sure that you can distinguish between incomplete metamorphosis and complete metamorphosis.

Metamorphosis is more dramatic in insects like butterflies, moths and flies. The young are called **larvae** and do not look at all like the adult. They moult and grow but then turn into **pupae**. The final changes into the adult take place inside the pupa. When the changes are complete the adult emerges from the pupa, dries off and flies away. This dramatic change from young to adult is called **complete metamorphosis** (see Figure 17.4G and Figure 17.4H).

It's a fact!

Metamorphosis allows a species (see p. 8) to occupy different niches (see p. 45) at different stages in its life. For example, the larvae of different species of butterfly eat leaves or other parts of a plant. The adults suck nectar (see p. 347). As a result competition for food between adults and their offspring is reduced.

Figure 17.4G 🔺 Larvae, pupae and adults – complete metamorphosis. The larvae of different insects do not look alike. Butterfly and moth larvae have a distinct head and short, stumpy legs. They are called caterpillars

Figure 17.4H 🔺 Blowfly larvae are much simpler; they are called maggots

It's a fact!

The heart beats roughly the same number of times during the lifetime of most mammals, but the heart rate varies. Small animals have a short life and fast heart rate; large animals live longer and have a slower heart rate. On this basis humans live three times longer than expected! We do not know why.

How humans grow

A **baby** grows to become a **child**; children develop into **adolescents** who become **adults** at around the age of 20 years. These are the stages signposting the route of human growth and development. Figure 17.4I shows how different parts of the body increase in size during the first twenty years of a person's life. Notice that different parts of the body grow at different rates because cell division occurs more quickly in some parts than in others. A child's head is bigger in proportion to the rest of the body compared with an adult's. Growth of arms and legs speeds up during adolescence resulting in the proportions of head and body we see in adults.

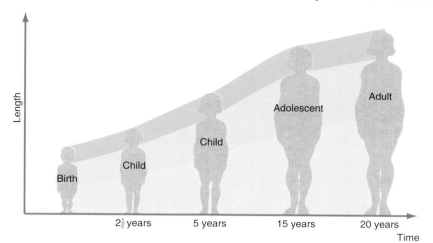

Figure 17.4I ◆ Human growth

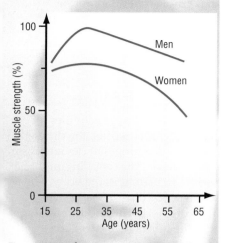

Figure 17.4J ◆ Muscle strength in men and women of different ages

Ageing begins in the middle twenties. By ageing we mean not only advancing years but also loss of **fitness**. Different aspects of fitness peak at different times, and individuals vary, especially between the sexes. For example Figure 17.4J shows the peak and decline of average muscular strength in men and women as they grow older.

Exercise helps to slow the ageing process so that early death from heart disease, for example, is less likely. Slowing down ageing does not mean living longer (although this may happen) but rather that good health is enjoyed for a much greater part of life.

CHECKPOINT

▶ 1 What are hormones and how are they transported round the body?

▶ 2 Complete the following paragraph using the words provided below. Each word may be used once, more than once or not at all.

oestrogen   target   long-term   testosterone   nervous   short-term

transports   progesterone   adrenalin   thyroxine   blood   sexual

The endocrine tissues secrete hormones directly into the _____ which _____ them all over the body. The tissues affected by hormones are called _____ tissues. The uterus, for example, responds to the hormones _____ and _____ . The action of most hormones is _____ . Female secondary _____ characteristics, for example, develop under the influence of _____ , male secondary _____ characteristics develop under the influence of _____ . The adrenal gland secretes _____ which prepares the body for sudden action. The rapidity of its effect is more like that of the _____ system than that of other hormones.

▶ 3 The diagram shows changes in the proportions of the human body from birth to adulthood. Analyse the changes in proportions of the head and the rest of the body in relation to total body length at each stage of development. Write a brief report of your analysis suggesting reasons for the changes.

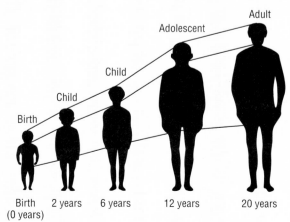

17.5 ▶ Maintaining a constant internal environment

The cells of the body work efficiently when they are at an appropriate temperature and pH (acidity/alkalinity) and supplied with a solution of all of the substances they need to maintain their activity. These conditions are part of the body's **internal environment**. Different mechanisms help regulate the body, keeping its internal environment fairly constant.

Homeostasis

Keeping conditions constant is called **homeostasis**. The process depends on **negative feedback** mechanisms which enable different activities to correct themselves when changes take place. In other words, the different processes of life are **self-adjusting**, returning something that deviates from a **set point** (or normal value) back to that set point. The set point might be blood sugar values or the levels of hormone circulating in the body. The menstrual cycle pictured in Figure 22.2C on p. 363 illustrates the principles. Rising levels of **follicle stimulating hormone** and **luteinising hormone** stimulate the ovaries to produce the hormone **oestrogen**. Increasing levels of oestrogen feed back negatively, inhibiting further release of follicle stimulating hormone. The level of the hormone returns to normal, its role of promoting the growth of egg follicles complete.

Regulating glucose

The different hormones which regulate the level of glucose in the blood are described on p. 301. Here we shall see how the hormones are a part of homeostasis. Figure 17.5A shows you the idea:

● *High* concentration of blood sugar promotes the release of insulin.

● *Low* concentration of blood sugar promotes the release of glucagon. The result of the interaction between the two hormones is the regulation of glucose at around 90 mg per 100 cm^3 of blood.

The hormones insulin and glucagon have opposite effects on blood glucose (sugar) levels.

See www.keyscience.co.uk for more about blood sugar regulation.

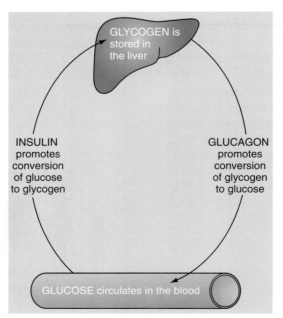

Figure 17.5A ▲ Homeostatic control of blood sugar levels

Each kidney consists of about one million tubules called nephrons.

Important functions of the nephrons are the removal of wastes (mostly urea) and regulation of the composition of the blood.

See www.keyscience.co.uk for more about excretion.

EXTENSION FILE
ACTIVITY

See www.keyscience.co.uk for more about the kidneys.

Regulating water

Metabolism produces waste substances which must be removed from the body. The removal of metabolic wastes is called **excretion**. Water and salts in excess of the body's needs must also be excreted. In humans and other mammals the kidney is the main excretory organ. It also plays a part in homeostasis by regulating the amount of water (osmoregulation – see p. 151) and concentration of salts and other useful substances in the blood.

The kidneys at work

Figure 17.5B shows you where the kidneys are in the human body. Each one consists of about one million tiny tubules called **nephrons**. The nephron is the working unit of the kidney (Figure 17.5C). Blood is brought to the kidneys by the **renal arteries**. It contains wastes (mostly **urea**) which the nephrons remove from the blood along with glucose, salts and other substances in solution. Some of these are useful and the nephron reabsorbs them into the blood. The nephron's contribution to homeostasis, therefore, is the part it plays in keeping the composition of the blood constant.

Look at Table 17.1. Notice the composition of blood in the **glomerulus**, before treatment by the nephron, and the composition of the liquid (called **urine**) at the end of its journey through the nephron. Blood cleaned of wastes is carried away from the kidneys by the **renal veins**.

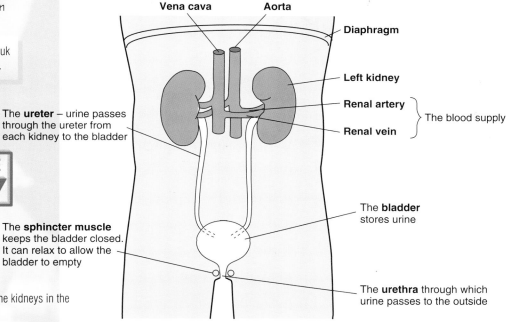

The **ureter** – urine passes through the ureter from each kidney to the bladder

The **sphincter muscle** keeps the bladder closed. It can relax to allow the bladder to empty

Figure 17.5B ◗ The position of the kidneys in the human body

Table 17.1 ▼ Composition of the blood in the glomerulus and the urine of a healthy adult

	Blood in the glomerulus (percentage by mass)	Urine
Water	91.7	96.5
Proteins	7.5	0
Urea	0.03	2
Ammonia	trace	0.05
Sodium ions (Na⁺)	0.3	0.6
Potassium ions (K⁺)	0.02	0.15
Chloride ions (Cl⁻)	0.36	0.6
Glucose	0.1	0

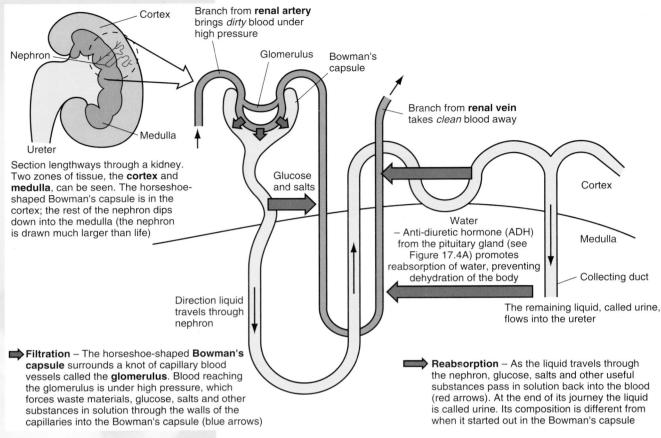

Section lengthways through a kidney. Two zones of tissue, the **cortex** and **medulla**, can be seen. The horseshoe-shaped Bowman's capsule is in the cortex; the rest of the nephron dips down into the medulla (the nephron is drawn much larger than life)

Branch from **renal artery** brings *dirty* blood under high pressure

Branch from **renal vein** takes *clean* blood away

Water
– Anti-diuretic hormone (ADH) from the pituitary gland (see Figure 17.4A) promotes reabsorption of water, preventing dehydration of the body

The remaining liquid, called urine, flows into the ureter

Direction liquid travels through nephron

Filtration – The horseshoe-shaped **Bowman's capsule** surrounds a knot of capillary blood vessels called the **glomerulus**. Blood reaching the glomerulus is under high pressure, which forces waste materials, glucose, salts and other substances in solution through the walls of the capillaries into the Bowman's capsule (blue arrows)

Reabsorption – As the liquid travels through the nephron, glucose, salts and other useful substances pass in solution back into the blood (red arrows). At the end of its journey the liquid is called urine. Its composition is different from when it started out in the Bowman's capsule

Figure 17.5C A kidney nephron

It's a fact!

Urea forms when **amino acids** (see Topic 10.3) break down in liver cells. The amino group (—NH$_2$) is removed from the amino acid molecule. This process is called **deamination**. The flow chart shows you what happens

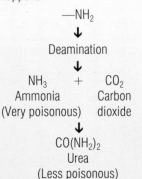

—NH$_2$
↓
Deamination
↓
NH$_3$ + CO$_2$
Ammonia   Carbon
(Very poisonous) dioxide
↓
CO(NH$_2$)$_2$
Urea
(Less poisonous)

Where do wastes come from?

Thousands of chemical reactions take place inside a living cell. These reactions make up the cell's **metabolism** (see p. 3).

Compounds are broken down and new ones are made. The waste products formed could be harmful to the body if they were allowed to accumulate (Table 17.2). Removal of these waste substances is part of the process of excretion.

Table 17.2 Waste substances produced by metabolism

Waste substances	How waste is made	Where waste is made	Where waste is excreted
Carbon dioxide	Cellular respiration (see p. 243)	All cells	From lungs or other gas exchange surfaces
Urea and other compounds containing nitrogen	Deamination of amino acids	Liver cells	From the kidney
Bile pigments	Breakdown of haemoglobin (see p. 259)	Liver cells	In bile which passes into the duodenum (see p. 237)
Oxygen	Photosynthesis (see p. 177)	Inside the chloroplasts of plant cells	From leaf cells through the stomata

It's a fact!

Proteins leaking from the blood, through the glomerulus (see p. 307) into the urine is one of the symptoms of kidney failure.

Kidney failure

The person in Figure 17.5D is attached to a machine that has taken over the work of her kidneys. This is one way of treating people whose kidneys are not working (**kidney failure**).

The **kidney machine** removes waste substances from the patient's blood. Figure 17.5E explains how it works.

Science at work

The benefits and limitations of dialysis and transplant surgery summarised

Benefits	Limitations
	Dialysis
The patient	The patient
● does not have to take drugs for a lifetime	● must have access to a kidney machine
● may be treated more quickly	● is restricted to eating a limited diet
	● must be treated regularly
	Transplant surgery
The patient	The patient
● does not depend on a kidney machine	● must wait for a suitable donor kidney
● can enjoy a normal diet	● is vulnerable to the body rejecting the new kidney
	● must take anti-rejection drugs for a lifetime
	● is vulnerable to infections (see p. 269)

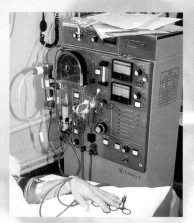

Figure 17.5D ⬆ The patient must use the kidney machine two or three times a week. Each treatment takes about ten hours

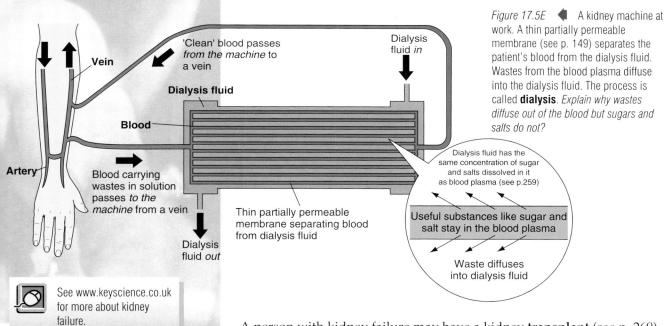

Figure 17.5E ◀ A kidney machine at work. A thin partially permeable membrane (see p. 149) separates the patient's blood from the dialysis fluid. Wastes from the blood plasma diffuse into the dialysis fluid. The process is called **dialysis**. *Explain why wastes diffuse out of the blood but sugars and salts do not?*

Vein

'Clean' blood passes *from the machine* to a vein

Dialysis fluid *in*

Dialysis fluid

Blood

Artery

Blood carrying wastes in solution passes *to the machine* from a vein

Thin partially permeable membrane separating blood from dialysis fluid

Dialysis fluid *out*

Dialysis fluid has the same concentration of sugar and salts dissolved in it as blood plasma (see p.259)

Useful substances like sugar and salt stay in the blood plasma

Waste diffuses into dialysis fluid

See www.keyscience.co.uk for more about kidney failure.

A person with kidney failure may have a kidney **transplant** (see p. 269). A healthy kidney is taken from a person (the donor) who has just died (Figure 17.5F), or from someone living who wants to help the patient. Live donors are very often close relatives of the patient because their body tissues are similar, which reduces the chances of the patient's body rejecting the transplant (see Topic 16.2). We have two kidneys, but it is possible to live a healthy life with only one.

It's a fact!

Treatment by dialysis restores the concentrations of substances dissolved in the blood to normal levels.

NHS Organ Donor Register
donorcard
I want to help others to live in the event of my death Please let your relatives know your wishes

Figure 17.5F ⬆ Donor card

CHECKPOINT

▶ **1** The diagram shows the position of the kidneys in the body and their connections with blood vessels and the bladder.

(a) Name the parts labelled A–G.

(b) Briefly explain why the concentration of urea in B is much less than in A.

(c) What is the role of E?

(d) Where is urine stored before it is released to the outside?

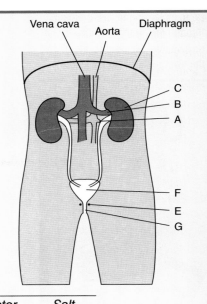

▶ **2** Copy the table below. It lists parts of the kidney and its connections to blood vessels and the bladder. Tick each substance present in the part listed in the left-hand column (assume the person is healthy).

	Protein	Glucose	Urea	Water	Salt
Blood in the renal artery					
Blood in glomerulus					
Fluid that passes into the Bowman's capsule					
Urine in the bladder					
Blood in the renal vein					

▶ **3** State three ways in which blood leaving the kidney is different from blood entering the kidney.

▶ **4** Match the lettered statements with the appropriate numbered parts.

A takes urine from the kidney to the bladder   **1** sphincter
B prevents urine leaving the bladder   **2** renal artery
C removes urea from the blood   **3** renal vein
D carries urine out of the body   **4** kidney
E carries blood to the kidney   **5** urethra
F stores urine   **6** ureter
G carries blood away from the kidney   **7** bladder

▶ **5** Complete the following paragraph about the kidney, using the words provided. Each word may be used once, more than once, or not at all.

Bowman's capsule   ureter   cortex   one million
medulla   renal vein   bladder   glomerulus

A kidney cut lengthways has two regions, an outer _____ and an inner . _____ Under a microscope many tiny nephrons are visible: there are about _____ in each kidney. Each nephron has a cup-shaped capsule called the _____ at one end. This surrounds a small bunch of capillaries called the _____ . Leading away from each capsule is a narrow tubule which twists and turns and eventually joins the _____ , which takes urine from the kidney to the _____ .

▶ **6** These questions refer to Table 17.1

(a) What is the mass of water in 100 cm³ of urine?

(b) Why do you think there are no proteins in urine?

(c) What percentage of glucose does the nephron reabsorb into the blood?

(d) Name the main waste product formed from the deamination of amino acids.

▶ **7** What is dialysis and why do some people need it?

Make sure you can answer an exam question which asks you to explain the significance of a graph similar to Figure 17.5G.

Regulating temperature

The **metabolism** of cells (see p. 3) releases heat which warms the animal body. Most enzymes which control metabolism work best at a temperature of about 37 °C. If an animal can maintain its body temperature at around 37 °C, even when the temperature outside changes, then metabolism will be at its most efficient. Birds and mammals (which include us) have evolved ways of keeping the temperature at the centre of the body (the **core temperature**) at around 37 °C. This is why we are described as **warm-blooded** even though the temperature at the body's surface may fluctuate a little with changes in the temperature outside.

Being warm-blooded means that we (and other mammals and birds) can live in places where other animals would soon perish because they cannot maintain a steady core temperature. Animals that cannot maintain a steady core temperature are called **cold-blooded**, not because they are 'cold' but because their body temperature fluctuates with the temperature of the environment. Their body temperature drops when the temperature of the environment drops and rises when the temperature of the environment rises (Figure 17.5G). Enzymes work less efficiently when body temperature drops. As a result metabolism slows down which helps explain why cold-blooded animals are usually slow moving in cold weather. Fish, amphibians and all invertebrates are cold-blooded. Many reptiles are cold-blooded but some species are able to achieve limited control of body temperature (Figure 17.5H).

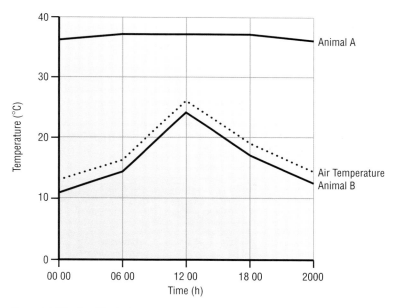

Figure 17.5G ▲ The body temperatures of two animals and the air temperature recorded over 24 hours. *Which animal is warm-blooded and which cold-blooded?*

■ Skin, hair and control of body temperature

Hair is made of the protein **keratin** (see p.168). It grows from pockets of cells in the skin called **hair follicles.** Figure 17.5I shows hair follicles and other structures in a section of human skin. Although parts of the human body appear hairless, we can see where the hair follicles are when we are cold because the follicle muscles contract covering the skin with small bumps which we call **'goose pimples'**.

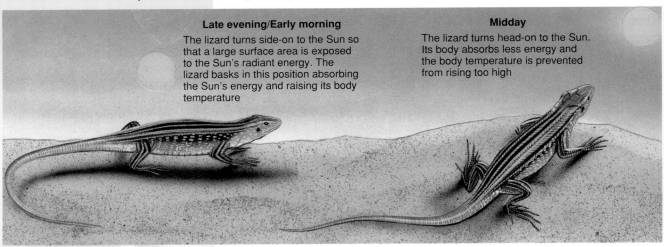

Late evening/Early morning
The lizard turns side-on to the Sun so that a large surface area is exposed to the Sun's radiant energy. The lizard basks in this position absorbing the Sun's energy and raising its body temperature

Midday
The lizard turns head-on to the Sun. Its body absorbs less energy and the body temperature is prevented from rising too high

Figure 17.5H ⬥ How lizards achieve limited control of core temperature

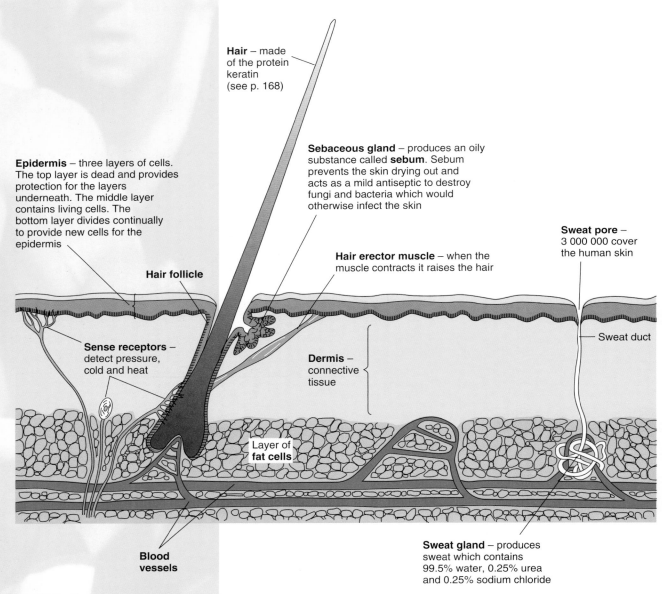

Hair – made of the protein keratin (see p. 168)

Sebaceous gland – produces an oily substance called **sebum**. Sebum prevents the skin drying out and acts as a mild antiseptic to destroy fungi and bacteria which would otherwise infect the skin

Epidermis – three layers of cells. The top layer is dead and provides protection for the layers underneath. The middle layer contains living cells. The bottom layer divides continually to provide new cells for the epidermis

Sweat pore – 3 000 000 cover the human skin

Hair follicle

Hair erector muscle – when the muscle contracts it raises the hair

Sweat duct

Sense receptors – detect pressure, cold and heat

Dermis – connective tissue

Layer of **fat cells**

Blood vessels

Sweat gland – produces sweat which contains 99.5% water, 0.25% urea and 0.25% sodium chloride

Figure 17.5I ⬥ Section through the human skin

Hair and other mechanisms help to control the body's temperature:

- **Hairs** raised by the **erector** muscles trap a layer of air which insulates the body in cold weather (air is a poor conductor of heat). In warm weather the hair is lowered and no air is trapped.
- **Fat** insulates the body and reduces heat loss.
- **Sweat** cools the body because it carries heat energy away from the body as it evaporates.
- Millions of temperature-sensitive **sense receptors** cover the skin. Nerves connect them to the **brain** which controls the body's response to changes in temperature in the environment.
- When it is warm, **blood vessels** in the skin dilate (**vasodilation**). More blood flows through the vessels in the skin and loses heat to the environment. In cold weather, the blood vessels in the skin constrict (**vasoconstriction**) and less heat is lost to the environment.
- **Shivering** helps warm the body when it is cold. Small muscles under the skin contract and relax repeatedly. The contractions and relaxations release heat.

The evaporation of sweat is not the only way the body loses heat energy. Conduction, radiation and convection play a part as well (see *Key Science Physics* p. 63).

- **Conduction** is the transfer of heat energy from a hot object to a cooler object, which it is touching. For example, a warm dog lying on a cool floor loses heat energy to the floor by conduction.
- **Radiation** is the transfer of heat energy from a hot object to a cooler object, *not* by touching. Infra-red waves (see *Key Science Physics* p. 66) account for most of the heat energy lost. For example up to 60% of the heat energy lost by a person sitting in a room at 22 °C may be caused by radiation.
- **Convection** is the transfer of heat energy by the movement of currents of water and air. The movements spread heat energy through the environment and speed up the loss of heat energy from the body by evaporation and conduction.

The body can also gain heat energy by conduction, radiation and convection if it is cooler than the surrounding environment. Heat transfer through evaporation, however, is a one-way process – the body can only lose it this way.

Body temperature is monitored and controlled by the hypothalamus of the brain (see Figure 17.1H). The hypothalamus has receptors (see p. 283) that are sensitive to the temperature of the blood flowing through the brain. Also, temperature receptors in the skin send nerve impulses to the hypothalamus giving information about skin temperature.

CHECKPOINT

▸ 1 Why does the concentration of sugar in the blood vary despite regulation which keeps the variations within acceptable limits?
▸ 2 Briefly explain:
  (a) Why raised body hair helps us keep warm
  (b) Why sweating helps us keep cool
▸ 3 Why does a person look pale when cold and flushed when warm?

Topic 18

Support and movement

18.1 ▶ Skeletons and muscles

All living things are supported in some way. Think what would happen to your body without the support of the skeleton. It would collapse into a shapeless heap.

There are three main types of skeleton (Figure 18.1A).

Skeleton: know the meaning of the prefixes:
● **endo** – inside
● **exo** – outside
● **hydro** – fluid

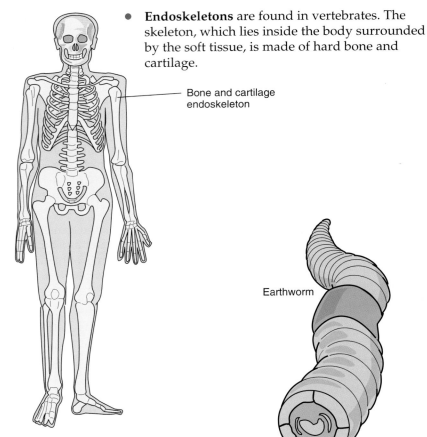

● **Endoskeletons** are found in vertebrates. The skeleton, which lies inside the body surrounded by the soft tissue, is made of hard bone and cartilage.

Bone and cartilage endoskeleton

Figure 18.1A 🔺 (a) The endoskeleton of a human

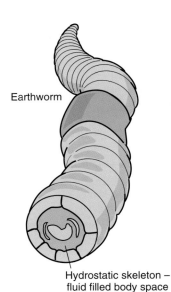

Earthworm

Hydrostatic skeleton – fluid filled body space

(c) The hydrostatic skeleton of a worm

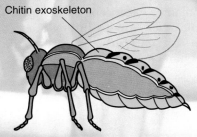

Chitin exoskeleton

(b) The exoskeleton of an insect

● **Exoskeletons** are found in insects and their relatives. The skeleton is made of hard **chitin** (see Topic 10.1) segments which surround the body like armour plate.

● **Hydrostatic skeletons** are found in the larger worms. They consist of body spaces filled with fluid under pressure. Hydrostatic skeletons are firm but flexible.

How muscles work

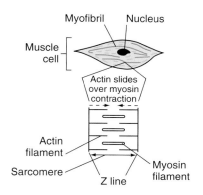

Figure 18.1B ⬆ The arrangement of muscle filaments

Up to 2000 parallel strands called **myofibrils** pack the cytoplasm of a muscle cell. Myofibrils consist of units called **sarcomeres**. Each sarcomere is composed of filaments of two types of protein parallel to one another. The thicker filament is made of the protein **myosin**; the thinner filament is made of the protein **actin**. Filaments from neighbouring sarcomeres link together in a region called the **Z line** (Figure 18.1B shows the arrangement.)

When nerve impulses stimulate the muscle cells, the thin actin filament slides over the thicker myosin filament and cross-bridges between the filaments rapidly form, break and reform. Because filaments are anchored at the Z line, the sarcomere shortens, the myofibrils shorten and the muscle cell as a whole **contracts**. The interaction between myosin and actin releases energy from **ATP** (see p. 245) for muscle contraction.

Antagonistic muscles

Figure 18.1C compares the arrangement of muscles in the locust leg and the human arm. The muscles are in pairs which stretch across the joint. One muscle of a pair has the opposite effect to that of its partner. We call them **antagonistic** pairs. For example, contraction of the biceps muscle lifts the lower arm (flexing); contraction of the triceps muscle straightens (extends) the arm (see Figure 18.1D). When the biceps contracts the triceps relaxes and vice versa. The antagonistic pairs of muscles in the locust leg work in a similar way.

The skeleton moves because muscles contract and relax across flexible joints.

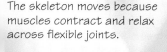

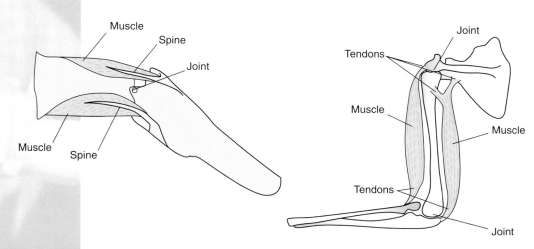

Figure 18.1C ⬆

(a) Locust leg – notice that the exoskeleton surrounds the muscles which are attached to its inside surface by inward-projecting spines

(b) Human arm – notice that the endoskeleton is surrounded by the muscles which are attached to the outside surface by tough **tendons**

What are antagonistic pairs of muscles? Check the question by studying Figure 18.1D carefully.

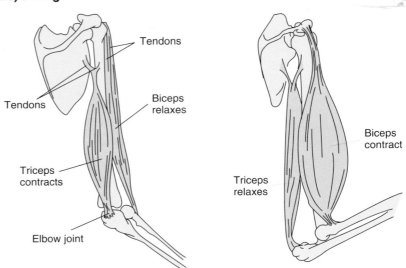

Figure 18.1D ◆ Moving the lower arm – the biceps and the triceps are an antagonistic pair of muscles

Each type of muscle tissue is adapted for a particular role in the body.

Types of muscle

Animals are able to move from place to place because of the action of muscles which pull on the skeleton. Moving from place to place is called **locomotion**. Muscular contractions are also responsible for other forms of movement. As heart muscle contracts and relaxes rhythmically it propels blood through the blood vessels. **Peristalsis** (see p. 240) is brought about by contraction and relaxation of the muscle in the wall of the intestine. There are three types of muscle in the human body – smooth muscle, skeletal muscle and cardiac muscle (see Figure 18.1E).

■ Smooth muscle

Smooth muscle contracts and relaxes slowly and steadily and does not become tired. These properties are ideal for the continuous movement of substances through the organs of the body. In mammals, smooth muscle is found in the walls of the intestine, blood vessels and air passages. Smooth muscle receives nerve impulses from the **autonomic nervous system** (which acts without conscious control from the brain). Contractions of smooth muscle therefore occur automatically. This is why smooth muscle is called **involuntary** (see p. 287) muscle.

■ Skeletal muscle

Skeletal muscle is often called **striated** (striped) muscle. It consists of fibres which are crossed with alternate light and dark bands. Skeletal muscle becomes tired after prolonged periods of activity but is otherwise ideal for moving parts of the skeleton. Its contractions are quick, strong and usually **voluntary** (see p. 288). Conscious messages from the brain control the strength and speed of contractions. The fibres of skeletal muscle receive branches from the axons of **motor neurones**. The muscle fibres contract when nerve impulses reach them.

■ Cardiac muscle

Cardiac (heart) muscle is striated, like skeletal muscle. Its fibres are branched and connect with one another. This structure lets nerve impulses spread throughout the whole tissue, co-ordinating its contractions. Cardiac muscle never becomes tired. Its action is **involuntary**.

SUMMARY

Muscular contractions are responsible for movement. Smooth muscle moves substances through the organs of the body; cardiac muscle propels blood through the blood system; skeletal muscle moves the skeleton and is responsible for locomotion (moving from place to place).

(a) Smooth muscle (×20)

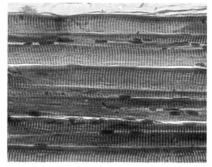

(b) Skeletal muscle (×20)

(c) Cardiac (heart) muscle (×20)

Figure 18.1E 🔼 The three types of muscle viewed under a microscope

18.2 ▶ Locomotion: how different animals move

FIRST THOUGHTS

Propulsion pushes an animal in the direction it wants to go. Support comes from the environment in which the animal lives (water gives more support than air because it is more dense). Stability is controlled by fins, wings and tail feathers or legs. Animals are adapted to overcome the problems of propulsion, support and stability.

How fish swim

Fish swim by pushing their bodies and fins against the water. Their **streamlined** shape helps to reduce resistance to the movement of their bodies through the water (see Figure 18.2A).

Contraction and relaxation of the blocks of muscle attached to each side of the fish's vertebral column move the body from side to side and drive the fish forward (see Figure 18.2B). The blocks of muscle on either side of the vertebral column form antagonistic pairs. The vertebral column itself is flexible and acts like a lever.

A gas-filled sac called the **swim bladder** helps to control buoyancy. When the swim bladder is full of gas the density of the fish decreases and the fish rises: when gas is removed, the density of the fish increases and the fish sinks. Cartilaginous fish such as the shark shown in Figure 18.2A do not have swim bladders. If they stop swimming they slowly sink to the bottom.

Figure 18.2A 🔼 Sharks are efficient swimmers – they have very streamlined bodies

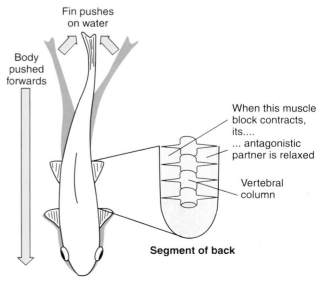

Fin pushes on water

Body pushed forwards

When this muscle block contracts, its....

... antagonistic partner is relaxed

Vertebral column

Segment of back

Figure 18.2B 🔼 How a fish swims

Fins control the direction of movement and stability of the fish rather than propel it through the water.

See www.keyscience.co.uk for more about muscles and movement.

It's a fact!

Flowing extensions of finger-like **pseudopodia** change the shape of the protist *Amoeba* (see p. 16) and help it move in search of food. Another protist called *Paramecium* is covered with hair-like **cilia**. The cilia beat against the water like oars, propelling the cell forwards or backwards.

Figure 18.2C shows how fish use their **fins** to control their direction of movement and stability. The dorsal and ventral fins (the medial fins) increase the vertical surface area of the fish and keep it upright by preventing **rolling** (rotation of the body about the long axis) and **yawing** (side to side movement of the front part of the body). The pectoral and pelvic fins prevent **pitching** (the tendency to nose dive), allowing the fish to swum up, down and even backwards!.

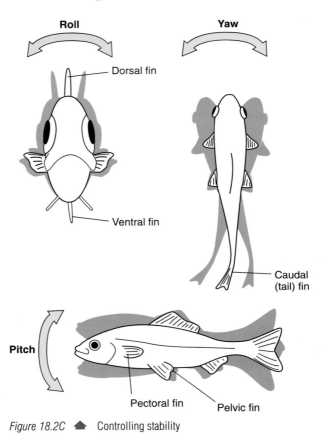

Figure 18.2C Controlling stability

How birds fly

Birds are adapted to fly:
- **Hollow bones** reduce weight.
- **Flight muscles** move the wings up and down.
- **Feathers** smoothly shape the body.

Feathers also **insulate** the body. At around 41 °C the body temperature of a bird is higher than that of most other warm-blooded animals (see Topic 17.5). The high temperature means that the flight muscles work more efficiently. Feathers make flying possible: they also keep in heat, keep out water and their colour is used either to attract mates or for protective camouflage. Feathers are arranged in layers over the body. The outer layers give a smooth shape to the body and are used for flight. The inner layers, called **down**, give extra insulation (Figure 18.2D).

Wing feathers are of different sizes. The longest are the stiff **flight feathers**. Figure 18.2E shows that those on the outer edge are called **primaries**; those on the inner edge are called **secondaries**. Rows of smaller feathers called **wing coverts** smoothly overlap the flight feathers on top and underneath the wing. The **alula** is formed from a group of small feathers on the outer front edge.

Skeletons and locomotion: Flexibility v rigidity

The fluid-filled hydrostatic skeleton of the earthworm is surrounded by an outer layer of circular muscle and an inner layer of longitudinal muscle (see Figure 14.5H and Figure 18.1A(c)). Contractions of the muscles send pressure waves through the fluid of the hydrostatic skeleton, causing changes in the shape of the body. These shape changes and the **chaetae** (bristles – see Figure 1.5K) projecting from the body help the earthworm to burrow through soil.

Limbs and a rigid skeleton make walking possible. An insect at rest is supported by all six legs attached to the rigid exoskeleton (see Figure 1.5N). When walking, the insect supports itself on three legs at a time: the first and third legs of one side, and the second leg of the other side. These form a tripod while the other three legs are carried forward.

EXTENSION FILE
ACTIVITY

The aerofoil shape of wings provides the lift for birds to fly.

Figure 18.2D 🔺 Flight feather and down

Figure 18.2E 🔺 Wing feathers

Figure 18.2F shows that the wing is in the shape of an **aerofoil**. The upper surface is more strongly curved than the lower surface. Air moving over the top has further to go than the air moving underneath and therefore travels faster. This increases air pressure beneath the wings and creates a reduced pressure above them. The result is an upward force and the wings lift.

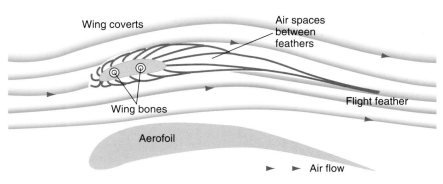

Figure 18.2F 🔺 Section of wing showing the shape of an aerofoil. Notice that the flow of air is smooth and free of turbulence

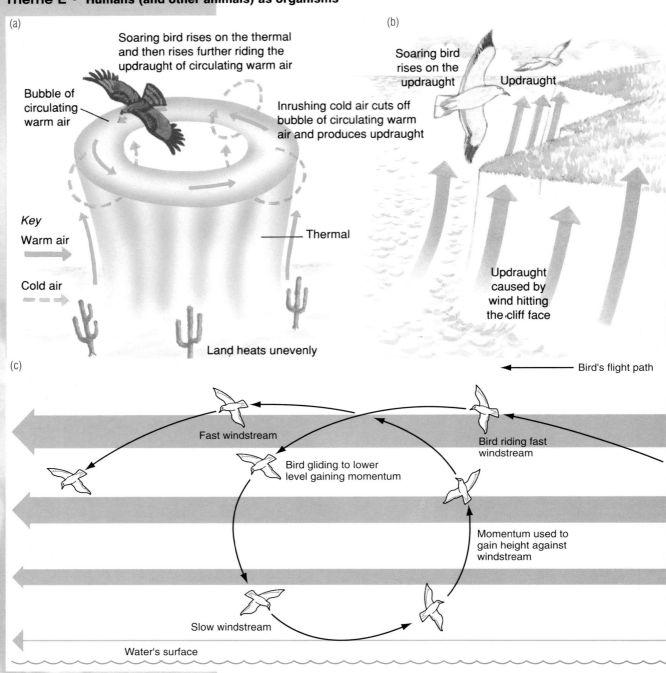

(a)

Soaring bird rises on the thermal and then rises further riding the updraught of circulating warm air

Bubble of circulating warm air

Inrushing cold air cuts off bubble of circulating warm air and produces updraught

Key

Warm air

Cold air

Thermal

Land heats unevenly

(b)

Soaring bird rises on the updraught

Updraught

Updraught caused by wind hitting the cliff face

Bird's flight path

(c)

Fast windstream

Bird riding fast windstream

Bird gliding to lower level gaining momentum

Momentum used to gain height against windstream

Slow windstream

Water's surface

(a) Land heats up unevenly producing rising columns of warm air called **thermals**. Vultures and other soaring birds circle on the updraught and rise with it

(b) Seagulls gain height by riding the **updraughts** when wind hits the cliffs

(c) Wind near the surface of the sea is slowed by friction with the waves. Wind speed increases with height reaching its maximum 15 m above the water's surface. The albatross uses the **variation in windspeed** to pick up momentum on the high, fast windstream, gliding down to a lower level then using its momentum to gain height against the wind. The bird travels thousands of kilometres across the southern oceans without flapping its wings

Figure 18.2G Gliding and soaring

Table 18.1 sets out the forces acting against a bird in flight and how the design of the body and wings is adapted to overcome them.

Table 18.1 ▼ *Counteracting forces*

Force	Adaptation
Drag caused by the resistance of air	Tips of feathers point backward in the same direction as the flow of air. The **leading edge** of the wing is blunt and rounded: wing feathers taper toward the **trailing edge**
Gravity causing a downward pull	Lift on both wings opposes the force of gravity and the bird stays aloft
Turbulence (uneven flow of air) caused by the bird's movements	The alula forms a slot in front of the wing, reducing turbulence

Have you ever watched a seagull gliding and soaring? Gliding is the simplest way of staying aloft. It makes the smallest demands on muscle power and energy reserves. Figure 18.2G gives you the idea.

Figure 18.2H shows that to climb higher a bird tilts the leading edge of each wing upwards. Notice, however, that the more the wings are tilted, the greater the **turbulence** as the smooth airflow over the wings breaks up into swirling eddies. The wings lose lift and the bird is in danger of **stalling**. At the critical moment the bird spreads the alula on the leading edge of each wing. Each alula acts as a slot through which air rushes, keeping the airflow over each wing smooth and turbulence free (Figure 18.2I).

Gliding and soaring require less effort than flapping flight.

Turbulence interrupts the smooth airflow across wing surfaces. Lift is lost and the bird is in danger of stalling.

Opening the alula keeps a smooth airflow over the wings. Lift is restored and a stall prevented.

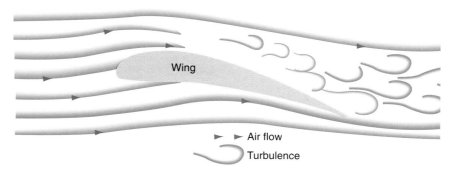

Figure 18.2H ▲ Gaining height – the angle between the wing and the airstream is increased. Notice the turbulence developing over the wing's surface

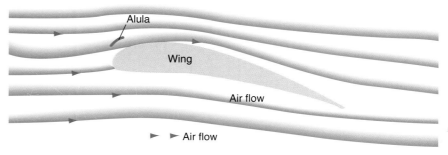

Figure 18.2I ▲ Controlling the stall – the alula opens, restoring a smooth airflow over the wing

Flapping flight requires more effort than gliding and soaring. Figure 18.2J shows how muscles and bones work together to move the wings. Multi-flash photography allows us to see the air movements as a bird flies (Figure 18.2K). The **downstroke** is the power stroke. Air is pushed downwards supporting the weight of the bird. The primary feathers lie flat against one another forming an unbroken surface which provides maximum resistance to the air underneath. As the wing moves downwards and forwards, the ends of the primaries curve back forming a propeller shape, pulling the bird forward.

The **upstroke** produces little air movement. The primary feathers twist open and air passes through the gaps between them. Minimum air resistance means that the upstroke needs less effort. It is a recovery stroke in preparation for the next powerful downstroke. Even so the upstroke still helps to drive the bird forward. The primaries curl back, each becoming a small propeller, and the wings tilt to maintain lift (see Figure 18.2H).

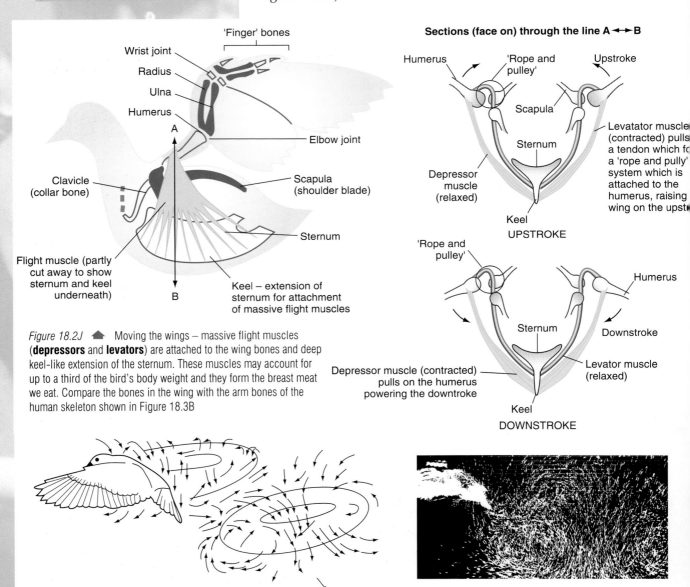

Figure 18.2J 🔺 Moving the wings – massive flight muscles (**depressors** and **levators**) are attached to the wing bones and deep keel-like extension of the sternum. These muscles may account for up to a third of the bird's body weight and they form the breast meat we eat. Compare the bones in the wing with the arm bones of the human skeleton shown in Figure 18.3B

Figure 18.2K 🔺 Soap bubbles filled with the gas helium are as dense as air (ordinary bubbles are denser than air and slowly sink). Movements of the bubbles show how the air is moving as the bird flies through the bubble cloud. Notice the circular vortex rings of moving air produced by the downstroke generating lift and thrust

How horses walk and run

Think how animals move from place to place. To swim, fish push against water; to fly birds push against air. *How do land animals move when walking and running?*

The hind legs of a horse push against the ground and the body moves forward. Analysing movement, therefore, is best begun with the hind legs because that is where the force is applied. A complete cycle of movement is called a **stride**. All four legs complete their motion and move the horse.

Figure 18.2L shows the arrangement of muscles and bones in the horse's hind leg. Muscles stretch across joints and pull on the bones. One group of muscles bends (flexes) a joint; an opposing group straightens (extends) it. *Which groups of muscles are antagonistic pairs (see Topic 18.1)?*

Legs push against the ground and the body moves forward.

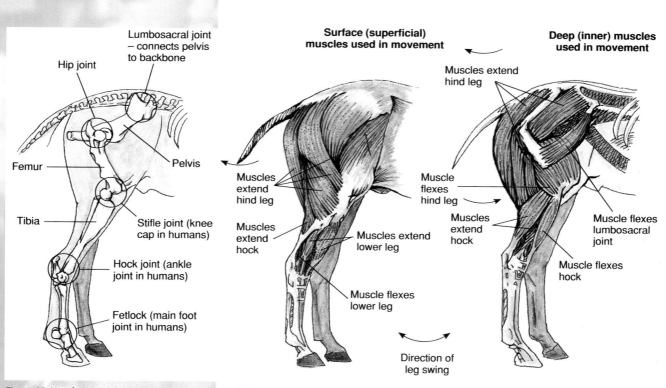

Figure 18.2L ▲ (a) Joints and bones of the hind leg of the horse

(b) The muscles used to move the hind leg of the horse. Two diagrams are needed to show the main muscles for movement because the muscles are layered over each other

Look at Figure 18.2M. Propulsion begins at the moment when the leg is vertically beneath the hind quarters and continues until the foot leaves the ground. Notice the **ligaments** which bind the bones together. Also notice the **tendons** which attach the muscles to the bones. At the end of each swing the leg briefly stops. At this moment the tendons store energy like a coiled spring, only to release it in the elastic recoil helping the leg to start moving the other way.

Next time you see a horse walking, listen to the cycle of its hoof beats. You should hear a repeating pattern of four distinct beats as each foot strikes the ground in succession. Figure 18.2N shows the sequence.

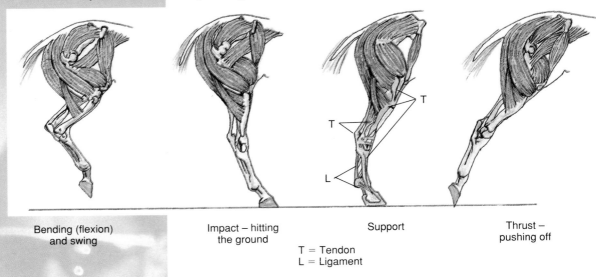

Bending (flexion) and swing	Impact – hitting the ground	Support	Thrust – pushing off

T = Tendon
L = Ligament

Figure 18.2M ⬆ Stages of the stride. Ligaments and tendons are shown at the support stage. They keep the bones in place even when the muscles are relaxed

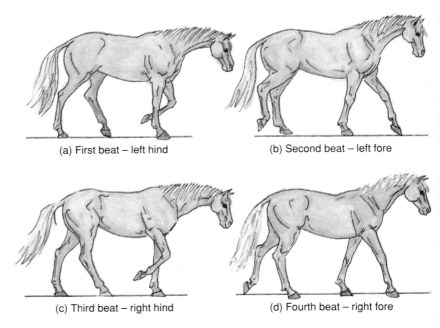

(a) First beat – left hind	(b) Second beat – left fore
(c) Third beat – right hind	(d) Fourth beat – right fore

Figure 18.2N ⬆ The walk is a cycle of four beats. Two or three legs are always on the ground giving stability. Although the foreleg appears to move first from the halt, the walk sequence starts with a hind leg because the power comes from the hindquarters

Distinguish between the strides of a horse walking and a horse galloping.

Galloping is also a four-beat cycle. The strides are long and the hind legs reach forward under the body. Figure 18.2O shows the sequence. Notice that when the horse pushes off with its foreleg it is suspended in the air for a moment before the hind foot strikes for the first beat. The head and neck move forward and back balancing the horse and carrying it on to the next stride. The speed of the gallop varies: a horse galloping cross-country can maintain a speed of 30 km h^{-1} for long distances; a racehorse can clock 70 km h^{-1} sprinting at top speed.

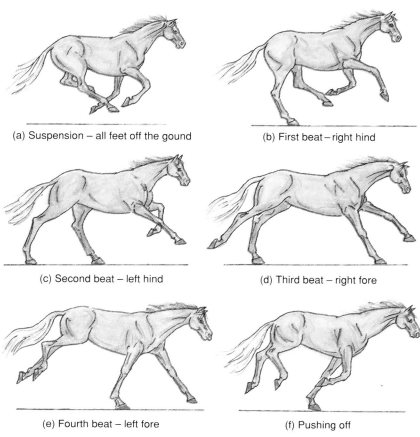

(a) Suspension – all feet off the gound

(b) First beat – right hind

(c) Second beat – left hind

(d) Third beat – right fore

(e) Fourth beat – left fore

(f) Pushing off

Figure 18.20 🔺 The gallop is a four-beat cycle of long strides with suspension. The abdominal muscles help the hind legs to reach forward under the body. Breathing is in rhythm with the strides

18.3 ▶ The skeleton

The vertebrate skeleton is made mostly of bone (see Figure 18.3A). It also contains **cartilage** which is softer than bone because it contains less **calcium** and **phosphate**. The amount of cartilage in the skeleton depends on the species and the age of an animal. The skeleton of sharks is made entirely of cartilage. In adult mammals, however, cartilage is found in only a few places. Cartilage covers the ends of limb bones where it helps to reduce friction in the joints as bones move upon one another.

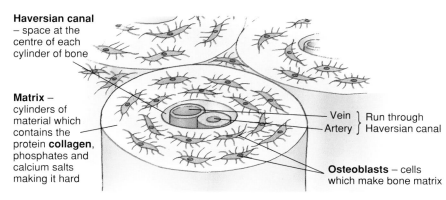

Haversian canal – space at the centre of each cylinder of bone

Matrix – cylinders of material which contains the protein **collagen**, phosphates and calcium salts making it hard

Vein } Run through
Artery } Haversian canal

Osteoblasts – cells which make bone matrix

Figure 18.3A 🔺 The structure of living bone (×200)

The human skeleton

The human skeleton is illustrated in Figure 18.3B. The bones of the skeleton are connected to one another at joints. Strap-like **ligaments** hold joints together. **Tendons** attach muscles to the skeleton. As muscles contract and relax across joints they move the different parts of the skeleton. If the bones of a joint separate, the joint is said to be **dislocated**.

The human skeleton consists of two parts: the **axial skeleton** and the **appendicular skeleton**.

Parts of the human skeleton:
● The axial skeleton protects major organs of the body.
● The appendicular skeleton is a system of levers used for lifting, walking and running.

Why is it that ligaments are elastic ('stretchy') but tendons are not?

Answer:
Elastic ligaments means that bones can move relative to each other at joints. If tendons were elastic some of the force generated by muscle contraction would stretch the tendons rather than move bones.

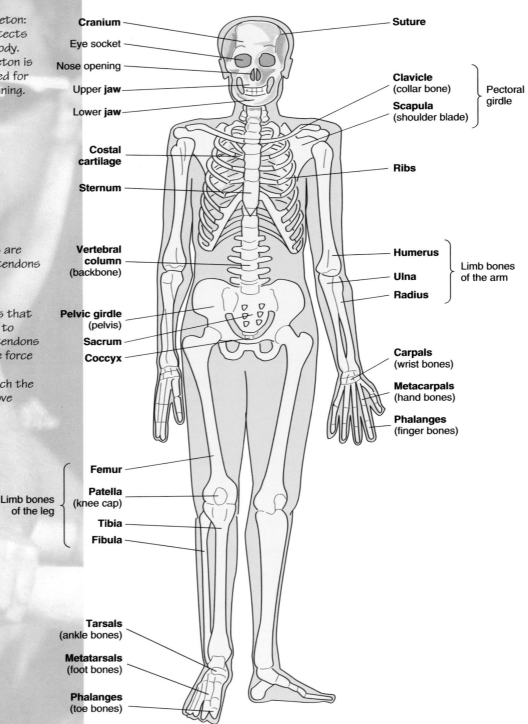

Figure 18.3B ⬆ The human skeleton

■ The axial skeleton

The axial skeleton consists of the skull, vertebral column and the ribs. These bones protect vital organs and tissues and form a strong support for the body. The skull and the vertebral column form a protective sheath of bone around the brain and spinal cord.

- **The skull** consists of plates of bone fused together to form the **cranium** which encloses and protects the brain. A hinged joint allows powerful muscles to move the lower jaw against the fixed upper jaw.

- **The vertebral column** (backbone) consists of a series of bones called **vertebrae** arranged like a curved rod (see Figure 18.3C). The vertebral column supports the skull and the limb girdles (**pectoral** and **pelvic** girdles). A cavity called the **neural canal** runs through the centre of each vertebra forming a continuous space in the vertebral column through which the spinal cord runs. At the top of the vertebral column the spinal cord passes through a hole in the skull and expands into the brain.

- **The ribs** form a curved bony cage around the heart and lungs (see Figure 18.3B). At the front they are attached by the **costal cartilages** to a bone called the **sternum**. At the back they form joints with the **thoracic** vertebrae. Each rib has ridges to which the **intercostal muscles** are attached. The costal cartilages and the joints make the ribs flexible. The intercostal muscles, the diaphragm and the muscles of the abdomen move them. Breathing depends on these movements (see Topic 15.2).

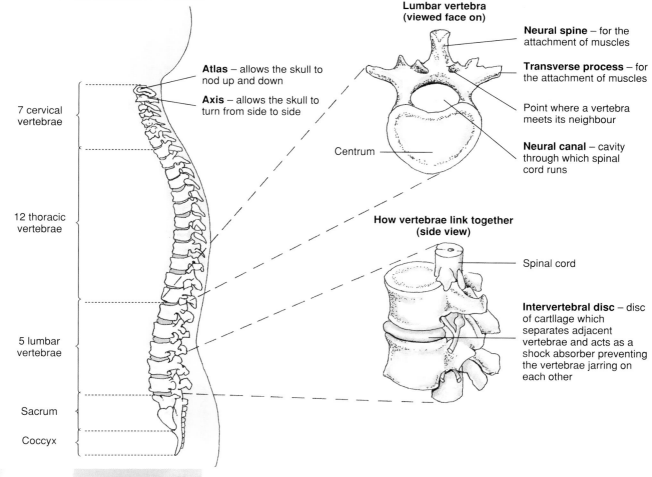

Lumbar vertebra (viewed face on)

Neural spine – for the attachment of muscles

Transverse process – for the attachment of muscles

Point where a vertebra meets its neighbour

Neural canal – cavity through which spinal cord runs

Centrum

Atlas – allows the skull to nod up and down

Axis – allows the skull to turn from side to side

7 cervical vertebrae

12 thoracic vertebrae

5 lumbar vertebrae

Sacrum

Coccyx

How vertebrae link together (side view)

Spinal cord

Intervertebral disc – disc of cartilage which separates adjacent vertebrae and acts as a shock absorber preventing the vertebrae jarring on each other

Figure 18.3C   The vertebral column

It's a fact!

The fossil remains of vertebrates are usually parts of skeletons. Bone is very hard and therefore preserves well.

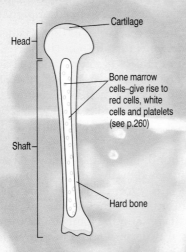

Figure 18.3D ▲ Section of a long bone cut down through its length

Make sure you can distinguish between a fracture and a dislocation.

A sprain occurs when the ligaments and other tissues of a joint are torn by a sudden wrench.

SUMMARY

The human skeleton is in two parts: the axial skeleton which protects vital organs and the appendicular skeleton which allows us to handle objects and move from place to place.

The appendicular skeleton

The appendicular skeleton consists of the limb girdles and limb bones. We use the hind limbs (legs and feet) for walking and running. Our upright position means that the forelimbs (arms and hands) are free for other activities.

- **The pelvic girdle** links the legs with the vertebral column (see Figure 18.3B). It is made of two bones fused together and gives a solid framework that helps to bear the weight of the body. Also it is joined rigidly to the base of the vertebral column. The rigid arrangement of bones allows forces on the leg to be transmitted to the rest of the body.
- **The pectoral girdle** links the arms with the vertebral column. It is made up of the **scapulas** (shoulder blades) and **clavicles** (collar bones). This arrangement is not as effective as the pelvic girdle in transmitting force from the limbs to the body but it gives the shoulders and arms great freedom of movement.
- **The limb bones** in humans consist of long bones (see Figure 18.3D) which form joints at the elbow in the arm and the knee in the leg. The upper long bone of each limb is attached to a limb girdle. The lower long bones are attached to the hand or the foot by a set of bones which form the wrist or the ankle. The joints with the pectoral and pelvic girdles and the joints at wrist, ankle, elbow and knee enable the limbs to move freely.

Broken bones

X-ray photographs help doctors locate broken (fractured) bones (Figure 18.3E). A plaster cast or supporting bandage holds the broken bone (broken end to broken end) in position. Bone cells divide and produce new bone tissue which hardens and heals the break. If the fracture is severe, the broken ends of the bones are lined up and pinned to help them grow together. Children's bones are softer, more flexible and less likely to break. However, an accident can splinter the bone causing a **green stick** fracture.

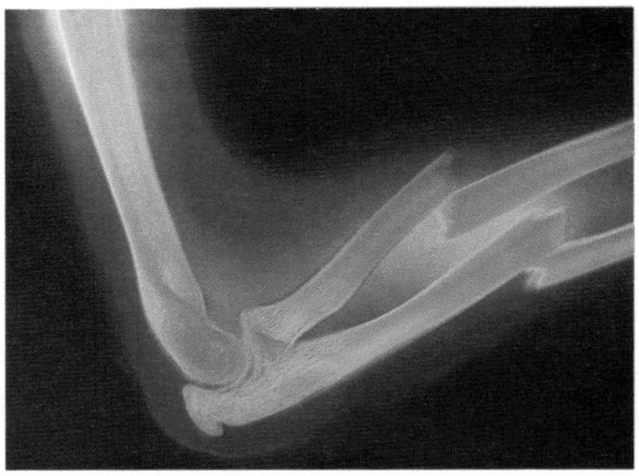

Figure 18.3E ▶ X-ray photograph showing a fracture of the lower arm bones (radius and ulna)

Joints

A joint is a meeting of two bones. Some joints are fixed. The bones of the skull, for example, meet at fixed joints called **sutures** (see Figure 18.3B). Other types of joint are not fixed:

- **Ball and socket joints** are formed where the upper long bones of the arms and legs meet their respective girdles. Figure 18.3F shows how the rounded end of the femur fits into a cup-shaped socket in the pelvic girdle.
- Figure 18.3G shows the **hinge joint** of the elbow. Hinge joints are also found in the fingers and knee. Notice how friction is reduced when the bones move against each other.

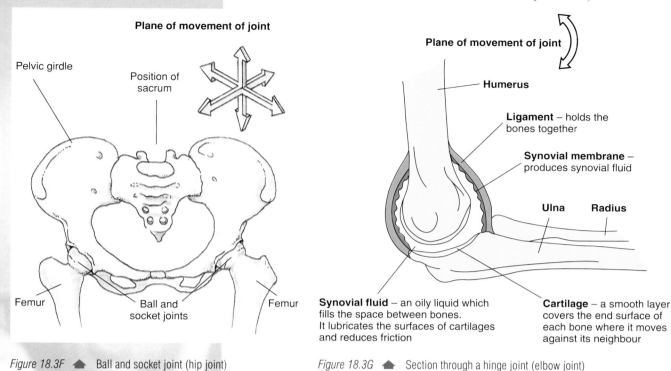

Plane of movement of joint

Pelvic girdle

Position of sacrum

Femur

Ball and socket joints

Femur

Figure 18.3F ▲ Ball and socket joint (hip joint)

Plane of movement of joint

Humerus

Ligament – holds the bones together

Synovial membrane – produces synovial fluid

Ulna **Radius**

Synovial fluid – an oily liquid which fills the space between bones. It lubricates the surfaces of cartilages and reduces friction

Cartilage – a smooth layer covers the end surface of each bone where it moves against its neighbour

Figure 18.3G ▲ Section through a hinge joint (elbow joint)

CHECKPOINT

▶ **1** Study the diagram of the arm below and answer the following questions.
  (a) Which muscle contracts when the arm is raised?
  (b) Which muscle contracts when the arm is straightened?
  (c) Why are the biceps and triceps called an antagonistic pair of muscles?
  (d) How are the arm muscles attached to the arm bones?
  (e) What type of joint is formed between the humerus and the scapula?

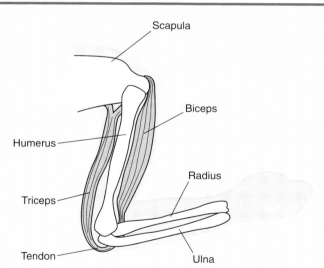

Scapula

Biceps

Humerus

Radius

Triceps

Tendon

Ulna

▶ **2** Match each of the bones in Column A with its description in Column B.

Column A	Column B
Cranium	Consists of vertebrae fused together
Rib cage	Upper arm bone
Sternum	Protects the brain
Sacrum	Moves as you breathe
Clavicle	Ribs are attached to it
Femur	Called the collar bone
Humerus	Upper leg bone

▶ **3** (a) Distinguish between the hydrostatic skeleton, exoskeleton and endoskeleton.

   (b) Say what features the hydrostatic skeleton, exoskeleton and endoskeleton have in common.

▶ **4** Give explanations for the following statements.

   (a) Pregnant women should be encouraged to drink milk.

   (b) The sutures of the skull are fixed joints.

   (c) The human femur is stronger than the humerus.

   (d) Breathing depends on the movement of the ribs.

   (e) Women usually have broader hips than men.

▶ **5** Distinguish between smooth muscle, cardiac muscle and skeletal muscle. Explain how each type of muscle is suited to the job it performs.

▶ **6** Complete the following paragraph using the words listed below. Each word may be used once, more than once or not at all.

   **levers   calcium   support   collagen   joint   rigidity   muscles   movement
   contract   food   bones   protect**

   The _____ of the skeleton _____ and _____ the soft tissues of the body. The limb bones are _____ which move because _____ attached to them _____ . Where bones meet a _____ is formed. Bone consists of the protein _____ and _____ salts. Blood vessels in the bone supply _____ and oxygen.

▶ **7** Name the bones of a human arm. Do they resemble the bones of a bird's wing?

▶ **8** List the different functions of feathers. Discuss two of the functions in more detail.

▶ **9** Different forces act against a bird in flight. Identify the forces and show with the help of diagrams how the design of the body and wings is adapted to overcome them.

▶ **10** Look at Figure 18.2J. Summarise how muscles and bones work together to move the wings.

▶ **11** Compare the arrangement of muscles and bones in the horse's hind leg with the arrangement of muscles and bones in the human. (Identify the similarities and comment on the differences.)

▶ **12** Describe the stages in a horse's stride.

▶ **13** What is the difference in the cycle of hoof beats between a horse walking and a horse galloping?

Topic 19   **Disease**

19.1 ▶

What is disease?

FIRST THOUGHTS

Think about these figures:

- In developing countries more than 4 million children die each year from the effects of diarrhoea.
- Parasitic worms infect around 2400 million people worldwide.
- Cancer accounts for more than 30% of deaths in developed countries but only 4% in developing countries.
- Influenza ('flu) and hepatitis B viruses infect enormous numbers of people world-wide. Millions die from these diseases each year.

This section explores the background to these startling facts.

See www.keyscience.co.uk for more about diseases.

It's a fact!

Cells may break away from a tumour and begin new cancerous growths elsewhere in the body. The tumour is said to be **malignant**. If cells do not break away, the cancer does not spread and the tumour is said to be **benign**.

For most of us the word 'disease' conjures up the idea of feeling unwell. Figure 3.7A on p. 72 shows a range of organisms that cause different diseases. We blame **bacteria** for most of our ailments but **viruses** are important disease-causing agents. Notice that **protists** and **fungi** are culprits as well, and that different animal **parasites** also cause disease as a result of their activities inside our bodies. Disease organisms are called **pathogens**. Diseases are said to be **infectious** if the organisms can be passed from one person to another.

Not all diseases are infectious. Many **non-infectious** diseases develop because the body is not working properly (see Table 19.1). The way we treat our bodies affects the onset of non-infectious diseases. Many disorders can be avoided or at least delayed providing we live sensibly. You can find more about disease using the *index* as well as good advice on looking after yourself.

Table 19.1 ▼ Non-infectious diseases

Type of disease	Description	Examples
Cancer	Cell division produces new cells to replace old cells as they wear out and die. If cell division runs out of control then a cancerous growth (called a tumour) develops	Lung cancer, leukaemia, cancer of the cervix are examples of malignant tumours
Degenerative ('wear and tear')	Organs and tissues 'wear' and work less well with age	Sight and hearing may deteriorate, heart attacks, arthritis
Deficiency	Poor diet may deprive the body of vitamins and other essential substances	Scurvy, rickets, kwashiorkor
Allergies	Reaction to a substance (antigen) which is normally harmless but which the body regards as harmful	Hay fever
Psychological/ mental illness	A wide range of disorders from feeling more than normally 'fed up' to more severe conditions due to brain damage or the brain not working properly	Depression, paranoia, schizophrenia
Metabolic	Defect (often inherited) that prevents an organ or tissue from working properly	Diabetes, dwarfism, haemophilia, sickle cell anaemia

Zoonosis

We share many pathogens and the diseases they cause with other animals. A pathogen which infects people and other animals is called a **zoonosis**. Animals may form a reservoir of zoonoses from which infection may spread to more and more people. Effective treatment depends on controlling the pathogen in animals as well as the human victims.

Fungi are a major cause of plant disease.

Rabies virus is a zoonosis. It infects warm-blooded animals and is found most often in wolves, jackals and other members of the dog family. Contact with domestic pets (biting, licking) is the most usual route by which rabies virus passes to people. However, in South America blood-sucking vampire bats are a common source of infection.

Rabies virus attacks the central nervous system (see p. 285) causing 'mad' behaviour. At this time the virus is present in the saliva. A rabid dog often runs and bites at random, passing on the infection to people and other animals.

The virus destroys nerve cells but takes time to establish itself in the central nervous system of the victim. If cuts and bite marks are washed with disinfectants and treatment with rabies serum (serum is plasma with fibrinogen removed – see p. 259) and vaccination is started immediately after infection, then there is a good chance of preventing the disease. However, once symptoms develop (from 2–3 weeks up to 5 months after infection) then death is nearly always inevitable within 10 days.

Strict quarantine laws have helped to keep Britain free of rabies. Quarantine means that animals brought into Britain from other countries are kept away from people and other animals until it is certain that they do not carry the disease. The quarantine time period is six months – the time it takes for symptoms of rabies to develop in an infected animal.

However, following a report by an expert group led by Professor Ian Kennedy of University College, London, changes have been made. Under the **Pet Travel Scheme (PETS)** cats and dogs vaccinated against rabies are able to travel to and from the UK, EU and other permitted countries. They will not be put into quarantine providing their owners follow the PETS rules. Quarantine will remain for animals from other countries.

Fungal diseases

Plant disease is caused by as wide a range of pathogens as animal disease. However, the impact of particular types of pathogen is different. Bacteria for example are a major cause of animal disease but pose few problems for plants. Fungi on the other hand are a major hazard. More than 1000 species cause plant disease with serious effects on crops and food production.

Phytophthora infestans causes **late blight** of potatoes. In the 1840s, the fungus wiped out the European harvest. Control measures were inadequate and potato plants had little resistance to the disease.

Phytophthora passes from plant to plant owing to:
- planting infected potatoes
- infected left-overs from the previous crop in the ground
- spores carried by wind or rain to other plants in the crop.

The spores germinate and fungal hyphae (see Topic 1.5) grow between the potato leaf cells. Side branches penetrate and feed on the cells, destroying them. The parts of the plant above ground eventually die. As a result little starch is produced through photosynthesis for storage by the potato tubers underground (see p. 356). The tubers fail to develop.

Before the potato dies, some hyphae grow through the stomata of the leaves to the outside. They carry spores which break off and infect other plants.

Good hygiene helps break the chain of infection. Burning infected plants and killing the plant above ground before harvest reduces the risk of

Infectious diseases are a major cause of death in developing countries but not in developed countries. Instead, non-infectious diseases are a major cause of death.

SUMMARY

Pathogens are organisms that cause infectious diseases. Non-infectious diseases arise when the body is not working properly. Zoonoses are pathogens which people share with other animals. Plant pathogens damage crops, reducing yields. Infectious diseases are responsible for the majority of deaths in developing countries.

infecting the potato tubers below ground. Other control measures include using disease-free seed potatoes and spraying plants with fungicide (see p. 92) to prevent spores that settle on the leaves from germinating.

Rust disease of wheat is caused by *Puccinia graminis*. This fungus lives in the stem and leaves. The weakened plant produces less grain and poor quality straw.

Wheat rust is persistent, helped by farming practices such as **monoculture** (see Figure 5.1A). Pathogens able to infect the crop spread rapidly through the hectares of identical plants causing huge losses of grain. Breeding rust-resistant strains of wheat is the best form of defence, but rust fungus quickly mutates. Scientists are in a race to discover new resistant varieties of wheat before new strains of fungus overwhelm the crop (see p. 439). *Do you think this is another example of the Red Queen Effect (see p. 405)?* Rust disease also affects other cereal crops.

Disease worldwide

In the developing countries of Africa and Asia infectious diseases are responsible for 40 per cent of all deaths. For example, each year more than 14 million children under the age of five years die from diarrhoea, measles and other diseases associated with inadequate medical services, poverty, poor housing and malnutrition.

In sharp contrast, relatively few people living in the developed countries of Japan, Europe and North America, where there are good medical services, die from infectious diseases. Instead many illnesses are related to our being better off. We can afford to smoke, drink too much alcohol and eat too much of the wrong sort of food. Unhealthy life-styles are partly responsible for the onset of heart disease and cancer, which are the leading causes of death in many developed countries.

Life expectancy and the infant mortality rate reflect the difference in causes of death between countries. In developed countries the low level of infectious diseases means that many people reach old age and most children survive to become adults. In developing countries infectious diseases make life shorter and the number of children who die is much greater. Despite the gap, progress is being made. Worldwide, for someone born in 1999, they can, on average, expect to live for 65 years. In the same year the infant mortality rate was 54 deaths per 1000 births. Education, good medical services, the relief of poverty and improvements in living conditions are all essential for success.

CHECKPOINT

▶ 1 Distinguish between infectious diseases and non-infectious diseases. Give examples of each category.

▶ 2 What is a zoonosis? Discuss how the biology of a named zoonosis helps its transfer between people and other animals.

▶ 3 Discuss how the biology of the fungus *Phytophthera infestans* makes it a serious pest of the potato.

▶ 4 Distinguish between the different causes of death in the developing countries of Africa and Asia and the developed countries of Japan, Europe and North America.

19.2 ▶ Influenza: a case study

Every few years influenza ('flu) breaks out in the winter affecting thousands of people. Every 10 to 20 years 'flu sweeps the world affecting millions. This section explains the outbreaks and why preventing them is difficult.

It's a fact!

Each year air travel moves millions of people at great speed around the world. Speed means that travellers may reach their destination infected with pathogens (see p. 331) before the symptoms of disease are apparent. Vaccination (see p. 262) of people travelling to and from countries where serious diseases are endemic (see p. 335) helps to prevent the spread of infections.

Few drugs are effective against viral diseases.

It's a fact!

How many times have you caught a 'cold'? Sneezing, a runny nose, a cough, a sore throat and a raised temperature make you feel miserable. Many different viruses cause these symptoms. Some resistance to colds may develop with frequent exposure to the viruses but their variety means that lifelong immunity is impossible.

A sneeze produces a jet of moisture droplets which shoot out of the nose and mouth. Microorganisms infecting the mucous membranes lining the nose and the rest of the **upper respiratory tract** (see Topic 15.2) are carried in the droplets and spread to other people nearby. This is how diseases like colds and 'flu pass from person to person, especially in crowded places like schools and hospitals. Remember to turn away from people when you sneeze or, if in time, sneeze into a handkerchief. It is not only good manners but also helps to cut down the spread of air-borne diseases.

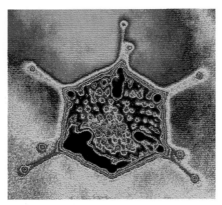

Figure 19.2A ▲ 'Flu virus particle (×200 000)

'Flu is caused by a virus (Figure 19.2A). Notice the spikes and knobs projecting from the surface of the virus. They are proteins which bind the virus to, and help it penetrate, the cells of the mucous membrane. The infected cells are destroyed, and after 48 hours the symptoms of 'flu appear. Muscles ache and overproduction of mucus causes a runny nose with sneezing and coughing. A sore throat and fever develop, and the victim feels generally unwell. Symptoms usually subside within 2–7 days unless the infection spreads to the lungs. Then viral pneumonia may set in making the patient seriously ill. The lungs fill with fluid and bacteria may set up other infections.

Treatment

Normally treating the patient amounts to little more than bed rest and good care. Drugs are available to treat only a few viral diseases. In the case of 'flu, **amantadine** can be given to patients who are at high risk of serious illness. 'High risk' includes people aged over 65 and people suffering from diseases of the upper respiratory tract and lungs (see p. 250), a lowered resistance to infection, or diabetes (see p. 301). Amantadine seems to prevent the 'flu virus from reproducing in the cells of the mucous membrane (see p. 249). **Zanamivir** (*Relenza*) does not cure 'flu but relieves the symptoms and reduces the time the patient suffers illness. The drug works by stopping the 'flu virus spreading between cells in the lung. A person who is ill with 'flu is easily attacked by bacteria. Antibiotics may be used to treat the bacterial infection.

Vaccines

The surface proteins of 'flu virus (see Figure 19.2A) are **antigens** against which an infected person produces **antibodies**. However, people may catch 'flu more than once during their lifetime. *Why are the antibodies ineffective against 'flu virus when it next invades the body?* After all you rarely catch diseases like chicken pox and measles more than once (see p. 262).

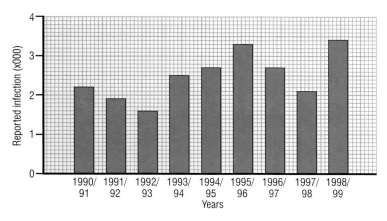

Figure 19.2B ▲ Infections due to different types of 'flu virus in England and Wales 1990–1999. (Source: Public Health Laboratory Service)

Antigenic drift and antigenic shift result in new types of 'flu virus and make it difficult to produce an effective vaccine.

SUMMARY

'Flu is caused by a virus. Epidemics and pandemics of the disease are related to frequent changes in 'flu antigens. To be effective, new vaccines must be developed quickly in response to changes in the virus.

Unfortunately, 'flu virus antigens often change shape. Minor changes are called **antigenic drift**. They produce new **strains** of virus which are probably responsible for the frequent occurrence of 'flu **epidemics**.

Major changes are called **antigenic shift** and result in new **types** of virus. They are less frequent and seem to be linked to the 10–20 year cycle of worldwide **pandemics**. Antigenic drift and antigenic shift means that antibodies produced against a particular type of 'flu virus do not protect the person from a new infection. They also mean that producing an effective vaccine is difficult.

Prevention

The **World Health Organisation** monitors the **antigenic changes** in 'flu virus and recommends the strains of virus which should be included in vaccines. The recommended strains are grown in fertilised chicken eggs, extracted and inactivated with methanal.

The vaccine usually consists of several strains and is used to protect young children, the old and other people who may be at risk of serious illness if they catch 'flu. Protection from a particular strain of 'flu virus is effective providing the vaccine for the strain is given each year and that its antigens do not change. When antigenic changes occur new techniques allow vaccine to be redeveloped quickly to keep pace.

CHECKPOINT

▶ 1 Describe the symptoms of 'flu and discuss how they are caused.

▶ 2 After many years of research why is an effective vaccine against the common cold still not available?

▶ 3 Why does the World Health Organisation monitor antigenic changes in the 'flu virus?

19.3 ▶ Fighting disease

Make sure you can answer an exam question which asks you to distinguish between disinfectants, antiseptics and asepsis.

Figure 19.3A shows the body's natural defences against disease. Notice the **physical** and **chemical** barriers to infection. They help to keep most of us healthy for most of our lives. Looking after ourselves and maintaining good hygiene help the defences to do their job. You can find out more about treating disease from the *index*.

Disinfectants and antiseptics

Disinfectants are chemicals that kill microorganisms; **antiseptics** stop them from multiplying. We use disinfectants and antiseptics to attack microorganisms outside the body.

Disinfectants help to keep surfaces clean and free from microorganisms. For example, bleach poured down the lavatory kills microorganisms lurking around the bowl. Bleach is a strong disinfectant which contains calcium chlorate(I). It is effective but corrosive, so use it carefully.

Antiseptics are usually weaker than disinfectants. They can be used to swab a wound or before an injection to clean the area of skin where the hypodermic needle is to be inserted.

The skin is one of the main barriers against infection. Washing regularly reduces the chances of infection. Some soaps and skin cleansers have antiseptic properties. They are particularly effective against the bacteria that cause body odour. They also help prevent fungal infections such as ringworm and athlete's foot (see Figure 3.7A, p. 72). Sweaty feet are an ideal environment for fungi. So, wash and dry carefully between your toes. Also, borrowing other people's footwear is not a good idea. The fungi are spread from person to person by physical contact.

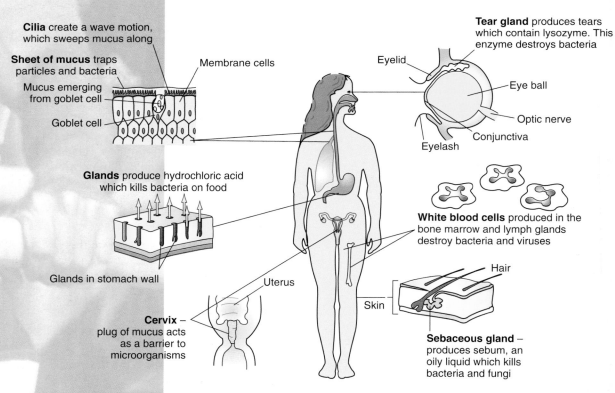

Cilia create a wave motion, which sweeps mucus along

Sheet of mucus traps particles and bacteria

Mucus emerging from goblet cell

Goblet cell

Membrane cells

Glands produce hydrochloric acid which kills bacteria on food

Glands in stomach wall

Cervix – plug of mucus acts as a barrier to microorganisms

Uterus

Tear gland produces tears which contain lysozyme. This enzyme destroys bacteria

Eyelid

Eye ball

Optic nerve

Conjunctiva

Eyelash

White blood cells produced in the bone marrow and lymph glands destroy bacteria and viruses

Hair

Skin

Sebaceous gland – produces sebum, an oily liquid which kills bacteria and fungi

Figure 19.3A ⬆ The body's natural defences

Developing bacterial resistance to antibiotics encourages research to find new drugs.

Chemotherapy

The word **chemotherapy** describes treatments which use drugs to attack pathogens. There is a lot more about drugs and treatment of disease elsewhere in the book. Turn to the *index* to look up the information.

In the fight against disease, **antibiotic drugs** have been the success story of the twentieth century. Before their discovery, bacterial diseases were major killers, especially of children. Now the majority of infectious diseases can be cured – thanks to antibiotics (see p. 412).

There are two sorts of antibiotic:

● **bactericides** like penicillin, which kill bacteria
● **bacteristats** like tetracycline, which prevent bacteria from multiplying.

Figure 19.3B shows how different antibiotics affect bacteria. Notice that those which damage the structure of the cell are mostly bactericidal. Those which disrupt metabolism (see p. 3) are mostly bacteristatic.

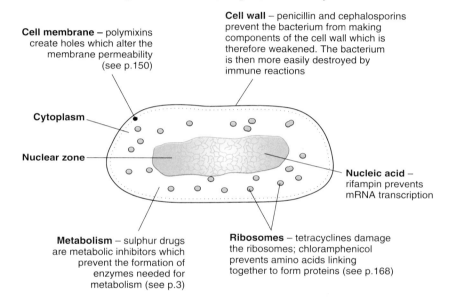

Cell membrane – polymixins create holes which alter the membrane permeability (see p.150)

Cell wall – penicillin and cephalosporins prevent the bacterium from making components of the cell wall which is therefore weakened. The bacterium is then more easily destroyed by immune reactions

Cytoplasm

Nuclear zone

Nucleic acid – rifampin prevents mRNA transcription

Metabolism – sulphur drugs are metabolic inhibitors which prevent the formation of enzymes needed for metabolism (see p.3)

Ribosomes – tetracyclines damage the ribosomes; chloramphenicol prevents amino acids linking together to form proteins (see p.168)

Figure 19.3B 🔺 Action of antibiotics

Resistance

Since the 1940s, widespread use of antibiotics has led to the development of **resistance** (see Figure 19.3C and p. 443). Bacteria multiply very quickly (in some species a new generation is produced every 20 minutes) and resistance spreads quickly. To be effective, dosage has to be increased step-by-step until the drug becomes so inefficient or poisonous to the patient that an alternative has to be found.

Even so, new drugs are only part of the answer to the problem. Resistance to each new drug soon develops. Precautions can be taken in an attempt to slow the development of resistance:

● *Reduce* the antibiotics given to farm animals to improve production. Meat contains the antibiotics. Eating meat, therefore, means that people take in traces of antibiotic.
● *Avoid* using antibiotics by practising good hygiene to prevent the spread of infection.
● Antibiotics should be used *sparingly* and not prescribed unnecessarily.
● A patient should always *finish* the prescribed course of antibiotics.

SUMMARY

The body's natural defences against disease are supported by advances in modern medicine. Disinfectants and antiseptics help to prevent the spread of infection. Antibiotic drugs, in particular, have had a major impact on treatment. Diseases which previously killed many people can now be cured. However, some bacteria have developed resistance to antibiotics. Sensible use of drugs and the discovery of new ones together help to combat the problem.

See www.keyscience.co.uk for more about penicillin.

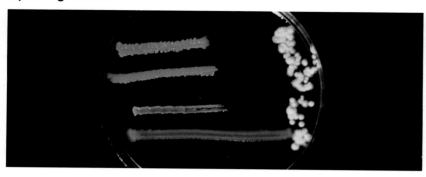

Figure 19.3C ◆ *Penicillium* mould is growing at one edge of the plate; four different bacteria have been streaked on the plate; three are sensitive to the penicillin diffusing from the mould and their growth stops short of the mould; the fourth bacterium is not sensitive and the streak therefore grows up to and in contact with the mould

Resistance develops in different ways. For example some bacteria are resistant to penicillin G (see p. 412 – one of the first of the family of penicillins, and still a useful antibiotic) because they produce **penicillinase**. This enzyme breaks down the penicillin, making the drug ineffective. A particular strain of the coccus bacterium *Staphylococcus aureus* (see Figure 1.5C, p. 14) is resistant to all antibiotics, except one with side-effects that can be lethal to the patient! These examples highlight the importance of the measures listed above, which aim to guard against the development of resistance.

CHECKPOINT

▶ **1** Discuss how the body's natural defences keep most of us healthy for most of our lives.

▶ **2** Distinguish between:

antiseptics	disinfectants
bactericides	bacteristats
antisepsis	asepsis

▶ **3** Discuss how different antibiotic drugs attack bacteria.

▶ **4** How do bacteria become resistant to antibiotic drugs? Discuss the precautions which can be taken to slow the development of resistance.

Theme Questions

1 Henry wanted to estimate the energy content of a biscuit. He placed 20 g of water in a large test tube. He then measured the temperature of the water.

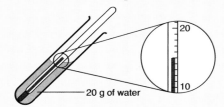

20 g of water

Henry burned the biscuit under a large test tube and measured the temperature of the water again.

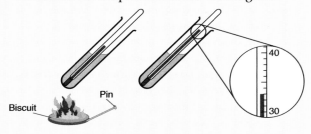

Biscuit   Pin

(a) Henry found out that 1 g of water needs 4.2 J of energy to increase its temperature by 1°C.
Calculate how much energy the water obtained from Henry's biscuit.
You **must** show how you work out your answer.
(b) Henry repeated his experiment using the following apparatus. He found that the water obtained more energy from the biscuit.

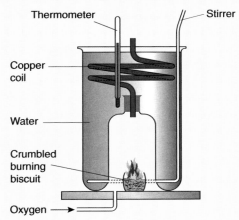

Thermometer   Stirrer
Copper coil
Water
Crumbled burning biscuit
Oxygen →

Write down **two** features of this apparatus which improved his result.
Explain how these features improved his result.
(OCR)

2 Photographic film is made of transparent celluloid. Silver compounds are fixed to it using gelatine. Pepsin releases the silver compounds by removing the gelatine. The release of the silver compounds changes the photographic film from dark to transparent.

Five test tubes were set up to investigate the effect of pH on the action of pepsin.
Each test tube contained the same volume of 5% pepsin solution

Each test tube contained the same volume of 5% pepsin solution

pH 2   pH 4   pH 6   pH 8   pH 10

Dark photographic film

The five test tubes were kept at 37°C for two hours. Changes in the transparency of the film were measured. The results are shown in the table.

	Percentage of light which passed through the photographic film				
	pH 2	pH 4	pH 6	pH 8	pH 10
At start	52	52	52	52	52
After 2 hours	72	96	62	53	52

(a) At which pH did the pepsin work best? Explain your answer.
(b) Pepsin removed the gelatine from the film in some of the tubes.
   (i) Suggest which type of substance gelatine is.
   (ii) Name TWO substances which would be present in solution in the test tubes after pepsin had removed all the gelatine from the film.
(c) The experiment was repeated at 70°C.
Explain the effect this would have on the result at pH 2.   (Edexcel) Abridged

3 Look carefully at the diagram of the alveoli and blood capillary in a lung .
(a) Which arrow represents (i) deoxygenated blood (ii) carbon dioxide?
(b) (i) By what process is oxygen able to pass from the air in the alveoli into the red blood cells in the capillary.

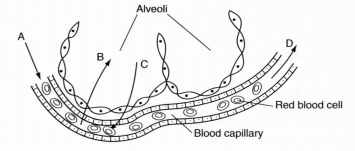

Alveoli
A
B   C
D
Red blood cell
Blood capillary

   (ii) Through how many layers of cells does the oxygen have to pass?
(c) Give two features of the alveoli that help this passage of gas.
(d) Which structures allow gas exchange to take place in (i) a leaf (ii) a fish?

Reproduction

All living things eventually die. They may be killed by accidents or other organisms or die of old age. Before dying some individuals produce new individuals of their own kind. The process is called reproduction.

Topic 20 Outlines of reproduction

20.1 ▶ Asexual reproduction

FIRST THOUGHTS

Why are offspring reproduced asexually identical to one another and their parent? Read this section and find out.

During asexual reproduction, mitosis means that the offspring are genetically identical with one another and with their parent.

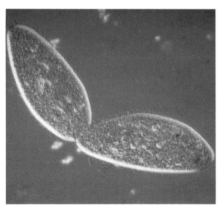

(a)

(e)

SUMMARY

Offspring reproduced asexually each receive an exact copy of the parent's genetic material because of mitosis. As a result the offspring are identical to each other and to their parent.

Reproduction more than anything else characterises living things. It continues the thread of life, unbroken from generation to generation.

Asexual reproduction gives rise to individuals which are genetically identical to each other and to their parent. The genetically identical individuals produced are called **clones**.

Figure 20.1A illustrates different ways in which asexual reproduction takes place. Notice that only *one* parent gives rise to new individuals. The offspring are identical to the parent because the parent's cells divide by mitosis. During mitosis DNA **replicates** (see Topic 9.4). The replication of DNA passes exact copies of the parent's genetic material to the daughter cells (see p. 154) which multiply and give rise to new genetically identical offspring. The sequence reads:

Parent cells →(replication of DNA)→ Mitosis →(exact copies of parent's genetic material inherited)→ Daughter cells →(develop)→ Offspring

Mistakes in DNA replication are called **mutations**, but they are rare (see p. 148).

(b) (c) (d)

Figure 20.1A ⬆ Ways of reproducing asexually (a) *Paramecium* dividing by **fission** – the parent cell divides into equal parts (b) *Hydra* **budding** – outgrowths of the parent's body (buds) separate from the parent. Each one becomes a separate individual. (c) The **vegetative parts** of the parent body grow into new individuals. For example parts of stems can sprout roots and grow into new plants. (d) **Regeneration** – the parent body breaks into pieces, each piece can grow into a new individual. (e) **Parthenogenesis** – a female aphid produces eggs which develop into new individuals without fertilisation

CHECKPOINT

> 1 What is asexual reproduction?
> 2 Briefly describe the different ways asexual reproduction takes place.
> 3 What is a clone?
> 4 Explain why the formation of a clone depends on mitosis.

20.2 ▶ Sexual reproduction

FIRST THOUGHTS

Why are offspring reproduced sexually different from one another (except identical twins) and their parents? Find out by reading this section.

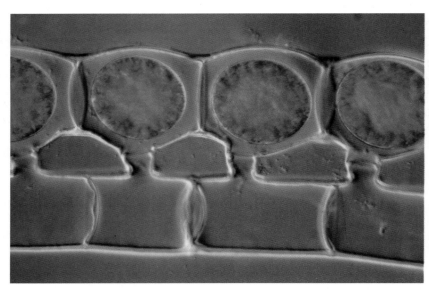

Figure 20.2A 🔺 Maleness and femaleness in *Spirogyra* – a new strand of *Spirogyra* develops from the fused nuclei

Figure 20.2A shows strands of the green alga *Spirogyra* (see Topic 1.5) lying side-by-side. A cell of one of the strands is joined by a tube to a cell of the other strand. The content of one of the joined cells passes through the tube and the nucleus fuses with the nucleus of the other cell. The genetic material in each nucleus combines with the genetic material of its opposite number. The combination of genetic material is *different* from the combination in each of the parent cells. This is an important feature of **sexual reproduction**; the production of much more **genetic variation** is possible than by mutation alone. Figure 20.2A illustrates another important feature of sexual reproduction – **maleness** and **femaleness**. The male cell of *Spirogyra* is the empty one. Its contents have moved toward the female cell.

■ Sex cells

There are two types of sex cell: male sex cells called **sperms** and female sex cells called **eggs**. Sperms and eggs are called **gametes**. Table 20.1 compares sperms and eggs.

Table 20.1 🔽 Comparing sperms and eggs

Sperms	Eggs
Small	Large
Move towards the egg	Do not move much
Have no food store	Have food store

The sperms of many types of organism are very similar. Human sperm is fairly typical (see Figure 20.2B). In fact, the sperm is little more than a mobile nucleus designed to bring genetic material from the male parent to the egg of the female parent.

Animal eggs are bigger than sperms because they contain yolk which is a food store. Some eggs have more yolk than others. This is why the eggs of different animals are different in size.

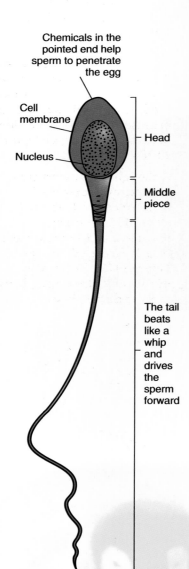

Chemicals in the pointed end help sperm to penetrate the egg

Cell membrane

Nucleus

Head

Middle piece

The tail beats like a whip and drives the sperm forward

Figure 20.2B 🔺 Human sperm

Fertilisation

Why are sex cells haploid (see Topic 9.4)? In other words why do they have half the chromosomes of the other cells of the body? During fertilisation (when sperm and egg join together) the sperm nucleus and the egg nucleus fuse. The chromosomes from each nucleus combine to produce new combinations of genes. Half come from the haploid sperm and half come from the haploid egg. The fertilised egg (called the **zygote**) is therefore **diploid** (see Topic 9.4) and develops into a new individual which inherits characteristics from both parents, not from just one as in asexual reproduction (see Figure 20.2C). *Can you think what would happen genetically in each generation of new individuals if sperms and eggs were diploid and not haploid?*

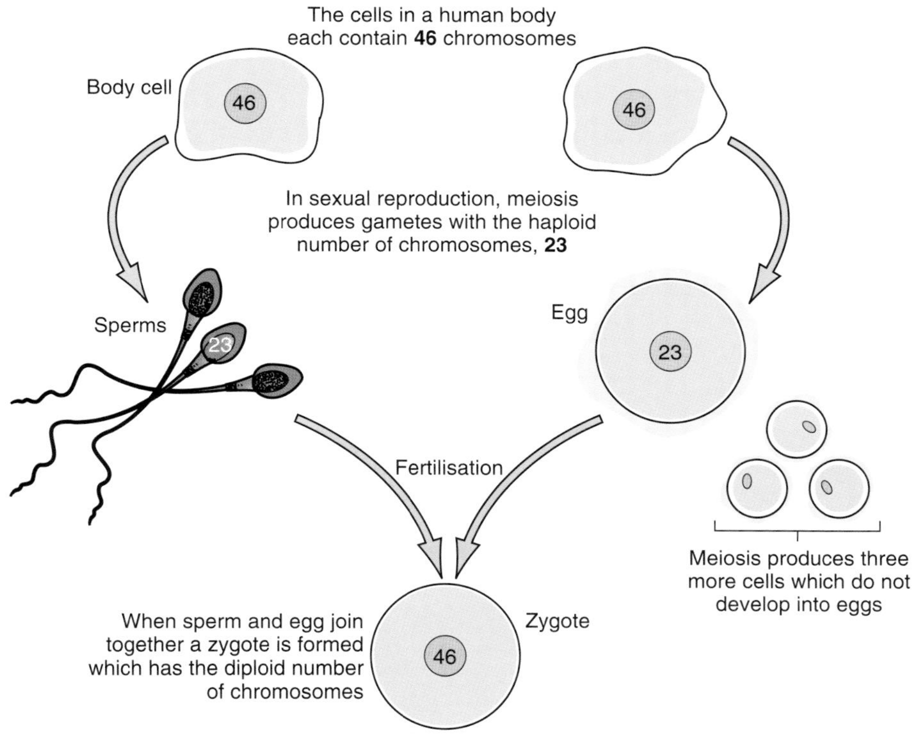

■ Patterns of fertilisation

Sperms swim. Swimming brings sperms to eggs. Sperms and eggs must therefore be in a liquid for fertilisation to take place. Most aquatic organisms release sperms and eggs into the surrounding water where fertilisation takes place. This is called **external fertilisation** because it takes place outside the parent's body (see Figure 20.2D).

Although moss plants grow on land, even they need water to complete their life cycle. Without it moss sperms cannot swim to moss eggs to fertilise them (see Figure 20.2E). This is why mosses and other such plants grow in damp places.

The cells in a human body each contain **46** chromosomes

Body cell
46

46

In sexual reproduction, meiosis produces gametes with the haploid number of chromosomes, **23**

Sperms
23

Egg
23

Fertilisation

Meiosis produces three more cells which do not develop into eggs

When sperm and egg join together a zygote is formed which has the diploid number of chromosomes
Zygote
46

Figure 20.2C 🔺 Fertilisation – restoring the diploid state

Figure 20.2D 🔺 Frogs mating. The male sheds sperms over the eggs which the female lays in the water

Figure 20.2E 🔺 Moss sperms are released in the film of water covering the plant. They swim towards the eggs

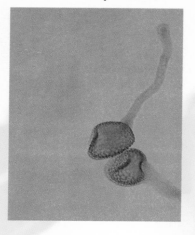

Figure 20.2F 🔼 Sprouting pollen tubes

SUMMARY

Offspring reproduced sexually each receive genetic material from both parents. As a result the offspring are different from one another (except identical twins) and their parents.

Without water sperms and eggs perish. *How do organisms that spend their lives on dry land overcome the problem?* **Seed-producing** plants can live in much drier places than moss plants. Their sperm is protected inside drought-resistant **pollen grains**. These are usually carried by animals or blown by the wind to the female part of the plant.

Each pollen grain sprouts a **pollen tube** which grows through the female tissues towards the eggs inside (see Figure 20.2F). Sperms pass down the pollen tube and one of them fertilises the egg. This is an example of **internal fertilisation**.

Internal fertilisation is one of the adaptations that insects, reptiles, birds and mammals have adopted for life on dry land. The male places sperm directly into the body of the female – a process that usually needs some kind of coupling, called **copulation**. Special organs help to transfer the sperm from male to female. In many insects, for example, sperm is transferred from male to female protected in a tiny package called the **spermatophore** (see Figure 20.2G).

In mammals the male has an organ called the **penis** which penetrates the opening to the reproductive system of the female (see Figure 20.2H). Sperms pass through the penis into her reproductive system where fertilisation takes place. The sperms swim towards the eggs in a liquid which is produced by the male reproductive system specially for the purpose and which is transferred to the female with the sperms.

Figure 20.2G 🔼 Bush crickets transferring a spermatophore

Figure 20.2H 🔼 Copulating zebras

CHECKPOINT

▶ **1** What are 'gametes'?

▶ **2** Explain, in simple terms, how sperms swim.

▶ **3** Why are eggs larger than sperms?

▶ **4** What is meant by the term 'fertilisation' and what is formed as a result of it?

▶ **5** 'The number of chromosomes in the nuclei of cells is kept constant from one generation to the next through meiosis and sexual reproduction.' Comment on this statement with reference to a named species.

Topic 21   **Plant reproduction**

21.1 ▶ **Flowers**

EXTENSION FILE ACTIVITY

EXTENSION FILE ACTIVITY

Flowers are **shoots** which are specialised for sexual reproduction. Although flowers come in all shapes and sizes, they are all made up of similar parts. The **sepals**, **petals**, **stamens** and **carpels** of the buttercup are shown in Figure 21.1A.

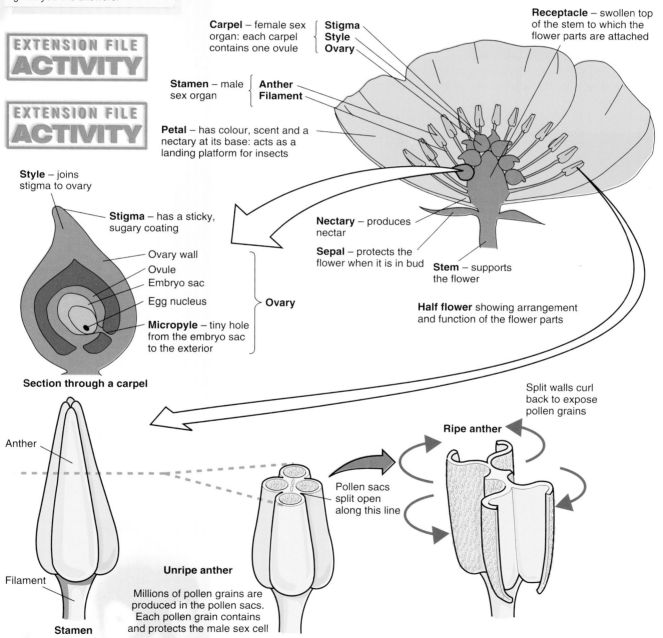

Carpel – female sex organ: each carpel contains one ovule { **Stigma Style Ovary** }

Receptacle – swollen top of the stem to which the flower parts are attached

Stamen – male sex organ { **Anther Filament** }

Petal – has colour, scent and a nectary at its base: acts as a landing platform for insects

Style – joins stigma to ovary

Stigma – has a sticky, sugary coating

Ovary wall
Ovule
Embryo sac
Egg nucleus

Micropyle – tiny hole from the embryo sac to the exterior

} **Ovary**

Section through a carpel

Nectary – produces nectar

Sepal – protects the flower when it is in bud

Stem – supports the flower

Half flower showing arrangement and function of the flower parts

Anther

Filament

Stamen

Unripe anther

Millions of pollen grains are produced in the pollen sacs. Each pollen grain contains and protects the male sex cell

Pollen sacs split open along this line

Ripe anther

Split walls curl back to expose pollen grains

Figure 21.1A ◆ The meadow buttercup

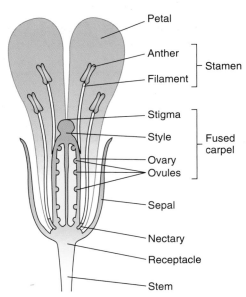

Petal
Anther — Stamen
Filament
Stigma
Style — Fused carpel
Ovary
Ovules
Sepal
Nectary
Receptacle
Stem

Types of flower

Flowers like the buttercup are simple flowers. Their parts are free and separate from each other and they have many stamens and carpels. The parts of other types of flower may be arranged differently with some parts joined together or have fewer parts than the buttercup has. Figure 21.1B illustrates examples of different arrangements.

(a) Half-flower of the wallflower. The one large carpel is formed from a number of carpels fused together. You can tell that the carpel is fused because it contains a number of ovules. In the buttercup each carpel contains one ovule

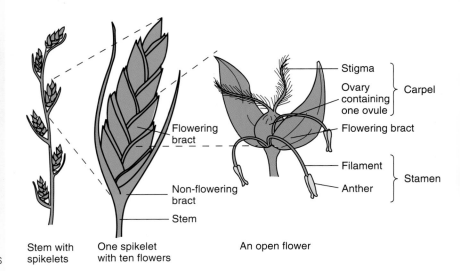

Stigma
Ovary containing one ovule — Carpel
Flowering bract
Filament — Stamen
Anther

Flowering bract
Non-flowering bract
Stem

Stem with spikelets   One spikelet with ten flowers   An open flower

(b) Rye grass flowers have no sepals or petals

Figure 21.1B ⬆ Arrangements of flower parts

Figure 21.1C ⬆ The anthers of hazel flowers produce large amounts of pollen. Notice that the leaves are in tight bud

Pollination

The pollen grain must pass from the anther to the stigma before the male sex cell inside it can fertilise the female sex cell in the ovule. This transfer of pollen is called **pollination**.

■ Wind pollination

Wind-pollinated flowers are adapted in ways that make sure that pollen is widely scattered. Some have long, slender anthers which hang clear of the sepals and petals, which otherwise might get in the way of the pollen being blown away. The flowers of different grasses (for example rye grass in Figure 21.1B) do not have sepals or petals at all. The dangling male catkins of the hazel tree produce a cloud of pollen which is blown away well before the leaves, which could get in the way, are fully open (see Figure 21.1C).

Figure 21.1D ⬆ Mallow flowers showing strongly marked honey guides

Insect pollination

Insects carry pollen from the anthers of one flower on to the stigmas of another flower. Insect-pollinated flowers, therefore, are adapted to attract insects to them. They are often large, brightly coloured and sweetly scented. They also produce a sweet liquid called **nectar**. Insects visit flowers to feed on pollen and nectar, attracted by the colour and scent. Marks on the petals called **honey guides** lead the insect to the nectar (see Figure 21.1D). As insects feed at the flower, their bodies become covered in pollen which is carried to the next flower the insect visits (see Figure 21.1E).

Figure 21.1E ⬆ Pollen is sticking to the body and legs of a honey bee on a dandelion. It is carried by the bee in pollen sacs, one on each of the back legs: one pollen sac can be seen. The curled stigmas of the dandelion florets can also be seen

The differences between wind-pollinated and insect-pollinated flowers are summarised in Table 21.1.

It's a fact!

Human eyes are damaged by ultraviolet light, but insects' eyes can see in ultraviolet light. The honey guides on the petals of flowers stand out when photographed on film sensitive to ultraviolet light. Bees searching for nectar can see the honey guides very clearly.

 See www.keyscience.co.uk for more about pollination.

It's a fact!

Pollination need not always lead to fertilisation. Chemicals in the stigma stop pollen from a different species growing pollen tubes. In the case of self-pollination the pollen tube may not grow as fast, if at all, as that of pollen from another flower.

EXTENSION FILE
ASSIGNMENT

Table 21.1 ⬇ Comparing insect-pollinated and wind-pollinated flowers

Part of flower	Insect-pollinated		Wind-pollinated	
Petals	Brightly coloured Usually scented Most have nectaries	Large flowers which attract insects	If present, green or dull colour No scent No nectaries	Small flowers
Anthers	Positioned where insects are likely to brush against them		Hang loosely on long thin filaments so that they shake easily in the wind	
Stigma	Positioned where insects can brush against them Sticky and flat or lobe-shaped		Long, branching and feathery to make a large area for catching wind-blown pollen grains	
Pollen	Small amounts produced Large grains Rough or sticky surface which catches on the insects' bodies		Large amounts produced Small, light grains with smooth surfaces – easily carried on the wind	

The wild arum is a trap for small insects. Insects that feed on dung and rotting flesh are attracted by its horrible smell. They fall down the slippery tube-like bract to the nectar at the bottom, brushing pollen on to the stigmas of the carpels as they fall. The insects cannot escape because of the downward pointing hairs. The anthers ripen within hours of pollination taking place, the hairs wither and the insect crawls out. As they brush past the anthers they are covered with fresh pollen which is transferred to another arum when the insect is trapped again.

Self-pollination and cross-pollination

(a) Self-pollination. Pollen is transferred from the anthers to the stigma(s) of the same flower or the anthers of one flower to the stigma(s) of another flower on the same plant

(b) Cross-pollination. Pollen is transferred from the anthers of a flower on one plant to the stigma(s) of a flower on a different plant. Cross-pollination increases **genetic variation** (see Topic 25.2)

Figure 21.1F ⬆

Figure 21.1F shows the difference between **self-pollination** and **cross-pollination** and how self-pollination can take place. Different adaptations make cross-pollination more likely than self-pollination. For example, self-pollination is less likely to occur if anthers and stigmas on the same flower or different flowers on the same plant are ripe at different times. The two possibilities are that:

- the anthers ripen and pollen falls before the stigma is ripe
- the stigma is ready to receive pollen before the anthers are ripe.

Figure 21.1G shows an example of the first possibility. However, pollination would not occur at all if all of the anthers (or stigmas) in flowers of a particular species were ripe at a time when all of the stigmas (or anthers) were not. Success depends on some anthers and some stigmas in different flowers being ripe at the same time.

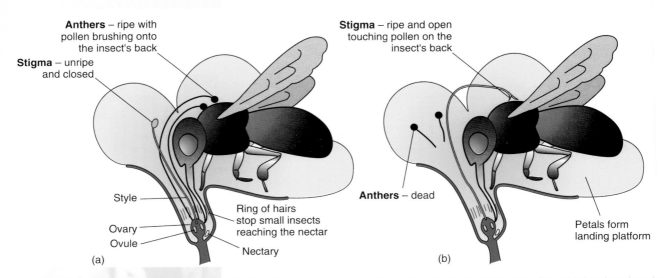

Figure 21.1G ⬆ Half-flower of the white deadnettle. The stigma and anthers touch the topside of the insect. In the young flower (a) the stigma is unripe and closed. The insect's weight brings down the ripe anthers which brush pollen on to the insect's back. In the older flower (b) the anthers have died and the stigma is ripe and open. The insect's weight brings down the stigma which picks up pollen from the insect's back

Different arrangements of
flower parts promote cross-
pollination.

Why should cross-pollination be
preferable to self-pollination?
Answer: cross-pollination
promotes genetic variation.

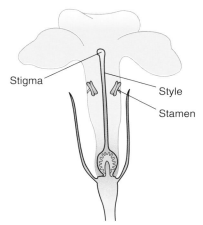

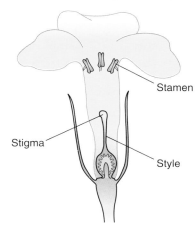

(a) Half-flower of the **pin-eye** primrose with
long style and stigma above stamens

(b) Half-flower of the **thrum-eye** primrose
with short style and stigma below stamens

Figure 21.1H 🔼

The primrose shown in Figure 21.1H illustrates another method for
making cross-pollination more likely. When a bee pushes into a thrum-
eyed flower in search of nectar, its head is dusted with pollen. If the next
flower it visits is pin-eyed some pollen will brush on to the stigma. When
a bee visits a pin-eyed flower, pollen sticks to its mouthparts at the place
where it will touch the stigma of a thrum-eyed flower. Also, thrum-eyed
flowers have large pollen grains and stigmas with small pits, pin-eyed
flowers have small pollen grains and stigmas with large pits. Since large
pollen grains fit best into large pits and small grains into small pits,
cross-pollination is more likely.

The only way to make sure of cross-pollination is for a plant to have
either all male or all female flowers. Few plants are like this but
examples are the poplar, ash and willow trees.

Fertilisation

Figure 21.1I 🔼 Pollen grains covering the tip
of the stigma (green) of a cress flower. Notice the
anthers (grey) carrying pollen ready for transfer to
the stigma of another flower

Pollination brings pollen grains to the stigma (see Figure 21.1I). A male
sex cell is inside each pollen grain. *How does it reach the egg cell in the ovule
inside the ovary?* The sugar coating the stigma's surface helps pollen
grains to stick to it. If conditions are right pollen grains begin to grow
pollen tubes (see Figure 21.1J). The pollen tube grows down through the

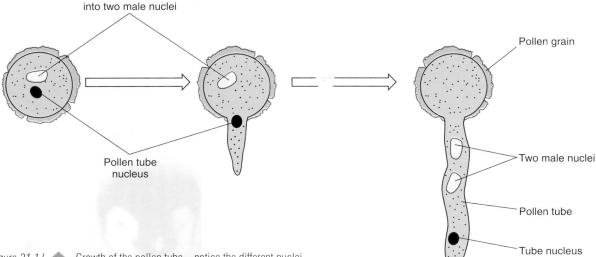

Figure 21.1J 🔼 Growth of the pollen tube – notice the different nuclei

style to the **micropyle**. The **tube nucleus** dies and the two **male nuclei** pass down the pollen tube and into the **embryo sac**. There, one nucleus fuses with the egg nucleus. This fusion of male and female nuclei is **fertilisation**. (see Figure 21.1K.)

Although more than one pollen grain may grow a pollen tube, it is a male nucleus of the first pollen tube to reach the egg nucleus which fertilises it. In an ovary which has more than one ovule each egg nucleus fuses with one male nucleus.

The fertilised egg divides and develops into the **embryo** which will become the new plant. The other male nucleus from the pollen tube fuses with two more nuclei in the embryo sac, developing into a special tissue which forms a **food store** for the embryo to use when it grows.

EXTENSION FILE
ACTIVITY

SUMMARY

Flowers are specialised for sexual reproduction. The carpel contains a female sex cell; the anther contains pollen grains, each enclosing a male sex cell. Pollination carries pollen to the stigma of a carpel. The pollen grains sprout pollen tubes each capable of carrying the male sex nucleus to the female sex nucleus. Fertilisation occurs when the nuclei fuse.

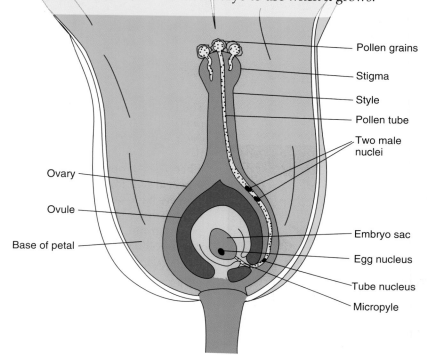

Labels:
- Pollen grains
- Stigma
- Style
- Pollen tube
- Two male nuclei
- Ovary
- Ovule
- Base of petal
- Embryo sac
- Egg nucleus
- Tube nucleus
- Micropyle

Figure 21.1K The male nucleus fertilises the egg nucleus

CHECKPOINT

▶ 1 Look at Figure 21.1A
   (a) Which parts are the male and female sex organs?
   (b) The male sex organs produce millions of tiny pollen grains. What is inside each pollen grain?
   (c) Which part of the female sex organ contains the ovule?
   (d) What does the ovule contain?

▶ 2 What is pollination?

▶ 3 Pollination usually happens by wind or by insects. Choose one wind-pollinated flower and one insect-pollinated flower from Figures 21.1A–D and study the pictures carefully. How is each flower you have chosen adapted for the way it is pollinated?

▶ 4 Briefly describe different ways flowers are adapted to make cross-pollination more likely than self-pollination.

▶ 5 Does pollination take place before or after fertilisation?

▶ 6 What is the name of the tiny hole through which the pollen tube goes to the ovule?

▶ 7 What happens when the pollen tube reaches the ovule?

21.2 ▶ Seeds and fruits

A **seed** is a fertilised **ovule**. Inside is the embryo plant and its store of food which develops following fertilisation. The seed is covered by a tough covering called the **testa**.

When the embryo is almost fully developed the tissues round it lose water, leaving the seed hard and dry. The seed can stay like this for a long time until conditions are right for it to grow.

Food may be stored in a thick, fleshy, wing-like structure called the **cotyledon**. Figure 21.2A shows that the seed of the broad bean has two cotyledons. Seeds like this are called **dicotyledonous** seeds. Figure 21.2A also shows which parts of the embryo will grow into which parts of a new plant.

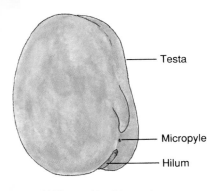

Testa

Micropyle

Hilum

(a) The outside of the whole seed

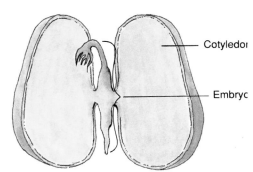

Cotyledon

Embryo

(b) The two cotyledons slightly separated showing the embryo plant between them

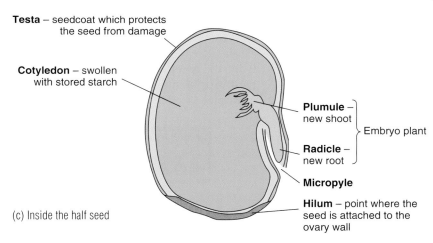

Testa – seedcoat which protects the seed from damage

Cotyledon – swollen with stored starch

Plumule – new shoot

Radicle – new root

> Embryo plant

Micropyle

Hilum – point where the seed is attached to the ovary wall

(c) Inside the half seed

Figure 21.2A ◆ The broad bean seed

Grasses produce seeds which store food in only one cotyledon. They are called **monocotyledonous** seeds.

After fertilisation it is usually the **ovary** which develops into the **fruit**. The wall of the ovary is then called the **pericarp**. As the fruit develops the pericarp becomes either dry and hard (examples are acorn, dandelion, wallflower, sycamore and poppy) or juicy and fleshy (examples are plum, tomato, blackberry, rosehip, honeysuckle and holly). Juicy fruits are more correctly called **succulent fruits** (see Figure 21.2B).

The fruit contains the seed or seeds. The number of seeds in a fruit depends on how many ovules there were in the ovary to begin with and how many were fertilised.

EXTENSION FILE
ACTIVITY

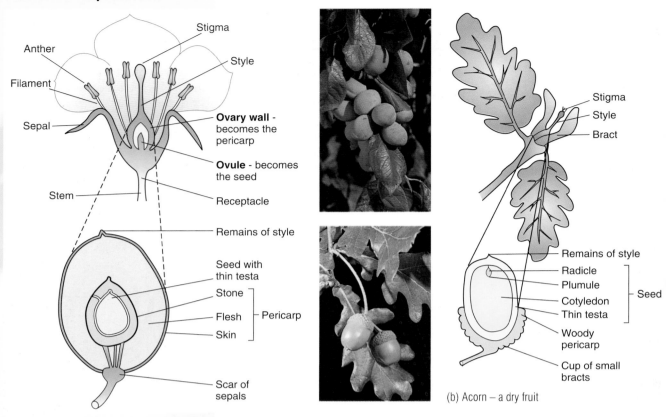

Ovary wall - becomes the pericarp

Ovule - becomes the seed

(a) Plum – a succulent fruit with one seed inside the stone

(b) Acorn – a dry fruit

Figure 21.2B ⬆ The parts of the flower that develop into parts of the fruit

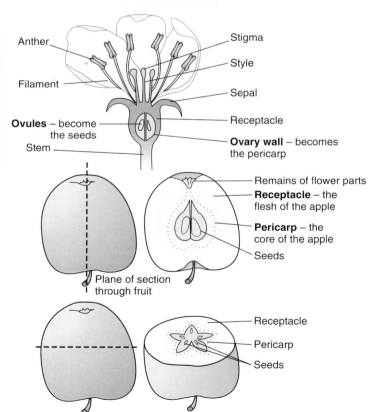

Ovules – become the seeds

Ovary wall – becomes the pericarp

Remains of flower parts

Receptacle – the flesh of the apple

Pericarp – the core of the apple

Seeds

Plane of section through fruit

Receptacle

Pericarp

Seeds

In some plants, parts of the flower other than the ovary develop into the fruit. These are called **false fruits**. In many false fruits the **receptacle** grows to form the fruit. Apples and strawberries are examples (see Figure 21.2C).

EXTENSION FILE
ASSIGNMENT

Figure 21.2C ⬆ The apple is an example of a false fruit. The receptacle forms the fleshy fruit around the pericarp which contains the seeds

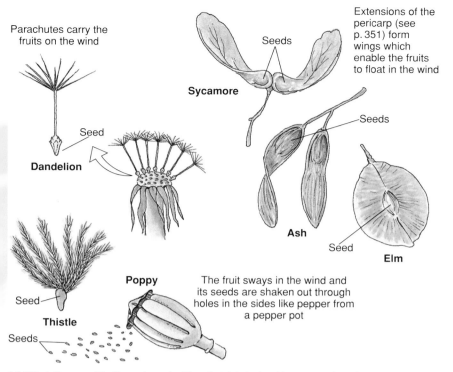

Parachutes carry the fruits on the wind

Dandelion

Seed

Thistle

Seed

Seeds

Poppy

The fruit sways in the wind and its seeds are shaken out through holes in the sides like pepper from a pepper pot

Sycamore

Seeds

Extensions of the pericarp (see p. 351) form wings which enable the fruits to float in the wind

Seeds

Ash

Seed

Elm

(a) Wind-dispersed fruits and seeds. 'Parachute', 'wing' and 'pepper-pot' mechanisms all help dispersal. The photograph shows the fruits of Old Man's Beard – the hairs make a large surface area which catches the wind

Dispersal of fruits and seeds

When a fruit is ripe it breaks away from the parent plant and its seeds are scattered. The scattering is called **dispersal**. It is important that seeds are dispersed far and wide so that the plants which grow from them will not be overcrowded.

The two main ways of dispersal are by wind and by animals. The pericarps of different fruits develop in different ways for different methods of dispersal (see Figure 21.2D).

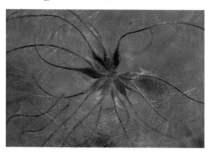

Hooks on the fruit wall cling to the bodies of passing animals

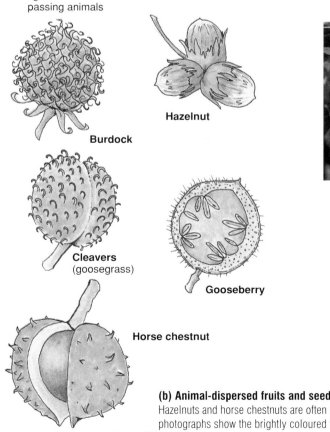

Burdock

Hazelnut

Cleavers (goosegrass)

Gooseberry

Horse chestnut

Bright colours attract birds and other animals to eat these fruits. The tough testa protects the seed from the digestive juices in the animals' intestine. Eventually the seeds pass out of the animal in the faeces

(b) Animal-dispersed fruits and seeds. Hooks and bright colours aid dispersal. Hazelnuts and horse chestnuts are often stored by squirrels and then forgotten. The photographs show the brightly coloured fruits of the honeysuckle and the wild rose

Figure 21.2D 🔺 Methods of fruit and seed dispersal

See www.keyscience.co.uk for more about seed dispersal.

The fruit wall of some plants dries and splits open. As it does so the seeds are thrown out. Examples of fruits that disperse seeds in this way are shown in Figure 21.2E. A few fruits are dispersed by water currents. Examples are the waterlily and coconuts (see Figure 21.2F).

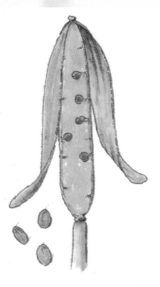

Figure 21.2E **Self-dispersal.** The fruit dries and splits and the seeds fall out. The photograph shows the fruit of lupin, split open with some of the seeds still in place

SUMMARY

The seed is a fertilised ovule. The fruit encloses the seed(s) and is usually produced from the ovary wall following fertilisation. Dispersal of fruits and seeds ensures that the plants that grow from them are not overcrowded.

Figure 21.2F Coconuts float and are dispersed by **water currents.** The outer fibres of the coconuts pictured have been removed. The coconut is thought to have come originally from South America. The water currents in the Pacific Ocean could have dispersed the coconut from South America to the South Sea Islands

CHECKPOINT

▶ 1 Which part of the flower usually develops into the fruit?

▶ 2 What is the fruit wall called?

▶ 3 Describe which parts of the flower have developed into the fruits illustrated in Figures 21.2B and 21.2C.

▶ 4 Why is the dispersal of seeds important?

21.3 ▶ Asexual reproduction in plants

Many different plants can reproduce asexually (see p. 341). The root, leaf or more often the stem may grow into new plants. These parts are called **vegetative parts** and asexual reproduction in flowering plants is sometimes called **vegetative reproduction**. Since new plants come from one parent plant, they are genetically the same. Remember that genetically identical individuals are called **clones** (see p. 341).

The strawberry plant shown in Figure 21.3A reproduces asexually by stems called **runners** which grow horizontally on the surface of the soil.

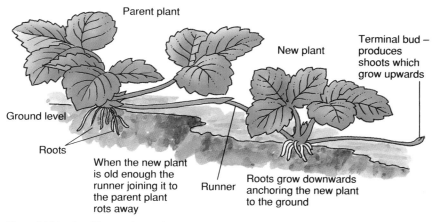

Parent plant

New plant

Terminal bud – produces shoots which grow upwards

Ground level

Roots

When the new plant is old enough the runner joining it to the parent plant rots away

Runner

Roots grow downwards anchoring the new plant to the ground

Figure 21.3A ▲ The strawberry plant

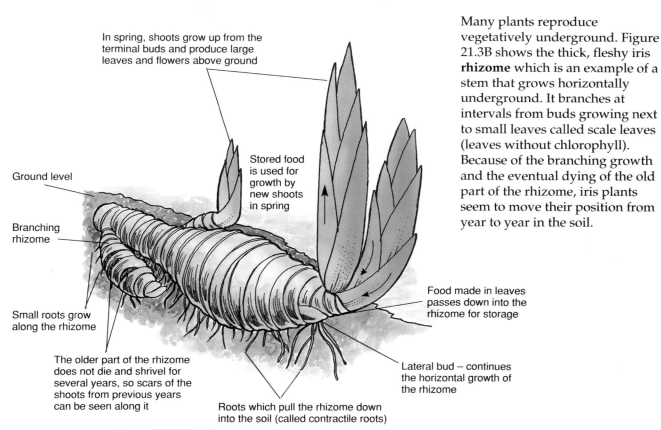

In spring, shoots grow up from the terminal buds and produce large leaves and flowers above ground

Stored food is used for growth by new shoots in spring

Ground level

Branching rhizome

Small roots grow along the rhizome

The older part of the rhizome does not die and shrivel for several years, so scars of the shoots from previous years can be seen along it

Roots which pull the rhizome down into the soil (called contractile roots)

Food made in leaves passes down into the rhizome for storage

Lateral bud – continues the horizontal growth of the rhizome

Many plants reproduce vegetatively underground. Figure 21.3B shows the thick, fleshy iris **rhizome** which is an example of a stem that grows horizontally underground. It branches at intervals from buds growing next to small leaves called scale leaves (leaves without chlorophyll). Because of the branching growth and the eventual dying of the old part of the rhizome, iris plants seem to move their position from year to year in the soil.

Figure 21.3B ▲ The thick, fleshy iris rhizome grows and branches horizontally underground. The arrows show the movement of stored food. *In which directions is the rhizome growing underground?*

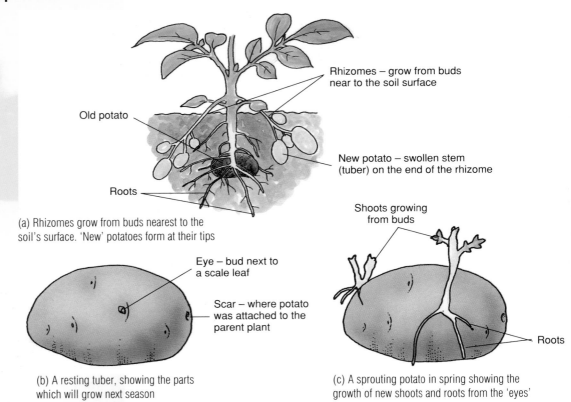

Rhizomes – grow from buds near to the soil surface

Old potato

New potato – swollen stem (tuber) on the end of the rhizome

Roots

(a) Rhizomes grow from buds nearest to the soil's surface. 'New' potatoes form at their tips

Eye – bud next to a scale leaf

Scar – where potato was attached to the parent plant

(b) A resting tuber, showing the parts which will grow next season

Shoots growing from buds

Roots

(c) A sprouting potato in spring showing the growth of new shoots and roots from the 'eyes'

Figure 21.3C ◆ The potato. Notice that an 'eye' is a bud next to a scale leaf

Figure 21.3C shows the potato plant. The potato is a swelling of stored food at the end of a rhizome. It is called a **stem tuber**. Leaves on shoots which grow above ground make starch by photosynthesis. The starch passes down the stem into the rhizomes. The old tuber rots away at the end of the growing season.

A **corm** is a short, swollen, underground stem and a **bulb** is a large underground bud. At the end of a growing season the leaves make starch by photosynthesis. The starch is stored in the corm or bulb underground until it is used the next year for the growth of new leaves and flowers. Daughter corms or bulbs develop from buds on the side of the parent organ and when they are large enough they break off and become independent plants. (See Figure 21.3D.)

Runners, rhizomes, tubers and corms are modified stems. Bulbs are modified buds. Onions are an example.

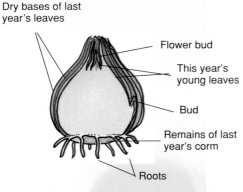

Dry bases of last year's leaves

Flower bud

This year's young leaves

Bud

Remains of last year's corm

Roots

(a) Resting corm, showing the parts which will grow next season

Figure 21.3D ◆ The crocus corm

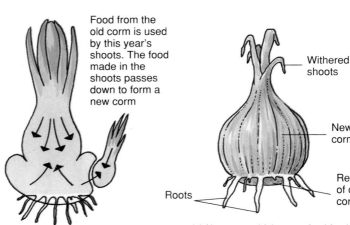

Food from the old corm is used by this year's shoots. The food made in the shoots passes down to form a new corm

Roots

(b) Growing leaves and flowers use up food stored in the old corm and a new corm begins to form on top of it (arrows show movement of stored food)

Withered shoots

New corm

Remains of old corm

(c) New corm which stores food for the next year – the new corm separates from the remains of the old corm at the end of the growing season

Bulbs, corms, tubers and most rhizomes are not only organs of asexual reproduction but also organs which store food. They fill up with starch during the summer when the plants which have grown from them are in leaf and making food by photosynthesis. In the autumn the plants die down but the organs underground, full of stored food, survive the winter. The stored food is used for the development of new plants the following year. Because these organs asexually reproduce new plants year after year they are called **perennials**.

Bulbs, corms, tubers and most rhizomes are not only organs of asexual reproduction but also organs which store food.

Artificial vegetative reproduction

Because vegetative propagation produces clones of new plants that are genetically the same as the parent plant, the desirable qualities of the parent are preserved in the offspring. Gardeners and farmers need to produce fresh stocks of plants with desirable characteristics like disease-resistance, colour of fruit or shape of flower. They exploit vegetative reproduction to guarantee plant quality from one generation to the next.

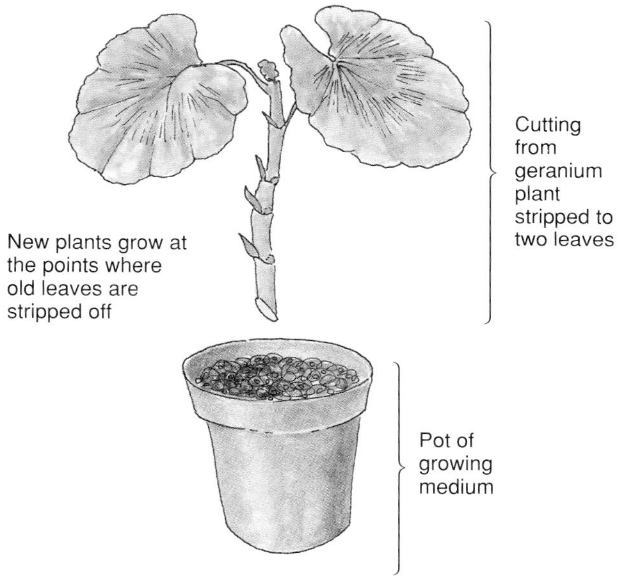

Taking **cuttings** is a method of vegetative reproduction often used. New plants are produced quickly and cheaply from older plants with characteristics that we wish to preserve. Figure 21.3E shows a geranium cutting. Short pieces of stem are cut just below the point (called the **node**) where a leaf joins the stem. Most of the leaves are stripped from the stem. New shoots grow at the points where the old leaves are stripped off. If the cutting is planted in a damp mixture of growing medium then roots will sprout from the cut surface of the stem. Roots that grow from part of a shoot are called **adventitious** roots. The cutting then becomes a complete, independent plant. Cuttings are more likely to grow successfully if they are first kept in a damp atmosphere until roots develop.

Grafting is often used for reproducing roses and fruit trees. A twig is cut from the tree to be reproduced and replaced in a slit in the stem of an already well-rooted tree so that the cut surfaces are brought together. The tissues of the twig (called the **scion**) and the tree (called the **stock**) join together and the **graft** grows on the rooted tree (see Figure 21.3F).

Cutting from geranium plant stripped to two leaves

New plants grow at the points where old leaves are stripped off

Pot of growing medium

Figure 21.3E ⬆ A cutting from a geranium plant

It's a fact!

Widespread use of clones (see p. 341) by gardeners and farmers reduces variation (see Topic 25.2). Fewer alleles (see p. 381) are available for the selective breeding of new varieties of crops and animals for food (see p. 424).

Figure 21.3F ⬆ Grafting a fruit tree

Cuttings, grafting and micropropagation using tissue culture are asexual methods used by gardeners, glasshouse owners and farmers to preserve the desirable characteristics of parents in their offspring.

Tissue culture under sterile conditions produces disease-free plants.

 See www.keyscience.co.uk for more about cloning.

Budding is another type of grafting. A bud is used instead of a twig. Several buds can be grown on one root stock, each growing into an individual plant.

The plants produced by these methods are the same genetically as the parents from which they are taken. This means that the desirable qualities of the parent are preserved in the offspring.

Tissue culture

Nowadays it is possible to grow plants of many kinds from small pieces of parent plant (**micropropagation**) using **tissue culture** of parent plant (Figure 21.3G). All the plants grown from pieces from one parent plant will be genetically alike (clones – see p. 341) and will grow in the same way if they are treated identically. The small fragments begin life growing in a liquid or gel which provides all necessary substances for their development. The conditions are **sterile** so the new plants are free of disease. This is an ideal way of propagating plants for glasshouse owners who want to produce many similar, healthy plants.

Special facilities are needed for tissue culture. The growing medium must be sterile. The fragments of plants are handled with sterile instruments too. Also, the correct temperature and other conditions have to be provided. The requirements for successful tissue culture are very precise.

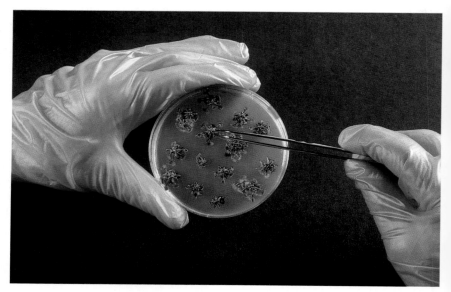

Figure 21.3G ⬆ Growing plants in tissue culture

■ Cloning from cultures

Some commercially important plants are mass-produced from tissue cultures. For example, large numbers of people rely on the oil palm for food and detergents either directly or because its products are exported. Clones are grown from palm trees that produce a lot of oil.

Figure 21.3H shows you the process. A small part of the growing point from the parent plant is collected. It is then grown in sterile conditions on agar jelly. Mitosis (p. 156) produces clusters of identical cells. These are then separated. Each can develop into a new plant with the same desirable genes found in the parent plant.

Selection
of hybrid (see p.8) variety of oil palm. Its fruit produces a lot of oil. A hybrid is the cross between two genetically unlike individuals

Tissue culture
Root tissue from the selected tree is placed in a medium which contains all the nutrients needed for growth. Auxin (p.197) is used to stimulate growth

Clone formation
Plants develop from the root tissue. They are genetically the same (clones) since they come from the same parent. All of the clones have fruit which produces a lot of oil

Figure 21.3H ◀ Cloning oil palms. Oil palms are produced which yield up to six times more oil than oil palms produced by traditional methods

Palm oil is an important source of food and detergents.

The process can be carried out in laboratories anywhere in the world and the young plants then exported to the tropics to grow into plantations of mature trees (see Figure 21.3I).

SUMMARY

Asexually reproduced plants are genetically identical to one another and their parent. Desirable characteristics of the parent, therefore, are preserved from one generation to the next.

Figure 21.3I ◀ A plantation of high-yielding palm trees

CHECKPOINT

▶ **1** What are the vegetative parts of flowering plants?

▶ **2** Briefly compare the similarities and differences between a runner and a rhizome.

▶ **3** What is a tuber?

▶ **4** Look at Figure 21.3D. Describe the movement of food in a corm from the beginning of the growing season to the end.

▶ **5** Grafting is a method of vegetative propagation. It is often used to reproduce roses and fruit trees. Find out how grafting is carried out. What are the advantages of reproducing a tree which produces sweet, crisp-eating apples by grafting?

▶ **6** List the conditions necessary for the successful culture of plant tissue to produce new plants.

Topic 22   Human reproduction

22.1 ▶   Getting to know you

Getting to know you charts the relationships between two imaginary people. Look at the different stages of their developing relationship.

Who is this person?
- Handsome • Pretty
What makes you notice someone of the opposite sex?

Attraction
- Caring • Easy to talk to
- Similar interests and opinions
Why do you want to see more of the other person?

Getting to know you
- You want to touch and kiss
- Are you nervous?
Why do you want to get closer to the other person?

Loving you
- Sharing your life with someone else
- Having children is a possibility
Why do you want to commit yourself to the other person?

- What points do you think are important at each stage of the developing relationship?
- Do you agree with the order of the different stages of *Getting to know you*? How would you describe each stage?
- What do you think the role of love is in a relationship?
- How do you think the responsibility of children should be shared between partners?

Sexual feelings for someone of the opposite sex are called **heterosexuality**. **Homosexuality** is having sexual feeling for someone of your own sex. Homosexual feelings are not uncommon in adolescents who, as a result, often feel different from people who are heterosexual. Talking about homosexual feelings to parents, friends or a counsellor can help to keep problems in proportion. What is important about relationships is that they should be based on trust, love and understanding of one another.

22.2 ▶ The human reproductive system

FIRST THOUGHTS

Parts of the male reproductive system hang outside the body. Most of the female reproductive system lies inside the body.

The visible parts of the reproductive system are called the **genitalia**. A man's genitalia consist of the **penis** and the **testes** which are contained in a bag-like **scrotum** which hangs down between the legs. This position protects the testes from injury. It also keeps them about 3 °C lower than body temperature. This is important because sperms only develop properly inside the testes in these slightly cooler conditions. The reproductive system of a man is illustrated in Figure 22.2A.

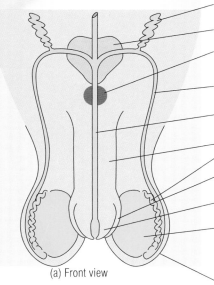

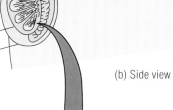

Seminal vesicle – opens into the sperm duct and produces seminal fluid

Bladder – stores urine

Prostate gland – opens into the urethra and produces an alkaline fluid which neutralises any urine in the urethra

Sperm duct – from the testis, opens into urethra

Urethra – passes down penis, both sperm and urine pass through it

Shaft of **penis**

Epididymis – leads to sperm duct

Foreskin – covers the glans (removed in circumcision)

Glans – head of penis, very sensitive to stimulation

Testis – bundle of sperm–producing tubules which join and lead to the epididymis

Scrotum

(a) Front view

(b) Side view

See www.keyscience.co.uk for more about human reproduction.

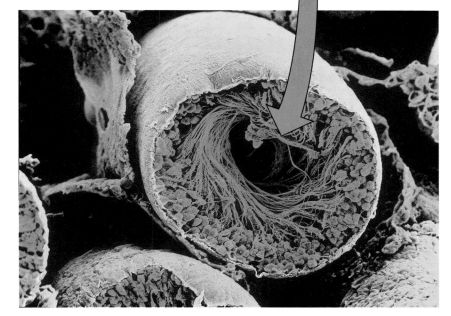

(c) Cross section through a sperm-producing tubule showing sperms clustered inside. Stretched out end-to-end the tubules inside each testis are more than 500 m long (×200)

Figure 22.2A ▲ The male reproductive system

EXTENSION FILE
ASSIGNMENT

The reproductive system of a woman is illustrated in Figure 22.2B. The genitalia cover and protect the opening to the rest of the reproductive system inside her body.

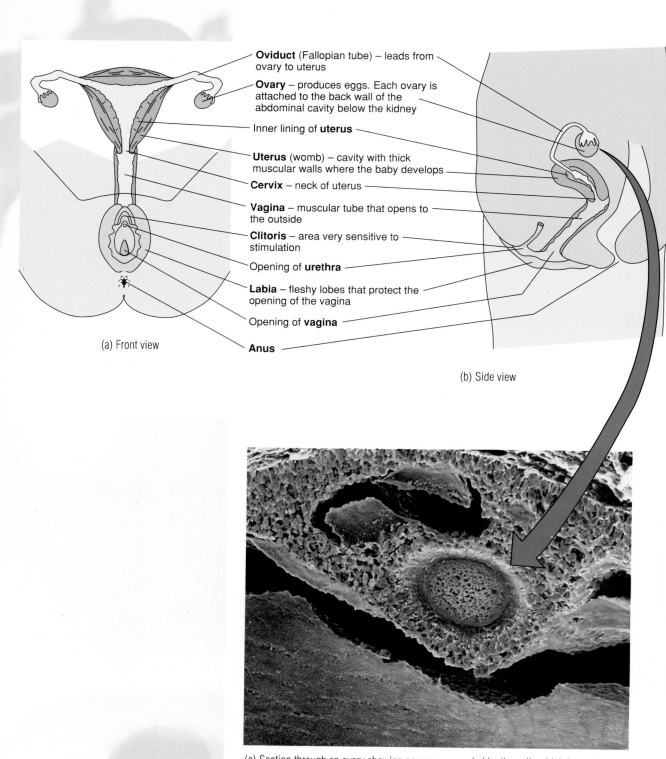

Oviduct (Fallopian tube) – leads from ovary to uterus

Ovary – produces eggs. Each ovary is attached to the back wall of the abdominal cavity below the kidney

Inner lining of **uterus**

Uterus (womb) – cavity with thick muscular walls where the baby develops

Cervix – neck of uterus

Vagina – muscular tube that opens to the outside

Clitoris – area very sensitive to stimulation

Opening of **urethra**

Labia – fleshy lobes that protect the opening of the vagina

Opening of **vagina**

Anus

(a) Front view

(b) Side view

(c) Section through an ovary showing an egg surrounded by the cells which form the follicle (×250). Each follicle contains one egg. At birth each ovary contains approximately 300 000 follicles. Of these, approximately 300 complete their development in a woman's lifetime

Figure 22.2B ◆ The female reproductive system

SUMMARY

The testes produce sperm. The ovaries usually each produce one egg on alternate months from the onset of puberty to the menopause. Egg production is controlled by hormones during the menstrual cycle.

Menstrual cycle

The human female usually produces one mature egg each month from the onset of **puberty** (age 11–14 years) to the beginning of the **menopause** (age about 45 years). This monthly cycle is called the **menstrual cycle** (from the Latin *mensis* meaning month). Egg production becomes more and more irregular during the menopause and stops altogether usually by about the age of 50. Figure 22.2C shows how the different events of the menstrual cycle fit together.

The changes occurring during the menstrual cycle prepare the uterus to receive an egg if it is fertilised. If the egg is not fertilised, the production of the **hormones** (see Topic 17.4) oestrogen and progesterone tails off and the thick lining of the uterus begins to break down as Figure 22.2C shows. The release of blood and tissue through the vagina is called **menstruation** and is what is meant by 'having a period'. It lasts for several days. A new menstrual cycle then begins.

Blood released during a period can be absorbed by a **sanitary towel**, which a woman wears as a lining to her underwear, or as an alternative she can put a tampon made of cotton wool in her vagina (see Figure 22.2D).

The menstrual cycle can affect a woman's emotions. How a woman feels depends on the individual. Some have few problems but others feel irritable and below their best just before and during menstruation.

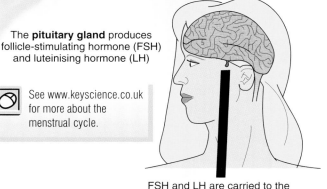

The **pituitary gland** produces follicle-stimulating hormone (FSH) and luteinising hormone (LH)

See www.keyscience.co.uk for more about the menstrual cycle.

FSH and LH are carried to the ovaries in the blood stream

Graafian follicle – contains the developing egg

The empty follicle is called the **corpus luteum**

Several follicles start to grow, stimulated by FSH and LH

Ovulation occurs when the egg is released from its follicle

The **corpus luteum** produces a hormone called *progesterone* which stimulates the uterus lining to thicken even more

The **ovary** produces a hormone called *oestrogen* which stimulates division of the cells lining the uterus. The uterus thickens and its blood supply increases. Rising levels of oestrogen feed back negatively (see p. 306), reducing the production of FSH and LH from the pituitary

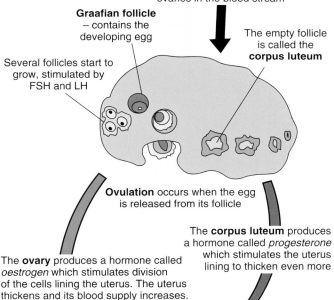

Menstruation

Thickness of uterus lining

0   14   28

Days

Figure 22.2D 🔺 Hygiene during a period. Santitary towels come in different thicknesses. The woman can choose which type suits her best depending on how much blood and tissue is released during her period. The tampon is removed from the vagina by its thread

Figure 22.2C 🔺 The menstrual cycle. A sharp increase in the level of luteinising hormone (LH) causes ovulation. The intervals of time for each stage may vary depending on the individual. For example, ovulation may occur earlier or later in the cycle than shown

22.3 ▶ How do sperms meet eggs?

An erect penis is a sign that a man is sexually excited. The penis stiffens and lengthens as blood fills the spongy tissue of the **shaft** (see Figure 22.3A). Signs of sexual excitement in women are less obvious. The labia (see Figure 22.2B) fill with blood and swell a little. All of these changes in a man and woman help to prepare them for **sexual intercourse**.

The swollen labia help to guide the erect penis into the vagina. The muscles of the vagina wall relax, helping entry. Fluid produced by the vaginal wall lubricates the movements of the penis during sexual intercourse. These movements stimulate the muscles in the scrotum and around the epididymis and sperm ducts (see Figure 22.2A) to contract, pushing sperm from the testes along the sperm ducts to the urethra. During the journey the sperm mix with fluids from the seminal vesicles and prostate gland. These fluids and the sperm form **semen**. Continuing stimulation results in **ejaculation** which is a series of contractions that propel semen through the urethra into the vagina (see Figure 22.3B).

Semen is white and sticky. It contains sugars which are an energy source for the sperms as they swim up through the vagina and uterus to the oviducts.

During ejaculation the man usually experiences a pleasant feeling called an **orgasm**. The woman may experience an orgasm as well. The muscles of the vagina gently relax and contract around the penis. The woman's orgasm is usually caused by gentle pressure stimulating the **clitoris**.

We have seen that ovulation releases an egg from the ovary into the opening of the oviduct (see Figure 22.2B). If a sperm is to meet an egg it

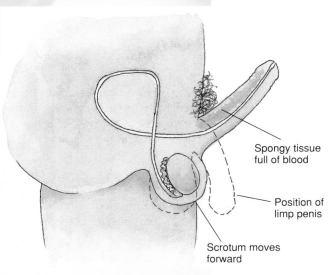

Spongy tissue full of blood

Position of limp penis

Scrotum moves forward

Figure 22.3A ▲ An erection

Sexual intercourse is a means by which sperm meet an egg.

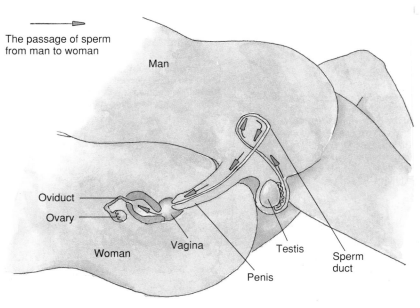

The passage of sperm from man to woman

Man

Oviduct

Ovary

Woman

Vagina

Penis

Testis

Sperm duct

Figure 22.3B ▲ Sexual intercourse

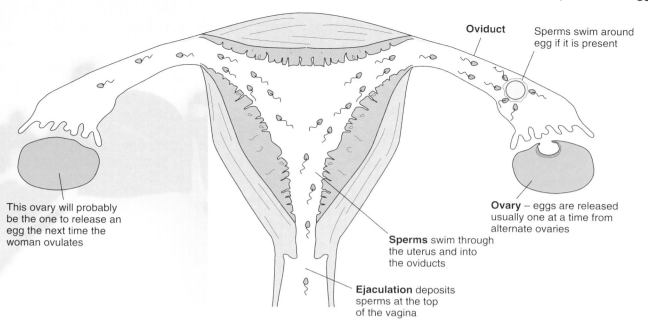

Oviduct

Sperms swim around
egg if it is present

This ovary will probably
be the one to release an
egg the next time the
woman ovulates

Ovary – eggs are released
usually one at a time from
alternate ovaries

Sperms swim through
the uterus and into
the oviducts

Ejaculation deposits
sperms at the top
of the vagina

Figure 22.3C ◆ How sperms meet an egg. Sperms reach the top of the oviduct approximately
40 minutes after ejaculation

SUMMARY

Once sperm are deposited in the
vagina following sexual
intercourse, they make their way
through the cervix and uterus to
the oviduct. Here they swarm
around the egg if one is present.
Of the millions of sperm that
start the journey only a few
thousand reach the oviduct.

must make its way from the
vagina, through the uterus to
the oviduct. Figure 22.3C shows
the distance a sperm must
travel. The journey is not an
easy one. Of the hundreds of
millions of sperm deposited in
the vagina only a million or so
make it through the cervix into
the uterus. Of these, only a few
thousand arrive at the end of
the opening of the oviduct. Here
they swarm around the egg if
one is present (see Figure
22.3D).

Figure 22.3D ◆ A sperm penetrating the egg
membrane. Only one succeeds: then the egg
membrane forms a barrier to other sperm

CHECKPOINT

▶ 1 Briefly describe the route taken by ejaculated sperm from where they are produced to the oviduct
of the woman.

▶ 2 Complete the following paragraph using the words below. Each word may be used once, more
than once or not at all.

**fertilisation   vagina   testes   weeks   semen   bladder   urethra   sperm   duct
seven   sperm   seminal   cervix   penis   uterus   sexual   ovaries   prostate   oviduct**

An egg is released by one of the two _____ about every four _____ . It passes into the
_____ . It may take up to _____ days to reach the _____ . If _____ is to take place
the egg must be met by a _____ before, or just after, it reaches the _____ . Sperm are
produced in the tubules of the _____ in vast quantities. Ejaculation forces the sperm from the
epididymis, into the _____ . The _____ vesicle, and _____ gland add their secretions to
the sperm, forming _____ . This leaves the body through the urethra running through the
_____ .

22.4 ▶ Fertilisation and development of the embryo

If an egg and sperm are present in the oviduct then fertilisation may occur. Read on and find out what happens afterwards.

When sperm meets an egg the sequence of events then reads:

Fertilisation

↓

Division of the zygote to form an embryo as it travels down the oviduct to the uterus

↓

Implantation of the embryo into the wall of the uterus

Although thousands of sperms may reach an egg only one enters it. The tail of the successful sperm is left outside as the head travels through the cytoplasm of the egg to the nucleus. Fertilisation occurs when the sperm nucleus fuses with the egg nucleus to form a **zygote** (see Figure 22.4A). This is the moment of **conception** and the woman is now **pregnant**.

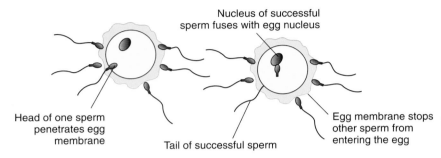

Nucleus of successful sperm fuses with egg nucleus

Head of one sperm penetrates egg membrane

Tail of successful sperm

Egg membrane stops other sperm from entering the egg

Figure 22.4A ◆ Fertilisation of the egg forms a zygote

Figure 22.4B shows what happens next. The zygote travels down the oviduct, dividing by mitosis (see Topic 9.4), as it goes, forming a ball of cells. The journey may take up to seven days. By the time the ball of cells reaches the uterus it has formed an **embryo**. Remember that at this stage of the menstrual cycle (see Figure 22.2C) the wall of the uterus has a thick lining. The embryo sinks into it – a process called **implantation**.

Finger-like extensions called **villi** project from the embryo into the lining of the uterus. The surfaces firmly bind together forming a region called the **placenta**. In the next few weeks the embryo develops into a **fetus** which is attached to the placenta by the **umbilical cord**. An artery and a

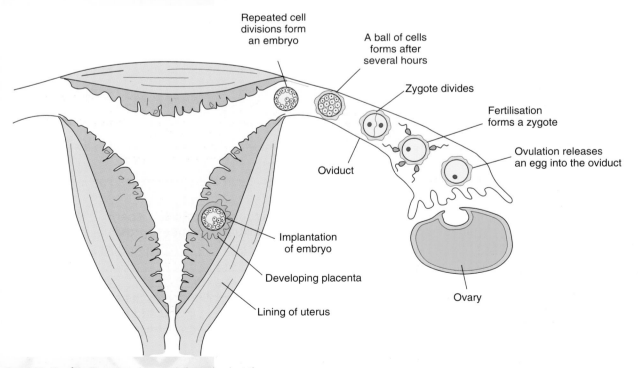

Repeated cell divisions form an embryo

A ball of cells forms after several hours

Zygote divides

Fertilisation forms a zygote

Ovulation releases an egg into the oviduct

Oviduct

Implantation of embryo

Developing placenta

Lining of uterus

Ovary

Figure 22.4B ◆ The stages from ovulation to implantation

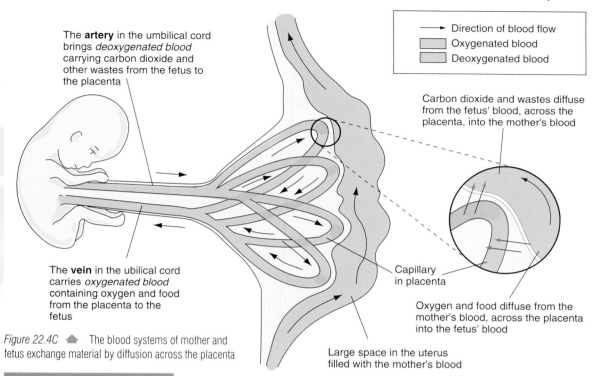

The **artery** in the umbilical cord brings *deoxygenated blood* carrying carbon dioxide and other wastes from the fetus to the placenta

Direction of blood flow
Oxygenated blood
Deoxygenated blood

Carbon dioxide and wastes diffuse from the fetus' blood, across the placenta, into the mother's blood

Capillary in placenta

The **vein** in the ubilical cord carries *oxygenated blood* containing oxygen and food from the placenta to the fetus

Oxygen and food diffuse from the mother's blood, across the placenta into the fetus' blood

Large space in the uterus filled with the mother's blood

Figure 22.4C ⬆ The blood systems of mother and fetus exchange material by diffusion across the placenta

It's a fact!

Gestation periods are:
- Mouse – 18 days
- Cat – 2 months
- Horse – 11 months
- Elephant – 20 months

Usually the larger the animal, the longer the gestation period.

vein run through the umbilical cord and connect the fetus' blood system to the placenta. Figure 22.4C shows that the fetus' blood system is not directly connected to the blood system of the mother. The exchange of oxygen, food and wastes between mother and fetus depends on diffusion across the thin wall of the placenta.

The time taken for the fetus to develop from conception into a baby is called the **gestation period**. It usually lasts nine months in humans. Figure 22.4D) shows the growth and development of the fetus from the early stages of pregnancy to just before the time the baby is due to be born.

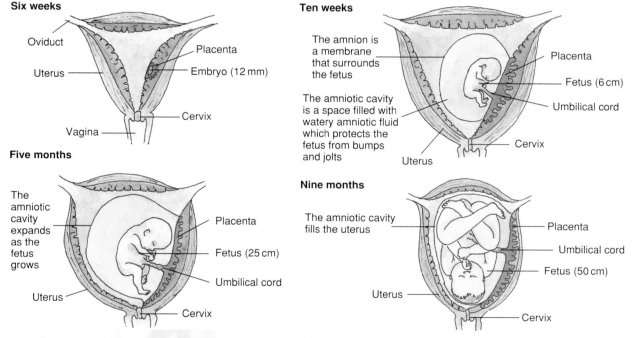

Six weeks

Oviduct
Uterus
Placenta
Embryo (12 mm)
Cervix
Vagina

Ten weeks

The amnion is a membrane that surrounds the fetus

The amniotic cavity is a space filled with watery amniotic fluid which protects the fetus from bumps and jolts

Placenta
Fetus (6 cm)
Umbilical cord
Cervix
Uterus

Five months

The amniotic cavity expands as the fetus grows

Placenta
Fetus (25 cm)
Umbilical cord
Uterus
Cervix

Nine months

The amniotic cavity fills the uterus

Placenta
Umbilical cord
Fetus (50 cm)
Uterus
Cervix

Figure 22.4D ⬆ Growth of the fetus in the uterus. As cells grow and divide, differentiation (see p. 178) increases the complexity of the tissues and organs of the developing body.

SUMMARY

Fertilisation is followed by implantation of the embryo in the wall of the uterus. It usually takes 9 months for the human baby to develop fully. The menstrual cycle stops during pregnancy.

Pregnancy and the menstrual cycle

After fertilisation, the placenta takes over the function of the corpeus luteum and produces progesterone. As a result:

● the lining of the uterus remains intact,

● production of FSH and LH is inhibited,

● the ovaries stop producing eggs and follicles.

When the baby is born progesterone levels fall, FSH and LH are released once more and the menstrual cycle restarts.

Looking again at Figure 22.2C will help you remember the events of the menstrual cycle.

22.5 ▶

Birth

FIRST THOUGHTS

Birth is a profound experience – for baby who enters the world, for Mum who gives birth and for Dad who has to wait for the outcome!

Figure 22.5A 🔺 Chartered physiotherapists teaching a group of antenatal mothers

It's a fact!

Animal mothers very often eat the afterbirth because the smell of blood might attract hungry predators. The afterbirth contains a lot of iron compounds and other nutritious substances.

The mothers pictured in Figure 22.5A are heavily pregnant. It will not be long before they give birth to their babies. They are visiting an **antenatal** clinic where a doctor will check that all is well with each mother and her baby. The mothers also receive advice on how best to prepare for the baby's birth.

When birth takes place (see Figure 22.5B), the membranes surrounding the baby break and the amniotic fluid escapes (see Figure 22.4D). Then the baby moves, usually head first through the vagina. Safely delivered, the baby starts to breathe, sometimes helped by a tap on the back which causes a surprised intake of breath. Now that the baby can breathe for itself, the placenta and umbilical cord are no longer needed.

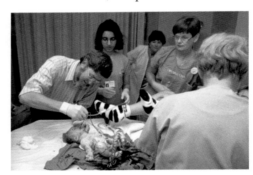

Figure 22.5B 🔺 Childbirth – the doctor is clearing the baby's mouth and airways of mucus

The placenta comes away from the uterus wall and passes out through the vagina as the **afterbirth**. The umbilical cord is clamped near to where it joins the baby and is cut. This does not hurt the baby because there are no nerves in the cord. The stump that remains becomes the baby's **navel**.

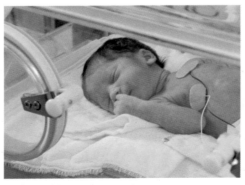

Figure 22.5C 🔺 Premature baby in an incubator

Babies born before the ninth month of pregnancy are described as **premature**. They have a good chance of survival providing they are not too small and weak. Premature babies are often kept in **incubators**. An incubator is a cabinet with a controlled environment that keeps the baby warm and provides extra oxygen to help with breathing. The baby stays in the incubator until he or she is strong enough to survive independently (see Figure 22.5C).

In humans, pregnancy usually results in the birth of only one baby. However, sometimes two babies are born one after the other. They are called **twins** and Figure 22.5D shows how this arises. The twins develop together in the uterus, each with its own placenta and umbilical cord.

Occasionally ovulation releases three or more eggs into the uterus at the same time, especially if the woman has been given a fertility drug to help her to become pregnant (see Topic 22.7). These **multiple** pregnancies can be difficult because of the space taken up in the uterus by the growing fetuses. Very often the mother gives birth early at around the seventh month and some of the babies may die.

Make sure you can answer an exam question which asks you to distinguish between identical twins and non-identical twins.

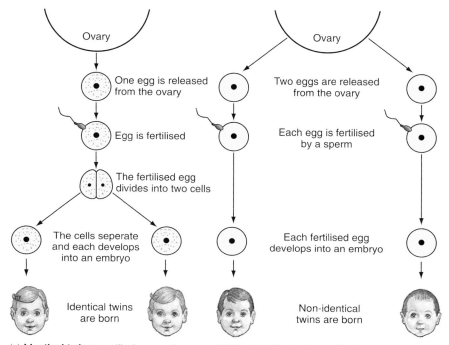

(a) **Identical twins** are alike because they each have the same genes (see p. 152)

(b) **Non-identical** or **fraternal** twins are different from one another because they do not have the same genes (see p. 152)

Figure 22.5D 🔺

SUMMARY

Following birth, the umbilical cord which connects baby and mother is cut. Twins are identical or fraternal depending on how they arise.

CHECKPOINT

▶ **1** Name three substances which show a net movement into fetal blood across the placenta, and three substances that show a net movement out of the fetal blood across the placenta.

▶ **2** The uterus can be the most powerful muscle in the body. Why does it need to be so powerful?

▶ **3** Why is birth a considerable shock to the baby?

▶ **4** Complete the following paragraph using the words provided below. Each word may be used once, more than once or not at all.

**cervix   uterus   placenta   oxygen   oviducts   oxygenated
wastes   amniotic   vagina   muscles   implants**

Once fertilisation has occurred, normally in one of the _____ the embryo grows, moves into the uterus and _____ into its wall. The _____ develops which provides a surface for the exchange of materials with the mother's blood. _____ and food cross into the fetal blood, whereas carbon dioxide and other _____ enter the mother's circulation. The developing fetus is surrounded and protected by the _____ fluid. At birth the _____ dilates, and powerful contractions of the _____ of the uterus push the baby out through the _____ .

22.6 ▶ Looking after baby

FIRST THOUGHTS

The newborn baby feeds on milk. Read on and find out how the baby's diet changes.

Mother's milk not only contains all the necessary nutrients but also provides antibodies which protect the baby from diseases until his/her immune system is mature.

A newborn baby will naturally suck at the **nipple** of the mother's breast. Figure 22.6A shows that glands (called **mammary glands**) inside the breast secrete **milk**. This happens soon after birth and is called **lactation**. Mother's milk is perfect food for the baby. It contains all the necessary nutrients as well as the mother's antibodies which help to protect the baby from diseases during the first few months of life (see Topic 16.1). Sometimes a mother does not produce enough milk, so the baby has to be **bottle-fed** (see Figure 22.6B).

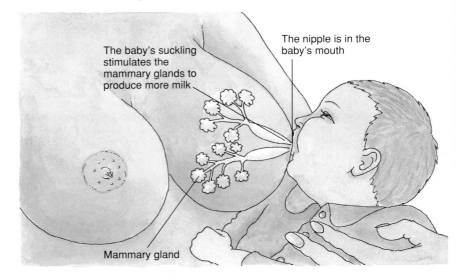

The baby's suckling stimulates the mammary glands to produce more milk

The nipple is in the baby's mouth

Mammary gland

Figure 22.6A A baby feeding from the mother's breast

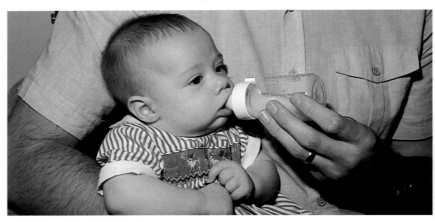

Figure 22.6B Bottle feeding

A newborn baby cannot take in solid food because he or she has no teeth to chew it with. Also the digestive system is unable to deal with solid food. After about six months the first teeth appear and solid food can now be added to the baby's diet. At this stage the baby's milk intake decreases. **Weaning** is the word used to describe the change from a diet of milk to one of solid food.

Looking after a baby is time-consuming and exhausting. Apart from feeding, the baby must be kept clean and warm. It is also very important that the emotional needs of the baby are cared for. Keeping the baby interested, stimulated and happy is just as important as looking after the baby's physical needs.

SUMMARY

Milk provides the newborn baby with all of the nutrients it needs. Later solid food is added to the diet.

22.7 ▶ Controlling fertility

FIRST THOUGHTS

Different methods of contraception prevent pregnancy. Hormones are used to treat infertility.

The fertility rate gives the average number of children a woman has throughout her childbearing life. In the UK, on average each woman bears 1.8 children. The '0.8' takes into account women who are unable to have children (**infertile**) or those who choose not to have them.

If a couple want to have sexual intercourse but do not want to have children they must use some form of **contraception** to prevent pregnancy. Using methods of contraception enables people to choose when they want children and how many children they want. This choice is called **family planning** or **birth control**.

Different forms of contraception prevent pregnancy.

Preventing pregnancy

To prevent pregnancy, the method of contraception must either:

- stop sperm from reaching the egg,
- stop eggs from being produced,
- or stop the fertilised egg from developing in the uterus.

Different methods of contraception are described below and illustrated in Figure 22.7A.

 See www.keyscience.co.uk for more about contraception.

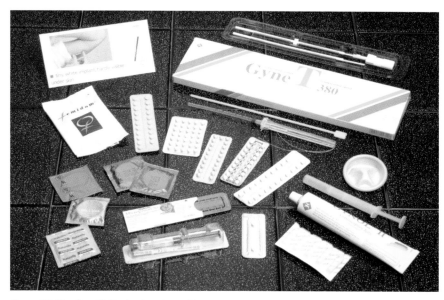

Figure 22.7A ◀ Methods of contraception

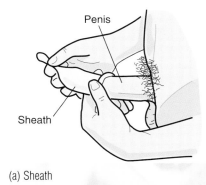

(a) Sheath

- **Sheaths (condoms).** The condom is a thin sheath which is rolled on to the erect penis before intercourse. The penis must be removed from the woman's vagina immediately after ejaculation to avoid any spillage of sperm. **The female condom** is a thin sheath which lines the vagina. It is closed at one end and open at the other. A ring at the closed end helps the woman to insert the condom before sexual intercourse. A ring at the open end remains outside the body, pushed flat against the labia. The penis is guided into the sheath and moves inside the lining during intercourse. When the penis is removed after ejaculation; the ring at the open end is twisted to make sure no sperm is spilt, and the sheath gently pulled from the vagina.

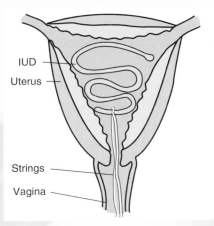

(b) Intrauterine device

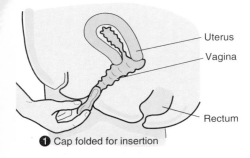

❶ Cap folded for insertion

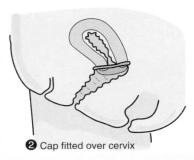

❷ Cap fitted over cervix

(c) Diaphragm or cap

Figure 22.7A 🔺 Methods of contraception (continued)

Distinguish between the 'pill' which stops ovaries from producing eggs and the 'morning after pill' which prevents implantation of the embryo.

- **Intrauterine devices (IUDs).** IUDs are fitted inside the uterus by a doctor. The IUD touches the inner wall of the uterus and prevents implantation of the embryo. The IUD can be removed by a doctor by pulling on the strings attached to it which pass through the cervix. IUDs are usually only used by women who have already had a child.
- **Diaphragms.** The diaphragm or **cap** is a dome-shaped device which fits over the cervix. They come in different sizes and a woman must be taught to insert the diaphragm by her doctor or at a family planning clinic. A diaphragm can be inserted immediately before intercourse but must remain in place for some time after intercourse. Diaphragms offer more protection when used with a **spermicide**.
- **The pill.** Hormonal contraceptives taken as pills are called **oral** (by mouth) contraceptives. There are different types of pill:
  - **The combined oral contraceptive (COC)** combines small amounts of oestrogen and progesterone. The combination is called the 'low dose' pill.
  - **The progesterone only pill (POP)** called the 'mini-pill' is suitable for women who are breast feeding or who may be at risk if they take the COC.

The concentration of the hormones stops the ovaries from producing eggs by reducing the secretion of FSH and LH from the pituitary (see Figure 22.2C). The woman takes a pill every day for 21 days of her menstrual cycle. When she stops taking the pill menstruation occurs. The woman begins taking the pill again on day one of her next menstrual cycle. If the woman forgets to take a pill on one day then the protection is not complete and another form of contraception must be used until the woman's next menstrual cycle begins.

Another type of pill is the **morning-after pill**: so called because it is an emergency contraceptive used after unprotected sex when the woman wants to avoid the risk of pregnancy. To be effective the pills can be taken right away (not necessarily the 'morning after') or up to three days after unprotected sex. The earlier the pills are taken within the three-day window the more effective they are. The pills deliver a large dose of hormones which prevent implantation of the embryo.

- **Spermicides.** Spermicidal creams kill sperms. A woman uses an **applicator** to put spermicide inside her vagina just before intercourse. Spermicides are not very effective on their own and are more often used as 'back-up' for other methods.

(d) Spermicide

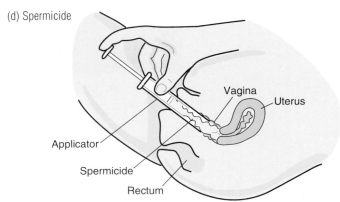

- **Injected contraceptives** contain the hormone progesterone. Once injected into the arm, the hormone is slowly released into the body over the next two or three months. Like the mini-pill, it stops the ovaries from producing eggs. Injected contraceptives are useful for women who find it difficult to take the pill or experience problems with other methods of contraception.

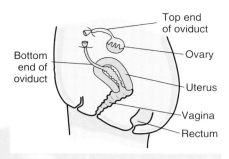

Top end of oviduct

Bottom end of oviduct

Ovary

Uterus

Vagina

Rectum

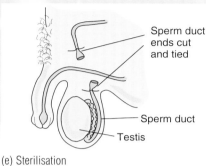

Sperm duct ends cut and tied

Sperm duct

Testis

(e) Sterilisation

Figure 22.7A ⬤ Methods of contraception (continued)

The rhythm method of contraception is used mostly by people whose religion does not allow other methods.

SUMMARY

Different methods of contraception prevent pregnancy. Oral contraceptive pills which contain hormone(s) are particularly effective. Hormones are also used to treat infertility. Understanding how the effects of hormones fit together in the events of the menstrual cycle helps us use hormones to our advantage.

● **Sterilisation**. Both men and women can be sterilised. Sterilisation involves a minor operation. In a man, the sperm ducts are tied off and cut by the surgeon. The man can still ejaculate as the ducts are cut below the **seminal vesicles** which produce seminal fluid, but his semen will not contain any sperms. In a woman, the oviducts are tied off and cut. This prevents the sperm from reaching the egg. Although some sterilisation operations can be reversed, this is not usually the case so a man or woman who is sterilised has to be very sure that he or she does not want any more children.

All the methods of contraception described so far, and shown in Figure 22.7A, depend on surgery, chemicals or a mechanical device to be effective. The **rhythm method** does not use any of these aids but depends on the woman (and possibly her partner) understanding how her menstrual cycle works. Look at Figure 22.2C once more and calculate at what time in the menstrual cycle intercourse is most likely to lead to pregnancy. *Why is pregnancy not possible at other times – the so-called 'safe period'?*

Figure 22.7B shows the changes in a woman's body temperature during her menstrual cycle. Notice the slight increase in body temperature when she ovulates. *How do you think the woman can use the information to help her prevent pregnancy?*

Unfortunately the menstrual cycle is not always predictable. The cycle can vary a good deal, especially in teenagers, which makes it difficult to predict whether or not it is safe to have intercourse. However, the rhythm method is a natural form of contraception and it is used mostly by people whose religion does not allow other methods.

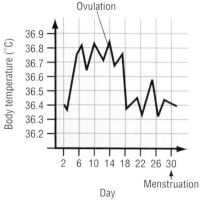

Ovulation

Figure 22.7B ⬤ The temperature rise at ovulation is about 0.5°C

Table 22.1 compares the reliability of different methods of contraception. However you should not think that it is a matter of just looking down the list, choosing one that suits best and having intercourse when you please and with whom you please. A close physical relationship is only part of the relationship you have with a member of the opposite sex. It means that you must be old enough to take responsibility not only for your own actions and feelings but also for your partner's actions and feelings as well. Look back at *Getting to know you* on page 360 and think again about your answers to the questions now that you have read most of this section. *Have your opinions changed?*

Table 22.1 ⬇ The reliability of contraceptive methods

Method	Percentage of pregnancies	How reliable is the method?
No method	54	Very unreliable
Rhythm method	17	Unreliable without expert help
Diaphragm with spermicide	12	Quite reliable when fitted well
Sheath	8	Quite reliable if used properly
IUDs	2	Reliable
The pill	0	Very reliable

Infertility

Couples are usually thought to be infertile if they have regular unprotected sexual intercourse for 12 months without pregnancy occurring. Of all couples who are infertile (10%) about 35% are due to female infertility. Insufficient FSH is one of the causes of infertility in women. Treatment aims to stimulate egg development by raising the level of FSH in the body (see Figure 22.2C).

Tablets of the antioestrogen drug **clomiphene** make tissues insensitive to oestrogen, including the pituitary gland. Remember that oestrogen inhibits the production of FSH (see Figure 22.2C). Treatment with clomiphene prevents this effect of oestrogen and the pituitary continues to produce FSH. Levels of FSH increase with the result that eggs develop normally. Drugs that increase the levels of FSH from the pituitary induce ovulation in 80% of treated women.

Artificial insemination (AI) helps some infertile couples to have a baby. Semen (see p. 364) is placed in the woman's vagina using a syringe. The semen may come from the husband or from an anonymous donor if the husband is infertile. The donor supplies his semen to a **sperm bank** where it is frozen and stored until needed. Some people feel that AI using donated sperm makes difficulties for the marriage partnership. Also, children resulting from AI may be upset when they discover their origin. Despite the problems, many couples use AI to produce children that otherwise they could not have.

In vitro **fertilisation (IVF)** also gives new hope to infertile couples. Hormonal treatment brings on ovulation (see p. 363). The eggs are then retrieved from the woman and prepared for fertilisation. Sperms are added to selected eggs, and two or three fertilised eggs are then transferred into the uterus. Repeated IVF results in around 40% of treated women giving birth to a healthy baby.

22.8 ▶ Sexually transmitted diseases

FIRST THOUGHTS

Different diseases may be passed from person to person by sexual activity. Some of these sexually transmitted diseases cause serious illness. Their effects are described in this section.

Sexually transmitted diseases (sometimes called **venereal diseases** or **VD**) are a group of diseases which can pass from person to person during sexual activity. **Syphilis** and **gonorrhoea** are examples.

Syphilis is caused by the bacterium *Treponema pallidum*. The bacterium can live for only a short time outside the body and is very quickly killed by heat, lack of water and antiseptics. The bacterium that causes gonorrhoea is quickly killed in similar fashion. This is why it is very difficult to pick up syphilis and gonorrhoea other than by the intimate contact between two people having sex. Table 22.2 summarises the symptoms of the two diseases in men and women.

Other diseases that may be passed from person to person by sexual activity include:

- **Herpes** – which is caused by a virus similar to the kinds that cause cold sores and chicken pox. Blisters appear, usually on the glans of the penis (see Figure 22.2A) and inside the vagina. Unfortunately, once infected a person remains infected for life and the blisters often recur.

- **AIDS** – which is caused by the human immunodeficiency virus (HIV) (see Topic 16.1).

Table 22.2 ▼ Syphilis and gonorrhoea in men and women

Syphilis		Gonorrhoea	
Men	**Women**	**Men**	**Women**
Sores appear on the genitals weeks or sometimes months after sexual intercourse		Becomes painful to pass urine; yellow discharge from penis	Many women show no symptoms but it may become painful to pass urine and there may be a yellow discharge from the vagina
Symptoms disappear		Symptoms disappear	
If syphilis is not treated, years later it can cause blindness, heart trouble, insanity and eventually leads to death	Same effects as in men In addition, babies can be badly affected in the uterus	In the long term, sperm ducts become blocked, resulting in sterility. May also lead to bladder problems	Oviducts become blocked resulting in sterility. Babies affected in the uterus may be born blind

These diseases are not always caught through having sex with an infected person. AIDS, for example, is spread when a person's blood infected with HIV mixes with someone else's blood. This is how many of the people suffering from **haemophilia** have become infected with HIV. They picked up the virus from the blood clotting agent factor VIII which had been donated by HIV-infected people. Now blood, and blood products like factor VIII, are screened for HIV before being given to patients (see p. 264).

Avoiding sexually transmitted diseases means avoiding sexual intercourse with a person who is infected. *How can you tell if someone has a sexually transmitted disease?* The short answer is, you cannot, but the chances of becoming infected are considerably reduced if you only have sex within the context of a stable relationship and do not have a lot of sexual partners.

People who are worried that they have been infected with a sexually transmitted disease can go to a special **clinic** (most large hospitals have one) where they are examined and if necessary treated. **Antibiotic drugs** like penicillin and streptomycin are used to treat syphilis and gonorrhoea. They will cure the disease providing treatment is started early enough. Nobody need know that treatment has been given; the hospital keeps the visit to the special clinic confidential.

Viral diseases like herpes cannot be cured with antibiotics. AIDS is a special problem for which there is no known cure at present. There are drugs that slow down the progress of HIV and scientists world-wide are trying to find new drugs and vaccines to fight the disease.

SUMMARY

The risk of infection with sexually transmitted diseases is reduced by not having sex with a lot of partners. Some of the diseases can be treated with drugs; others are much more difficult to deal with.

CHECKPOINT

▶ 1 (a) A diaphragm is fitted over the cervix: how does it work as a contraceptive?
    (b) How do condoms (or sheaths) work as contraceptives?
▶ 2 Briefly explain how the contraceptive pill affects the menstral cycle to prevent pregnancy.
▶ 3 Describe how drugs affect the menstrual cycle to improve fertility.

Theme Questions

1 The diagram shows a section through a flower.

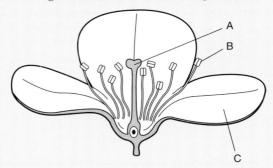

Name the parts of the flower labelled **A**, **B** and **C**. What are their functions?

2 The mass of living things plus its contained water is called **fresh mass**. The relationship between the fresh mass of lupin seeds and the percentage seed germination, percentage seedling survival and seedling fresh mass is shown in the table.

Fresh mass of seeds (mg)	Percentage of seeds germinated	Percentage of seedlings surviving to one week	Mean seedling fresh mass after five weeks (mg)
Below 16	41.9	84.6	24.3
Between 16–25	90.2	96.8	44.2
Between 26–35	95.6	98.1	60.7
Between 36–45	97.5	100.0	85.4
Above 45	100.0	100.0	106.4

(a) Refer to the table and comment on **two** relationships between seed fresh mass and percentage seed germination

(b) State the evidence provided by the figures that the seedlings produced from large seeds grow more rapidly than the seedlings produced from small seeds. Suggest an explanation for the more rapid growth

(c) What **three** conditions would it be necessary to provide for the seeds to ensure germination?

(d) Name two additional conditions necessary for the healthy growth of the seedlings.

3 (a) The diagram shows a bean seed opened up. Name the labelled parts of the seed, and briefly explain the function of each part.

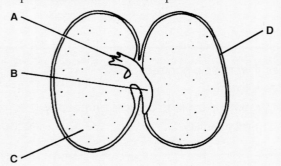

(b) The diagram shows a broad bean seedling. Name the labelled parts, and briefly explain the function of each part.

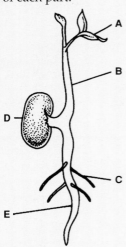

(c) Here are drawings of fruits and seeds. For each one describe how **two** features are adapted to the method of dispersal that the plant uses. The first one has been done for you.

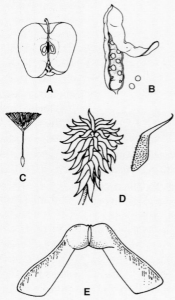

Plant A
Feature 1 Succulent fruit encourages animals to feed.
Feature 2 Tough seed to resist digestion.

4 The diagram at the top of the next page shows one way carnation plants are grown in commercial nurseries.

(a) Explain how the method produces new plants.

(b) What **two** advantages does the method have for the commercial grower?

(c) Explain the main disadvantages of the method for the grower.

Side shoot pegged down

Soil level

5 (a) Describe the events which occur after the release of sperms into the vagina of a human female until implantation of the embryo.

(b) The chart below shows how the thickness of the uterus lining and the levels of two hormones A and B, made in the ovaries, vary during the menstrual cycle. The first month is a normal menstrual cycle but fertilisation occurs during the second month.

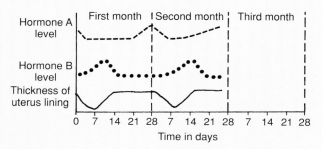

Hormone A level

First month | Second month | Third month

Hormone B level

Thickness of uterus lining

0 7 14 21 28 7 14 21 28 7 14 21 28
Time in days

(i)   Say when was the most likely time of ovulation in the first month.
(ii) During which period in the second month is fertilisation likely to have occurred?

(c) Copy and complete the chart above to show what happens, following fertilisation, to:
(i)   the uterus lining in the third month,
(ii) the levels of hormones A and B in the third month.

(d) One type of contraceptive pill contains a mixture of hormones A and B.
(i)   Explain briefly how this pill works as a contraceptive.
(ii) If she is using the contraceptive pill, it is usual for a woman to take a hormone pill each day for 21 days and then to take a pill without any hormones for the next 7 days. What is the advantage of taking the hormones for 21 days only

6 (a) If the oviducts (Fallopian tubes) are blocked, a woman cannot have a baby in the normal way, but may be able to have a 'test tube' baby. In order for this to happen, a doctor pushes a fine tube through the body wall and takes several eggs from the ovary.
Why can't the eggs be obtained through the vagina and uterus?

(b) The eggs are then put into a small glass dish and sperm are mixed with them. After a few days the developing zygotes are put back into the woman's uterus through the cervix.
(i)   Why are sperm mixed with the eggs before they are put back into the woman?
(ii) Explain why the zygotes are kept for a few days before they are put back into the woman.

(c) Give **two** reasons why the term *test-tube baby* is misleading.

(d) Name **three** changes which take place in the uterus to help protect and nourish the developing embryo.

7 (a) External human fertilisation (*in vitro* fertilisation) may be done to help a woman with blocked oviducts to become pregnant.
(i)   Suggest **one** reason why the blockage of the oviducts would cause infertility.
(ii) Some of the main stages of *in vitro* fertilisation are:
• the woman is given an injection of follicle-stimulating hormone (FSH) shortly after menstruation;
• she is also given an injection of luteinising hormone (LH);
• several eggs are then removed from her body;
• the eggs are placed in a solution very similar in composition to the fluid inside the female reproductive tract before the sperm are added to them.
Suggest the reasons for the inclusion of each of these stages in external fertilisation.

(b) The diagram shows a fetus in the uterus shortly before birth.

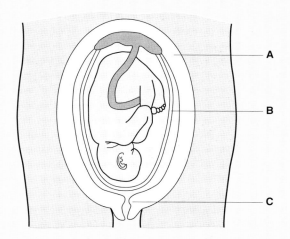

A

B

C

During birth changes occur to many of the structures shown in the diagram. Describe the changes, and the reasons for them, which occur to the structures labelled **A**, **B**, and **C**.   (SEG)

Genetics and biotechnology

Geneticists help our understanding of heredity by investigating how offspring inherit characteristics from their parents. Biotechnologists turn our understanding to good use. New techniques such as the use of monoclonal antibodies and genetic engineering help fight diseases, make new products, improve food production and protect the environment.

Topic 23   Inheritance and genetics

23.1 ▶ Gregor Mendel

The study of the patterns of **heredity** (the ways in which offspring resemble or differ from their parents) is called **genetics**.

Like any other branch of science, genetics has its own vocabulary. Before you start reading about Gregor Mendel's experiments you need to know the meaning of a few words. Checking the genetics glossary will help you.

Genetics glossary

- **monohybrid inheritance** – the processes by which a single characteristic is passed from parents to offspring, e.g. flower colour, eye colour
- **pure-breeding** – characteristics that 'breed true' appearing unchanged generation after generation
- **parental generation** (symbol **P**) – individuals that are pure-breeding for a characteristic
- **first filial generation** (symbol F_1) – the offspring produced by a parental generation
- **second filial generation** (symbol F_2) – the offspring of the first filial generation
- **gene** – a length of a strand of DNA which codes for the whole of one protein (see Topic 10.4)
- **allele** – one of a pair of genes which control a particular characteristic
- **homozygote** – an individual with identical alleles controlling a particular characteristic. Individuals which are pure breeding for a particular characteristic are **homozygous** for that characteristic
- **heterozygote** – an individual with different alleles controlling a particular characteristic
- **dominant** – any characteristic that appears in the F_1 offspring of a cross between pure breeding parents with contrasting characteristics, e.g. tallness and shortness in pea plants OR any characteristic expressed by an allele in preference to the form of the characteristic controlled by the allele's partner. 'Expressed' describes the processes which produce a protein because of the activity of a gene.
- **recessive** – any characteristic present in the parental generation that misses the F_1 generation but which reappears in the F_2 generation OR any characteristic of an allele which is not expressed because the form of the characteristic of the allele's partner is expressed in preference OR any characteristic of an allele which is only expressed in the absence of the allele's dominant partner
- **genotype** – describes the genetic make-up (all of the genes) of an individual
- **phenotype** – describes the outward appearance of an individual

You may come across other ways of describing 'dominant' and 'recessive'. Here the alternative explanations will help you understand alternative descriptions. An individual's phenotype is the result of those genes of the genotype which are actively expressing characteristics.

Make sure that you check the *Genetics glossary* . It gives the meanings of words which regularly crop up in questions.

Genetics which traces the inheritance of characteristics according to the principles established by Mendel, is sometimes called Mendelian genetics (or inheritance).

Remember that peas are seeds and that pods are the fruits of the pea plant.

Remember also that the stamens and carpels are the sex organs of the flower (see Topics 21.1 and 21.2).

Mendel's experiments

The work of Gregor Mendel marks the beginning of modern genetics. At the monastery where Mendel was a monk, he kept a small garden plot where he experimented with breeding the garden pea (*Pisum sativum*). He observed the way in which its characteristics were inherited from one generation to the next.

Mendel's choice of the garden pea for his experiments was fortunate:

- Pea plants are easy to grow.
- Pea plants have different characteristics which breed true, appearing unchanged generation after generation (for example height of plant, colour of seed).
- The petals of the pea flower enclose the stamens and the carpels (see Figure 23.1A). Cross pollination, therefore, does not usually occur. If pea plants could naturally cross pollinate, Mendel's experiments would have been unsuccessful. *Can you think why?*

Mendel's success also lay in the methodical way he planned his work. He only studied one characteristic at a time (**monohybrid inheritance**) in thousands of pea plants and his mathematical training allowed him to analyse his results.

To prevent self-pollination in the plants in his experiments, Mendel prised open the flower buds before the pollen matured and removed the anthers. He then pollinated the flowers with pollen from other mature plants and tied small, muslin bags over the pollinated flowers to prevent any stray pollen settling on the flowers.

Mendel took pollen from **pure-breeding** short plants and dusted it onto the stigmas of **pure-breeding** tall plants and vice versa. These plants were the **parental generation**. He collected and grew the seeds they produced (see Figure 23.1B).

It did not matter if Mendel took pollen from tall or short parent plants, the F_1 **generation** were always tall. He called 'tallness' a **dominant** characteristic.

Mendel then bred from the F_1 generation. He collected and grew their seeds. The 'shortness' characteristic that skipped the F_1 generation reappeared in the F_2 **generation**. Mendel called 'shortness' a **recessive** characteristic (see Figure 23.1C).

Figure 23.1A 🔺 A pea plant with peas visible against the light shining through the pods. The petals of a pea flower enclose the stamens and carpels

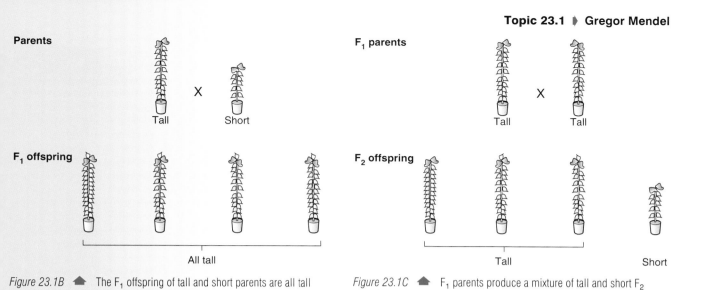

Parents

Tall   X   Short

F₁ offspring

All tall

Figure 23.1B ⬆ The F₁ offspring of tall and short parents are all tall

F₁ parents

Tall   X   Tall

F₂ offspring

Tall   Short

Figure 23.1C ⬆ F₁ parents produce a mixture of tall and short F₂ offspring

EXTENSION FILE
ACTIVITY

Make sure you can answer an exam question which asks you to make the link between meiosis and Mendel's *law of segregation*.

Remember: a recessive characteristic only appears if the individual's genotype (see p. 387) is homozygous for the recessive alleles. In the case of peas, short plants appear only if the individuals are **tt**. If **T** is present in the genotype, then the plants will be tall because the dominant **T** masks the effect of the recessive **t**.

How characteristics are inherited

From his results Mendel realised that parents passed 'something' on to their offspring which made them look like their parents. When these offspring became parents they passed on the 'something' to their offspring, and so on from generation to generation. Today, we call the 'something' which parents pass to offspring **genes** (see p. 152).

Mendel reasoned that sexually reproduced offspring receive the same number of genes from each parent and that any particular characteristic, therefore, must be controlled by a pair of genes. Paired genes controlling a particular characteristic are called **alleles**. They may be identical to one another or different. An individual with identical alleles controlling a particular characteristic is called a **homozygote** (*homo*- means the same); an individual with different alleles controlling a characteristic is called a **heterozygote** (*hetero*- means different).

Mendel concluded that alleles must separate when **gametes** form. He called the general rule the **law of segregation**. We know that the separation of alleles occurs at **meiosis** and that only one allele goes to each gamete (see Topic 9.4). Mendel, however, did not know this. Nearly 30 years went by after his experiments with peas before meiosis was discovered.

Mendel used letter symbols to simplify his observations. Capital letters symbolise alleles for dominant characteristics and small letters symbolise alleles for recessive characteristics The letter used to symbolise the recessive allele is the same letter used to symbolise the dominant allele. For example, he used **T** for the allele which produced tallness in pea plants and **t** for the allele which produced shortness, and set out the crosses between plants diagrammatically (see Figure 23.1D).

Mendel repeated his experiments using other pairs of characteristics. For example, he crossed purple flowered plants with white flowered plants. He always found that the flowers in the F₁ generation were purple. In other words, purple was a dominant characteristic; white was a recessive characteristic. Table 23.1 summarises the different pairs of dominant and recessive characteristics studied by Mendel in his experiments.

It's a fact!

*How do you know if a tall plant is homozygous (**TT**) or heterozygous (**Tt**)?*

One way of finding out is to do a **backcross**. The tall plant is crossed with a short plant which we know must be homozygous (**tt**) (see p. 379). If the tall plant is homozygous (**TT**), then all the offspring of the cross will be tall and heterozygous (**Tt**). If the tall plant is heterozygous (**Tt**) then 50% of the offspring will be tall (**Tt**) and 50% short (**tt**). By doing the backcross, and counting the proportion of tall offspring to short offspring (if any), you will know whether the tall parent is homozygous (**TT**) or heterozygous (**Tt**).

The Cambridge statistician RA Fisher thought that the data providing evidence for Mendel's ratio of 3 : 1 was too good to be true. The comment does not mean that Mendel's conclusions were wrong. Just that research data varies within limits. One source of variation is 'experimental error'.

Parental generation

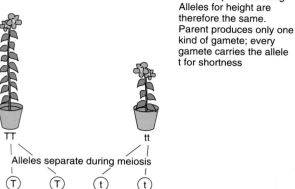

Parent is pure-breeding. Alleles for height are therefore the same. Parent produces only one kind of gamete; every gamete carries the allele T for tallness

Parent is pure-breeding. Alleles for height are therefore the same. Parent produces only one kind of gamete; every gamete carries the allele t for shortness

TT   tt

Alleles separate during meiosis

Gametes   T   T   t   t

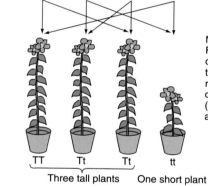

F₁ generation

Tt   Tt   Tt   Tt

Alleles in each F₁ individual are different. All F₁ plants are tall because T is dominant. Each F₁ plant produces two kinds of gamete: 50% carry the allele T, 50% the allele t

Alleles separate during meiosis

Gametes   T   t   T   t

F₂ generation

Not all tall plants in the F₂ generation have the same combination of alleles. Two thirds have dominant and recessive alleles (Tt), the other third is pure breeding (TT). The short plants are also pure-breeding (tt)

TT   Tt   Tt   tt

Three tall plants   One short plant

Figure 23.1D ⬆ How alleles controlling a characteristic pass from one generation to the next. *Which plants are homozygotes and which are heterozygotes? Why are only homozygotes pure-breeding?*

Table 23.1 ⬇ Dominant and recessive characteristics in pea plants. (**Note:** axial flowers sprout along the stem; terminal flowers at the end)

Characteristic	Dominant	Recessive
Seed shape	Round seed	Wrinkled seed
Seed colour	Yellow seed	Green seed
Seed coat colour	Coloured seed coat	White seed coat
Pod shape	Smooth pod	Wrinkled pod
Pod colour	Green pod	Yellow pod
Flower position	Axial flowers	Terminal flowers
Plant height	Tall stem	Short stem

KEY SCIENTIST

The Cambridge geneticist RC Punnett was the first to set out a cross between parents with contrasting characteristics as a table. Called a **Punnett square**, the alleles of the gametes of one parent are written along the top; the alleles of the gametes of the other parent down the side. The possible combinations of alleles are written in the appropriate boxes, avoiding the criss-cross of lines in Figure 23.1D.

CROSS	TT × tt	
Parental gametes	t	t
T	Tt	Tt
T	Tt	Tt

Pure breeding recessive parent

F$_1$ generation

Pure breeding dominant parent

Genetics of human blood groups

Sometimes a characteristic is controlled by more than two alleles. Human blood groups, for example, are controlled by three, A, B and O (see Topic 16.1). An individual has two out of the three alleles.

The A and B alleles control production of the antigens which determine a person's blood group. The A and B alleles are dominant to the O allele but not to each other. So, if the A and B alleles are both present, then the person's blood group is AB. If neither the A nor B alleles are present, then the person's blood group is O (see Table 23.2).

Because the A and B alleles are not dominant to each other, they are said to be **co-dominant**. However, they are both dominant to the O allele.

Table 23.2 ▼ The genetics of blood groups (Notice that blood groups A and B each have two possible combinations of alleles)

Alleles	Antigen on red blood cells	Blood group
AA or AO	A	A
BB or BO	B	B
AB	A and B	AB
OO	None	O

See www.keyscience.co.uk for more about inheritance of sex.

Inheritance of sex

Figure 23.1E shows the chromosomes of a man and a woman. In each case a photograph of all the chromosomes from the nucleus of a body cell has been cut up and the chromosomes arranged into **homologous pairs** and in order of size. The arrangement is called a **karyotype**.

Of the 23 pairs of chromosomes in each photograph, the chromosomes in each of 22 pairs are similar in size and shape in both the man and the woman. Notice, however, that the chromosomes of the 23rd pair in the man are different from the 23rd pair in the woman. These are the sex chromosomes. The larger chromosomes are called the **X chromosomes**; the smaller chromosome is called the **Y chromosome**.

Since the body cells of a woman each carry two X chromosomes, meiosis can only produce eggs containing an X chromosome. Each

SUMMARY

Genetics is the study of the ways in which characteristics are inherited by offspring from their parents. Gregor Mendel pioneered the science and established some fundamental principles of genetics.

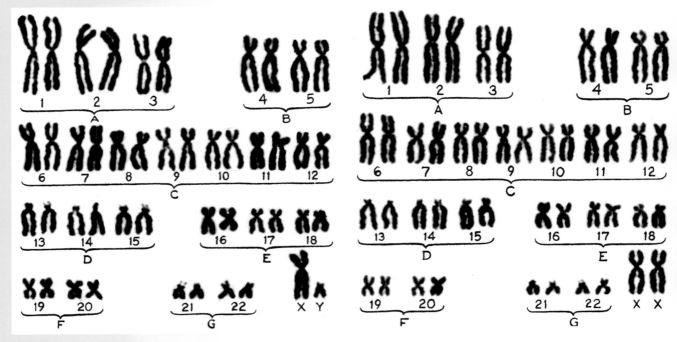

Figure 23.1E ⬆ (a) The chromosomes of a man

(b) The chromosomes of a woman

Meiosis determines the 50:50 ratio of girls to boys.

body cell of a man, however, carries an X chromosome and a Y chromosome, so meiosis produces two types of sperm. Of the sperms produced, 50% carry an X chromosome and 50% carry a Y chromosome. A baby's sex depends on whether the egg is fertilised by a sperm carrying an X chromosome or one carrying a Y chromosome (see Figure 23.1F). The birth of almost equal numbers of girls and boys is governed by the production of equal numbers of X and Y sperms at meiosis.

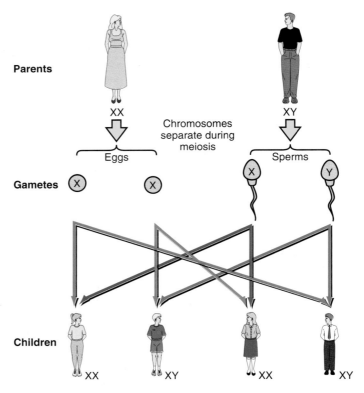

Figure 23.1F ⬆ Inheritance of sex in humans

CHECKPOINT

◗ **1** The table shows the results of breeding experiments with pea plants beginning with parents pure-breeding for tallness and shortness.

Cross	Original parental cross	F₁ plants from parental cross	F₂ plants from F₁ cross
Height	Tall × short	All tall	779 tall: 268 short

(a) Explain how you can tell from the results that tallness is dominant and shortness is recessive.

(b) To the nearest whole number, what is the ratio of tall to short plants in the F₂ generation?

(c) All the F₁ plants are tall. Explain how the combination of their alleles is different from that of the parent tall plant.

(d) Copy and complete the diagram opposite to explain the F₂ results obtained from crossing within the F₁ generation.

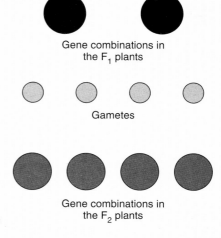

Gene combinations in the F₁ plants

Gametes

Gene combinations in the F₂ plants

◗ **2** What is the distinction between genes and alleles?

◗ **3** Match the terms in Column A with their definitions in Column B.

Column A	Column B
Gene	A gene which can express itself despite the presence of another version of the gene
Recessive gene	Portion of DNA molecule which controls a specific characteristic such as eye colour in humans and height in pea plants
Dominant gene	A plant which will always breed true for a particular characteristic
Pure-breeding plant	A gene not expressed in the presence of another version of the gene

◗ **4** Explain why the sex of a baby is determined by the father and not the mother. How does this account for the birth of almost equal numbers of boys and girls?

23.2 ◗ Human genetic disease

FIRST THOUGHTS

In this section you will find out about some examples of human genetic diseases.

Down's syndrome is one of the more familiar conditions resulting from a genetic abnormality. It is called a syndrome because it involves a number of disorders. People with Down's syndrome are usually short and stocky. They are more likely to suffer from infectious diseases, find speech difficult and have abnormalities of the heart and other organs. Most are also slightly mentally handicapped, which often improves with treatment. Some are above average in intelligence. Some genetic diseases caused by abnormalities in the number of chromosomes can be detected in the karyotype (see p. 383).

Figure 23.2A shows the chromosomes of a person with Down's syndrome. Carefully examine the 21st set of chromosomes and compare them with the 21st set in Figure 23.1E. *Can you see the extra one making 47*

385

47 XX +21

(a) A child with a variety of Down's syndrome working alongside pupils in a normal school class

Figure 23.2A ⬆

(b) The chromosomes of a person with Down's syndrome

chromosomes in all? A defect occurring during cell division produces the extra copy of chromosome 21. It is the presence of the extra chromosome that causes Down's syndrome.

Defects occurring during cell division can also affect the numbers of sex chromosomes (set 23). In humans a combination of an X and a Y chromosome (XY) produces male children. However, XXY, XXXY and XXXXY combinations also produce males. The combinations XXX and X0 (only one X chromosome present) as well as the normal combination XX produce female children. Men and woman with these unusual combinations of sex chromosomes are usually sterile and suffer from other abnormalities.

Other genetic diseases are caused by changes (mutation – see p. 154) in the genes themselves. Most of us carry some defective genes. Fortunately we usually have a normal copy of the allele which masks the effect of its abnormal recessive partner. However, if the defective allele is dominant or a person inherits two copies of a defective recessive allele (one from each parent) then the affected individual inherits the disorder. The various possibilities have a clear pattern of Mendelian inheritance (linked to the X(sex) chromosome or recessive/dominant) so that the risk of someone inheriting a particular disorder can be predicted.

Human genetic diseases are caused either by abnormalities in the number of chromosomes or by mutations in the genes themselves.

Human genetic diseases caused by mutations show a pattern of Mendelian inheritance: linked to the X (sex) chromosome, recessive or dominant.

See www.keyscience.co.uk for more about sex linkage.

Haemophilia – a sex-linked problem

Haemophilia is a genetic condition in which the blood does not clot properly after an injury. In its most common form the person fails to produce the **clotting agent factor VIII** (see Topic 16.1). The allele responsible for haemophilia is recessive and is located on the X chromosome. Therefore, haemophilia is a **sex-linked** disease.

Although women may carry the defective allele on one of the X chromosomes, they do not usually suffer from haemophilia. This is because the normal allele on the other X chromosome is dominant. The dominant allele masks the effect of its recessive partner and ensures enough factor VIII is made for normal blood clotting to take place. A woman who carries the recessive allele on one of the X chromosomes is called a **carrier**. Although she does not suffer from the disease she is able to pass it on to her children. For a woman to have haemophilia she would have had to receive the recessive allele for the characteristic from both her mother and her father. Since the recessive allele is rare, this only happens very occasionally.

Men have one X chromosome and one Y chromosome. The Y chromosome does not carry as many genes as the X chromosome. If a man inherits the recessive allele for haemophilia on the X chromosome, there is no dominant allele on the Y chromosome to mask the effect of the recessive allele. No factor VIII is produced and the man suffers from haemophilia.

Haemophilia is a genetic disease with a royal connection. Queen Victoria was a carrier of the haemophilia allele. One of her sons and two of her daughters inherited the defective allele. Although haemophilia is usually rare, various intermarriages caused the defective allele to spread among the royal families of Europe.

Figure 23.2B is a **pedigree chart** of Queen Victoria's family. A pedigree chart shows genetic data about related individuals through a number of generations.

It's a fact!

The word pedigree comes from the French **pied de grue** meaning a crane's foot. The crane is a type of fish-eating bird. The branching lines of a pedigree chart reminded geneticists of the way the toes of a crane splay out from each of its ankles.

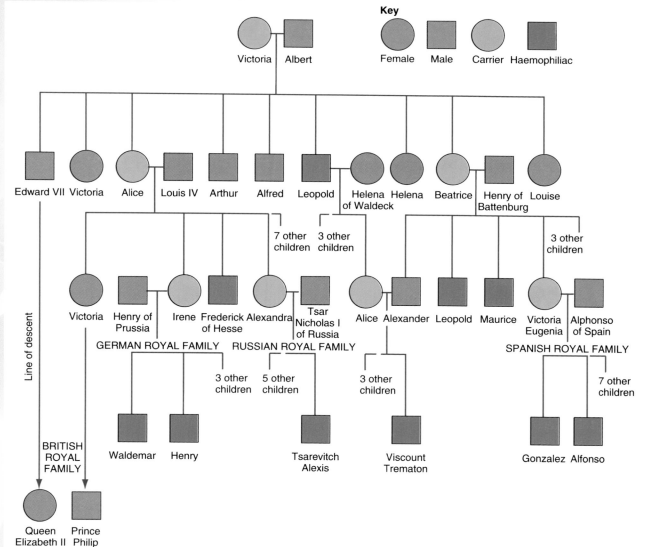

Figure 23.2B ⬆ Distribution of the haemophilia allele in Queen Victoria's descendants. (Notice that there are no female haemophiliacs and that the British Royal Family escaped the disease because Edward VII did not inherit the defective allele)

Red-green **colour-blindness** is another sex-linked disorder caused by a recessive allele on the X chromosome. As with haemophilia, women can be carriers but rarely suffer from the disorder. Red-green colour-blindness occurs in 8% of men but only 0.04% of women.

Cystic fibrosis – a recessive problem

Cystic fibrosis is an inherited condition which affects the pancreas and the bronchioles of the lungs. It is one of the most common fatal diseases of childhood and is inherited as a recessive gene in the following way.

Let C represent the allele for a normal pancreas and bronchioles and c be the recessive allele for cystic fibrosis. A person suffers from the disease only if he or she has two alleles for cystic fibrosis (the genotype is cc). A person with the genotype Cc is a **carrier** of the disorder but does not suffer from the disease. There is, therefore, a one-in-four chance of a child suffering from the complaint if two carriers have a child (see Figure 23.3C and p. 250).

Parents Cc × Cc

♀ gametes ♂	C	c
C	CC	Cc
c	cC	cc

Figure 23.2C ⬆ Inheritance of cystic fibrosis

In people suffering from cystic fibrosis the bronchioles become blocked with mucus and have to be cleared regularly (see Figure 23.3D).

Figure 23.2D ⬆ Regular physiotherapy helps clear the lungs of a child with cystic fibrosis of mucus

Huntington's chorea – a dominant problem

Huntington's chorea is an inherited disease characterised by involuntary muscular movement and mental deterioration. The age of onset is about 35 years. The majority of those affected can therefore have a family before being aware of their own condition. It is transmitted by a dominant gene and both sexes can be equally affected.
Let HC be the allele for Huntington's chorea and hc the normal allele. Notice in Figure 23.3E that affected people are heterozygous.

Gene therapy (see p. 416) is being used to treat cystic fibrosis.

See www.keyscience.co.uk for more about cystic fibrosis.

It's a fact!

Gene therapy (see p. 416) is one approach used in attempts to cure cystic fibrosis. The healthy gene for normal pancreas and bronchioles is taken up into fatty droplets called **liposomes**. A fine aerosol containing the liposomes is then sprayed deep into the lungs of the patient. If the liposomes fuse with the cells of the patient's alveoli (see p. 249), then the healthy genes may enter the cells and do their work. However, gene transfer from liposomes to cells is not always successful, and when it does occur benefits are often short lived.

Parents HChc × hchc

♀ gametes ♂	hc	hc
HC	HChc	HChc
hc	hchc	hchc

Figure 23.2E Inheritance of Huntington's chorea

**EXTENSION FILE
ASSIGNMENT**

Figure 23.2F (a) A genetics laboratory where the results from genetic probes are analysed

Theoretically two parents could be affected and a homozygous child could be produced. The effect of this would probably be lethal. The gene is so rare that it is most unlikely to occur in the homozygous condition.

Shall we start a family?

A couple thinking of starting a family can be especially anxious that any children they have are fit and healthy. If either partner has a family history of genetic disease, the couple may want to know what the chances are of handing on the problem to any children they have. A pregnant woman may want to know if the baby she is carrying has inherited a particular disease. An older woman may wish to evaluate the risk of having a baby with Down's syndrome (the likelihood of having a child with Down's syndrome increases with the mother's age). These are just some of the worries that people can have about starting a family.

Scientists and doctors have developed techniques which allow them to diagnose genetic disorders in prospective parents and in babies developing in the uterus.

- **Genetic probes.** For some types of genetic disorder, differences between the defective allele responsible for the disorder and its normal partner can be detected with genetic probes. The probes detect differences in the sequences of bases on the **DNA** and therefore reveal changes in the codons which determine the way proteins are made (see Topic 9.3). Genetic probes give very accurate results (see Figure 23.2F(a)).

- **Amniocentesis. Amniotic fluid** (see Topic 22.4) contains living cells from the fetus. These cells, once removed, can be grown in the laboratory and examined for genetic disorders. A thin needle is used to withdraw fluid from the amniotic cavity. Doctors work out where the fetus is in the uterus by using an **ultrasound scanner** (see Figure 23.2F(b)).

- **Statistical evidence.** The risk of having a child with a genetic disorder is well documented for some disorders such as Down's syndrome (see Figure 23.2F(c)). Karyotyping (see p. 383) helps identify the problem.

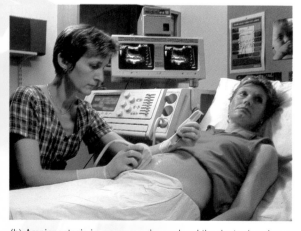

(b) Amniocentesis in progress – in one hand the doctor is using a syringe to draw fluid from the amniotic cavity through a thin needle piercing the abdomen of the pregnant woman. About 30 cm³ of fluid is removed. In the other hand the doctor is holding an ultrasound scanner to detect the position of the fetus in the uterus, preventing damage to it by the needle. Images of the fetus can be seen on the screens in the background

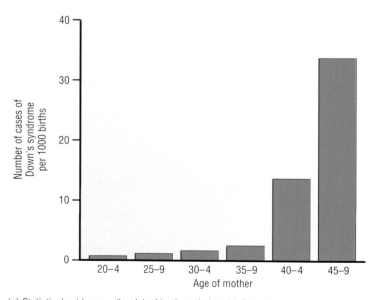

(c) Statistical evidence – the risk of having a baby with Down's syndrome

389

Any prospective or expectant parents who have reason to be worried about genetic disorders can attend **genetic counselling**. During genetic counselling, statistical evidence and results from karyotyping, amniocentesis and genetic probes are discussed by the couple and a specially trained genetic counsellor. The counselling helps couples to make informed choices about whether to start a family if there is a chance of having children with genetic disorders. It also helps expectant mothers and their partners to decide whether to continue with a pregnancy when they know their baby has a genetic disorder.

SUMMARY

Occasionally inheritable abnormalities arise in humans. Such genetic disorders may be caused by altered chromosomes or altered genes. Genetic counselling helps couples to make informed decisions about the risk of their producing children with a genetic disorder.

CHECKPOINT

▶ 1 Defective alleles on the X chromosome are the cause of red–green colour-blindness. How do you account for the difference in the occurrence of red–green colour-blindness between men and woman?

▶ 2 Women carry the defective allele for haemophilia but do not usually suffer from the disease. Explain why this is so.

23.3 ▶ The growth of modern genetics

FIRST THOUGHTS

Developments in cell biology and molecular biology have helped to explain the results of Mendel's breeding experiments. Discovering the structure of DNA and the nature of genes has opened up the possibility of manufacturing processes based on biotechnology. This section tells you about these exciting prospects.

Mendel reported the results of his experiments in 1865 at a meeting of the Brünn Natural History Society. Nobody really understood what he was talking about. His paper was published the following year but was ignored (see Figure 23.3A), probably because most scientists at the time were unfamiliar with the mathematical analysis of biological problems. **Remember** that Mendel was a mathematiciain as well as a natural historian (see p. 380) – an unusual combination of skills in the mid-nineteenth century. Mendel continued his breeding experiments with pea plants until 1871 when he was made Abbot of the monastery. His new duties gave him little time for further research. He died in 1884.

There matters rested until 1900 when Mendel's work was rediscovered by three biologists working independently of one another. The growth of modern genetics dates from their recognition of the importance of Mendel's work.

In the interval between the publication of Mendel's paper and its rediscovery, important advances had been made in cell biology. New ways of staining cells showed that the nucleus contained strands of material that took up dyes very strongly. They were called **chromosomes**, which literally means 'coloured bodies'. **Mitosis** and **meiosis** were seen for the first time.

Early work on genetics depended on breeding experiments to establish the effect of genes on the appearances of organisms. Questions about what genes are and how they produce their effects were not tackled until new techniques in **molecular biology** were developed. Figure 23.3B shows how developments in genetics, cell biology and molecular biology were brought together to give a full explanation of Mendel's observations. Today the techniques of cell biology and molecular biology have revolutionised genetics. New discoveries are stimulating the growth of biotechnology bringing benefits to agriculture, industry and medicine (see Topic 24).

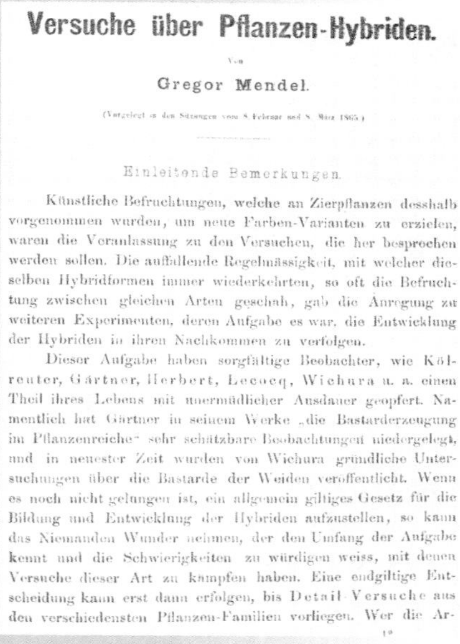

Versuche über Pflanzen-Hybriden.

Von

Gregor Mendel.

(Vorgelegt in den Sitzungen vom 8. Februar und 8. März 1865.)

Einleitende Bemerkungen.

Künstliche Befruchtungen, welche an Zierpflanzen desshalb vorgenommen wurden, um neue Farben-Varianten zu erzielen, waren die Veranlassung zu den Versuchen, die hier besprochen werden sollen. Die auffallende Regelmässigkeit, mit welcher dieselben Hybridformen immer wiederkehrten, so oft die Befruchtung zwischen gleichen Arten geschah, gab die Anregung zu weiteren Experimenten, deren Aufgabe es war, die Entwicklung der Hybriden in ihren Nachkommen zu verfolgen.

Dieser Aufgabe haben sorgfältige Beobachter, wie Kölreuter, Gärtner, Herbert, Lecocq, Wichura u. a. einen Theil ihres Lebens mit unermüdlicher Ausdauer geopfert. Namentlich hat Gärtner in seinem Werke „die Bastarderzeugung im Pflanzenreiche" sehr schätzbare Beobachtungen niedergelegt, und in neuester Zeit wurden von Wichura gründliche Untersuchungen über die Bastarde der Weiden veröffentlicht. Wenn es noch nicht gelungen ist, ein allgemein giltiges Gesetz für die Bildung und Entwicklung der Hybriden aufzustellen, so kann das Niemanden Wunder nehmen, der den Umfang der Aufgabe kennt und die Schwierigkeiten zu würdigen weiss, mit denen Versuche dieser Art zu kämpfen haben. Eine endgiltige Entscheidung kann erst dann erfolgen, bis Detail Versuche aus den verschiedensten Pflanzen-Familien vorliegen. Wer die Ar-

Figure 23.3A ⬆ The title page of Mendel's paper published in 1866. Almost all of the notebooks containing the results of his experiments were destroyed soon after his death

It's a fact!

The fruit fly *Drosophila* is a small fly which feeds on the sugar it finds in rotting fruit. The fly has been studied intensively and today more is known about its genetics than any other animal. It is ideal for genetic experiments because:

- It reproduces quickly (a generation every two weeks),
- It is easy to keep,
- Its cells each have only four pairs of chromosomes in the nucleus (the fewer chromosomes the better for genetic experiments),
- It has giant chromosomes in the salivary glands (giant chromosomes are easy to see under the microscope).

SUMMARY

Since Mendel's discoveries in genetics and cell biology have come together, driving forward the development of molecular biology. One result is genetic engineering which underpins the growth of modern biotechnology.

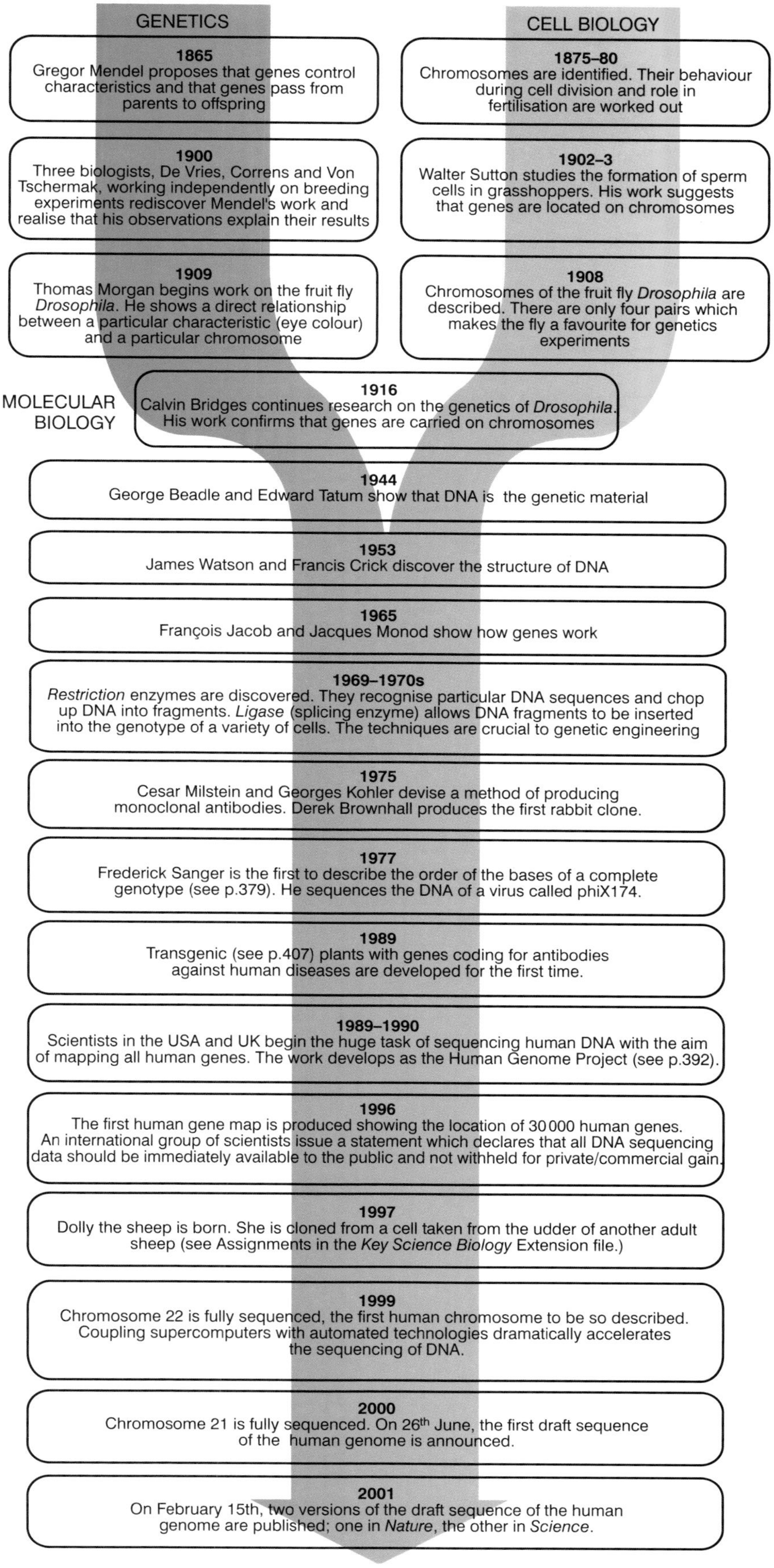

GENETICS

1865
Gregor Mendel proposes that genes control characteristics and that genes pass from parents to offspring

CELL BIOLOGY

1875–80
Chromosomes are identified. Their behaviour during cell division and role in fertilisation are worked out

1900
Three biologists, De Vries, Correns and Von Tschermak, working independently on breeding experiments rediscover Mendel's work and realise that his observations explain their results

1902–3
Walter Sutton studies the formation of sperm cells in grasshoppers. His work suggests that genes are located on chromosomes

1909
Thomas Morgan begins work on the fruit fly *Drosophila*. He shows a direct relationship between a particular characteristic (eye colour) and a particular chromosome

1908
Chromosomes of the fruit fly *Drosophila* are described. There are only four pairs which makes the fly a favourite for genetics experiments

MOLECULAR BIOLOGY

1916
Calvin Bridges continues research on the genetics of *Drosophila*. His work confirms that genes are carried on chromosomes

1944
George Beadle and Edward Tatum show that DNA is the genetic material

1953
James Watson and Francis Crick discover the structure of DNA

1965
François Jacob and Jacques Monod show how genes work

1969–1970s
Restriction enzymes are discovered. They recognise particular DNA sequences and chop up DNA into fragments. *Ligase* (splicing enzyme) allows DNA fragments to be inserted into the genotype of a variety of cells. The techniques are crucial to genetic engineering

1975
Cesar Milstein and Georges Kohler devise a method of producing monoclonal antibodies. Derek Brownhall produces the first rabbit clone.

1977
Frederick Sanger is the first to describe the order of the bases of a complete genotype (see p.379). He sequences the DNA of a virus called phiX174.

1989
Transgenic (see p.407) plants with genes coding for antibodies against human diseases are developed for the first time.

1989–1990
Scientists in the USA and UK begin the huge task of sequencing human DNA with the aim of mapping all human genes. The work develops as the Human Genome Project (see p.392).

1996
The first human gene map is produced showing the location of 30 000 human genes. An international group of scientists issue a statement which declares that all DNA sequencing data should be immediately available to the public and not withheld for private/commercial gain.

1997
Dolly the sheep is born. She is cloned from a cell taken from the udder of another adult sheep (see Assignments in the *Key Science Biology* Extension file.)

1999
Chromosome 22 is fully sequenced, the first human chromosome to be so described. Coupling supercomputers with automated technologies dramatically accelerates the sequencing of DNA.

2000
Chromosome 21 is fully sequenced. On 26th June, the first draft sequence of the human genome is announced.

2001
On February 15th, two versions of the draft sequence of the human genome are published; one in *Nature*, the other in *Science*.

Figure 23.3B ⬆ The growth of modern genetics

23.4 ▶ Discovering the human genome

FIRST THOUGHTS

The first draft readout of the sequence of all the bases of human DNA was announced on June 26th, 2000. The project continues and is the largest ever undertaken in biological research.

Estimates of the number of human genes currently vary between 30 000 and 40 000.

It's a fact!

Genes are mostly the same in all people, but 'silent' DNA varies from person to person and is nearly always unique to the individual. This uniqueness is the basis for **genetic fingerprinting** which can help to identify criminals in cases where some of the criminal's body cells are found at the scene of the crime (see p. 400).

The word genome refers to all of the DNA in each cell of an organism. Scientists working on the human genome aim to:

- work out the order (sequence) of bases of the lengths of DNA (see p. 152 which form our genes (see p. 153)
- identify where individual genes are located on chromosomes (see p. 148).

Nearly all human cells each have thousands of genes which determine appearance, vulnerability to disease, patterns of behaviour and all of the other characteristics we inherit from our parents. However, genes account for around 3% of the 3.1 billion nucleotides (see p. 170) which join together to make the DNA of our genome. The rest of the genome consists of 'silent' or 'junk' DNA, so called because its role in the cell remains a mystery. However, working out its base sequence is also part of the work of the scientists engaged in discovering the human genome.

Genes are found in the cell nucleus (see p. 152), and in humans they are distributed between 23 pairs of chromosomes. In women, each pair is similar in structure, because the 23rd pair – the sex chromosomes – are XX. In men, however, the X chromosome is paired with a dissimilar Y chromosome (see Figure 23.1E, p. 384), so there are 24 different kinds of human chromosome in all

Setting up genomic research: an issue of funding

In 1986 scientists in the USA and UK made moves to set up the **Human Genome Project**. Two years later the **Human Genome Organisation** was established to co-ordinate the work. Now more than 1000 scientists are employed in laboratories undertaking genomic research in Germany, France, Japan and China, as well as the USA and the UK.

In 1992 the UK contribution to the project centred on the Genome Campus near Cambridge (see Figure 23.4A). Its scientists are responsible for one-third of all of the DNA sequencing work, with particular focus on chromosomes 1, 6, 9, 10, 13, 20, 22 and X. They receive public funds to the tune of millions of pounds each year.

Figure 23.4A ⬆ The Genome Campus: its complex of buildings houses the Sanger Centre (named after Frederick Sanger who discovered the technique of DNA sequencing), the European Bioinformatics (see p. 395) Institute and the UK Human Genome Mapping Project Resource Centre

It's a fact!

In 1995 Craig Venter announced the DNA sequence of the genes of the bacterium *Haemophilus influenza*. It was the first time that all of the genes of an organism had been sequenced.

The American scientist Craig Venter set up the privately funded *Institute of Genomic Research* soon after his application in 1991 for patents on more than 300 human genes. *What were his motives?* The hoped-for prize is the new drugs and treatments which may come from knowing the genetic causes of disease (Figure 23.4B).

Genetic key to personal medicines

Genetic pioneers herald medical revolution

Figure 23.4B ▲ Headlines predict that new medicines and treatments will soon be available because of genomic research. *Do you think the newspapers are right?*

The issue of public versus private funding of genomic research has sparked fierce debate. Scientists receiving private funds say that the investment of millions of pounds and dollars into work on the human genome must be protected by **patents**. They argue that it is the investors who have risked their money, making it possible to discover the DNA sequences of genes. Any financial benefits from the work therefore belongs to the investors, and patents allow their interests to be protected.

Scientists receiving public funds say that the genetic information which is in effect about our biology should be available to everyone. In 1996 they co-ordinated their work in response to the challenge from privately funded genomic research by issuing the 'Bermuda Statement'. The statement declares that all DNA sequencing data should be immediately available for everyone to read and not withheld for private or commercial gain.

It's a fact!

A patent is a legal document which registers a person's (or organisation's) claim for an invention. It does not establish ownership but gives the holder of the patent the right to stop others from making use of the invention. In the case of genomic research, the invention refers to the base sequence of the DNA which makes up the genome.

Many find it shocking that parts of the human genome should be in the hands of a few individuals who run genome companies and the shareholders who invest in the companies. Spokespeople for the companies make the point that unless the money invested to make the discoveries in the first place is protected by patents there is no encouragement for investors to make the money available for scientific progress. Without the possibility of profits from new medicines and treatments, why should people risk their money? *Whose genome is it – ours, the government's, the shareholders of genome companies? What are your views?*

Reading the genome

First things first! The genes on chromosomes carry the code (a set of instructions) which cells need to make protein (see Topic 9.3). Notice the strings of letters on the computer shown in Figure 23.4C. The letters are ATC and G. They represent the bases of the nucleotides which make up a length of DNA (see Topic 10.4). On screen you are looking at a tiny part of the human genetic code. The order of the letters – GGGTGC and so on – is part of a gene, setting instructions for making a particular type of protein. Another sequence of the letters instructs the cell to make a different protein. In this way the proteins needed for growth and for building and repairing bodies are made.

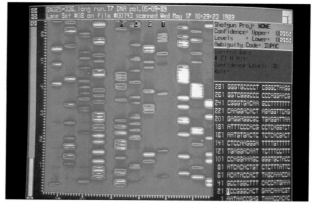

Figure 23.4C The screen shows part of the human genetic code

How do scientists read the letter sequence of the human genome? Samples of cells (blood and sperm) are taken from anonymous volunteers. Scientists break up the chromosomes of the cells into pieces to get at their DNA. Thousands of copies of the pieces of DNA are made (the procedure is called **amplification**) to give scientists enough material to work on. **Decoding** the DNA begins. By 'decoding' we mean that the order (sequence) of the letter symbols representing the bases of the DNA is worked out. Figure 23.4D shows you the results of sequencing methods at work. Each letter symbol is represented by a colour. Seeing the colours gives a reading of the sequence of bases for a particular piece of DNA. Computers analyse the pieces and assemble them into the complete sequence for a whole gene.

In 1999 the work of DNA sequencing speeded up with the introduction of new technology which couples automated sequencing equipment with supercomputers. At the same time the first human chromosome – chromosome 22 – was fully sequenced at the Sanger Centre, England, and a working draft of the human genome completed. On 26th June 2000 the first draft sequence of the human genome was announced. In a joint communique the then US President Bill Clinton and UK Prime Minister Tony Blair confirmed the principles of the Bermuda Statement issued in 1996 (see p. 391 and Figure 23.4E).

Figure 23.4D Reading the genome. Remember that the order of the letters is part of the code of a gene for making a particular protein (see Topic 9.3)

Figure 23.4E On 14th March 2000 US President Bill Clinton and UK Prime Minister Tony Blair declare that genomic data should be made available to the public and that patents should be issued only for inventions based on genes

The working draft of the publicly funded version of the human genome was published on 15th February 2001 in the journal *Nature*. The privately funded version was published in *Science* on the same day. Publication was the result of the joint work of all of the scientists involved who shared in the triumph of what is the biological equivalent of sending men to the moon (Figure 23.4F).

Exploiting the genome

The sequencing of the human genome is expected eventually to lead to more effective treatment of a range of different diseases. Mistakes in genes are the cause of a number of disorders such as Duchenne muscular dystrophy, cystic fibrosis and Huntington's disease (see Topic 23.2). Genetic errors combined with factors in the environment may increase the risk of some people dying of heart disease and cancer.

Being able to read the messages of our genes means that treatment can be tailored to the individual rather than the population as a whole. For example, the drug **salbutamol** is used to treat people suffering from **asthma** (see p. 251). It acts on the muscles which control the opening and closing of the tubes through which air passes into and out of the lungs (see Figure 15.2B). There are different versions of the muscles. One version is unresponsive to salbutamol. For people with the unresponsive version, another form of treatment for their asthma is called for. Reading the genes responsible for the different versions of the muscles in question means that people unresponsive to salbutamol can quickly be identified and provided with alternative drugs.

Cancer kills large numbers of people each year (see p. 155). It is a disease of DNA. Throughout life our cells are exposed to radiation, chemicals and viruses which may bring about changes in the sequence of our DNA (p. 154). The changes result in abnormal genes which accumulate and may reach a stage when the affected cells become cancerous. Tumours form and cancerous cells break away from the tumours and spread through the body (see p. 331). Many different genes are abnormal for any particular cancer and different genes are abnormal in different types of cancer.

In 1999 the **Cancer Genome Project** began work at the Sanger Centre. The aim is to study variations in the sequence of human genes and involvement of the variations in tumour formation (see Figure 23.4G). Finding cancer genes will allow scientists to identify the causes of cancer, understand the mechanisms by which cancer develops and help pinpoint more precisely where and how treatment can be most effective.

Figure 23.4F ▲ Reading the human genome: a student studies the draft outline of the instructions for building the human body. The place is the courtyard of *The Eagle* in Cambridge where Watson and Crick often discussed their work on DNA (see p. 172)

It's a fact!

Sequencing the genome is like producing the words in a dictionary but leaving out their meanings. The meanings in the genomic dictionary are the proteins for which the base sequences of the genes are the code (see Topic 10.3). The question is: *which gene(s) code(s) for which protein(s)?* **Bioinformatics** is helping to answer the question. It involves the use of powerful computers which help match the base sequences of the genome with the proteins for which they are the code.

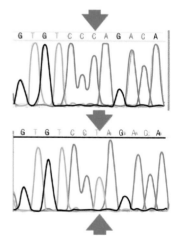

Figure 23.4G ▶ Analysis of the base sequence of genes helps to identify the genetic basis of disease. In this case, a mutation substitutes the base cytosine (C) by the base thymine (T) of a gene called SH2D1A found on the X chromosome. The substitution causes the uncontrolled multiplication of B and T white blood cells of the immune system in a person infected with a type of virus called Epstein–Barr.

In the next few years more and more will be discovered about the contribution of our genes to haemophilia, diabetes, high blood pressure, multiple sclerosis, schizophrenia and a host of other diseases where treatments at the moment are unsatisfactory or not even available. The continuing success of human genome research seems set to revolutionise medicine in the coming decades.

A brave new world

Is there a downside to success? The short answer to the question is almost certainly 'yes'. Many people are worried that the brave new world of molecular medical genetics could lead to genetic discrimination and 'designer' babies. The question *Whose genome is it?* asked on page 373 raises yet more important issues.

If we can read our genome, should companies that provide life insurance be allowed to require people to be tested for diseases that may develop later but for which there is no cure? Genetic tests may also play a part in the success of an individual's application for a mortgage. *What if the genetic readout shows that their health is at risk?* An unfavourable genetic profile may make it difficult for people to obtain insurance and a mortgage. Genetic discrimination is the backdrop for the development of a genetic underclass at the margins of society. *Should genetic testing of a fetus go beyond establishing its well-being?* Testing for intelligence, choosing eye colour and hair colour – all will soon be possible so that parents can opt for a 'designer' baby.

These are just a few of the issues causing concern. As with most developing technologies, benefits must be weighed against dangers – the good against the bad.

Morals establish the values and reasons lying behind the choices we make. They help us to distinguish between good and bad. **Ethics** is the code of behaviour by which morals are put into action. Science presents us with the challenge. We must decide. *What are your views?*

24.1 ▶ Introducing biotechnology

FIRST THOUGHTS

Biotechnology has a long history. This section explains the modern industrial process and the biology behind some traditional products of biotechnology.

The word **biotechnology** describes the way we use plant cells, animal cells and microorganisms to produce substances that are useful to us. Although the word biotechnology is new, the processes of biotechnology have a long history. The use of moulds to make cheese, and bacteria to make vinegar are early examples. For thousands of years we have exploited **yeast** to make wine, beer and bread. The ancient civilisations of Egypt and China knew that lactic acid bacteria preserved milk by turning it into yogurt. Figure 24.1A shows different ways of making traditional products of biotechnology.

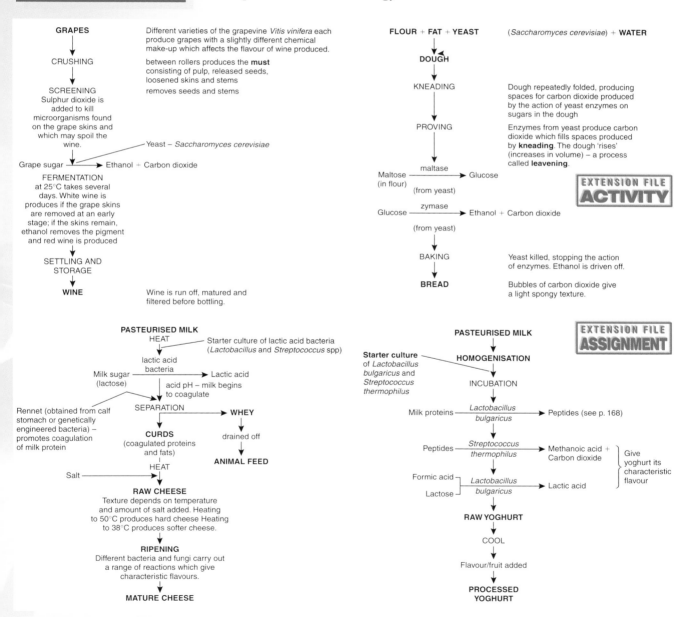

Figure 24.1A ◆ Making the traditional products of biotechnology has a long history

Figure 24.1B ⬆ Demi-johns of wine – each demi-john has a volume of about 10 litres. The cork is protected by a trap which allows carbon dioxide gas produced during fermentation to escape but prevents unwanted air-borne microorganisms from contaminating the mixture. The demi-johns and traps are sterilised before pouring in the mixture of grape-juice, water and yeast

Production techniques

Some people make wine at home. If so, the glass containers (called demi-johns) full of grape-juice, water and yeast may be kept next to a radiator or in the airing cupboard.

Figure 24.1B shows the set-up. You may not realise it, but making wine at home illustrates the principles behind biotechnology: demi-johns are **fermenters**, grape-juice is a **nutrient solution** and yeast is the **cell culture** that ferments the sugar in grape-juice into ethanol (alcohol) – the desired **product**. Warmth from the radiator provides the best temperature for the **fermentation** reactions to take place (see Topic 15.1). Fermenter, nutrient, cell culture and product come together in an industrial process which is illustrated in Figure 24.1C.

Substances are produced either by:

- **batch culture** which produces substances in a fermenter for a limited period. The fermenter is then emptied of product and nutrient solution. **Sterilisation** using super-heated steam prepares the fermenter for the next batch.

- **continuous culture** which produces substances in a fermenter as an ongoing process. Product is drawn off and nutrients replaced as they are used.

What are the advantages of continuous culture processing compared with batch culture? Continuity means that the process can run for a long time without the cost of stops for cleaning and loss of product while cleaning. However, some products of biotechnology cannot be produced by continuous culture. Technical difficulties mean that batch culture is the only cost-effective option.

Figure 24.1C ⬇ Industrial biotechnology

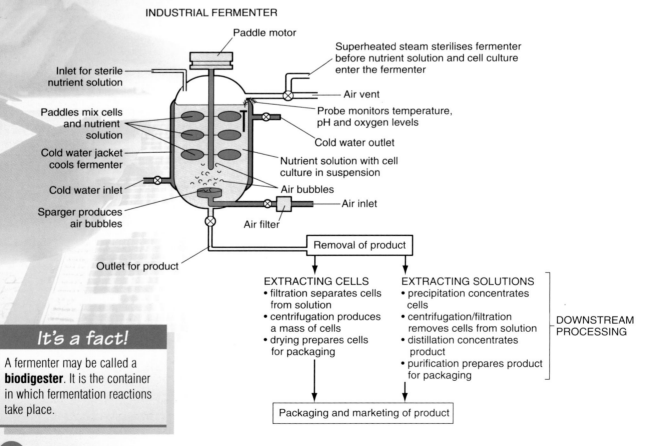

INDUSTRIAL FERMENTER

Paddle motor

Superheated steam sterilises fermenter before nutrient solution and cell culture enter the fermenter

Inlet for sterile nutrient solution

Air vent

Probe monitors temperature, pH and oxygen levels

Paddles mix cells and nutrient solution

Cold water outlet

Cold water jacket cools fermenter

Nutrient solution with cell culture in suspension

Air bubbles

Cold water inlet

Air inlet

Sparger produces air bubbles

Air filter

Outlet for product

Removal of product

EXTRACTING CELLS
- filtration separates cells from solution
- centrifugation produces a mass of cells
- drying prepares cells for packaging

EXTRACTING SOLUTIONS
- precipitation concentrates cells
- centrifugation/filtration removes cells from solution
- distillation concentrates product
- purification prepares product for packaging

DOWNSTREAM PROCESSING

Packaging and marketing of product

It's a fact!

A fermenter may be called a **biodigester**. It is the container in which fermentation reactions take place.

Scaling up

There is a big jump from home wine-making to industrial biotechnology. Fermentation in a 10-litre demi-john usually looks after itself, with perhaps an occasional shake to help air to circulate. Scaling up to an industrial fermenter perhaps 25 000 times the size creates considerable problems:

- **Sterile** conditions in the fermenter and in the pipelines leading to and from the fermenter must be maintained. **Superheated** steam at around 120 °C is pumped through the system to kill unwanted microorganisms.

- **Oxygen** levels at the centre of the fermenter may drop, slowing the growth of the cell culture. Sterile air is fed into the bottom of the fermenter through a perforated metal ring or disc called a **sparger**. Air bubbles rise through the liquid. Motorised paddles stir the mixture.

- **Heat** generated by the fermentation reactions and motorised paddles can quickly raise the temperature inside the fermenter. Temperatures of more than 60 °C would kill the cell culture and damage the product. A cooling system, therefore, is vital.

Notice the label 'downstream processing' in Figure 24.1C. When fermentation is complete, **downstreaming** is the process that collects and purifies the product, and packages it. The product may be cells, substances that cells produce or substances that cells ferment from nutrient solutions. Some of the products and uses of biotechnology are investigated in the following sections.

SUMMARY

Ancient civilisations used the process of biotechnology to produce food. Modern biotechnology produces many useful substances and operates on an industrial scale.

CHECKPOINT

▶ 1 Describe the mode of action of three types of microorganism used to produce food (see *Key Science: Biology* Extension File).

▶ 2 A demi-john is a glass container used for making home-made wine. 'In principle an industrial fermenter is a large version of a demi-john'. Do you think this statement is correct? Give reasons for your answer.

▶ 3 What is 'downstream processing'?

24.2 ▶ Using biotechnology

FIRST THOUGHTS

Biotechnology is creating new industries which earn millions of pounds annually. This section tells you about some of the developments.

Before the First World War, glycerol and propanone (which are used in the manufacture of explosives) were made by using bacteria. After the war, the petrochemical industry expanded and the manufacture of glycerol and propanone from oil replaced these early examples of biotechnology. However, rising oil costs have revived interest in the production of these chemicals by biotechnology

In 1928, Alexander Fleming reported that the mould *Penicillium notatum* made a substance that killed bacteria. The substance was called penicillin: the first of a family of antibiotic drugs made by biotechnology.

See www.keyscience.co.uk for more about genetic engineering.

It's a fact!

Did Nicholas II, Russia's last Tsar, and his family escape the Bolshevik firing squad 75 years ago? Some people think so, but British scientists are virtually certain that the skeletons discovered in 1991 in a grave at Ekaterinberg in the Ural Mountains are those of Nicholas II, his wife the Tsarina, three of their children, the royal doctor and three servants. Using DNA technology, samples from bones of the skeletons were analysed for their DNA and genetic fingerprints prepared. The Duke of Edinburgh is a descendant through the female side of the Russian royal family. He provided blood and hair samples for DNA fingerprinting. The DNA taken from the Tsarina and the three children and the Duke's DNA matched. The Tsar's DNA also matched the DNA obtained from two of his descendants. Further DNA tests showed that the children in the grave were related to the Tsar and Tsarina. The chances that the people from whom the samples of DNA were taken are not related are millions to one against.

Science at work

The bacterium *Thiobacillus ferrioxidans* is used to extract copper from low grade ores (material which contains a small proportion of copper compounds). Water containing *Thiobacillus* is sprayed over the copper-containing ore. The bacterium converts the ore to copper sulphate. Copper is extracted by reacting the copper sulphate with scrap iron.

Modern biotechnology is branching out in new and exciting ways. Central to success is our understanding that the products of biotechnology are the result of the action of **genes** (see p. 152). We now know what genes are, how they work and how to manipulate them to our advantage.

The breakthrough occurred in the early 1970s when scientists developed the techniques of **genetic engineering** (see p. 391) which introduced the modern era of biotechnology. New methods of manipulating genes were possible following the discovery of different enzymes in bacteria.

- **Restriction enzymes** cut DNA into pieces, making it possible to isolate specific genes.
- **Ligases** (splicing enzymes) allow desirable genes to be inserted into the genetic material of host cells. Genetic engineering makes it possible to create organisms for producing products that we need and want.

DNA technology

A person's DNA is as unique as their fingerprints (see p. 391). DNA 'fingerprinting' can help to identify criminals in cases where the criminal's body cells are found at the scene of the crime. Except in the case of identical twins, the chances of two people having the same DNA 'fingerprint' are millions to one against (see p. 392).

Figure 24.2A shows how DNA 'fingerprints' can help to identify criminals. The X-ray film shows the DNA maps made from a semen stain found at the scene of a rape, the body cells of a suspect and the body cells of the victim. *Do you think the police have caught the right man?*

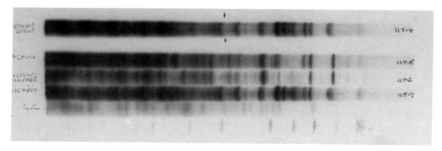

Figure 24.2A 🔺 DNA fingerprinting

Biosensors

An **antibody** will attach itself only to a particular antigen (see p. 262); an enzyme will catalyse only a particular reaction or group of reactions (see p. 168). **Biosensors** use the sensitivity of these reactions and microelectronic circuits to detect minute amounts of chemicals. Biosensors help scientists to diagnose disease and to monitor pollution in the environment.

Figure 24.2B shows a biosensor that is able to detect glucose levels in the blood. If this is coupled to a portable insulin pump, the glucose level in the blood of a **diabetic** can be continuously monitored, and the correct level of insulin maintained without the need for daily injection (see Topic 17.4).

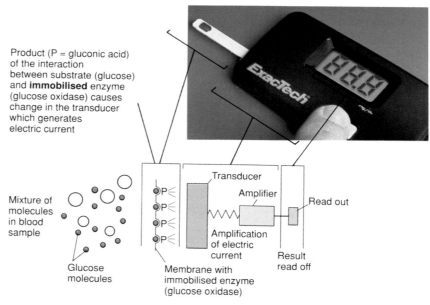

Product (P = gluconic acid) of the interaction between substrate (glucose) and **immobilised** enzyme (glucose oxidase) causes change in the transducer which generates electric current

Figure 24.2B ▲ How a biosensor works

Enzymes

Whoever wore the clothes about to go into the washing machine shown in Figure 24.2C is a careless eater! Notice the egg stains, grease marks and sticky jam. Also notice that the washing powder to be used is biological. It contains enzymes! The enzymes are sealed in **capsules** to protect them from the action of the detergent powder. Encapsulation also protects the people making the washing powder from possible harmful side-effects caused by the enzymes. *What do you think such side-effects might be? Give reasons for your answer. What are enzymes doing in washing powder? Which types of enzyme are best for washing food-stained clothing? Table 14.9 on p. 235 will help you answer these questions.*

Enzymes are **protein catalysts** which speed up chemical reactions taking place inside cells (see p. 169). Biotechnology depends on them! Most industrial enzymes come from microorganisms. Fungi and bacteria are grown in a nutrient solution inside large fermenters. Enzymes are secreted into the nutrient solution, then filtered off, concentrated and packaged for sale either as liquids or as powders (see p. 398).

Enzymes are very useful industrial catalysts because:

- only a particular reaction is catalysed, making it easier to collect and purify the product
- activity is high at moderate temperature and pH
- only small amounts are required
- the enzyme is not used up in the reaction – at the end of the reaction it is as good as new and can be used again.

Table 24.1 summarises the sources and uses of some important industrial enzymes.

The enzyme added to the substances whose reaction it catalyses may be lost when the product is collected. Dilution also occurs, making it difficult to recover the enzyme when the extraction is complete. Developments in the manufacture of enzymes have overcome these

Figure 24.2C ▲ Wash time with biological washing powder

Table 24.1 ▼ Some important industrial enzymes (B) = Bacterium, (F) = Fungus. Notice that carbonic anhydrase is derived from red blood cells

Use	Enzyme	Source
Industrial Food production, leather making, brewing and washing powder manufacture	*Chymosin:* • clots milk, producing curds which is used to make cheese *Amylase:* • converts starch into glucose producing syrup for the food industry • used in biological washing powders *Glucose isomerase:* • converts glucose into fructose, which is used as a sweetener in foods *Proteases:* • 'clear' beer of yeast haze • coagulate milk for cheesemaking • remove hairs from skins used to make leather	*Escherichia coli* (B) *Bacillus subtilis* (B) *Aspergillus oryzae* (F) *Bacillus coagulans* (B) *Streptomyces spp* (B) *Bacillus subtilis* (B)
Medical Diagnosis and treatment	*Glucose oxidase:* • indicates glucose levels in blood *L-asparaginase:* • breaks down the amino acid L-asparagine required for the growth of cancers *Streptokinase:* • dissolves blood clots and cleans wounds	*Aspergillus niger* (F) *Escherichia coli* (B) *Streptomyces* spp (B)
Analysis Environmental pollution and crime detection	*Carbonic anhydrase:* • detects insecticides *Urease:* • indicates amount of urea in urine and blood	Red blood cells *Bacillus pasteurii* (B)

Check the advantages of using immobilised enzymes.

problems. Various insoluble materials are used to bond to the enzyme, so that the enzyme can still catalyse reactions but remains attached to the insoluble support and is not lost when the products are collected. The enzyme is said to be **immobilised**.

Immobilised enzymes are:

- easily recovered and can be used again and again
- often active at temperatures that would destroy the unprotected enzyme. For example, immobilised *glucose isomerase* (see Table 24.1) is stable at 65 °C whereas unprotected *glucose isomerase* is soon destroyed at 45 °C
- not diluted and therefore do not contaminate the product.

Remember that the substance which an enzyme helps to react is called the **substrate**. The substance formed in the reaction is the **product**.

Immobilised enzymes make it easier to produce products by continuous culture (see p. 393). Substrate flowing over immobilised enzyme inside a fermenter is converted into product. The product is drawn off from the fermenter as an *ongoing* process.

Immobilised enzymes are the biological heart of different types of biosensor. For example immobilised **glucose oxidase** is a vital part of the biosensor shown in Figure 24.2B. The enzyme catalyses the reaction:

$$\text{Glucose} + \underset{\text{Oxygen}}{O_2} + \underset{\text{Water}}{H_2O} \xrightarrow{\textit{Glucose oxidase}} \text{Gluconic acid} + \underset{\substack{\text{Hydrogen}\\\text{peroxide}}}{H_2O_2}$$

The acid produced conducts an electric current which is proportional to the amount of glucose in solution – an important measurement for people suffering from diabetes.

It's a fact!

Most hard cheese is made using the enzyme chymosin produced by genetically engineered bacteria, in place of rennet taken from calves' stomachs. The genetically modified cheese is therefore suitable for vegetarians (see p. 220).

Biofuel

The sun floods the Earth with light energy. Some of the light energy is absorbed by chlorophyll and converted into the chemical bond energy of sugars by the reactions of photosynthesis (see Topic 11.1). The sugars (mainly sucrose) formed are the source of energy which plants use to live and grow.

Although they are a source of energy, sugars are little use as a fuel. Sucrose (and glucose) is very unreactive (see *Key Science Chemistry* p. 233) and does not burn easily. Easy burning (combustion) is a key feature of fuels (see *Key Science Chemistry* p. 146). Ethanol does burn easily and has great fuel potential! Yeast cells ferment sugar to ethanol (see p. 244). Sugar, therefore, is a raw material for fuel which in theory could replace petrol in the engines of cars, lorries and other vehicles.

Sugar cane contains around 12–14% sucrose. It is therefore a rich source of sugar for yeast to ferment into ethanol. Brazil is sunny and warm and therefore an ideal place for growing lots of sugar cane. Since the 1930s successive governments have worked on turning the theory of ethanol-based fuels into practice. Low sugar prices and rising fuel costs in the 1970s encouraged development of ethanol-based fuel and the engines to run on it. The work was supported through the National Ethanol Programme (called Proálcool). However, the discovery of off-shore oil fields changed their priorities. With low-cost crude oil now readily available, government encouragement for Proálcool fell away.

There are around 12 million cars on the roads of Brazil. Exhaust emissions add greenhouse gases to the atmosphere and pollute the air of Brazil's towns and cities (see p. 103). Smog and the threat of global warming (see Topic 6.7) has helped to re-awaken government support for Proálcool. Ethanol-based fuels burn more 'cleanly' than petrol alone. Ethanol is now being blended with petrol as a fuel called 'gasohol' (see Figure 24.2D). *Will gasohol help reduce the problems of smog in cities?* Proálcool hopes so! The success of gasohol depends on its environmental benefits. Less pollution will also help Brazil contribute to the global reduction of greenhouse gases (see Topic 6.7).

Figure 24.2D

SUMMARY

Biotechnology is creating products for industrial, medical and analytical purposes. The new processes depend on the activities of microorganisms.

CHECKPOINT

▶ 1 How does a biosensor work? Explain why biosensors help scientists diagnose disease and monitor pollution in the environment.

▶ 2 Why are enzymes useful industrial catalysts?

▶ 3 What is an immobilised enzyme?

▶ 4 'Plants represent a store of energy'. Explain this statement and describe how biotechnology converts the stored energy in plants into fuel.

24.3 ▶ Biotechnology on the farm

Around 6 billion people populate the world. Feeding everybody is difficult. Biotechnology, however, is solving some of the problems. Developments include:

- genetically engineering crops to grow in places where at present there is little chance of success
- altering nitrogen-fixing bacteria so that they can live in the roots of cereal crops

FIRST THOUGHTS

Strawberries growing in winter, kid goats born to sheep, potato and tomato plants joined to form a 'pomato' – what other changes are taking place on the farm because of biotechnology? Read this section and find out.

See www.keyscience.co.uk for more about gene transfer.

EXTENSION FILE
ASSIGNMENT

Agrobacterium tumefaciens is a useful bacterium. It can be used to introduce desirable genes into host cells.

- designing insecticides, produced by bacteria, which are selective for particular insect pests
- producing plants resistant to disease
- developing livestock to produce more and better quality meat and milk.

Can we look forward to a future where crops tolerate cold, flourish in drought conditions and resist insects and disease? Think carefully about your vision of what is possible. Make a list of your ideas.

Genetic engineering and *Agrobacterium tumefaciens*

Scientists use genetic engineering (see p. 400) to isolate desirable genes and insert them into crop plants. The bacterium *Agrobacterium tumefaciens* is an ideal cell for introducing desirable genes into host cells. Figure 24.3A shows the technique.

Notice that cells of *Agrobacterium* each have a loop of DNA called the **Ti plasmid**. Also notice that a desirable gene has been inserted into the *Ti* plasmid. In other words, the *Ti* plasmid has been **genetically engineered**. As a result the bacterial cell itself is **genetically modified (GM)**. The plant is infected with modified *Agrobacterium* and, because of the *Ti* plasmid, produces a cancerous growth (**tumour**) called a **crown gall**. Cells in the gall each contain the engineered *Ti* plasmid with the desirable gene in place. Plantlets can be cultured from small pieces of tissue (**explants**) cut out of the gall. The plantlets are genetically identical, forming a **clone** (see p. 341). Each one carries the engineered *Ti* plasmid with its desirable gene and is therefore genetically modified. Modified plantlets are transferred to soil where they grow into a mature crop, helped by the characteristic (in this case, resistance to herbicide) which the gene inserted into the *Ti* plasmid controls.

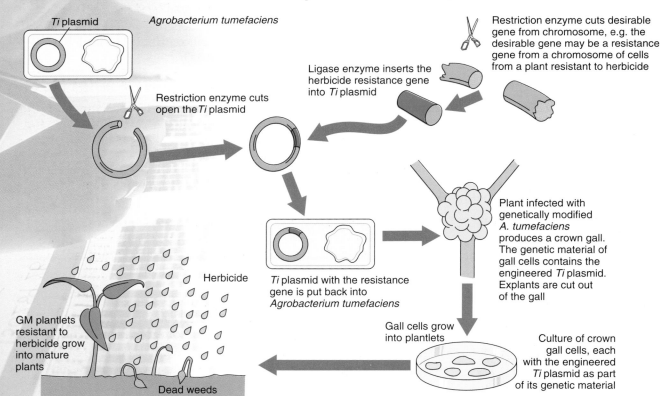

Figure 24.3A ⬆ Engineering *Agrobacterium tumefaciens*

If nitrogen-fixing bacteria could be made to live in the roots of cereal plants, the huge amounts of nitrogen fertiliser used to boost crop productivity would not be needed. Pollution of the environment would also be reduced.

Nitrogen fixation

Nitrogen (N_2) is an essential element of proteins, DNA and RNA. Crop growth depends on the availability of nitrogen. Different species of the bacterium *Rhizobium* live in swellings (called **nodules**) on the roots of **leguminous** plants (beans, peas, soybeans, clover). Each nodule contains millions of bacteria which convert nitrogen into nitrates. This process is called **nitrogen fixation** (see p. 55). The host plant uses the nitrates (as a source of nitrogen to make protein) and provides the bacteria with sugar.

Cereal crops account for 50% of food supplies world-wide. Unfortunately they do not have nitrogen-fixing bacteria. As a result, huge amounts of nitrogen fertiliser are applied to crop plants to boost growth and productivity. Surplus fertiliser runs off the land and pollutes drinking water, with serious consequences for the environment (see p. 115).

Biotechnology offers alternative strategies. For example, increasing the range of plants which associate with *Rhizobium* is one idea under investigation. Success depends on understanding the mechanism by which *Rhizobium* recognises legume root tissue and using this knowledge to achieve the same result with non-leguminous plants like cereals.

Other ideas involve manipulating the set of about 12 genes (called the **nif genes**) which control nitrogen fixation in *Rhizobium*. The options include inserting the nif genes into:

- suitable cereal plants
- bacteria which already live in association with cereal plants
- the *Ti* plasmid of *Agrobacterium tumefaciens*, and then infecting suitable plants with the modified bacterium, causing crown gall formation.

Looking at Figure 24.3A once more will remind you of the idea. *How do you think food production would benefit from this type of technology? Do you think there are risks to human health and the environment? Make a list of your ideas.*

Bacterial insecticides

Insects quickly develop resistance to conventional insecticides (see Topic 5.5). More and more insecticide has to be used until its inefficiency and expense forces the farmer to switch to another compound. Resistance to the new insecticide soon appears and the cycle starts again. Scientists are in a race between resistance developing and finding new chemicals with which to control insect pests. Even so, insect damage to crops costs $billions and £billions per year worldwide.

Bacterial insecticides are safer to use and environmentally more acceptable than conventional insecticides. New products are available for farmers to use but conventional insecticides are likely to remain an important weapon against insect pests for some time.

The bacterium *Bacillus thuringiensis* kills leaf-eating caterpillars and the larvae of flies and mosquitoes. For example, it kills the codling moth, a pest of apples and pears, and the cabbage looper moth, a pest of lettuces, broccoli, cabbages and potatoes. The damage is caused by a poison called **insecticidal crystal protein (ICP)**, which is produced by the bacterium. ICP attacks the gut lining, the caterpillar stops feeding and eventually dies.

Batches of *Bacterium thuringiensis* are fermented in large quantities. The product is drawn off, mixed with a sticky substance and sprayed on to crops as a mixture of bacterium and poisonous ICP.

Unfortunately *Bacterium thuringiensis* is a poor survivor when exposed to the weather. As an alternative to spraying, therefore, plants are modified to carry the gene for the ICP poison. The gene is inserted into the *Ti* plasmid (see p. 404). *Draw a flow diagram to show the method. Look once more at Figure 24.3A to help you with your task.*

Cold tolerance

Late frosts in spring-time often damage crops. The bacterium *Pseudomonas syringae* is nearly always found on the surfaces of leaves. It feeds on frost-affected plants and damages them. It contains a gene which is the code (see p. 152) for the production of a protein around which ice crystals form at temperatures between 0 °C and −7 °C.

If the bacterium is not present the plants are unaffected by ice crystals unless the temperature drops below −7 °C. Scientists have produced a strain of *P. syringae* without its 'ice-protein' gene. This strain competes and replaces wild-type 'ice' bacteria when it is sprayed over crops. Frost-sensitive crops like strawberries are less vulnerable to damage after the treatment and can be grown at a time earlier in the year when frosts are more likely and would kill untreated plants.

Protoplasts take up genes more easily than plant cells with their tough walls intact

Protoplasts

Protoplasts are plant cells with their walls removed. Figure 24.3B shows the idea.

Protoplasts will take up substances added to the medium in which they are growing. For example, scientists have incorporated genes into protoplast DNA by adding genes to the growth medium. The protoplasts grow into complete cells which then divide. Treatment with **auxins** (see Topic 13.1) stimulates shoot and root development and whole plants can be raised carrying the genes engineered into the original protoplasts. *Think of some of the benefits of this technique. Make a list of your ideas.*

Protoplasts can also be fused together (Figure 24.3C) and grown to produce hybrid (see p. 8) plants with new mixtures of characteristics. The **pomato** mentioned in *FIRST THOUGHTS* was produced in the late 1970s as an early demonstration of the technique. The possible benefits of protoplast fusion include overcoming diseases, such as potato leaf roll disease, shown in Figure 24.3D.

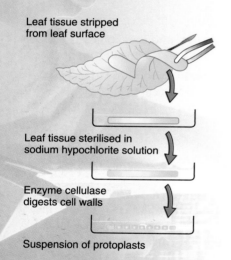

Leaf tissue stripped from leaf surface

Leaf tissue sterilised in sodium hypochlorite solution

Enzyme cellulase digests cell walls

Suspension of protoplasts

Figure 24.3B ⬆ Preparing protoplasts

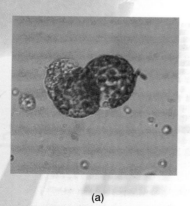

(a)   (b)

Figure 24.3C ⬆ Fusing protoplasts: (a) Early stage (b) Late stage

Figure 24.3D ⬆ Potato leaf roll disease

It's a fact!

A **transgene** is a gene which is 'foreign' to the cell into which it is transferred. A transgenic organism contains a transgene which is 'foreign' to its normal genotype (see p. 379). **Microinjection** is often used to produce transgenic animals. A very fine glass-needled syringe is used to inject the transgene into a fertilised egg. With luck the transgene becomes part of the egg's DNA. Mitosis (see p. 156) means that the transgene is inherited by all cells of the embryo as it develops into a new individual.

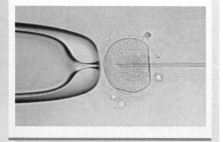

Improving animals

The sheep shown in Figure 24.3E are **transgenic**. Genetic engineering has inserted the human gene for **Factor VIII** (an essential blood-clotting protein) into their DNA. The sheep will make Factor VIII which can be drawn off in the milk, purified and used to treat **haemophilia**. *Why are new sources of blood products important? Think about blood-borne diseases such as AIDS before making a list of your answers.*

New techniques which manipulate embryos are also helping to improve animals to human advantage. In simple terms, eggs from donor animals are fertilised in the laboratory and the young embryos which develop are split up before the cells become specialised. The cells are transplanted into different regions of the uterus of a **surrogate** mother – so-called because the transplanted cells are not her own. The cells divide and embryos develop normally. Eventually the surrogate mother gives birth to a number of youngsters where only one or perhaps two offspring would have been normal. In this way rare breeds of a particular type of animal may be conserved, or new breeds with desirable characteristics developed reliably and quickly using surrogate mothers of a common breed.

A variation of the technique allows surrogate mothers to carry embryos of a different species. Figure 24.3F shows a surrogate eland with her baby bongo. The technique is useful because the fertilised eggs of rare species can be cooled and preserved for years after the original parents have died. When a suitable surrogate is available the fertile eggs can be transplanted.

Figure 24.3E ▲ Transgenic sheep

Figure 24.3F ▲ Eland (a common species of antelope) has given birth to a baby bongo (a rare species of antelope)

Using surrogate mothers of common species helps conserve rare species.

These methods are an increasingly important source of **rare** species. Zoos are developing the ideas to help **conservation** (see Topic 8). Their breeding programmes build up numbers of rare stock and arrange for their release in the wild.

Embryo cloning produces many genetically identical copies of an animal. Although not yet common practice for farm animals, laboratory studies point the way for future developments. Imagine that a mutation that occurs in a cow makes her an exceptional milk producer. Conventional breeding techniques would reshuffle her genes with the risk that the desirable mutation would be lost. Cloning to produce identical copies of the cow will conserve her unique milk-producing abilities for future generations.

Biotechnology is also producing healthier farm animals which produce more meat, milk and fibre for clothes manufacture. For example, the gene for **bovine somatotropin (BST)** has been genetically engineered into bacteria. As a result large amounts of this hormone are available for injection into cows.

BST stimulates growth and increases milk production. The meat from treated cows carries less fat, but there are worries about the long-term effects on human health. Analysis shows traces of BST in the meat and milk. Although we break down BST in the gut and the hormone seems to be active only in cattle, public concern remains. *Where do you think the balance lies? In Britain, the use of BST is not allowed. Even though there is no evidence of ill-effects, do you think people's worries are sufficient reason for abandoning developments? Do you think it cruel to treat animals in this way? Carefully consider the arguments before listing your answers.*

CHECKPOINT

▶ **1** How is biotechnology helping to solve some of the problems of feeding the world's growing human population?

▶ **2** Describe the discoveries made in the early 1970s which marked the beginning of genetic engineering.

▶ **3** Why is the bacterium *Agrobacterium tumefaciens* an ideal cell for introducing desirable genes into host cells?

▶ **4** How does the bacterium *Bacillus thuringiensis* kill insect pests?

▶ **5** What is the advantage of developing crops that are able to tolerate cold weather?

▶ **6** Describe how biotechnology is helping zoos to conserve rare animal species.

24.4 ▶ Single-cell protein

FIRST THOUGHTS

Eating microorganisms – bacteria, algae and fungi – may seem odd, but it could help to feed the world's growing population. Biotechnology is the key to new types of food based on microorganisms.

Food! We all need it, but how can we produce enough food to feed the world's growing population? Farmers try to produce as much food as possible by farming land intensively. Eating microorganisms produced by biotechnology is another option.

Think about these facts:

● Microorganisms double their mass within hours. Plants and animals may take weeks.

● Microbial mass is at least 40% protein and has a high vitamin and mineral content.

Eating microorganisms may not seem so odd after all! In the light of the facts, do you agree with this statement? Give reasons for your answer.

High-protein food produced from microorganisms is called **single-cell protein (SCP)**. The idea is not new. During the First World War, food shortages in Germany were overcome with the help of scientists who discovered how to grow large amounts of yeast to fill out sausages and soups. 'Marmite' and similar spreads made from yeast are firm favourites at tea-time. Yeast left over from brewing beer is turned into animal feed for livestock on the farm.

SCP production

In the 1960s, oil was used as a nutrient for growing SCP microorganisms. However, massive increases in the oil prices in the 1970s made SCP production based on oil uneconomical. Today glucose syrup, fruit pulp, waste from paper making, agricultural waste and sewage are just some examples of the wide range of nutrients used to grow SCP microorganisms.

Large fermenters capable of producing thousands of tonnes of SCP per year under sterile conditions are run in continuous culture (see p. 398) for months at a time. Nutrients are replaced as they are used up, and the temperature and pH carefully controlled. Microorganisms are harvested at regular intervals and processed.

Quorn – a case study

Figure 24.4A shows the mould *Fusarium graminearum*. With 45% protein content and 13% fat content, *Fusarium* is as nutritious as meat with the added advantage that, unlike meat, it is high in fibre and cholesterol-free. Hyphae of the mould *Fusarium* form a highly nutritious food called **mycoprotein**.

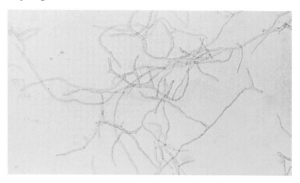

Figure 24.4A 🔺 Hyphae (see p. 18) of the mould *Fusarium graminearum*

'Quorn' is the market name for mycoprotein. It is high in fibre, has a high protein content, low fat content and is cholesterol-free.

Mycoprotein is marketed as 'Quorn'. The product looks like sheets of moist uncooked pastry. Careful control of hyphal length during fermentation and the addition of colours and flavours results in a range of food including biscuits, soups and drinks, as well as substitutes for ham and chicken (Figure 24.4B). Quorn also blends well with meat products such as sausages and readily absorbs flavours from any sauces used for cooking.

Figure 24.4B 🔺 Quorn, pineapple and coconut curry. The chunks which look like meat are Quorn

Is SCP a sign of the future? Food can be grown in large quantities in fermenters which take up little space and in a controlled environment independent of the weather. The technology is established and can be adapted for developing countries that experience food shortages. *Do you think people might be reluctant to eat food made from microorganisms? Devise ways of making SCP acceptable. Discuss the contribution that SCP production could make to solving the world's food crisis.*

24.5 ▶ Biotechnology and medicine

Biotechnology has enormous potential in the field of medicine. Cancer and heart disease, for example, are responsible for more than 50 per cent of deaths in the developed countries of the world. Biotechnology will play a key role in the prevention, diagnosis and treatment of these and other diseases.

The human hormone **insulin** (see Topic 17.4) was the first substance made by genetic engineering to be given to humans. In 1980, volunteer diabetics successfully tried out genetically engineered insulin and by 1982 it was in general use. Before then, insulin was obtained from slaughtered cattle and pigs. It was expensive to produce and in limited supply. The chemical structure of animal insulin is different from human insulin; some diabetics reacted allergically to it. Genetically engineered insulin is cheaper, available in large quantities and chemically the same as human insulin. Figure 24.5A explains how scientists use genetic engineering to make insulin.

Genetic engineering is used to make hormones and different medicines.

Other genetically engineered hormones, including **human growth hormone** and **calcitonin** (the hormone that controls the absorption of calcium into bones), are being produced by methods similar to those for producing insulin. Using human genes to produce substances like hormones helps to prevent the harmful side-effects that can come from products obtained from animal tissues. It also reduces the use of animals for medical research.

Monoclonal antibodies

White blood cells produce millions of antibodies to defend the body from attack by bacteria, viruses, fungi and other potentially dangerous antigens. It is difficult to separate different antibodies into pure samples of the antibodies required to fight specific antigens (see p. 262). The problems can be overcome by fusing B cell lymphocytes that produce a particular antibody with a type of rapidly dividing cancer cell. The fused cells only produce the antibody required. Pure samples of antibodies made in this way are called **monoclonal antibodies**.

Monoclonal antibodies are made by fusing B cell lymphocytes that produce a particular antibody with a type of rapidly dividing cancer cell.

Monoclonal antibodies have a wide range of uses. Scientists hope to be able to use monoclonal antibodies to treat cancer. Some types of cancer cell make proteins (antigens) that are different from the proteins made by healthy cells. Monoclonal antibodies that attach to only the abnormal proteins have been made. They target the cancer cells with drugs without affecting the healthy cells (Figure 24.5B). Monoclonal antibodies can be put to other uses:

- The urine of pregnant women contains **human chorionic gonadotrophin (hCG)**. This hormone, produced by the placenta, can be detected with monoclonal antibodies as early as 12 days after fertilisation. A pregnancy test kit consists of a dipstick impregnated with antibodies. The position of the coloured band after dipping the stick into a urine sample tells the woman whether or not she is pregnant. If hCG is present in the urine, the molecules bind to the antibodies at the end of the dipstick. The hCG–antibody complex diffuses up the dipstick and meets a band of antibodies which also bind hCG. Colour change at this level shows that the woman is pregnant. If there is no hCG in the urine, antibodies alone diffuse further up the dipstick to meet a second band of different antibodies. The antibodies bind together producing a colour change which shows that the woman is not pregnant.

It's a fact!

Restriction enzyme cuts DNA, with the desired gene in place, into fragments of different length. The sequence (order) of codons (see p. 152) on the fragment which is the desired gene is identified by the sequence of complementary codons (see p. 152) on a length of DNA which forms the **gene probe**. The codons of the probe join with the codons of the desired gene according to the rules of base pairing (see p. 171). The probe is labelled with a radioactive tag or fluorescent dye. The labelling identifies the gene/probe complex.

See www.keyscience.co.uk for more about gene transfer.

The same restriction enzyme used to cut out the insulin gene is used to cut open a loop of bacterial DNA. Loops of DNA are called **plasmids**, and plasmids are often found in bacterial cells (see p. 14). When a gene from another type of organism is inserted into a plasmid, the DNA is a mixed molecule (part from the bacterium, part from the other type of organism). We call the DNA **recombinant**. The plasmid itself is said to be genetically engineered (see p.404).

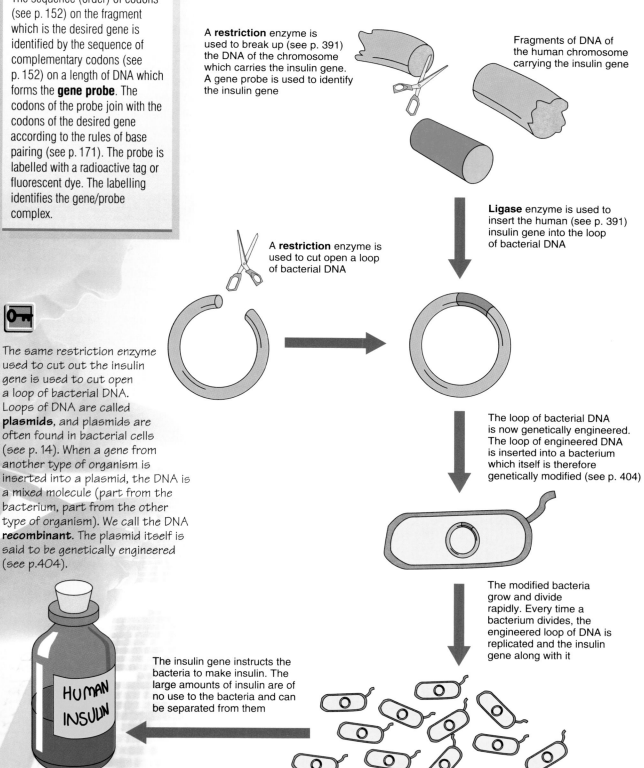

A **restriction** enzyme is used to break up (see p. 391) the DNA of the chromosome which carries the insulin gene. A gene probe is used to identify the insulin gene

Fragments of DNA of the human chromosome carrying the insulin gene

Ligase enzyme is used to insert the human (see p. 391) insulin gene into the loop of bacterial DNA

A **restriction** enzyme is used to cut open a loop of bacterial DNA

The loop of bacterial DNA is now genetically engineered. The loop of engineered DNA is inserted into a bacterium which itself is therefore genetically modified (see p. 404)

The modified bacteria grow and divide rapidly. Every time a bacterium divides, the engineered loop of DNA is replicated and the insulin gene along with it

The insulin gene instructs the bacteria to make insulin. The large amounts of insulin are of no use to the bacteria and can be separated from them

HUMAN INSULIN

Figure 24.5A ◀ Making genetically engineered insulin

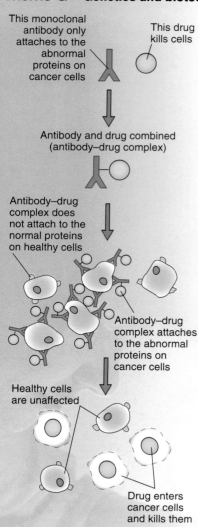

This monoclonal antibody only attaches to the abnormal proteins on cancer cells

This drug kills cells

Antibody and drug combined (antibody–drug complex)

Antibody–drug complex does not attach to the normal proteins on healthy cells

Antibody–drug complex attaches to the abnormal proteins on cancer cells

Healthy cells are unaffected

Drug enters cancer cells and kills them

Figure 24.5B How monoclonal antibodies are used to treat cancer

It's a fact!

The word antibiotic means 'against life'. Antibiotic drugs kill bacteria. However, they do not affect viruses.

Antibiotics are drugs used to combat diseases caused by bacteria.

- Monoclonal antibodies bind with poisons, inactivating them. Tetanus toxin (see Topic 16.1) and overdosage of digoxin (a drug used to treat heart disease) have been inactivated with different types of monoclonal antibody.

- The success of transplants depends on matching the tissue of donor and patient as nearly as possible. The closer the match, the less chance there is of **tissue rejection** (see Topic 16.2). Monoclonal antibodies, produced against the cell surface proteins of the donor's and patient's tissues, make tissue matching more accurate. *Explain how monoclonal antibodies improve the accuracy of tissue matching. Reading p. 269 will help you.*

Developing different types of monoclonal antibody for medical and other uses is expensive. However, the potential for profit is considerable. As a result, there are many new companies developing products in response to the demand.

Antibiotics

Penicillin was the first of a family of antibiotic drugs made by biotechnology. Today hundreds of different antibiotics are used to combat diseases caused by bacteria. Many more substances with antibiotic activity have been discovered. However, most are too expensive for commercial production and/or have harmful side-effects. Nevertheless, some have their uses outside medicine, in research, agriculture and food production.

There are four groups of antibiotic:

- **penicillins** – commercially produced penicillin is a mixture of compounds. The main component is penicillin G which can be converted into other forms, each with a slightly different activity. For example, penicillin G is broken down by stomach acid, and is therefore best given by injection; **ampicillin** is unaffected by stomach acid and can therefore be given by mouth as tablets or in a syrup – something children may prefer! The range of penicillins allows medical staff to choose which type is the best for treating a particular disease. Choice also helps to combat the development of drug **resistance** (see p. 337).

- **cephalosporins** – cephalosporins made by the mould *Cephalosporium* were discovered in 1948. All are active against a similar range of bacteria. Newer cephalosporins are effective against bacteria which have developed resistance to penicillin.

- **tetracyclines** – tetracycline is made by the bacterium *Streptomyces aureofaciens*. Its various forms are active against a similar range of bacteria, although the development of widespread resistance has reduced effectiveness. Tetracyclines bind to calcium and are deposited in growing bones and teeth. *Why should tetracyclines not be given to children and pregnant women?*

- **erythromycins** – the activity range of erythromycin is similar to that of penicillin. Erythromycin, therefore, is useful against bacteria resistant to penicillin or where the patient is allergic to penicillin.

Sales of antibiotics make £billions annually world-wide. Some, like streptomycin, destroy a limited number of species of bacteria. They are **narrow-spectrum** antibiotics. Others, such as tetracycline, are **broad-spectrum** antibiotics. They act against a wide range of bacteria.

The story of penicillin

In the 1880s, new ways of staining cells and further development of the light microscope boosted research on microorganisms. Bacteria were clearly visible for the first time. The way was open for fresh discoveries, which eventually led to our conquest of many of the diseases caused by bacteria.

At the time, scientists noticed that laboratory cultures of bacteria did not grow on culture plates on which colonies of mould were also growing. Nobody followed up the observations until 1928 when the British bacteriologist Alexander Fleming (Figure 24.5C) noticed that some of his bacteria cultures were contaminated with mould. Figure 24.5D shows what happened. Fleming identified the mould as *Penicillium notatum* (Figure 24.5E) and reasoned that it produced a substance which killed bacteria. He isolated the substance and called it **penicillin**.

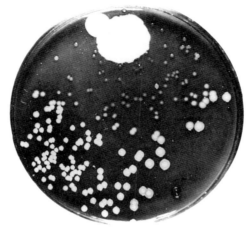

Figure 24.5D ⬆ The dish in which Fleming grew penicillin mould. Notice that colonies of bacteria do not grow near the mould because it secretes penicillin into its surroundings

Figure 24.5E ⬆ Hyphae and spore capsules (see p. 18) of *Penicillium*

Figure 24.5C ⬆ Alexander Fleming (1881–1955) in his laboratory at St Mary's hospital, London. He was awarded the Nobel Prize with Florey and Chain for working on penicillin

See www.keyscience.co.uk for more about penicillin.

Fleming thought that penicillin might help fight disease, but extracting it from the mould was very difficult. There matters rested until 1938 when Howard Florey and Ernst Chain, working at Oxford, made commercial production of penicillin possible. *Penicillium notatum* was replaced by *P. chrysogenum* which produces more penicillin. The scientists developed methods to produce enough penicillin for clinical trials, using all kinds of vessels (including milk bottles!) to contain the mould–sugar nutrient mixture.

By then (1940) the Second World War (1939–45) was in progress and there was an urgent need to treat wounded soldiers. Before penicillin, wounds often became infected with bacteria which caused **gangrene**, **blood-poisoning** and other fatal diseases. Penicillin prevented infections taking hold and troops could return to duty once they had recovered.

The work moved to the USA where large-scale production techniques were developed, using 'cornsteep liquor' (a waste product of the manufacture of starch from maize) as the nutrient solution for the growth of the penicillin mould. By 1944, enough penicillin was being produced to treat all the British and American casualties during the Normandy landings and invasion of Germany.

Penicillium mould begins to make penicillin after it has used up most of the nutrients in solution for growth.

Today the massive demand for penicillin world-wide is met by the development of new high-yielding strains of the mould. The yield is improved through genetic manipulation, selection, mutation and protoplast fusion (see Topic 24.3). Huge fermenters holding up to 200 000 dm³ of mould–nutrient solution produce penicillin by **batch culture** (see p. 393). The mould is then filtered off and the penicillin extracted.

Interferons

Have you ever suffered from the symptoms of 'flu? If so, **interferon** was responsible for your shivers, aches, fevers and tiredness. It was produced in response to the virus causing your illness.

Interferons are proteins produced by virus-infected cells. Interferons help healthy cells to break down viruses and prevent them from multiplying. Figure 24.5F shows you the idea.

Interferons break down viruses and prevent them from multiplying.

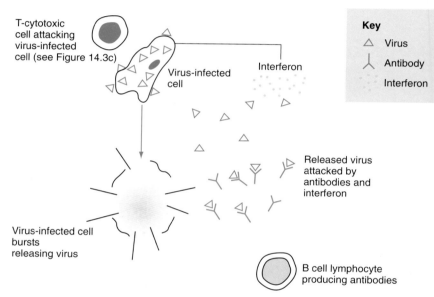

Figure 24.5F ▲ Interferon does not directly attack viruses, but interacts with host cells, helping them to protect themselves from the virus

Different types of human interferon have been isolated so far. Other animal interferons have also been identified. However, their action is **specific**. Interferon is only effective in the animal species that produces it.

Originally, interferon was extracted from white blood cells. Thousands of dm³ of blood were processed but yields of interferon were minute. At a cost of £10 million per gram, research designed to find out how interferon worked was slow and very expensive.

In 1980, the gene for one type of human interferon was isolated and inserted into the bacterium *Escherichia coli*. Prices fell sharply as much more genetically engineered interferon became available at a fraction of the previous production costs.

Hopes have run high that interferon would be the treatment for viral diseases that penicillin has proved to be for diseases caused by bacteria. Hopes remain, but turning interferon into the answer for viral infections is proving very difficult. Although the effect of interferon is outlined in Figure 24.5F, its actions are very complicated and incompletely known.

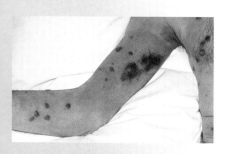

Figure 24.5G ⬆ Kaposi's sarcoma is a type of cancer. It is one of the diseases symptomatic of AIDS, as a result of HIV infection

Different interferons are available for medical use. Some products seem to be effective against certain types of viral disease. For example, the drug **Alferon**, produced by Interferon Sciences Inc. in the USA, shows promising results against the virus which causes **genital herpes**.

Other products treat different forms of cancer. For example, interferons are used to treat **HIV** infected patients who develop **Kaposi's sarcoma** (Figure 24.5G) *Do viruses cause some types of cancer? How does the effect of interferon on Kaposi's sarcoma help you to answer the question?*

Vaccines

Vaccines made by the methods described on p. 262 protect millions of people world-wide from serious diseases. However, there is a minute risk that:

- microorganisms used to make vaccines may survive the production process
- microorganisms used to make vaccines may regain the ability to cause disease
- patients may have an allergic reaction to the cellular debris that remains from vaccine production despite purification processes
- people working in production may come in contact with the dangerous organisms used to make vaccines despite strict safety precautions.

Genetic engineering is helping to reduce even these minute risks. The principles are:

- isolate genes from the disease organism responsible for producing the antigens that stimulate lymphocytes to produce antibodies
- insert the genes into a less dangerous organism
- culture the engineered organism, which makes antigen in large quantities
- extract the antigen, which is then used to make vaccine.

Difficulties arise in locating the genes which contain the code for the antigens most useful in stimulating lymphocytes to produce antibodies, and in identifying the antigens themselves. However, progress is being made. *Salmonella* bacteria have been engineered to carry the genes responsible for the cell surface proteins of the malaria parasite *Plasmodium* (see Topic 4.2). If the cell surface proteins are effective antigens then a vaccine made from them will be safer than one made from the whole organism.

Hepatitis B virus causes serious liver disease. Medical workers and patients in places where there may be human blood contaminated with the virus and families of hepatitis B carriers are at risk from the disease. They are high-priority groups for hepatitis B vaccine.

The vaccine is made from the blood plasma of people carrying the virus, but genetic engineering is taking over! Genes responsible for the protein coat (antigen) surrounding the virus have been isolated and inserted into yeast cells. Culturing the engineered yeast produces large quantities of virus coat protein which is used to make the vaccine. Figure 25.5H shows the method.

It's a fact!

Ten million people worldwide were infected with smallpox in the early 1970s. Since then a mass immunisation programme organised by the World Health Organisation has eliminated the disease.

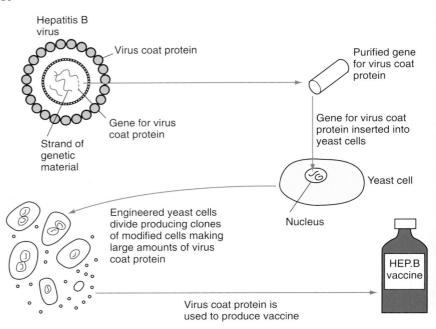

Hepatitis B virus

Virus coat protein

Strand of genetic material

Gene for virus coat protein

Purified gene for virus coat protein

Gene for virus coat protein inserted into yeast cells

Yeast cell

Nucleus

Engineered yeast cells divide producing clones of modified cells making large amounts of virus coat protein

Virus coat protein is used to produce vaccine

HEP.B vaccine

Figure 24.5H ⬆ Genetic engineering makes hepatitis B vaccine – conventional techniques include the genetic material of the virus in vaccine production. The genetic material is responsible for the virus reproducing itself, causing disease. Genetic engineering eliminates the risk

It's a fact!

People with sickle-cell trait (carriers of the recessive allele for sickle-cell anaemia) have greater resistance to malaria than individuals who are not carriers. This is why the sickle-cell allele persists in human populations living in regions of the world where malaria is endemic.

Gene therapy

First things first! Before continuing you need to *remember* that:

- any particular characteristic (e.g. height of pea plants) is controlled by pairs of genes called **alleles**
- a pair of alleles may be the same (**homozygous**) or different (**heterozygous**)
- a dominant allele masks the effect of its recessive partner (see p. 379).

Check the facts by reading pp. 380–383. They will help you to understand the exciting possibility of treating diseases by **gene therapy**.

We all carry a few faulty genes. Most of us are unaware of them because their dominant partners mask their harmful effects (see Topic 23.2). For example, **sickle-cell anaemia** is caused by a single mutation on the gene controlling the synthesis of the blood pigment haemoglobin (Figure 24.5I). In different regions of Africa up to 30 per cent of the population may carry the recessive allele for sickle-cell anaemia but its effect is masked by its normal partner. Carriers of the faulty allele are heterozygous and do not develop the disease. However, an individual who inherits the allele for sickle-cell anaemia from each parent is homozygous and soon develops the disease. He or she is unlikely to survive childhood.

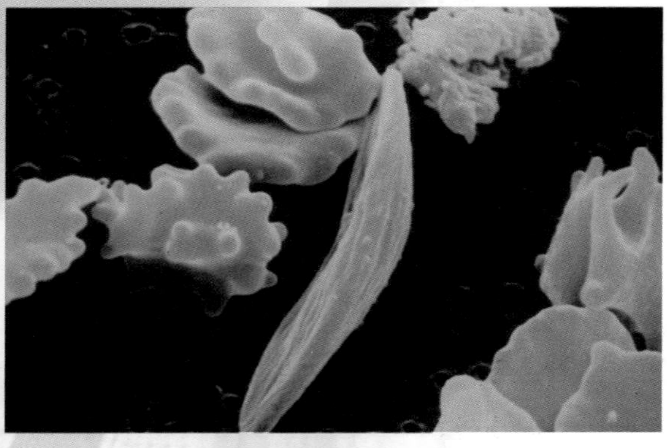

Gene therapy to control sickle-cell anaemia would aim to add normal genes to the patient's genetic make up. Before gene therapy can begin, the faulty gene must be identified. **Gene probes** (see p. 411) are powerful tools used in the hunt. They are designed to bind to specific portions of DNA and can distinguish between faulty genes and their normal partners.

Figure 24.5I ⬆ The disc-shaped red blood cells carry normal haemoglobin. The cells shaped like the blade of a sickle contain abnormal haemoglobin. Sickle cells do not carry as much oxygen as normal cells, so the affected person soon becomes breathless and tires easily (×1650)

CHECKPOINT

▶ 1 What are monoclonal antibodies? Describe how monoclonal antibodies could be used to treat cancer.

▶ 2 Penicillin is an antibiotic drug. Different types of penicillin each have a slightly different activity. Why is it useful for medical staff to be able to choose between different types of penicillin for treating patients?

▶ 3 How does interferon work?

▶ 4 Vaccines protect millions of people world-wide from serious diseases. However, there are minute risks attached to their use. What are the risks? How is genetic engineering helping to reduce the risks?

▶ 5 Briefly discuss the principles behind treating disease by gene therapy.

SUMMARY

Biotechnology is advancing our understanding and treatment of disease. New products and techniques of diagnosis will help to prevent and possibly eliminate many serious diseases.

24.6 ▶ Treating sewage

Have you thought where wastes and water go when you flush the toilet, pull the plug after a bath or finish the washing up? Where do you think rainwater runs to off the streets? What happens to waste chemicals and other materials from factories? Make a brief summary of your ideas.

Before there were drains and sewage works, sewage went directly into streams or open ditches which emptied into the nearest river. Drinking water supplies were often contaminated with microorganisms causing diseases such as cholera, typhoid (see p. 82) and diphtheria. Microorganisms decompose the organic matter in sewage, but in doing so use up the water's oxygen (called the **Biological Oxygen Demand – BOD** – see p. 115) and release poisonous substances like ammonia. Wildlife dies through lack of oxygen and/or poisoning. Anaerobic bacteria, which function without oxygen, continue the decomposition of waste organic matter, blackening the water and producing evil-smelling gases. Figure 24.6A gives you the idea.

Managing sewage

Raw sewage is a health hazard. Microorganisms help us to manage sewage by breaking it down into harmless and even useful substances. Treatment occurs in sewage works. Biotechnology, engineering and computer control are harnessed to deal with the wastes that pour in from homes, factories and gutters and drains which collect water and debris from every street. Figure 24.6B shows what happens.

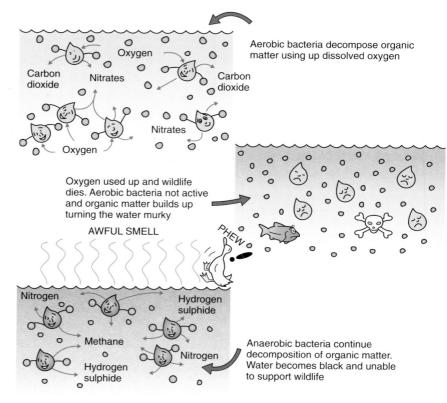

Figure 24.6A 🔼 Aerobic and anaerobic bacteria decomposing waste organic matter

Key to Figure 24.6B

1. **Sewers** – waste from homes, industry and the streets passes into a network of underground sewers which run to the treatment works

2. **Pumphouse and screen** – sewage passes through closely spaced metal bars which remove rags, paper, wood and large objects that would otherwise damage the machinery

3. **Grit removal** – sewage trickles through channels. Grit and sand in suspension settle out, are washed, collected and used to fill holes in the ground

4. **Primary settling tanks** – sewage flows into large tanks where solids settle to the bottom forming **crude sludge**. Electrically powered scrapers load crude sludge for transfer to sludge 'digestion' tanks (7)

5. **Aerobic processing** – liquid (called **primary effluent**) passes from the primary settling tanks to the **secondary treatment plant**. Here a great variety of microorganisms (bacteria, protists, fungi) break down the organic material in the primary effluent into minerals, gases and water. Their activities require a lot of oxygen. The main reactions are:

$$\text{Carbon compounds} \rightarrow \underset{\text{Carbon dioxide}}{CO_2}$$

$$\underset{\text{Hydrogen sulphide}}{H_2S} \rightarrow \underset{\text{Sulphates}}{SO_4^{2-}}$$

$$\underset{\text{Ammonium compounds}}{NH_4^+} \rightarrow \underset{\text{Nitrates}}{NO_3^-}$$

There are two systems:

a) **Activated sludge process** – diffusers bubble air through the primary effluent. It takes about 8 hours for the microorganisms to convert most of the organic material present into simpler substances

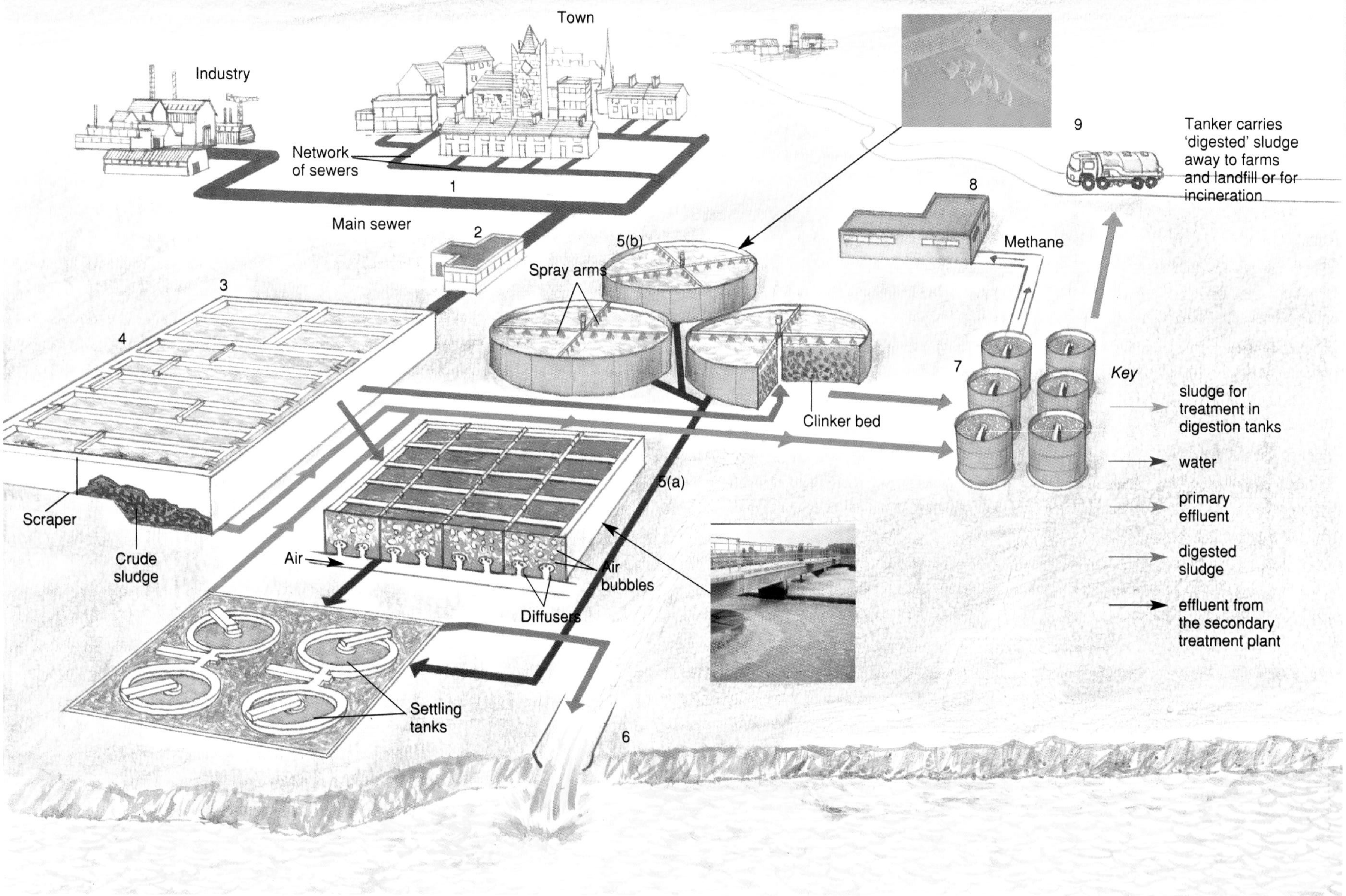

Industry

Town

Network of sewers

Main sewer

1

2

3

4

Scraper

Crude sludge

Air

Diffusers

Air bubbles

Settling tanks

5(a)

5(b)

Spray arms

Clinker bed

Methane

8

7

6

9

Tanker carries 'digested' sludge away to farms and landfill or for incineration

Key

sludge for treatment in digestion tanks

water

primary effluent

digested sludge

effluent from the secondary treatment plant

Figure 24.6B ◀ Managing sewage – large sewage works use the activated sludge process to treat primary effluent; the trickling filter process is used in smaller sewage works

FIRST THOUGHTS

The earliest human settlements were always beside rivers. The settlers needed water to drink and used the river to carry away their sewage and other waste. Obtaining clean water is more difficult now.

b) **Trickling filter process** – primary effluent is sprayed from slowly rotating horizontal arms on to a filter bed of clinker, gravel or moulded plastic contained in a concrete tank. A film of microorganisms covers the filter bed. As effluent slowly trickles over and through the filter bed, the microorganisms break down the organic matter. The reactions are similar to those of the activated sludge process. Protists such as *Vorticella* feed on the bacteria and fungi preventing them from smothering the filter and slowing down the process of breakdown.

Effluent from the activated sludge process or the trickling filter process passes into **settling tanks** where microorganisms and any remaining organic material separate from the water, forming a sludge. Some of the sludge is used for the next batch of aerobic processing, the rest is sent to the sludge 'digestion' tanks (7)

6. **Outfall** – water from the settling tanks is usually clean enough to go into a river. If the water needs to be especially clean before discharge it is filtered through beds of fine sand and activated charcoal, and treated with chlorine to prevent the growth of any remaining microorganisms

7. **Anaerobic processing** – sludge from the primary settling tanks and aerobic processing is treated for 2 to 3 weeks in tanks in the absence of oxygen at a temperature between 30 and 40 °C. Anaerobic 'digestion' of the sludge by bacteria such as *Methanobacterium* converts the organic material into methane (CH_4), carbon dioxide (CO_2) and hydrogen (H_2) gases along with water and minerals

8. **Powerhouse** – methane gas produced during anaerobic processing is burned to heat the 'digestion' tanks and generate electricity, which is used to power machinery and/or supply the National Grid

9. **Dumping** – 'digested' sludge may be incinerated (burnt) or spread on land as a soil 'conditioner', providing nitrates and phosphates and improving water retention. 'Digested' sludge is also sent for landfill. Decomposition by bacteria in the soil produces methane which can be piped to nearby houses as cheap fuel.

Water treatment

The water that we use is taken mainly from lakes and rivers. It is purified at treatment plants, making it safe to drink (see Figure 24.6C). Purification includes:

- **filtration** to remove solid matter followed by . . .
- **bacterial oxidation** to get rid of organic matter and . . .
- **treatment** with chlorine to kill microorganisms which cause disease (pathogens – see p. 331).

Biogas fermenters

Notice in Figure 24.6A that the anaerobic stage in sewage treatment produces the fuel gas **methane**, which is burned and used to generate power to run the sewage works. In China, India and other countries, village communities have used **biogas fermenters** to produce methane for hundreds of years. Human and animal faeces, leafy waste from crops, paper and cardboard are all grist to the bacterial mill inside the fermenter. Figure 24.6D shows one in action.

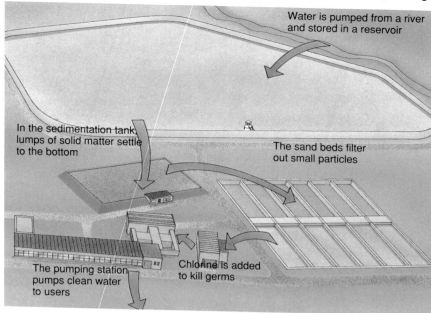

Figure 24.6C 🔺 A water treatment works

Water is pumped from a river and stored in a reservoir

In the sedimentation tank, lumps of solid matter settle to the bottom

The sand beds filter out small particles

Chlorine is added to kill germs

The pumping station pumps clean water to users

Biogas fermenters decompose the waste organic material of local communities, producing the gas methane which is a valuable fuel.

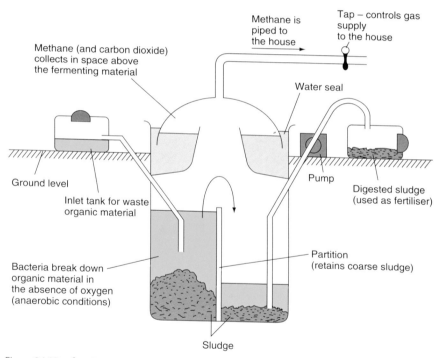

Figure 24.6D 🔺 Biogas fermenter – 'digestion' of organic material by *Methanobacterium* and other anaerobic bacteria produces methane

Methane is piped to the house

Tap – controls gas supply to the house

Methane (and carbon dioxide) collects in space above the fermenting material

Water seal

Ground level

Pump

Inlet tank for waste organic material

Digested sludge (used as fertiliser)

Bacteria break down organic material in the absence of oxygen (anaerobic conditions)

Partition (retains coarse sludge)

Sludge

SUMMARY

Raw sewage is a health hazard. Bacteria break down sewage waste into harmless and even useful substances. Sewage works are designed to deal with large volumes of waste. Biogas fermenters digest small amounts of waste and serve the needs of local communities.

The biogas fermenter is an example of small-scale technology serving local needs. In developing countries where wood is scarce, the methane produced is a valuable fuel for cooking, lighting and powering refrigerators. Left-over sludge is used as a soil conditioner. Biogas fermenters can be expensive. Cheaper designs are being developed so that more people can afford them.

Make the link between these developments and benefits to the environment. For example, how can biogas fermenters help conserve trees and soil? Report your ideas in a way that would help people decide to buy a biogas fermenter for their village.

CHECKPOINT

▶ **1** Discuss how the activities of microorganisms that decompose the organic matter in sewage use up dissolved oxygen.

▶ **2** Summarise the stages in the treatment of sewage in a modern sewage works.

▶ **3** Distinguish between the activated sludge process and the trickling filter process in sewage works.

▶ **4** The table shows the world consumption of water over the past 40 years.

Year	World consumption of water (millions of tonnes per day)
1960	10.0
1970	11.5
1975	13.0
1980	15.0
1985	17.0
1990	20.0
1995	21.0
2000	22.5

(a) On graph paper, plot the consumption (on the vertical axis) against the year (on the horizontal axis).

(b) Say what has happened to the demand for water over the past 40 years.

(c) Suggest three reasons for the change.

(d) From your graph, predict what the consumption of water will be in the year 2020.

▶ **5** Describe how a biogas fermenter works.

Theme Questions

1 The diagram shows some of the features of the inheritance of sex.

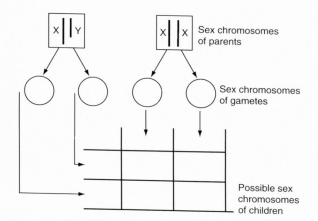

(a) Copy the diagram and indicate the sex of each parent.
(b) Complete your copy of the diagram by drawing in the spaces provided the chromosomes present in the gametes and in the children.
(c) The diagram shows the chromosomes taken from a cell of an adult human.

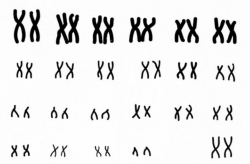

(i) State whether the chromosomes are from a male or female.
(ii) Give one reason for your answer.

2 Busy Lizzies are small plants which are easily grown and produce large numbers of white or red flowers all through the summer.

In Busy Lizzies the allele for red flowers (R) is dominant to the allele for white flowers (r). A red-flowered Busy Lizzie (RR) was *cross-pollinated* with a white-flowered Busy Lizzie (rr). The seeds produced by this cross were grown to produce an F_1 generation of plants all with red-flowers.

(i) What is meant by *cross-pollination*?
(ii) Why did **all** the F_1 generation of plants have red flowers?

(iii) Two of these F_1 red-flowered plants were cross-pollinated. The resulting seeds were grown to produce an F_2 generation of plants. Use a genetic diagram to explain the genotypes and phenotypes of the F_2 plants.

3 The diagram below shows part of the equipment used in the **activated** sludge treatment of sewage.

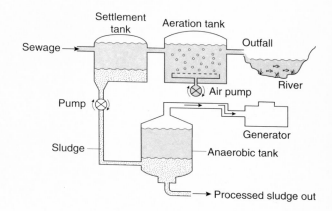

(a) (i) What is the function of the settlement tank?
(ii) Describe and explain what happens in the aeration tank.
(iii) In which tank is methane produced?
(b) Biological oxygen demand (BOD) is the amount of oxygen needed by bacteria to remove organic material from $1000\ cm^3$ of water or effluent. The more polluted a sample is, the higher its BOD. At a sewage works, the BOD was measured for raw sewage entering the works and for effluent entering the river.
The graphs below show the results for each month.

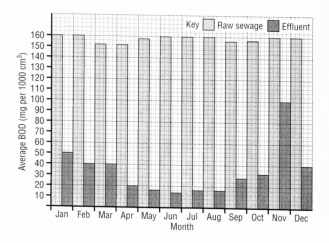

(i) In which month did the sewage works operate most efficiently? Give a reason for your answer.

(ii) In which month did the sewage works break down?

(iii) During this breakdown, effluent with a high content of organic matter entered the river. Suggest the effect this may have had on the population of bacteria and of fish. Give a reason for each answer. (Edexcel)

4 (a) A plasmid is a circle of DNA found in the cytoplasm of bacterial cells. The plasmid in the diagram contains a gene for resistance to antibiotic **A** and a gene for resistance to antibiotic **T**.

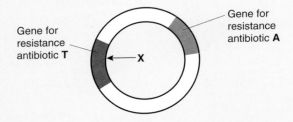

Explain how having this plasmid would help a bacterium to survive.

(b) A genetic engineering technique is used to make large amounts of human growth hormone. Read the passage and answer the questions following.

Genetic engineering to make human growth hormone involves cutting the DNA of the plasmid at **X**, inserting the gene for making human growth hormone and joining the cut ends together to form a complete plasmid again. Sometimes this process fails and the cut ends rejoin without including the gene for human growth hormone.

Engineered plasmids are then mixed with bacteria so that they may enter bacterial cells. The process results in three types of bacteria:

• bacteria containing no plasmid;
• bacteria containing plasmid without the human growth hormone gene;
• bacteria containing plasmid with the human growth hormone gene.

Each type responds to antibiotics in different ways and this fact can be used to select the type of bacteria to grow in a fermenter.

(i) Name the type of chemical used to cut DNA.

(ii) Write out and complete the table below to show if each type of bacterium would grow (✓), or would not grow (✗), when each antibiotic is added to a culture of the bacteria.

Type of bacterium	Antibiotic A added to culture	Antibiotic T added to culture
Containing no plasmid		
Containing plasmid without the human growth hormone gene		
Containing plasmid with the human growth hormone gene		

(iii) Describe how large amounts of human growth hormone can be made from genetically engineered bacteria.

(c) A vaccine can be used to treat hepatitis B. The diagram shows stages in the production of this vaccine.

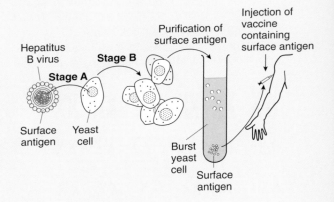

(i) What happens at stage **A**?

(ii) What happens at stage **B**?

(iii) Why is a vaccine, rather than an antibiotic, needed to treat hepatitis B? (Edexcel)

The diagram below shows the main stages of a biotechnological process which uses a fermenter. The process was designed to produce single-cell protein (SCP) from bacteria which grow on methanol. The bacteria and raw materials are put into the fermenter which is kept at a constant temperature of 40 °C.

(a) Using **only** the information provided in the diagram, answer the following questions:

  (i)   What other raw materials, in addition to methanol, are used in the process?

  (ii) What process ensures that the raw materials entering the fermenter are free of disease organisms?

  (iii) Why is the process in (ii) necessary?

  (iv) What substance is recycled into the fermenter after being separated from the bacteria?

  (v) State **two** forms in which SCP is sold.

(b) Explain the importance of keeping the fermenter at a constant temperature of 40 °C.

(c) State **three** advantages of using bacteria to produce protein food.

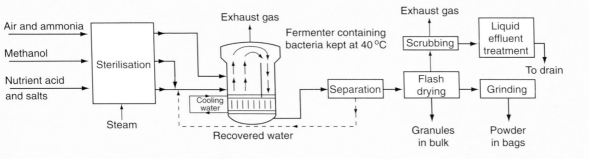

Variation and evolution

E volution is the process whereby living things become different from their ancestors through thousands of generations of change. Its mechanism is natural selection acting on the genetic variation inherited from one generation to the next. The variety of life today is a result of evolution from the origins of life on Earth around 4000 million years ago.

Topic 25

Genes and the environment

25.1 ▶ Continuous and discontinuous variation

Make sure that you can distinguish between continuous and discontinuous variation. It's all to do with the presence or absence of intermediate forms of the characteristic in question.

The ability to roll the tongue is an example of discontinuous variation. Another example is the A, B, AB and O blood groups in humans (see p. 264)

(see p. 264)

SUMMARY

Characteristics which have a range of intermediate forms between extremes vary continuously. Characteristics which have either/or categories without intermediate forms vary discontinuously.

Look at your classmates, the members of your family and your friends. They have different coloured hair, different coloured eyes, different shaped faces. They all show variations in the different characteristics that make up physical appearance.

The variations shown by some characteristics are spread over a range of measurements. All **intermediate** forms of a characteristic are possible between one extreme of the characteristic and the other. We say that the characteristic shows **continuous variation**. The height of a population of people is an example of continuous variation (Figure 25.1A). Notice that the average height of the adult population lies at the centre of the curve. Most people have heights close to the average. Only a few individuals are really short or really tall compared with the majority.

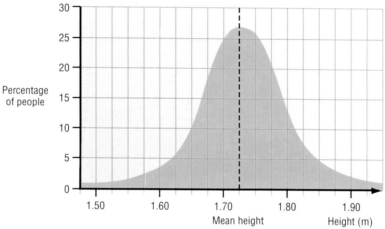

Figure 25.1A ▲ Continuous variation

Other characteristics do not show a continuous trend in variation from one extreme to another. They show categories of the characteristic without any intermediate forms. The ability to roll the tongue is an example – either you can do it or you can't. There are no half-rollers (Figure 25.1B)! We say that the characteristic shows **discontinuous variation**.

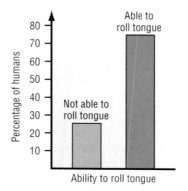

Figure 25.1B ▲ Ability to roll the tongue – an example of discontinuous variation

427

CHECKPOINT

▶ **1** Phenylthiocarbamide is a chemical which has a bitter taste to some people. It tastes like the rind of a grapefruit. However,'non-taster's' do not experience its sensation. The table shows the frequency of 'non-tasters' in different groups of people.

(a) Is people's response to phenylthiocarbamide an example of continuous or discontinuous variation?

(b) Can you think of reasons for the differences in response to phenylthiocarbamide between different groups of people?

▶ **2** In a populations of 300 goldfish variations in two characteristics were measured and the results displayed as charts. Chart A shows variation in the length of the fish; chart B shows variation in their colour.

(a) Which chart shows
   (i) continuous variation
   (ii) discontinuous variation?

Briefly give reasons for your answers

(b) Using chart B calculate the per cent number of yellow goldfish in the population.

(c) Albino goldfish are relatively rare compared with the colours of the other fish. Give a genetic explanation for the occurrence of albino goldfish.

Group	'Non-tasters' (%)
Hindus	33.7
Danish	32.7
English	31.5
Spanish	25.6
Portuguese	24.0
Japanese	7.1
Lapps	6.4
West Africans	2.7
Chinese	2.0
S.American Indians	1.2

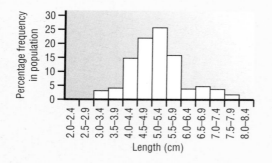

A

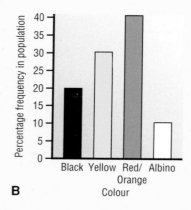

B

25.2 ▶ Sources of variation

FIRST THOUGHTS

Variation arises from genetic and environmental causes. Both affect the phenotype but only variation resulting from the genotype is inherited. Read on but also remember the principles of genetics discussed in Topic 23.1.

The outward appearance of an organism (the **phenotype**) depends on a combination of inherited characters (the **genotype**) and characteristics due to the effects of the environment (see p. 379).

A simple example gives the idea. The litter of labrador puppies shown in Figure 25.2A have inherited genes from their parents. If we compare the growth of the puppies, we might see that one of the litter does not develop as well as the rest. The reason could be that it has not been given the correct diet and exercise. In other words, the effects of the environment have influenced its phenotype. By altering the environment we can alter the phenotype.

Figure 25.2A ▲ Different environments will affect some of the pups' characteristics

On the other hand the litter of corgis shown in Figure 25.2B will grow up to be small dogs no matter how well fed they are. The genes determine the phenotype if the environment is kept the same.

Figure 25.2B ▲ These corgis will always be small, regardless of how well fed they are

Where does variation come from?

Variation arises from genetic causes through:

- **Sexual reproduction** which results in the fusion of the nucleus of a **sperm** (male sex cell) with the nucleus of the **egg** (female sex cell). The fusion (fertilisation) recombines the genetic material from each parent in new ways within the **zygote** (fertilised egg) (see p. 343).
- **Mutations** which arise as a result of mistakes in the **replication** of DNA (see p. 154). Occasionally the wrong base adds to the growing strand of DNA, making the new DNA slightly different from the original. **Ionising radiation** and some chemicals increase the probability of gene mutation:

Make sure you can answer an exam question which asks you to distinguish between phenotype and genotype.

See www.keyscience.co.uk for more about variation.

The greater the dose of ionising radiation the greater is the chance of mutation occurring.

♦ Ionising radiation strips **electrons** (see *Key Science Physics* p. 97) from matter exposed to it (ionisation). Emissions from radioactive substances are ionising radiations. They may cause mutations by damaging DNA directly or by generating highly active components of molecules called **free radicals** which cause the damage indirectly. Free radicals are caused by the action of radiation on water molecules in cells.

♦ Chemicals such as the **carcinogens** (substances that cause cancer) in tobacco may lead to mutations of the genes which inhibit cell division. A cancer develops because cell division runs out of control (see p. 155).

♦ **Crossing over** during **meiosis** (see Topic 9.4) exchanges a segment of one chromosome (and the genes it carries) with the corresponding segment of its **homologous** chromosome (see Figure 9.4D). As a result the sex cells produced by meiosis have a different combination of genes from the parent cell.

Variations that arise from genetic causes are inherited from parents by offspring who pass them on to their offspring and so on from generation to generation. Inherited variation is the raw material on which **natural selection** acts, resulting in **evolution** (see p. 437).

Variation also arises from **environmental** causes. Here 'environmental' means all of the external influences affecting an organism. Examples are:

● **Nutrients** in the food we eat and minerals that plants absorb in solution through the roots. For example, in many countries children are now taller and heavier, age for age, than they were 50 or more years ago because of improved diet and standards of living (see Figure 25.2C).

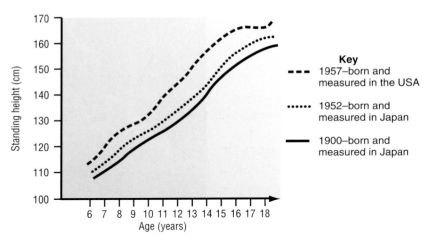

Figure 25.2C ⬆ The effects of improved diet on the growth of Japanese boys aged 6–18 years

Distinguish between the genetic causes of variation and the environmental causes. Remember that natural selection acts on the genetic causes, resulting in evolution. It does not act on the environmental causes.

● **Drugs** which may have a serious effect on appearance. For example, **thalidomide** was given to pregnant women to prevent them feeling sick and help them sleep. The drug can affect development of the fetus and some women prescribed thalidomide gave birth to seriously deformed children.

● **Temperature** affects the rate of enzyme-controlled chemical reactions. For example, warmth increases the rate of photosynthesis and improves the rate of growth of plants kept under glass (see Topic 11.1 and the Assignment 'Growing crops under glass' in *Key Science Biology* Extension File).

● **Physical training** uses muscles more than normal, increasing their size and power. For example, weight-lifters develop bulging muscles as they train for their sport (see Figure 25.2D).

Figure 25.2D ▲ Special training and a high-protein diet have produced this weight-lifter's bulging muscles. Compare with the other person who has a less strenuous life-style

SUMMARY

Genetic variation is inherited; variation due to the environment is acquired. Only genetic variation affects the course of evolution.

Variations that arise from environmental causes are *not* inherited because the sex cells are *not* affected. Instead the characteristics are said to be **acquired**. Because the weight-lifter has developed bulging muscles does not mean that his/her children will have bulging muscles unless they take up weight-lifting as well! Variations as a result of acquired characteristics are not acted upon by natural selection and therefore they do *not* affect evolution.

CHECKPOINT

▶ **1** Inherited characteristics and acquired characteristics are sources of variation in species.
   (a) Distinguish between inherited and acquired characteristics
   (b) Which type of characteristic is important for the evolution of new species?
       Try to give reasons for your answer.
▶ **2** List the different sources of variation in living things.
▶ **3** Why does sexual reproduction produce much more genetic variation than asexual reproduction?

Topic 26   Evolution

26.1 ▶   Evolution: the growth of an idea

FIRST THOUGHTS

Species change over the course of time. The change is called evolution. How evolution takes place was a puzzle until Charles Darwin discovered the principles of natural selection. In this section you can find out how the idea developed in Darwin's mind.

Figure 26.1A is a photograph of the night sky taken by the Hubble Space Telescope circling the earth. Light from the dimmest of the galaxies (see *Key Science Physics* p. 18) emitted at around the time of the origins of the Universe has taken 12–15 000 million years to reach us. By comparison the planets of the solar system are relative newcomers. Interstellar dust and gases circling a star we call the Sun condensed into the Earth and its companions around 4500 million years ago. Evidence suggests that soon after, the first stirrings of life on Earth occurred.

The variety of life today almost certainly owes its origins to a scenario billions of years ago similar to the one in Figure 26.1B. *How do the events in Earth's history take us from such beginnings to the present day?* Present-day living things are descended from ancestors that have gradually changed through thousands of generations. We call the process **evolution**.

Figure 26.1A ⬆   The specks and swirls of light are galaxies viewed from space by the Hubble telescope. Millions of light years distant from Earth, each galaxy consists of thousands of millions of stars formed soon after the birth of the Universe.

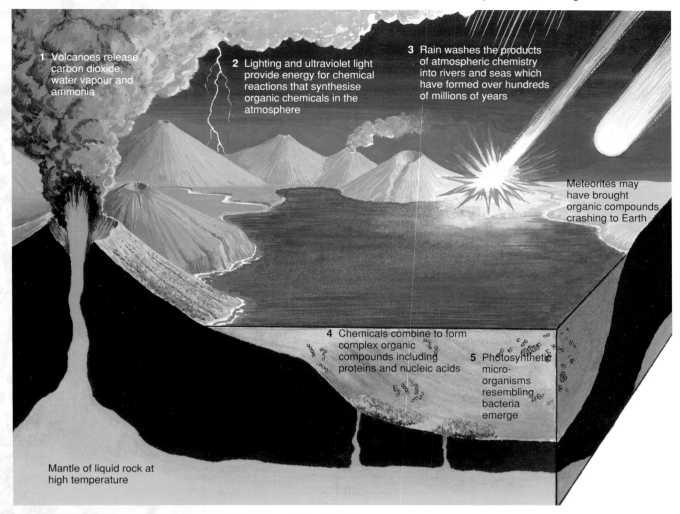

Figure 26.1B 🔺 Did events like these lead to the beginnings of life on Earth?

Life first appeared on Earth about 3 900 million years ago.

Charles Darwin was *not* the first person to propose that species evolve. However, he was the first (with Alfred Russel Wallace) to explain *how* species evolve. The process is called natural selection.

Figure 26.1C is a timetable of life on Earth. Its time line shows major events during the evolution of life from its earliest beginning. Notice that we humans are relatively recent. Our nearest ancestors first gazed at the world around them about 2.5 million years ago. Think of it like this. Humans appeared on Earth just a few seconds ago if we condense all of time since the origins of the Universe into a calendar year.

Charles Darwin

The British naturalist Charles Darwin (Figure 26.1D) was the first person to work out a theory explaining *how* species can evolve. In 1831 Darwin was invited to join the company of HMS *Beagle* on a world voyage which included a survey of the South American coast. The journey took five years. During the *Beagle's* time around South America, Darwin took over as the ship's naturalist (see Figure 26.1E). He collected fossils and specimens of plants and animals on expeditions inland. He made many observations about the wildlife he encountered.

On one April day in 1832 Darwin collected 68 different species (see p. 8) of small beetle from the rain forest around Rio de Janeiro in Brazil. What puzzled him was that there should be so many different species of one kind of insect. *How did such variety come about?*

433

Figure 26.1C 🔺 A timetable of life on earth

Figure 26.1D 🔺 Charles Darwin (1809–82)

It's a fact!

Space is filled with dust and gases made of carbon, hydrogen and nitrogen; some of the elements necessary to build organic compounds (see Topic 10). Some scientists believe that dust and gases entering the Earth's atmosphere 4000 million years ago helped to trigger the appearance of life on Earth.

As the *Beagle* sailed up the Pacific coast of South America, Darwin noticed that one type of organism gave way to another. He also noticed that many of the animals were different from those he had seen on the Atlantic coast.

Its South American survey complete, the *Beagle* set sail for the Pacific islands of the Galapagos on 7 September 1835. Once ashore, Darwin observed that the wildlife of the islands was similar to the wildlife he had seen on the South American mainland. However, he noticed differences in detail. The birds shown in Figure 26.1F are cormorants.

Figure 26.1E 🔺 HMS *Beagle* during her voyage around South America

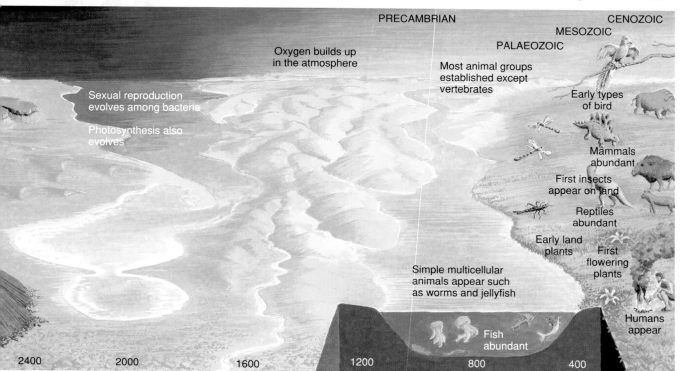

PRECAMBRIAN

PALAEOZOIC

MESOZOIC

CENOZOIC

Oxygen builds up
in the atmosphere

Sexual reproduction
evolves among bacteria

Photosynthesis also
evolves

Most animal groups
established except
vertebrates

Early types
of bird

Mammals
abundant

First insects
appear on land

Reptiles
abundant

Early land
plants

First
flowering
plants

Simple multicellular
animals appear such
as worms and jellyfish

Humans
appear

Fish
abundant

2400 2000 1600 1200 800 400

One comes from the Galapagos islands, the other from Brazil. *What differences can you see between them?*

Each of the Galapagos Islands Darwin visited was inhabited by tortoises which were similar to the mainland forms but much larger. He noticed something else: the tortoises on each island were slightly different, so he could tell which island particular tortoises came from (Figure 26.1G). Darwin wondered why the tortoises were different from island to island.

(a) Galapagos Islands cormorant

(b) Brazilian cormorant

Figure 26.1F 🔺 Flying demands a lot of energy. The absence of predators on the Galapagos Islands means that cormorants do not have to fly away to escape danger. Birds with small wings save energy and therefore are at an advantage. However, predators threaten Brazilian cormorants. Birds with large wings can fly away and are at an advantage even though they use a lot of energy

(a) Isla Isabela tortoise

(b) Isla Espanola tortoise

Figure 26.1G Tortoises from Isla Isabela (formerly Albemarle) and Isla Espanola (formerly Hood), two of the Galapagos islands. Isla Isabela is well watered and the tortoises wallow in the pools. Isla Espanola is arid and here the tortoises browse on the water-filled stems of cacti and juicy leaves and berries. Notice the shape of the shell of the tortoises from each island. *Can you account for the differences?*

Figure 26.1H Sir Charles Lyell (1797–1875) greatly influenced Darwin's view of Earth and its history. Darwin met Lyell on his return to England and they became great friends

Figure 26.1I Reverend Thomas Malthus

In Darwin's day, most people believed that each species was fixed and unchangeable, but Darwin could not agree. The variety of species discovered on his expeditions in South America and the Galapagos islands convinced him that species are not fixed but change through time; that is, they evolve. *The puzzle was, what mechanism brought about their evolution?*

■ The influence of Lyell and Malthus

Just before Darwin set sail in the *Beagle*, he was given the first volume of a newly published book *Principles of Geology* by Charles Lyell (Figure 26.1H). The second and third volumes were sent to him during the voyage.

Lyell believed that the Earth's rocks were very old and that natural forces had produced continuous geological change during the course of the Earth's history. Lyell explained that the age of rocks could be estimated from the types of fossil they contained. He also stated that the fossil record was laid down over hundreds of millions of years.

Darwin reasoned from Lyell's work that if rocks and rock formations have changed slowly, over long periods of time, living things might have a similar history.

Another piece of the jigsaw fell into place soon after Darwin's return to England. One day in 1838 he was reading *An Essay on the Principle of Population* written in 1798 by the Reverend Thomas Malthus (Figure 26.1I). In the essay Malthus stated that a population would increase indefinitely unless it was kept in check by shortages of resources such as food, water and living space. Darwin reasoned that in nature a 'struggle for existence' must occur. In modern language we say that organisms **compete** for resources in short supply (see Topic 3.5).

Darwin understood that living organisms produce more offspring than can normally be expected to survive. A beech tree, for example, produces thousands of seeds each year. Hundreds of seedlings sprout from the seeds but only one or two grow into mature trees (Figure 26.1J). He also understood that the supply of food, amount of space and other resources in the environment are limited. Only those organisms with the structures and the way of life that suit (**adapt**) them to make best use of these limited resources survive long enough to reproduce (see Figures 26.1F

and 26.1G). Less well-suited organisms leave fewer offspring or do not survive to reproduce at all. Therefore, **variations** (see Topic 25) which are favourable for survival accumulate from one generation to the next, while less favourable variations die out (became **extinct**). Darwin called this process **natural selection** (see p. 65). He realised that this is the mechanism of evolution. At last he understood how species change through time.

Figure 26.1J ◀ Hundreds of seedlings are growing in an area about 10 m². Very few will grow to maturity. Most perish, crowded out by the two or three that grow the fastest

See www.keyscience.co.uk for more about natural selection.

26.2 ▶ Artificial selection

FIRST THOUGHTS

By choosing particular characteristics, breeders have developed new varieties of plants and animals. The examples described in this section illustrate the process of selective breeding and its importance in support of Darwin's idea.

Darwin was a cautious man. He knew that other people who had proposed theories of evolution before him had been ignored and disliked. After the *Beagle's* voyage, he spent the next twenty years gathering more evidence to support his ideas. His search for evidence led him to investigate the work of breeders of animals and plants. Figure 26.2A shows two very different breeds of dog. Centuries of selecting dogs with particular characteristics such as size, colour, length of coat and shape of ear have resulted in a wide variety of breeds. The corgi, spaniel, labrador and bulldog are just a few examples. *How many other breeds can you think of?*

Dog breeders have taken advantage of the large amount of variation in the characteristics of the members of generations of dogs and chosen the characteristics which they want to be passed on to the next generation. The dogs with these characteristics have been allowed to reproduce and so the genes controlling the selected characteristics have been passed to their offspring. This process of choosing which characteristics should pass to the next generation and which should not is called **artificial selection** (see Figure 26.2B).

Figure 26.2A ◆ The same species?

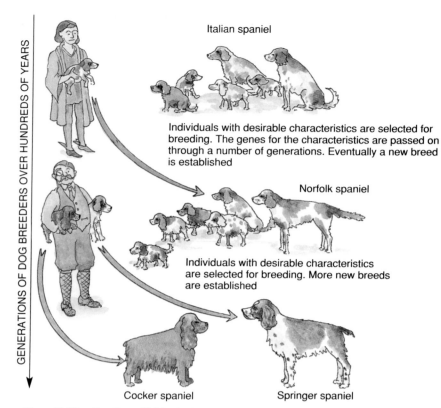

Italian spaniel

Individuals with desirable characteristics are selected for breeding. The genes for the characteristics are passed on through a number of generations. Eventually a new breed is established

Norfolk spaniel

Individuals with desirable characteristics are selected for breeding. More new breeds are established

Cocker spaniel   Springer spaniel

Figure 26.2B ⬆ The artificial selection of desired characteristics produces breeds of dog that look very different from their ancestors

Figure 26.2C ⬆ The many varieties of domestic pigeon

See www.keyscience.co.uk for more about selective breeding.

It's a fact!

Selective breeding greatly reduces variation (see Topic 25.2) in populations of crops and animals raised for food, because the number of different alleles (see p. 381) is reduced. Should conditions alter, reduced variation may mean that selective breeding of new populations that are able to survive the changed environment will not be possible.

Artificial selection is used to improve the productivity of farm animals and crops.

Generations of artificial selection have resulted in the great variety of dog breeds we know today. All dogs belong to the same species, *Canis familiaris*, and are descended from the same ancestral species even though they look very different from one another.

Darwin added the experience of breeding pigeons to his observations of other people's work. He saw that by mating pigeons with different characteristics, new varieties of pigeons could be produced. All the pigeons in Figure 26.2C belong to the species *Columba livia*.

Darwin reasoned that if artificial selection produces change in domestic animals and plants, natural selection should have the same effect on wildlife. The only difference is that artificial selection by human beings produces new varieties in a relatively short time, selection by nature takes much longer.

Artificial selection helps farmers

For thousands of years the selected **pedigree** (see p. 387) animal has been prized. Such animals have a recorded ancestry over many generations. The dairy person wants selected pedigree cattle from which to breed milking cows. The hunter wants a pedigree dog. Similarly, pure lines of cultivated plants are needed by the farmer. Characteristics such as grain yield, fruit yield and disease resistance are desirable qualities that can be maintained through artificial selection (**selective breeding**).

Figure 26.2D shows the variety of vegetables which have been produced through artificial selection from one species of plant, *Brassica oleracea* (a relative of the mustard plant). By choosing different characteristics which make good eating, different vegetables have been developed to suit different tastes.

Rice

Maize

Barley

Wheat

Figure 26.2E ▲ Important cereal crops

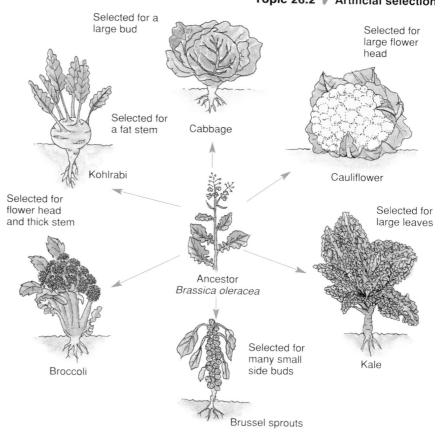

Figure 26.2D ▲ A variety of vegetables artificially selected and bred from a single ancestral species

Increasing yields of cereals

The grasses of the world provide the staple diet for most of its population. We know them as **cereal** crops. Humans have cultivated cereals since the earliest records of history. More people eat rice than any other food. Wheat, maize, oats, barley, rye and millet make up the bulk of the rest of the world's cereals (Figure 26.2E).

Today an enormous variety of cereals are available to the farmer as a result of artificial selection over many generations. Planned **hybridisation** (crossing two different varieties) forms the basis of the development of new varieties. The aim of hybridisation is to concentrate as many desirable features as possible in a single variety. One variety of wheat, for instance, may have good milling properties but poor resistance to fungal disease. Another may resist fungi but provide a low yield of grain. A third might not thrive in dry conditions but may give good baking flour when well irrigated (see p. 91). New varieties are constantly being developed by plant breeders which retain the favourable properties and eliminate the less favourable ones. To the basic properties already mentioned are added others that vary according to the climate and soil in which the wheat is to be grown. Also the season of planting may differ; some varieties are planted in autumn, others in spring. Figure 26.2F shows varieties of wheat. The improvement in varieties is largely responsible for the enormous increase in world production of wheat during the last 50 years. Hybridisation and selection of varieties of rice has also boosted yields by 25 per cent. It is a slower process than wheat breeding but promises to give even higher yields in the future.

Figure 26.2F Six of 14 varieties of wheat are shown: 7-chromosome einkorn, 14-chromosome emmer, macaroni wheat, Polish wheat, 21-chromosome club wheat and spelt

26.3 ▶ *The Origin of Species*

Alfred Russel Wallace concluded that natural selection was the mechanism of evolution independently of Darwin. However, he acknowledged that Darwin's ideas were more thoroughly worked out.

Darwin hesitated from publishing his ideas on natural selection, but in 1858 he had an unpleasant surprise. He received a letter and an essay from Alfred Russel Wallace who was a naturalist living and working in Malaya (Figure 26.3A). In the letter, Wallace asked for Darwin's opinion as a respected naturalist on the essay which set out the principles of natural selection as the mechanism for evolutionary change. This was a terrible blow to Darwin who had worked patiently on the same ideas for 20 years.

Like Darwin, Wallace was led to the idea of natural selection through reading Malthus' essay on population. Over the years, Darwin had discussed natural selection with only a few close friends and fellow scientists. They had urged him to publish his ideas. When Wallace's letter arrived, Darwin was stung into action.

Both scientists recognised the other's contribution and acted most generously over the matter. They agreed to the joint publication of their ideas which appeared in the *Journal of the Linnean Society* for 1858.

Although Wallace was a distinguished naturalist in his own right, he realised that Darwin's ideas were more thoroughly worked out than his own. For the rest of his life Wallace insisted that Darwin take most of the credit. After the publication of the joint paper, Darwin hastily prepared a fuller account of his work. *The Origin of Species* was published in 1859 (Figure 26.3B).

In the book Darwin brought together so much evidence that evolution was established as a viable theory. However, the position of natural selection was less secure. Many scientists thought it likely that natural selection was the mechanism of evolution but wanted more evidence. Not until the development of modern genetics, beginning with the discovery of Mendel's work (see Figure 23.3B), was natural selection generally accepted as the mechanism for evolutionary change. Understanding genetics allows us to understand why organisms vary. Natural selection works on these variations to produce evolutionary change.

SUMMARY

Darwin's key ideas

So far we have tracked the key ideas which were crucial to Darwin's understanding of how species evolve. They are the components of evolution:

Variation – Darwin's work during the voyage of HMS *Beagle* and his experience of selectively breeding pigeons provided evidence for the large amount of variation in the characteristics of different species.

Natural selection – Malthus' work contributed to Darwin's idea of a 'struggle for existence' (competition for resources). The result is natural selection of those organisms best suited (**adapted**) to survive.

Time – Lyell's work showed Darwin that the Earth is very old, giving time for the evolution of species to occur.

The summary will help you keep the different ideas in mind as you read on.

THE ORIGIN OF SPECIES

BY MEANS OF NATURAL SELECTION,

OR THE

PRESERVATION OF FAVOURED RACES IN THE STRUGGLE FOR LIFE.

By CHARLES DARWIN, M.A.,

FELLOW OF THE ROYAL, GEOLOGICAL, LINNEAN, ETC., SOCIETIES;
AUTHOR OF 'JOURNAL OF RESEARCHES DURING H. M. S. BEAGLE'S VOYAGE
ROUND THE WORLD.'

LONDON:
JOHN MURRAY, ALBEMARLE STREET.
1859.

Figure 26.3A ⬅ Alfred Russel Wallace

Figure 26.3B ⬅ Title page of *The Origin of Species*

CHECKPOINT

▶ 1 Briefly explain your understanding of evolution.

▶ 2 Why do you think studying life in the past helps us to understand biology today?

▶ 3 What is an ancestor?

▶ 4 What is a descendant?

▶ 5 What is meant by the term 'natural selection'?

EXTENSION FILE
ASSIGNMENT

26.4 ▶ Evolution in action

FIRST THOUGHTS

In this section you can read about examples of evolutionary change.

Members of a species do not live in isolation. They are part of a **population** (see Topic 3). No two individuals of a population are the same genetically (except identical twins). Some individuals have genes controlling characteristics which are better suited for survival than the genes of other members of the population. The individuals with the favourable characteristics are more likely to reproduce successfully and have offspring more likely to survive than the individuals with less favourable characteristics. The result is evolution (see Figure 26.4A).

Selection pressure

Offspring reproduced asexually are genetically identical with each other and with their parent. Mutation (see p. 154) is the only source of genetic variation. Sexually reproduced offspring inherit genetic material from both parents. They are genetically different from each other (except in the case of identical twins) and from their parents (see Topic 20).

Natural selection works on the genetic variation in populations of organisms selecting the variations which enable individuals to adapt to a

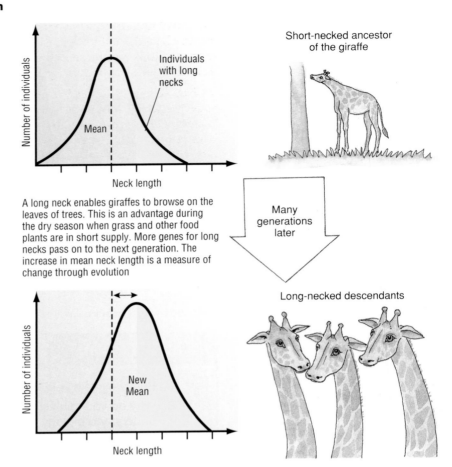

Figure 26.4A ◆ Passing on favourable characteristics

It's a fact!

In rare cases, a mutation (see p. 154) may increase an organism's chances of survival. The mutant gene will be favoured by natural selection and so inherited by future generations. Selection of favourable mutations is an important cause of evolution. Some mutations do not affect an organism's chances of survival one way or the other. We say that the mutations are **neutral** in their effects.

changing environment. In stable environments the **selection pressure** for organisms to adapt is low. Asexual reproduction, therefore, has advantages. One parent reproducing identical copies of itself can more easily populate and exploit the environment than two parents reproducing sexually.

In rapidly changing environments, however, the balance tips in favour of sexual reproduction. Selection pressure is high and favours individuals which can change in response to changing circumstances. Only sexual reproduction is able to produce enormous amounts of genetic variation. Natural selection can then select the change which enables organisms to adapt to the new environment.

Evolution is still happening

When a new **insecticide** is used against an insect pest, a small amount is enough to kill most of the population. However, a few individuals are not affected because they have genes which code for enzymes that help to break down the insecticide into harmless substances. These individuals are **resistant** to the insecticide. These are the insects that survive and reproduce. The majority of their offspring inherit the genes for the enzymes that break down insecticide. In this way the number of resistant insects in the population increases.

Insects reproduce quickly so resistance spreads quickly through the population. More and more insecticide has to be used until it becomes so

It's a fact!

The insecticide DDT was used to control insect pests during the Second World War (1939–1945). In 1947 the first cases of resistance to the new compound were reported in houseflies. Since then resistance in houseflies has developed so that some varieties are resistant to one hundred times the dose of DDT needed to kill unresistant flies. Today, hundreds of insect species are resistant to a range of insecticides.

It's a fact!

Resistance in bacteria means that antibiotics become less effective for the treatment of disease. Doctors try to get round the problem by using different types of antibiotic and by making sure that patients are given an antibiotic only if it is absolutely necessary (see Topic 19.3). In hospitals, the selection pressure by antibiotics is so intense that strains of bacteria resistant to a range of antibiotics rapidly evolve.

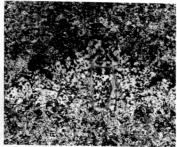

Figure 26.4C ◀ Can you see the moths?

inefficient and so expensive that an alternative insecticide has to be found. Even then, the effect is only temporary because the insect pest soon develops resistance to the new insecticide (Figure 26.4B). The insecticide acts as the *agent of selection*.

Resistance to **antibiotics** has evolved in bacteria in much the same way as resistance to insecticides has evolved in insects. Populations of bacteria always contain a few individuals with genes that make them resistant to an antibiotic. These individuals survive and reproduce. The new generation inherits the genes for resistance. Resistance develops rapidly in bacteria because they reproduce rapidly (in some species a new generation is produced every 20 minutes). The antibiotic acts as the *agent of selection*.

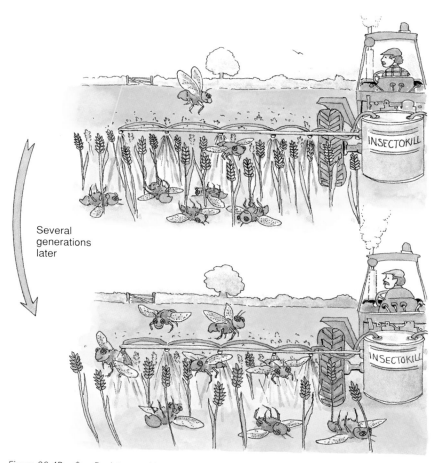

Figure 26.4B ▲ Resistance of insect pests to insecticide

Figure 26.4C shows two forms of the peppered moth *Biston betularia*. The speckled, pale moth (called the **peppered variety**) is found most often in the countryside where it merges into the light-coloured background of lichen-covered trees and rocks. In towns and cities air pollution kills the lichens leaving trees and other surfaces bare and sooty. The black moth (called the **melanic variety**) is the most common form in this environment where it merges into the dark background.

Birds eat the moths whose colour does not merge into the background because the moths are more easily seen. The birds act as the *agent of selection* favouring the melanic variety in urban areas and the peppered variety in the countryside (Figure 26.4D).

The environment and buildings of cities such as Manchester and Leeds in the north of England grew up during the Industrial Revolution. Buildings and the surrounding environment were blackened with the soot and grime of smoke-stack industries. The numbers of melanic moths in the cities increased dramatically. Non-melanic moths were rarely seen. In the past 40 years smoke-stack industries have given way to new technologies and the cities are much cleaner. Populations of non-melanic moths are increasing while the numbers of melanic moths are declining.

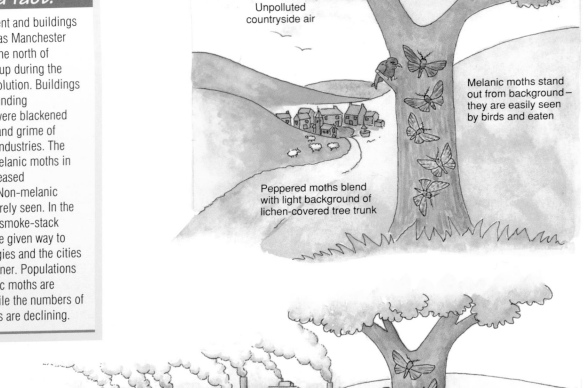

Unpolluted countryside air

Melanic moths stand out from background — they are easily seen by birds and eaten

Peppered moths blend with light background of lichen-covered tree trunk

Melanic moths blend with black background of soot-covered tree trunk

Peppered moths stand out from black background — they are easily seen by birds and eaten

Figure 26.4D ⬆ Maintaining the balance between the peppered and melanic varieties of the moth *Biston betularia*

SUMMARY

Natural selection acts on the genetic variation in populations. Individuals with genes for characteristics that adapt them to the environment survive in preference to individuals less well adapted. Individuals that survive are more likely to reproduce and pass on favourable genes to their offspring. The accumulation of favourable genes over many generations results in evolution: the change in species over time.

CHECKPOINT

▶ 1 Look at Figures 26.1F and 26.1G on pp. 435 and 436. Explain in your own words the evolution of flightless cormorants and different varieties of tortoise on the Galapagos Islands. In your explanation mention variation, favourable characteristics, ancestors and descendants.

▶ 2 Look at Figure 26.4A. The graphs show the variation in length of neck in giraffes and their ancestors. What do the graphs tell you about the process of evolutionary change? Why do long necks in giraffes favour survival?

▶ 3 Briefly summarise the reasons for the differences in abundance of peppered moths and melanic moths in the countryside and industrial areas.

▶ 4 Why is the bird in Figure 26.4D an agent of natural selection?

26.5 ▶ The vertebrate story

See www.keyscience.co.uk for more about evolution.

When rock layers containing fossils are undisturbed, they are a record of the sequence of the evolution of life on Earth.

Fossils

Fossils are the remains or impressions made by dead organisms. They are usually preserved in **sedimentary** rocks which are formed layer on layer by the deposition of mud, sand, silt and calcium carbonate over millions of years. Much of the Earth's surface consists of layers of sedimentary rock. Provided it remains undisturbed, the more recently formed the layer, the nearer it is to the Earth's surface. Undisturbed rock layers, therefore, follow on like the chapters of a book. The fossils in each rock layer are a record of life on Earth at the time when the layer was formed. The sequence of fossils therefore traces the history of life on Earth.

The rock layers of the Grand Canyon have lain undisturbed for hundreds of millions of years (Figure 26.5A). They have been exposed by the Colorado River eroding the soft rock. Today the river flows through a gorge 1.6 kilometres deep. The layers of rock at the bottom of the canyon are 2000 million years old. To travel down one of the trails from the rim to the bottom is to travel back in time. The fossils in each layer represent a particular stage in the evolution of life (Figure 26.5B).

Figure 26.5A 🔼 The Grand Canyon. The rocks are mostly sandstones and limestones deposited in shallow seas that once covered the western USA

What do the rocks of the Grand Canyon tell us about the evolution of life? Careful study of the fossils in successive layers of rock reveals that:

- representatives of the major invertebrate groups (see Topic 1.5) were in existence by 600 million years ago
- fish were the first vertebrates to appear about 500 million years ago
- reptiles appeared about 300 million years after the fish
- birds and mammals appeared more recently, too late to be represented as fossils in the rocks of the Grand Canyon.

The sequence of fossils also tells us that the older the fossils, the more likely they are to differ from present-day living things. From the fossil record we can conclude that organisms change through time. In other words, fossils are evidence that present-day living things are a product of evolution.

If selection pressure is low, then organisms show little evolutionary change through time.

A few organisms change very little through time. Descendants therefore look much like their distant ancestors. The coelacanth is a case in point (Figure 26.5C). Everyone believed that the fish had been extinct (see p. 450) for 200 million years. Then in 1938 a fisherman caught one in the Indian Ocean. Since then a number of specimens have been caught. The skeleton and structure of modern coelacanths differ little from the remains of fossil coelacanths 400 million years old, but are different from those of other fish living today. It seems that coelacanths have existed for all of this time in an environment where there is little selection pressure for evolutionary change.

Figure 26.5B ⬆ Animals representative of those living at the time when the rock layers of the Grand Canyon were deposited, and which have been exposed in the strata by the Colorado River

Figure 26.5C ⬆ Coelocanths resemble the fossils of vertebrates that were among the first to live on land

Fossil formation

Dead organisms rapidly decompose (see p. 54). Fossil formation therefore depends on conditions which replace organic material as it decays away with something more permanent, or which prevent decomposition altogether. Bones and shells are best preserved as fossils because they do not decompose easily. As the organic material gradually decays away, the bone or shell becomes porous. Mud and mineral particles infiltrate the structure filling up the spaces. If the mud turns into rock, the result is a fossil which may be preserved for hundreds of millions of years (Figure 26.5D).

Figure 26.5D ⬆ Fossilised bones and tusk of a mammoth

Similar processes may also fossilise plants. For example, the 'Petrified Forest' of Arizona, USA consists of tree trunks where the original organic material of the wood has been replaced by the mineral silica in a process of petrification. Originally the trees were buried under sediment, and water infiltrated silica particles into the trunks. The result was logs of rock (Figure 26.5E). Fine details have been preserved. For example, xylem tubes (see p. 188) and annual rings are clearly visible.

Figure 26.5E ⬆ Arizona's petrified forest

447

Fossil footprints show us that 3.75 million years ago our ancestors walked upright and were free-striding, just as we are today. The footprints are those of an individual walking across damp ash released from an active volcano nearby. Later, the ash dried and set like cement, preserving the footprints for fossil hunters to discover at the end of the twentieth century.

Fossils form when:
- mud and mineral particles infiltrate the remains of dead organisms, preserving structural details in rock,
- an organism is rapidly buried in mud or volcanic ash and a cast formed,
- a dead organism is rapidly frozen.

Sometimes when an organism is rapidly buried in hardening mud or cooling volcanic ash and then decays, the space it occupied is filled with another material. The space acts as a **mould**. The replacement material takes on the shape of the original organism, forming a **cast**. Figure 26.5F shows that fossils can preserve fine details which scientists then interpret as whole structures.

Figure 26.5F (a) The fossil preserves fine structural details of an ancient fish (b) Reconstruction from fossil remains

Do you think you could eat mammoth steaks? Whole mammoths (a type of woolly haired elephant) 10 thousand years old have been found preserved in the icy wilderness of northern Siberia (Figure 26.5G). After the animals died, rapid freezing in subzero temperatures prevented decomposition. The bodies were in such a good state of preservation that it was possible for the scientists who discovered them to cook and eat mammoth steaks!

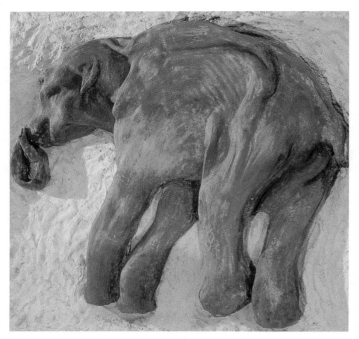

Figure 26.5G Plaster cast of the carcass of a baby woolly mammoth, whose frozen body was found in Siberia in 1977

Figure 26.5H ⬆ *Eusthenopteron*

The invasion of land

Coelacanths (see p. 447) resemble the fossil remains of *Eusthenopteron*, which was a type of fish that lived about 375 million years ago (see Figure 26.5H). These fish had lungs and were able to gulp air. Breathing air enabled them to survive periods of drought when the lakes and marshes in which they lived dried out. They also had strong limbs which allowed them to move on land. There was abundant invertebrate food on the banks of the pools in which they lived; insects and plants had moved on to land about 25 million years before the appearance of *Eusthenopteron*.

As *Eusthenopteron*, and other fish like it, could live for part of the time out of water they were more likely to survive and leave behind new generations of air-breathing descendants. The vertebrate invasion of land had begun.

Amphibians

Frogs, newts and salamanders and all other amphibians are descendants of the earliest land dwellers (Figure 26.5I). The adults usually live on land, but need water for breeding. The female lays her eggs in water. They are fertilised in water and hatch into tadpoles which swim and have gills. The tadpole grows: legs develop, it loses its tail and its gills are replaced by lungs. The change from tadpole to adult is called a **metamorphosis**.

The amphibian body is covered with soft skin through which the body quickly loses water in a dry atmosphere. This is why you find frogs and toads on land only where it is damp.

Figure 26.5I ⬆ *Ichthyostega* – an early type of amphibian that lived about 350 million years ago

Independence from water for reproduction, the ability to prevent the loss of water through the skin, to breathe air and move on land were adaptations essential for the evolution of animals able to live on land.

Reptiles

Reptiles, descended from an ancient type of amphibian, completed the move from water to land around 200 million years ago. Over the next 130 million years an enormous variety of reptiles appeared: land giants, agile runners, fliers, swimmers. They dominated the environment (Figure 26.5J).

449

Figure 26.5J ⬆ 155 million years ago reptiles dominated the land

Dimorphodon – although the reptile probably lacked the muscles for flapping flight, it had a wingspan of two metres and probably glided

Megalosaurus – a large carnivorous dinosaur

Scelidosaurus – herbivorous dinosaur

Compsognathus – chicken-sized dinosaur with powerful legs for running

Plesiosaurus – long necked reptile well adapted for swimming in the sea

Ichthyosaurus – reptile adapted for swimming. Specimens have been found with the skeletons of embryos inside them. It seems possible that the female *Icthyosaurus* gave birth to live young

Reptiles are true land animals. The most important features which allow them to be so are:

- horny scales which waterproof the skin
- well developed legs
- an egg which is surrounded by a hard impermeable shell. The liquid inside the egg provides the developing embryo with its own 'private pool'. This allows reptiles to breed away from water.

About 70 million years ago many species of reptile became **extinct**.

Extinction

We often read that species die out (become extinct) because of the harmful effects of human activities on the environment. Such extinctions usually occur over a relatively short period of time (see Topic 8). Natural extinction is more long term. It makes room for new species to evolve and replace the previous ones.

Competition between species (see p. 65) and environmental change are the causes of naturally occurring extinctions. For example, tortoises, turtles,

It's a fact!

The extinction of most reptiles occurred in a relatively short period of time. The extinction of dinosaurs in North America, for example, happened over 12 million years. This seems like a very long time to us but it is an instant compared with the 4000 million years of life on Earth. Think of it like this: if the 4000 million years of life on Earth occupied one day, the reptiles would have died out in the space of five minutes.

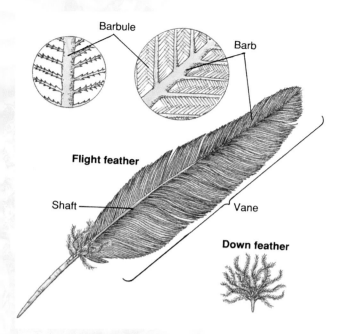

Figure 26.5K 🔺 Feathers

Figure 26.5L 🔺 *Archaeopteryx* was about the size of a crow.
Notice the impressions of feathers (characteristic of birds).
Archaeopteryx also had teeth in its beak (characteristic of reptiles)

snakes, lizards and crocodiles are virtually all that remain of the reptiles which once dominated the land, water and air. Why so many species of reptile became extinct is not clear. Explanations include that at the time:

- appearance of the first flowering plants replaced the vegetation on which herbivorous reptiles depended. The food webs (see p. 46) of the reptilian world therefore collapsed.

- some mammals which evolved from a group of reptiles about 200 million years ago were efficient thieves of reptilian eggs.

- a meteor plunged to Earth causing catastrophic damage. The impact threw up dust and smoke that triggered global cooling which most reptile species did not survive.

Whatever the cause, the mass extinction of most reptile species made way for mammals and birds to fill the vacant spaces. Pause and think for a moment. If the reptiles had not given way to mammals, we probably would not have evolved to read about these extraordinary events!

Birds

A bird's body is covered with feathers. Its forelimbs are modified for flying (see Topic 18.2). Feathers make **flight** possible and help to retain **body heat** (Figure 26.5K). The colour of a bird's feathers also helps to attract **mates** and provide **camouflage**.

Birds spend a great deal of time **preening** themselves to keep their feathers in good condition. Feathers are fluffed out and the bird nibbles them, drawing each one through its beak to remove fleas, lice and mites. The **preen gland** (called the parson's nose) lies just above the tail. It produces **oil** which the bird works into the feathers to waterproof them. Preening also repairs feathers and puts them back into place. When the fossil remains of an early type of bird called *Archaeopteryx* were found in 1861, it might have been mistaken for a reptile if it had not been for the beautifully preserved impressions of feathers which surrounded the specimen (Figure 26.5L).

Figure 26.5M 🔺 Hoatzin chick clinging to a branch. Notice the claw on the leading edge of its left wing

(a) Duck-billed platypus

Figure 26.5N ⬆ (b) Spiny anteater

Figure 26.5O ⬆ A grey kangaroo carrying her offspring in her marsupium

Flying

How did flight evolve? One idea is that the reptile ancestors of birds lived in trees and that the evolution of feathers enabled them to glide from branch to branch. In Figure 26.5L you can see a **claw** on the leading edge of the wing. Scientists think that this helped *Archaeopteryx* to clamber along the branches. The chicks of the hoatzin have claws like this and lead a similar sort of life in the tangled branches of the South American rain forest (see Figure 26.5M).

Another idea suggests that the evolution of flight began with a feathered carnivorous reptile running after its prey in hops and bounds, helped by its feathered arms. According to this theory, flight developed from long leaps after food.

No one knows the origins of flight. All we do know is that some long extinct group of reptiles gave rise to the feathered, warm-blooded, creatures we call birds, and that *Archaeopteryx* is an early stage in their evolution.

Mammals

One of the early groups of reptile which lived 200 million years ago was the ancestor of mammals. The fossil evidence shows that at first the mammal line flourished but later it nearly died out in the 'age of reptiles' that followed. However, mammals became abundant after most of the reptile species had become extinct 70 million years ago.

Of present-day mammals, the duck-billed platypus and spiny anteater most resemble the animals that were the ancestors of the mammal line (Figure 26.5N). Both animals have a mixture of reptile and mammal features. For example, they lay eggs as reptiles do. However, they are covered with hair and suckle their young with milk as mammals do.

Present-day mammals

Apart from the duck-billed platypus and the spiny anteater, the class mammals includes two groups: the **marsupials** and the **placentals** (see p. 10). Marsupial mammals include the kangaroo, wallaby and koala bear. They give birth to offspring which are at an early stage of development. The offspring crawl into their mother's pouch (called the **marsupium**) and attach themselves to the nipples of her milk glands. Inside the pouch they grow and develop until they can look after themselves (Figure 26.5O).

Placental mammals include moles, bats, rodents, whales, monkeys and humans (Figure 26.5P). They give birth to offspring which are at a later stage of development than marsupials. The embryo is attached to the wall of the uterus by the **placenta** through which it obtains nourishment (see p. 366).

Mammals are successful animals. They colonise every environment: from polar ice caps to tropical forests to hot deserts. They fill the air, land and sea.

Humans are relative newcomers to the scene. Our ancestors first appeared approximately 2.5 million years ago. They were as we are – **curious**. We use **words** to communicate our curiosity, construct codes of **moral behaviour** and ask questions about the workings of the natural world. We have even invented a way of investigating such questions. It is called **Science**!

SUMMARY

It is possible that there is life somewhere else in the Universe, but we do not know of it yet. For the time being, therefore, for us Earth is unique. It is filled with life – a product of 4000 million years of evolution.

Figure 26.5P ▲ Some of the many placental mammals

Theme Questions

1 The passage below is about Charles Darwin.

Who inspired Darwin?

Thomas Malthus lived in the late eighteenth century. He wrote 'An Essay on the principle of Population'. In this essay he pointed out that human beings produce far more offspring that ever survive. However, the adult population tends to remain stable from generation to generation.

Darwin realised that this idea applied to other animals. For example, one fish, which lays thousands of eggs in a year, would overpopulate an area with its offspring if they all survived.

The work of Malthus helped Darwin to develop his own ideas of how a species changes. He produced his theory of natural selection. Darwin realised that there must be a reason why some offspring survived but others did not. He suggested that small variations between individuals of a species might give certain individuals a better chance of survival. For example, those organisms with characteristics that made them better at escaping from predators or finding food would have a better chance of survival.

(a) (i) What is meant by the phrase "the adult population tends to remain stable from generation to generation"?
   (ii) Suggest why fish lay thousands of eggs rather than just a few.
   (iii) What can cause "small variations between individuals of a species"?
   (iv) What is meant by the phrase **natural selection**?

(b) Here are four statements about evolution. Which statement is false. Explain your answer.

The theory of evolution through natural selection was developed by Darwin.

DNA is the genetic material that transfers information from generation to generation.

Acquired characteristics **cannot** be passed on from parent to offspring.

Nature plays an important part in artificial selection.

(c) Suggest **two** ways that scientists can let other groups of scientists know about their ideas.

(Edexcel)

2 Snails are eaten by thrushes. Some snails have shells that are very striped, others unstriped. Every September for seven years a scientist counted all the snails he could find in an area of grassland. Here are the results:

Year	% ground covered by grass	Number of snails with ... very striped shells	unstriped shells
1	98	58	13
2	25	24	22
3	5	2	33
4	97	34	10
5	96		
6		9	43
7	98	68	13

(a) State:
   (i) a probable number of snails you would have expected to find in year 5,
   (ii) a probable percentage cover of grass in year 6.

(b) (i) All the snails were of the same type (species). During the seven years of study a single specimen was found with a completely black shell. What word could be used to describe this unusual form of the species?
   (ii) Choose the best answer. The term which best **explains** the results in the table is
   1 heredity   3 conservation
   2 natural selection   4 artificial selection.

3 Busy Lizzies are small plants that normally produce large numbers of white, pink or red flowers. Imagine that on a remote, uninhabited island scientists discovered colonies of Busy Lizzies growing and that most of these plants produced **blue** flowers. The plants could have evolved as a result of *natural selection* acting on a *mutant* **blue**-flowering Busy Lizzie plant.
(a) Explain how a *mutant* (genetic mutation) can arise by referring to the way DNA replicates.
(b) Explain how *natural selection* could have produced the colonies of **blue**-flowering Busy Lizzie plants.

4 *Hydra* is an animal which can reproduce sexually or asexually. If *Hydra* lives in an unchanging environment, it reproduces asexually. If the environment changes it begins to reproduce sexually. Explain why *Hydra* changes its pattern of reproduction.

(SEG)

5 The table shows how, for two species of gull, various features of their characteristics, behaviour and lifestyle have evolved.

Feature	Black-headed gull	Kittiwake
Nesting site	On the ground	Narrow ledges on cliffs
Type of nest	Shallow hole in ground	Made from mud and grass plastered to the cliff
Number of eggs	Three	Two
Colour of eggs	Match background	Match background
Chicks	Camouflaged	Not camouflaged
Reaction to predators	Adults give alarm calls, chicks often run away	Adults rarely give alarm calls, chicks do not run away

The two species of gull have evolved to nest in very different sites. Give **two** ways the characteristics, behaviour and lifestyle of each species have evolved to cope with the consequences of this. For **each** of the ways you mention give **one** advantage.

6 The diagram shows two types of peppered moth found on the trunks of trees in both city and countryside areas.

Dark moth

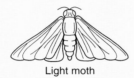
Light moth

Read the following passage about the peppered moth.

The peppered moth is often used to provide evidence for evolution. Before the Industrial Revolution the moths found were light in colour. They were harder for birds to see on light coloured tree trunks. Fewer moths are eaten by birds if they are well hidden. During the Industrial Revolution, many trees in large cities became black with soot. In the daytime the moths rest on tree trunks. Dark coloured moths were better hidden on the trees than the light coloured ones.

Records show that few light coloured moths survived in cities and few dark coloured moths survived in countryside areas. In 1956 the Clean Air Act gradually reduced the amount of smoke pollution. After 1956 the number of light coloured moths in city areas gradually increased.

(a) Use this information to answer the following question.
  (i) In which area were the greatest number of light coloured moths found after the Industrial Revolution?
  (ii) Describe and explain the effect of the Clean Air Act on the numbers of the two types of moth in city areas.
  (iii) Name the process of evolution described in the passage.

(b) Some students counted the number of moths they found resting on five trees in a city and five trees in a countryside area. The results are shown in the table below. Copy and complete the table by filling in the total number of moths counted and calculating the mean (average) number of moths found.

Tree	City trees		Countriside trees	
	Dark moths	Light moths	Dark moths	Light moths
A	10	2	3	8
B	8	3	4	7
C	8	4	3	8
D	7	2	2	6
E	8	2	4	7
Total				
Mean				

(c) Which group of trees showed the highest mean (average) number of moths?

7 Galapagos is a group of islands in the Pacific Ocean. The islands are several hundred miles from the mainland of South America. Few animals have reached the islands. On one of the islands there are three types of finch.
- One type feeds on soft seeds and some small insects.
- Another type feeds on large seeds and large insects.
- The third type feeds on small insects which it picks off leaves and twigs.

The drawings below show the heads of these three types of finch.

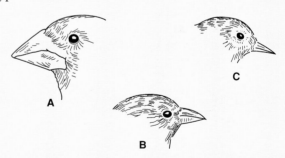

(a) Which type of finch:
  (i) feeds on soft seeds and some small insects;
  (ii) feeds on large seeds and large insects;
  (iii) feeds on small insects which it picks off leaves and twigs?
(b) Charles Darwin suggested that the three types of finch on the island had all evolved from one type of finch which had arrived from South America.
  (i) Name the process by which this evolution may have taken place.
  (ii) The finch that came from South America probably had a thick beak. Suggest how the thin beak could have evolved.

Index

A

abdominal cavity 236, 237, 238
abiotic environments 35, 40
absorption 238–9
abuse, drug 289, 291
accommodation 296
acid rain 103, 105–7
acquired characteristics 431
Acquired Immune Deficiency Syndrome (AIDS) 266, 374, 375
actin 315
active transport 151, 183, 189, 192
adaptation 35, 436
addiction, drug 289
adenine 152, 170, 171
adenosine triphosphate (ATP) 245, 315
adolescents 304
adrenal glands 300
adrenalin 300
adventitious roots 357
aerobic respiration 3, 39, 50, 243–4, 245
aerofoil 319
aerosols 109
afterbirth 368
AIDS (Acquired Immune Deficiency Syndrome) 266, 374, 375
air 38, 90
alcohol (ethanol) 214–6, 244, 245, 290, 403
alleles 379, 381–2, 386–9, 415
alula 318, 321
alveoli 249, 253, 254
amino acids 152–3, 167–8, 170, 183, 206, 238, 308
amniocentesis 389
amniotic fluid 368, 389
amphibians 6–7, 10, 32, 447
amplification (DNA) 394
amylase 168–9
anabolic steroids 301–2
anaemia 206, 207, 208
anaerobic respiration 3, 244, 245
analgesic drugs 337
angina 278
animals 9, 12, 24–34
annual rings 22, 186, 447
annuals 23
anorexia nervosa 224
anterior end 26
anthers 345, 346, 347, 348
antibiotics 19, 337–8, 375, 412–13, 443
antibodies 259, 261–2, 264, 267–8, 334–5, 337–8, 400, 410–12
antigens 261, 262, 264, 267, 268, 269, 270, 334–5, 399, 410
antiseptics 336
antitoxin 262
aphids 76, 96, 194
appendix 237, 241
arteries 270–1
arterioles 271, 273
artificial selection 437–9
asbestosis 251
aseptic surgery 337
asexual reproduction 341, 355–9, 441, 442
assimilation 238
association cortex 289
association pathways 289
asthma 251, 395
atheroma deposits 216
atmosphere 38, 55
auditory nerves 292
autism 263

autotrophs 45
auxin 196, 197, 198, 199, 406
axons 284, 287, 288, 316

B

bacilli bacteria 15
backbone 32, 326, 327
backcross (test) 382
bacteria 12, 14–15, 225, 226–8, 232, 241, 331, 337–8, 443
    denitrifying 55
    nitrifying 55
    nitrogen-fixing 55, 74–5, 90, 403, 405
    putrefying 55
balance canals 293
basal metabolic rate 204, 212–13
base pairing 153, 154, 171
bases 152, 153, 171
battery hens 88
Beagle, HMS 433–4
Bermuda Statement 393, 394
biennials 23
bilateral symmetry 26, 28
bile 236
bilharzia 74
binominal system 8
biodigester (fermenter) 399, 400
biodiversity 44
biofuel 402
biogas fermenters 420–21
biological control 76–7, 80, 98–9
biological oxygen demand (BOD) 115, 417
biomass 49, 50–1
biosphere 35
biotechnology 397–9, 407–8, 410–17
biotic environments 35, 40, 42
birds 10, 318–22, 451
blackfly 76
bleeding 265
blight 332
blood 259–62
blood antigens 264
blood donors 259, 264, 265
blood groups 264–5, 383
blood pressure 217, 276, 281
blood proteins 259
blood transfusion 259, 264
blood vessels 158, 270–3, 313
bloom, algal 61, 115
bomb calorimeter 203, 204
bones 325–8
botanic gardens 140
botulin 227
bovine somatotropin (BST) 408
Bowman's capsule 308
Braille, Louis 285
brain 285, 287, 288–9, 297
    cerebellum 288
    cerebrum 288
    hypothalamus 313
    medulla 288
    midbrain 288
    pons 288
bread mould 19
breathing 243–4
breathing movements 250
bronchioles 248, 249, 250
bronchitis 103, 250
budding 358
bulbs 356, 357
bulimia 224

C

caecum 241
caffeine 289–90
calcium 39, 183, 206, 207, 213
camouflage (prey) 69
cancer 155, 331, 410
capillaries 271–3
carbohydrases 235
carbohydrates 162–4, 203
carbon cycle 56
carbon dioxide 3, 39, 56, 111, 112, 124, 177–81, 243, 245–6, 249, 261
carbon monoxide 103, 108, 253
carcinogens 253, 430
cardiac arrest 278
cardiac muscle 274, 316
carnassial teeth 231
carnivores 9, 10, 45, 231, 241
carotenoids 179–80
carpels 326, 345
carriers 387, 388
cartilage 248, 325, 329
catalysts 168, 177
    enzymes 235
    proteins 402
cell division 154–7
cell surface membrane *see* plasma
    membrane
cell walls 12, 147, 149, 150, 151, 337
cells 145, 147–8, 149–51, 186, 302, 337
    bone 159
    ciliated 159
        fat 159
        female (egg) 159
        male (sperm) 159
        muscle 159
        nerve 159
        skin 159
        surface area to volume ratio 160–1
cellulose 163, 241
central nervous system 285, 286, 287
centromere 155
cephalosporins 412
cereal crops 405, 439
cerebral haemorrhage (stroke) 217, 276
CFCs (chlorofluorohydrocarbons) 109–10, 112
chaetae 319
Chain, Ernst 412
characteristics of life 3
cheese making 402
chemotherapy 337
chitin 163, 314
chlorophyll 177, 178, 179–80
chloroplasts 12, 17, 21, 23, 147, 177, 178, 191
cholera 82, 97
cholesterol 210, 216–17, 221
chromatids 155
chromosomes 148, 154, 155, 156, 157, 343, 390
    X and Y 383–4, 386–7
chyme 236
cilia 17, 159, 247, 249, 253, 318
cirrhosis 214
CITES (Convention on International Trade in Endangered Species) 133, 134, 135
classification (species) 9–10
Clean Air Acts 102, 103, 104
Clear Lake 117
climax communities 67
clonal expansion 268
clones 341, 355, 357, 358–9, 404

bacteria 14
cochlea 292, 293
'cold turkey' 291
collagen 325
colon 237, 238, 239
colour-blindness 388
colour vision 297
commensalism 72, 73, 75
communities 42, 45
companion cells 192
competition 58, 65–6, 73, 436
competitive exclusion 66
compost 93
concentration gradient 148, 151
cones, pine 22
conservation 133–41, 407
constipation 239
consumers 45, 46, 47, 49
contact inhibition 155
continuous variation 427
contraception 371–3
Convention on Biological Diversity 135
co-ordination 285
copper mining 122
copulation 344
corms 356, 357
cornea 296
coronary arteries 274, 281
coronary thrombosis 278
cortex 188
cotyledons 185, 351, 352
counter-current effect 257, 258
crenation 151
Crick, Francis 170, 172–3
crop rotation 333
cultivation 92
cutting 357–8
cycles 54–6
cyclosis 17, 178
cystic fibrosis 250–1, 388
cytoplasm 147, 149, 150, 153, 154, 170, 192, 366
cytosine 152, 170, 171

D

Darwin, Charles 128, 431–6, 440–41
daughter cells 154, 155, 156, 157
DDT 97–8, 117, 443
deactivated enzymes 170, 181
deafness 292
deamination 238, 308
deciduous trees 22
decoding (DNA) 394
decomposers 36, 37, 39, 45, 46, 47, 54
decomposition (decay) 37, 46, 53–4, 445
defaecation 239
deficiency diseases 207, 208
deforestation 139
denatured enzymes 170, 181
dental caries 232
dental formula 230
dentine 229, 232
dentition 230–1
deoxygenated blood 260
deoxyribonucleic acid (DNA) *see* DNA
detritus 39, 42, 43, 45
developed nations 63, 333
developing nations 63, 333
diabetes 301, 400
dialysis 309
diarrhoea 81, 82, 331, 333
diastema 230
diastolic pressure 276
dichotomous keys 6
diet 211–14, 216–17
differentiation 186, 302, 367
diffusion 148, 151, 160, 367
digestion 235–8
dioxin 96

diploid number 155, 156, 157, 343
disaccharides 162, 163
discontinuous variation 427
diseases
    infectious 262–3, 331–3, 334–5
    non-infectious 331, 333
    see also diseases by name
disinfectants 336
dispersal (fruits/seeds) 353–4
DNA 152–54, 155, 170–3, 341, 389, 392–4, 400, 411
dominant characteristics 379, 380, 382
donors
    blood 259, 264, 265
    organ 269–70
dorsal surfaces 26
double helix (DNA) 153, 154, 171, 172
Down's syndrome 385–6, 389
downstream processing 399
drugs, effects of 289–91
dysentery 16, 81

E

E Numbers 218–19
ear 292–4
ear wax 292
eardrum 292
Earth 3, 35, 36, 38, 111–12, 119–20, 431–2, 433
ecdysis (insects) 303
ecdysone 303
ecological pyramids 48–51
ecology 35–41
ecosystems 40–1, 99–101, 119, 133, 136
ectoparasites 73
effectors 283
eggs 342–4, 362, 363, 364–5, 366, 372, 373, 374
ejaculation 364, 371
electrons 430
elements 183, 207
embryo 154, 185, 350, 351, 366, 407
emigration 61
emphysema 253
enamel 229, 233
endangered species 135, 140
endocrine glands 300, 301, 363, 372
endodermis 188, 189
endoparasites 73, 74
endoskeletons 32, 314
endotoxins 227
energy 99–101, 203–6, 212, 213
energy flow 50–1
energy nutrients 203–4, 212, 213
Environmental Impact Statement 134
environmental pressures 119–20
enzymes 152, 167, 168–70, 178, 181, 235, 400, 401–2, 404
epidemics 335
epidermis 177, 188, 312
epiphragm 28
epiphytes 44
erosion 91–2, 136
erythromycins 412
essential amino acids 168, 206
ethanol (alcohol) 214–6, 244, 245, 290, 403
ethene 198
ethics 395
eutrophication 115–16, 418
evergreens 22
evolution 426, 430, 430–1, 441–44, 445
excretion 3, 307, 308
exercise 281
exoskeleton 30, 31, 163, 303, 314, 319
exponential increase (population) 58
extinction 132–3, 135, 136, 140, 437, 450–1
Exxon Valdez 128
eye 295–8

F

factor VIII 266, 375, 386, 387, 407
factory farming 88–9
faeces 239
families (species) 9–10
family planning 371
fast food 213
fat layer 164, 312
fats 203
    saturated 165–6, 216–17, 221
    unsaturated 166, 216
feathers 33, 318–19, 451
fermentation 244, 398, 399
fermenter (biodigester) 399, 400
fertilisation 157, 343–4, 349–50, 366–8
fertilisers 55, 87, 91, 93, 94, 115–17
fertility, control of 371–3
fetal alcohol syndrome 216
fetus 253, 366–7
fibre 203, 210, 221, 239
fibrinogen 259, 265
filariasis 78
fins 318
first filial generation (F_1) 379
fish 10, 32, 247, 257–8, 317–18, 446, 449
fishing 121
    factory ships 70
    nets 70, 71
    sonar 70
    sustainable 70–1
flagellum 148
Fleming, Alexander 399, 413
Florey, Howard 413
flowers, structure 345–6
flue gas desulphurisation (FGD) 107
fluoride 233–4
flying 318–22, 451, 452
food additives 218–19
food chains 46–7, 51, 114
food groups 212
food nutrients 203, 206–9, 212
food poisoning 219, 225–7
food preservation 227–8
food vacuoles 16, 17
food webs 47
fossil formation 447–8
fossils 445, 449, 451, 452
fracture (bones) 328
Franklin, Rosalind 172
free radicals 430
free range 87, 88
fronds (ferns) 21
fructose 162, 163
fruit
    dispersal 353–4
    false 352
    structure 352
    succulent 351, 352
FSH (follicle stimulating hormone) 306, 363, 368, 372, 374
fungi 12, 18–19, 159, 225, 226, 228, 331

G

Galapagos Islands 435–6
Galton, Francis 440
gametes 342, 381
gaseous exchange 3, 177, 243, 247–50, 256–8
gastric juice 236
gene probes 411, 416
gene therapy 388, 416
genes 153, 170, 302–3, 379, 381, 341, 443, 440
genetic code 153
genetic diseases 266, 385–90
genetic engineering 262, 400, 403, 404, 407–8, 410–17
genetic fingerprinting 392, 400
genetic modification (GM) 404, 415–16
genetics 379–84, 390–1

Index

genomes, human 392–5
genotype 379, 381, 428
genus 8
germination 186
gestation period 367
gibberelin 199
gills
    fish 32, 247, 257–8
    fungi 18, 159
glands 336
    endocrine 300, 301, 363, 372
gliding (flight) 320, 321
global warming 111–12, 403
glomerulus 307, 308
glucagon 300, 301, 306
glucose 162, 163, 244, 245, 300, 301, 306
glycogen 163, 238
goblet cells 247
goitre 302
gonorrhoea 374, 375
graafian follicle 363
grafting
    scion 357–8
    stock 357–8
grazing 139
green revolution 94
greenfly 76
greenhouse effect 110–12, 132
greenhouses 110
groups (species) 9–10
growth 3, 185–6, 196–8, 301–5
guanine 152, 170, 171
guard cells 159, 177, 191
gut 236–41

H

habitats 42–4, 45, 122
haemoglobin 108, 206, 207, 259, 260, 261
haemophilia 266, 375, 386–7, 407
hair 34, 311–13
hair erector muscle 312
hair follicle 311, 312
haploid number 155, 156, 157, 343
Harvey, William 272
healthy diet 211–12, 216–17
heart 158, 274–6
heart attack 278
heart disease 166, 210, 214, 216–17, 221, 278–81
heart muscle 274, 316
heart rate 280, 304
hepatitis B 331, 414–15
herbaceous plants 23, 186, 187
herbicides 87, 94–6, 198, 200
herbivores 45, 50–1, 230–1, 241
heredity 379–84, 390–1
hermaphrodite parasites 74
herpes 374, 375, 415
heterotrophs 45
heterozygote 379, 381, 382, 388
heterozygous alleles 379, 416
HIV (Human Immunodeficiency Virus) 266, 268, 375, 416
homeostasis 300, 306, 307
homologous pairs 383–4
homozygote 379, 381, 382, 389
homozygous alleles 379, 416
honey guides 347
honey sac 83
honeybees 82–4
Hooke, Robert 145
hormones 167, 300–5, 306, 363, 368, 372, 374
horses, movement of 323–5
hosts 26, 27, 73, 74
human genome 392–5
Human Genome Organisation 392
Human Immunodeficiency Virus (HIV) 266, 268, 375, 414

humus 36–7, 90
Huntington's chorea 388–9
hybridisation 439
hybrids 8
hydrolysis 235
hydroponics 184
hydrostatic skeletons 314, 319
hyphae 18, 19, 159, 226, 332

I

'ice protein' gene 406
immigration 61
immobilised enzymes 401–2
immune reactions 73, 262
immunisation 262–4
immunological memory 267, 268
immunology 267–70
immunosuppressive drugs 270
implantation 366
incubators 368
infectious diseases 262–3, 331–3, 334–5
infertility 374
inflammatory action 268
influenza 331, 334–5
insecticidal crystal protein (ICP) 405–6
insecticides 80, 87, 97–8, 117, 404, 405–6, 442–3
insects 30, 76–7, 98–9, 303–4
insulin 300, 301, 306, 400, 410–11
intelligence 289
intensive farming 87–9
interferons 413–14
International Convention for the Regulation of Whaling 135
intestinal juice 236
intestines 236, 237, 239
involuntary actions 287, 316
iron 183, 206–7, 208
irradiation (food) 228
Islets of Langerhans 300

J

Jenner, Edward 269
Jessica 128
joints 314
    ball and socket 328, 329
    hinge 328–9
    suture 328

K

Kaposi's sarcoma 415
karyotype 383, 389
Kendrew, John 172, 173
keratin 229, 311, 312
keys 5–7
kidney failure 309
kidneys 307, 309
    cortex 308
    medulla 308
kingdoms 9, 11–34
kwashiorkor 168
Kyoto agreement 112

L

lactation 370
Lamarck, Jean-Baptiste de 441
land use, pressures on 122–6
landfill 123, 138
larvae 304
laterite 124
law of segregation 381
leaching 105
leaf mosaic 177, 178
legumes 55, 405
leisure 127

leukaemia 266
LH (luteinising hormone) 306, 363, 368, 372
lichen 19, 104
life expectancy 63, 333
ligaments 323, 326, 329
ligases 400, 404, 411
light 177, 180–1
lignin 186, 189
limiting factors 180–1
Linnaeus, Carolus 7, 8, 9
lipids 164–6
liposomes 388
Lister, Joseph 337
litter layer 37
liver 236, 237, 238, 239
loam 89
low-sulphur fuels 107
lung cancer 251, 252
lungs 247, 248, 249, 250
Lyell, Sir Charles 436
lymph 273
lymph vessels 272, 273
lymphocytes 260, 261, 262, 264, 266, 267–8, 269
lysis 151

M

maggots 77
magnesium 183, 207
major elements 183, 207
malaria 78–80, 97
Malpighian tubules 307
Malthus, Reverend Thomas 436, 440
mammals 10, 34, 452–3
mammary glands 34, 370
mandibles (insects) 30
manure 93
marrow 260
marsupials 10, 452
McKay, Fred 234
meiosis 154, 155–7, 381, 383, 430
melanic moths 443, 444
Mendel, Gregor 380–2, 390–1, 440
meninges 286
meningitis 286
menopause 363
menstruation 363, 366, 368, 373
mercury 114
metabolic rate 204, 212–13, 238, 313
metabolism 3, 50, 203, 208, 259, 307, 308, 311, 337
metal reserves 122
metamorphosis 32, 303–4, 449
methane 112, 420–21
micropropagation 358–9
microscopes
    optical 145, 146
    transmission electron 145–6, 147
milk 34, 370
Minamata Bay 114
minerals 183, 206
mitochondria 146, 147, 183, 245
mitosis 154, 155–7, 186, 341, 358, 366
MMR vaccine 263
monoclonal antibodies 410–12
monocultures 87, 333
monohybrid inheritance 379, 380
monosaccharides 162, 163
monotremes 10
Montreal agreement 110
morals 395
mosquitoes 78–9
mosses 20–1, 343–4
moulting (insects) 303
movement 3, 196–8
multicellular organisms 157
muscles 3, 315–16
mutations 154, 341, 386, 429, 430, 441, 440

mutualism 72, 73, 74–5
mycelium 18, 19, 159
myofibrils 315
myosin 315
myxomatosis 59, 332

N

nastic movements 196
natural selection 65, 430, 437, 440, 441–42
nectar 82, 83, 84, 347, 348, 349
nephrons 307
　　Bowman's capsule 308
　　collecting duct 308
　　glomerulus 307, 308
nerve impulses 297, 283, 284, 285
nerves 285
neurones 283, 284, 287
neurotransmitters 284, 285
niches 45, 65–6, 304
nitrates 55, 105, 117
nitrites 117
nitrogen 3, 38, 54, 55, 183
nitrogen cycle 55
nitrogen-fixing bacteria 55, 74–5, 90, 403, 405
nitrogen oxides 103, 112
nodes 357
nodules 405
non-essential amino acids 168, 206
non-infectious diseases 331, 333
non-renewable resources 122, 132, 137
non-selective herbicides 95
North Sea 121
NPK fertilisers 94, 183
nuclear membrane 147, 152
nuclear waste 129–31
nucleic acid 13, 170–3
nucleotides 170–1
nucleus 147, 152–3, 154, 170, 366
nutrients 90–1, 93, 115, 121, 184, 203, 206–9, 212, 430
nutrition 3
nymphs 303

O

obesity 221–2
oedema 273
oesophagus 237
oestrogen 300, 301, 306, 363, 372, 374
oil pollution 128–9
opercula 257
optimum (enzymes) 169–70
oral rehydration 82
order (species) 9
organs 157, 158, 161
Origin of Species, The 440, 441
osmoregulation 151, 307
osmosis 149–50, 151, 191, 193, 228
osteoporosis 214
ovaries 300, 351, 352, 362, 363, 368, 372
overfishing 70–1, 121
ovules 351, 352
oxygen 3, 39, 54, 55, 108, 124, 177, 180, 186, 243, 249, 308
oxygen debt 244
oxygenated blood 260
oxyhaemoglobin 260
ozone layer 38, 109–110

P

pacemaker (heart) 275
palisade cells 159, 177, 178, 179
palps (insects) 30
pancreas 236, 237, 300, 301
pancreatic juice 236

pandemics 335
parasites 14, 16, 19, 26, 27, 49, 72–4, 76–7, 78–80, 331
parent cells 154, 155, 156
parental generation 379, 380
partially permeable membranes 149
passive smoking 254
Pasteur, Louis 228
pasteurisation 227, 228
patents 393
pathogens 73, 331–3, 334, 337
Pavlov, Ivan 287
pedigree 436
penicillin 19, 337, 338, 375, 399, 412, 413–14
penis 344, 361, 364, 371
peppered moth 443, 444
pepsin 168–9
peptides 168
perennials 22, 357
pericardium 274
pericarp 351, 352, 353
peripheral nervous system 285
peristalsis 240, 316
persistence 97–8, 114
Perutz, Max 172, 173
pesticides 76–7, 87, 97–8, 121
　　see also herbicides; insecticides
Pet Travel Scheme (PETS) 332
pH 39, 91, 105, 107, 168, 169–70
phagocytes 260, 261, 262, 267–8, 269
phagocytosis 16
pharmacogenomics 395
phenotype 379, 428–9
pheromones 301
phloem 159, 186, 187, 188, 192–4
phosphates 116, 171
phospholipid molecules 164–5
phosphorus 54, 183
photomicrographs 145, 146
photosynthesis 3, 39, 50, 56, 177–81, 245–6
pigments 179–80
pinna 34, 292
pinnae (ferns) 21
pioneers 67
pituitary gland 300, 301, 363, 372
placenta 366–7, 368
placentals 10, 452, 453
plankton 46
plant growth substances 197–200
plants 12, 20–3, 90–1, 185, 187, 332–3, 355–9
plaque 232
plasma 259, 264, 272, 273
plasma membrane 145, 147, 149, 150, 151, 164–5
plasmids 404, 405, 411
plasmolysis 150
platelets 260
plumule 185
pneumonia 250
pollen 83, 344, 346, 347, 348, 349, 380
pollen comb 83
pollen grains 159, 344, 345, 346, 349, 350
pollen sacs 83, 345, 347
pollen tubes 344, 349–50
pollination
　　cross 348–9
　　insect 347
　　self 348
　　wind 346, 347
pollutants 103–12
pollution 113–17, 121, 127, 128–31, 403
polypeptides 168
polysaccharides 163
'pomato' 403, 405
populations 58–64, 87, 441
posterior end 26
potassium 94, 183, 207
predators 60–1, 68–9, 70–1, 73, 98
preen gland 451
preening 451

pregnancy 216, 253, 263, 366
prey 68–69
primary consumers 45
primary hosts 74
primary tissues 186
proboscis 30, 81
producers 45, 46, 47, 49
product 397, 399
productivity (farming) 101
progesterone 300, 363, 368, 372
proglottids 26
proteins 55, 152, 153, 167–70, 203, 206, 302
　　first/second class 168
protists 12, 16–17, 46, 74, 75, 318, 331
protoplasts 406
pseudopodia 16, 318
PSSAs (particularly sensitive sea areas) 128
puberty 301, 363
pulse 276
pulverised fluidised bed combustion (PFBC) 107
Punnett, R.C 383
Punnett square 383
pupil 296
pure breeding 379, 380
pyramids 48–51
　　biomass 49
　　energy 50
　　numbers 49
　　population 63–4

Q

quadrats 61, 62
Quorn 409

R

rabies 332
radial symmetry 24
radiation 129–31, 228, 429–30
radicle 185
radula 28
Ramsar Convention 135
Ray, John 9
recessive characteristics 379, 380, 382
recipients (blood) 264, 265
recipients (organs) 269–70
recycling 138–9
red blood cells 159, 259, 260–1, 264
reflex arc 286, 287
reflex responses 286–7
renal blood cells 307–8
renewable resources 121, 137
replication 13, 154, 341, 429
reproductive organs 361–2
reptiles 10, 33, 449–51
resistance 337–8, 405, 442–3
resources
　　non-renewable 122, 132, 137
renewable 121, 137
respiration 3, 39, 50, 162, 203, 243–6, 261
response 195–8
restriction enzyme 400, 404, 411
retina 296, 297, 298
rhizomes 355, 356, 357
ribonucleic acid (RNA) see RNA
ribosomes 146, 152, 153, 337
rickets 206, 207, 208–9
risk factors (heart disease) 279–81
RNA 13, 152–3, 170–1
　　messenger RNA (m-RNA) 153
　　transfer RNA (t-RNA) 153
roots 74, 185, 186, 188, 189, 195, 355, 357
rumen 241
ruminants 241
runners 355
Ruska, Ernst 146

Index

S

saliva 236
Salmonella 226–7
Sanger Centre 392, 394
Sanger, Frederick 391, 392
sanitary towels 363
sap 76
saprobionts 14, 19, 46
sarcomeres 315
saturated fats 165–6, 216–17, 221
scales, fish 32
scaling up 399
scavengers 54
scolex 26
screw-worm flies 77
scurvy 208, 209
sebaceous glands 312
secondary consumers 45
secondary growth 186
secondary hosts 26, 74
sedimentary rocks 445
seed banks 140–1
seed dispersal 353–4
seedling 185
seeds 185, 351–4
    dicotyledonous 351
    monocotyledonous 351
segmentation 240
segments (worms) 27
selection pressure 442, 443, 446
selective breeding 437–40
selective herbicides 95, 198
Sellafield 130–1
semen 364, 373, 374
sense receptors 312, 313
senses 283
sensitivity 3
serum 265
sewage 127
sewage treatment 417–20
sex hormones 363
sexual reproduction 342–50, 429, 441, 442
sexually transmitted diseases 15, 374–5
shoots 185
sickle-cell anaemia 260, 416
sieve cells 192
sieve tubes 76, 192, 193, 194
sight defects 298
single-cell protein (SCP) 408–9
skeletal muscle 316
skeleton, human 325–8
skin 311–13, 336
slime capsule 148
slimming 223–4
smog 102, 403
smoking 251–5
smooth muscle 316
soaring (flight) 320, 321
soil 36–7, 89–92, 93, 136, 183, 258
species 4, 8, 9–10
species diversity index 44
sperm 342–4, 361, 363, 364–5, 366, 373, 374
spermatophore 344
sphygmomanometer 276
spirilla bacteria 15
spleen 273
spongy mesophyll cells 177, 179
spores 18, 19, 20–1
staining (specimens) 147
starch 163, 356, 357
stems 188, 189, 195
sterilisation (food) 227
stimuli (plant growth movements) 195–8
stomach 237, 238
stomata 90, 159, 177, 178, 191
stratosphere 38

strike 77
substrates 168, 169
succession 67
succulent fruit 351
sugars 162, 171, 177, 178, 185, 216, 232, 403
sulphur dioxide 103–4, 107
sustainable use 70–1, 138
sweat 313
sweat gland 312
swim-bladder 32, 317
swimming (fish) 317–18
symbiosis 73
synapses 284, 287, 288
syphilis 374, 375
systemic insecticides 76
systolic pressure 276

T

tapeworms 26, 72, 74
target tissue 300
tartar 232
teeth 229–34
temperature 168, 169–70
temperature inversion 39
temperature regulation 311–13
tendons 315, 316, 323, 326
teratogenic effects 96
tertiary consumers 45
testes 300, 361, 363, 364
testosterone 300, 301, 302
tetracyclines 337, 411
thalidomide 430
thermal pollution 115
thiomersal 263
thoracic cavity 249, 250
thrombosis 278
thrombus 278
thymine 152, 170, 171
thyroid 300
thyroxin 300, 301, 302, 303
tissue 157, 158
tissue culture 358–9
tissue fluid 272, 273
tissue rejection 412
tissue typing 269
topsoil 91
trace elements 183, 207
trachea 248, 250
transamination 238
transducers 283
transgene 407
translocation 192–3
transpiration 38, 90, 189, 191
transpiration stream 177, 178, 189, 190
transplants 269–70, 309, 412
trophic levels 45, 49–51
tropical rain forests 67, 124–6
tropism
    geotropism 196
    hydrotropism 196
    negative 196
    phototropism 196
    positive 196
troposphere 38
tubers 356, 357
tumours 251
turgor 149–50
turgor pressure 150, 185
twins 369
    2,4-D (synthetic auxin) 198
typhoid fever 82

U

ultrasound scanner 389
umbilical cord 366–7, 368

unsaturated fats 166, 216
upper respiratory tract 247, 248, 249, 250–1, 334
uracil 152, 170
urea 271, 272, 273, 307, 308
uterus (womb) 362, 363, 364, 365, 366, 368, 372

V

vaccination 262, 334–5
vaccines 262, 263–4, 415–16
vacuoles 147, 149, 150
vagina 362, 363, 364, 365, 371, 374
valves 274, 275
variation 427–31, 437
vascular bundles 188
vasoconstriction 313
vasodilation 313
vectors 26, 74, 78, 79, 80, 97
vegetarianism 51, 220–1, 402
vegetative reproductions 341, 355–9
veins 270–1
venereal disease (VD) 15, 374–5
Venter, Craig 393
ventral surfaces 26
vents 50
venules 271, 273
vertebrae, human 327
vertebrates 6, 9, 10, 32–4, 445
vibrios 15
villi 238–9, 366
viruses 13, 331–2, 334–5
vitamins 208
voluntary reactions 288

W

Wallace, Alfred Russel 433, 440, 441
warning colours (prey) 69
water cycle 38
water potential 149
water table 90
water treatment 420–21
Watson, James 170, 172–3
weaning 370
weathering 36, 89
whales 71, 135, 164
white blood cells 159, 260, 261
wildlife conservation 133–6, 140
Wilkins, Maurice 172–3
wine making 397–9
wood 186, 187, 189
World Conservation Strategy 134–5
World Health Organisation 335
World Heritage Convention 135
Worldwide Fund for Nature (WWF) 128, 133, 134

X

xylem 159, 177, 186, 187, 188–90, 192, 193

Y

yeasts 19, 244, 245, 396, 408

Z

zoonoses 331–2
zoos 140, 407
zygote 157, 343, 366